Conservation Translocations

Conservation translocation – the movement of species for conservation benefit – includes reintroducing species into the wild, reinforcing dwindling populations, helping species shift ranges in the face of environmental change, and moving species to enhance ecosystem function. Conservation translocation can lead to clear conservation benefits and can excite and engage a broad spectrum of people. However, these projects are often complex and involve careful consideration and planning of biological and socio-economic issues. This volume draws on the latest research and experience of specialists from around the world to help provide guidance on best practice and to promote thinking on how conservation translocations can continue to be developed. The key concepts cover project planning, biological and social factors influencing the efficacy of translocations, and how to deal with complex decision-making. This book aims to inspire, inform and help practitioners maximise their chances of success and minimise the risks of failure.

MARTIN J. GAYWOOD is a Senior Research Associate at the University of the Highlands and Islands, and Species Project Manager at NatureScot, Scotland's nature conservation agency. He has led a wide range of species conservation projects, including conservation translocations, and has been closely involved in beaver reintroduction to Scotland since 2000. He has provided the secretariat role to the National Species Reintroduction Forum since its inception, has managed the production of the Scottish Code for Conservation Translocations, and is a member of the IUCN SSC Conservation Translocation Specialist Group.

JOHN G. EWEN is a Senior Research Fellow at the Institute of Zoology, Zoological Society of London, and a member of the IUCN SSC Conservation Translocation Specialist Group. His research focuses on conservation translocations, often through providing decision support to recovery programmes. He is Co-chair of New Zealand's Hihi Recovery Group and Chair of the United States Fish and Wildlife Service's Sihek Recovery Team, and is involved in several conservation translocation projects globally.

PETER M. HOLLINGSWORTH is Director of Science and Deputy Keeper at the Royal Botanic Garden Edinburgh, a Visiting Professor at the University of Edinburgh, University of Johannesburg, and Heriot-Watt University, and an Honorary Professor in the Kunming Institute of Botany at the Chinese Academy of Sciences. His research focuses on understanding and conserving plant biodiversity. He has a strong interest in linking research to practical conservation outcomes and guidance, including conservation translocations and the integration of genetic and genomic data into conservation planning. He is a member of the IUCN SSC Conservation Translocation Specialist Group.

AXEL MOEHRENSCHLAGER is the Chair of the IUCN SSC Conservation Translocation Specialist Group, pursuing its mission 'to empower responsible conservation translocations that save species, strengthen ecosystems, and benefit humanity'. He also serves IUCN more widely as a member of the IUCN Species Survival Commission Leadership and Steering Committee. Moehrenschlager is an Adjunct Professor at the University of Calgary in Canada, Adjunct Associate Professor at Clemson University in the United States, Erskine Fellow at the University of Canterbury in New Zealand, and Research Associate at Oxford University's Wildlife Conservation Research Unit where he received his PhD. Aligned with additional research interests to innovate and implement sustainable synergies for biodiversity conservation and improved human livelihood, he serves on the Technical Advisory Committee of the United Nations Equator Prize and as Board Trustee of the St Andrews Prize for the Environment.

ECOLOGY, BIODIVERSITY AND CONSERVATION

General Editor
Michael Usher, University of Stirling

Editorial Board
Jane Carruthers, University of South Africa, Pretoria
Joachim Claudet, Centre National de la Recherche Scientifique (CNRS), Paris
Tasman Crowe, University College Dublin
Andy Dobson, Princeton University, New Jersey
Valerie Eviner, University of California, Davis
Julia Fa, Manchester Metropolitan University
Janet Franklin, University of California, Riverside
Rob Fuller, British Trust for Ornithology
Chris Margules, James Cook University, North Queensland
Dave Richardson, University of Stellenbosch, South Africa
Peter Thomas, Keele University
Des Thompson, NatureScot
Lawrence Walker, University of Nevada, Las Vegas

The world's biological diversity faces unprecedented threats. The urgent challenge facing the concerned biologist is to understand ecological processes well enough to maintain their functioning in the face of the pressures resulting from human population growth. Those concerned with the conservation of biodiversity and with restoration also need to be acquainted with the political, social, historical, economic and legal frameworks within which ecological and conservation practice must be developed. The new Ecology, Biodiversity and Conservation series will present balanced, comprehensive, up-to-date and critical reviews of selected topics within the sciences of ecology and conservation biology, both botanical and zoological, and both 'pure' and 'applied'. It is aimed at advanced final-year undergraduates, graduate students, researchers and university teachers, as well as ecologists and conservationists in industry, government and the voluntary sectors. The series encompasses a wide range of approaches and scales (spatial, temporal and taxonomic), including quantitative, theoretical, population, community, ecosystem, landscape, historical, experimental, behavioural and evolutionary studies. The emphasis is on science related to the real world of plants and animals rather than on purely theoretical abstractions and mathematical models. Books in this series will, wherever possible, consider issues from a broad perspective. Some books will challenge existing paradigms and present new ecological concepts, empirical or theoretical models, and testable hypotheses. Other books will explore new approaches and present syntheses on topics of ecological importance.

Ecology and Control of Introduced Plants
Judith H. Myers and Dawn Bazely

Invertebrate Conservation and Agricultural Ecosystems
T. R. New

Risks and Decisions for Conservation and Environmental Management
Mark Burgman

Ecology of Populations
Esa Ranta, Per Lundberg, and Veijo Kaitala

Nonequilibrium Ecology
Klaus Rohde

The Ecology of Phytoplankton
C. S. Reynolds

Systematic Conservation Planning
Chris Margules and Sahotra Sarkar

Large-Scale Landscape Experiments: Lessons from Tumut
David B. Lindenmayer

Assessing the Conservation Value of Freshwaters: An International Perspective
Philip J. Boon and Catherine M. Pringle

Insect Species Conservation
T. R. New

Bird Conservation and Agriculture
Jeremy D. Wilson, Andrew D. Evans, and Philip V. Grice

Cave Biology: Life in Darkness
Aldemaro Romero

Biodiversity in Environmental Assessment: Enhancing Ecosystem Services for Human Well-Being
Roel Slootweg, Asha Rajvanshi, Vinod B. Mathur, and Arend Kolhoff

Mapping Species Distributions: Spatial Inference and Prediction
Janet Franklin

Decline and Recovery of the Island Fox: A Case Study for Population Recovery
Timothy J. Coonan, Catherin A. Schwemm, and David K. Garcelon

Ecosystem Functioning
Kurt Jax

Spatio-Temporal Heterogeneity: Concepts and Analyses
Pierre R. L. Dutilleul

Parasites in Ecological Communities: From Interactions to Ecosystems
Melanie J. Hatcher and Alison M. Dunn

Zoo Conservation Biology
John E. Fa, Stephan M. Funk, and Donnamarie O'Connell

Marine Protected Areas: A Multidisciplinary Approach
Joachim Claudet

Biodiversity in Dead Wood
Jogeir N. Stokland, Juha Siitonen, and Bengt Gunnar Jonsson

Landslide Ecology
Lawrence R. Walker and Aaron B. Shiels

Nature's Wealth: The Economics of Ecosystem Services and Poverty
Pieter J. H. van Beukering, Elissaios Papyrakis, Jetske Bouma, and Roy Brouwer

Birds and Climate Change: Impacts and Conservation Responses
James W. Pearce-Higgins and Rhys E. Green

Marine Ecosystems: Human Impacts on Biodiversity, Functioning and Services
Tasman P. Crowe and Christopher L. J. Frid

Wood Ant Ecology and Conservation
Jenni A. Stockan and Elva J. H. Robinson

Detecting and Responding to Alien Plant Incursions
John R. Wilson, F. Dane Panetta, and Cory Lindgren

Conserving Africa's Mega-Diversity in the Anthropocene: The Hluhluwe-iMfolozi Park Story
Joris P. G. M. Cromsigt, Sally Archibald, and Norman Owen-Smith

National Park Science: A Century of Research in South Africa
Jane Carruthers

Plant Conservation Science and Practice: The Role of Botanic Gardens
Stephen Blackmore and Sara Oldfield

Habitat Suitability and Distribution Models: With Applications in R
Antoine Guisan, Wilfried Thuiller, and Niklaus E. Zimmermann

Ecology and Conservation of Forest Birds
Grzegorz Mikusiński, Jean-Michel Roberge, and Robert J. Fuller

Species Conservation: Lessons from Islands
Jamieson A. Copsey, Simon A. Black, Jim J. Groombridge, and Carl G. Jones

Soil Fauna Assemblages: Global to Local Scales
Uffe N. Nielsen

Curious About Nature
Tim Burt and Des Thompson

Comparative Plant Succession Among Terrestrial Biomes of the World
Karel Prach and Lawrence R. Walker

Ecological-Economic Modelling for Biodiversity Conservation
Martin Drechsler

Freshwater Biodiversity: Status, Threats and Conservation
David Dudgeon

Joint Species Distribution Modelling: With Applications in R
Otso Ovaskainen and Nerea Abrego

Natural Resource Management Reimagined: Using the Systems Ecology Paradigm
Robert G. Woodmansee, John C. Moore, Dennis S. Ojima, and Laurie Richards

The Species–Area Relationship: Theory and Application
Thomas J. Matthews, Kostas A. Triantis, and Robert J. Whittaker

Ecosystem Collapse and Recovery
Adrian C. Newton

Animal Population Ecology: An Analytical Approach
T. Royama

Why Conserve Nature? Perspectives on Meanings and Motivations
Stephen Trudgill

Invading Ecological Networks
Cang Hui and David Richardson

Hunting Wildlife in the Tropics and Subtropics
Julia E. Fa, Stephan M. Funk, and Robert Nasi

The Life, Extinction, and Rebreeding of Quagga Zebras
Peter Heywood

Impacts of Human Population on Wildlife
Trevor J. C. Beebee

Conservation Translocations

Edited by

MARTIN J. GAYWOOD
University of the Highlands and Islands

JOHN G. EWEN
Zoological Society of London

PETER M. HOLLINGSWORTH
Royal Botanic Garden Edinburgh

AXEL MOEHRENSCHLAGER
IUCN SSC Conservation Translocation Specialist Group

Shaftesbury Road, Cambridge CB2 8EA, United Kingdom

One Liberty Plaza, 20th Floor, New York, NY 10006, USA

477 Williamstown Road, Port Melbourne, VIC 3207, Australia

314–321, 3rd Floor, Plot 3, Splendor Forum, Jasola District Centre, New Delhi – 110025, India

103 Penang Road, #05–06/07, Visioncrest Commercial, Singapore 238467

Cambridge University Press is part of Cambridge University Press & Assessment, a department of the University of Cambridge.

We share the University's mission to contribute to society through the pursuit of education, learning and research at the highest international levels of excellence.

www.cambridge.org
Information on this title: www.cambridge.org/9781108494465

DOI: 10.1017/9781108638142

© Cambridge University Press & Assessment 2023

This publication is in copyright. Subject to statutory exception and to the provisions of relevant collective licensing agreements, no reproduction of any part may take place without the written permission of Cambridge University Press.

First published 2023

Printed in the United Kingdom by TJ Books Limited, Padstow Cornwall

A catalogue record for this publication is available from the British Library.

Library of Congress Cataloging-in-Publication Data
Names: Gaywood, Martin J., editor. | Ewen, John G., editor. | Hollingsworth, Peter M., editor. | Moehrenschlager, A., editor.
Title: Conservation translocations / edited by Martin J. Gaywood, University of the Highlands and Islands, John G. Ewen, Zoological Society of London, Peter M. Hollingsworth, Royal Botanic Garden Edinburgh, Axel Moehrenschlager, Wilder Institute/Calgary Zoo.
Description: Cambridge, United Kingdom : Cambridge University Press, 2023. | Series: Ecology, biodiversity and conservation | Includes bibliographical references and index.
Identifiers: LCCN 2022036229 (print) | LCCN 2022036230 (ebook) | ISBN 9781108494465 (hardback) | ISBN 9781108714570 (paperback) | ISBN 9781108638142 (epub)
Subjects: LCSH: Animal introduction. | Plant translocation. | Wildlife conservation. | Wildlife management–Environmental aspects. | BISAC: NATURE / Ecology
Classification: LCC QL86 .C68 2023 (print) | LCC QL86 (ebook) | DDC 333.95/4–dc23/eng/20220816
LC record available at https://lccn.loc.gov/2022036229
LC ebook record available at https://lccn.loc.gov/2022036230

ISBN 978-1-108-49446-5 Hardback
ISBN 978-1-108-71457-0 Paperback

Cambridge University Press & Assessment has no responsibility for the persistence or accuracy of URLs for external or third-party internet websites referred to in this publication and does not guarantee that any content on such websites is, or will remain, accurate or appropriate.

Dedicated to our children Poppy, Luca, Liam, Izzy, Lauren, Lance, Tatyana, and Kaden. We hope this book inspires people to help nature during your lifetimes, and enjoy the many benefits from doing so.

MG, JE, PH, and AM

Contents

List of Contributors		page xv
Foreword		xxi
Razan Al Mubarak		
Preface		xxiii
Acknowledgements		xxvi

Part I	**Conservation Translocations: Getting Started**	1
1	**Moving Species: Reintroductions and Other Conservation Translocations** Martin J. Gaywood and Mark Stanley-Price	3
2	**Conservation Translocations: Planning and the Initial Appraisal** Sarah E. Dalrymple and Joe M. Bellis	43

Part II	**Conservation Translocations: The Key Issues**	75
3	**Conservation Translocations and the Law** Arie Trouwborst, Andy Blackmore, Sally Blyth, Floor Fleurke, Phillipa McCormack, and Martin J. Gaywood	77
4	**Decision-Making in Animal Conservation Translocations: Biological Considerations and Beyond** John G. Ewen, Stefano Canessa, Sarah J. Converse, and Kevin A. Parker	108
5	**Animal Disease and Conservation Translocations** Anthony W. Sainsbury and Claudia Carraro	149

6 **Animal Welfare, Animal Rights, and Conservation Translocations: Moving Forward in the Face of Ethical Dilemmas** 180
Lauren A. Harrington, Natasha Lloyd, and Axel Moehrenschlager

7 **Conservation Translocations for Plants** 212
Joyce Maschinski and Matthew A. Albrecht

8 **Plant Health, Biosecurity, and Conservation Translocations** 241
Ruth J. Mitchell, Sarah Green, and Peter M. Hollingsworth

9 **Genomics and Conservation Translocations** 271
Linda E. Neaves, Rob Ogden, and Peter M. Hollingsworth

10 **The Human Dimensions and the Public Engagement Spectrum of Conservation Translocation** 303
Jenny A. Glikman, Beatrice Frank, Camilla Sandström, Samantha Meysohn, Michelle Bogardus, Francine Madden, and Alexandra Zimmermann

11 **Assisted Colonisation and Ecological Replacement** 331
Maria Hällfors and Sarah E. Dalrymple

12 **The Role of Conservation Translocations in Rewilding and De-extinction** 354
Philip J. Seddon

Part III Conservation Translocations: Looking to the Future 379

13 **From Genes to Ecosystems and Beyond: Addressing Eleven Contentious Issues to Advance the Future of Conservation Translocations** 381
Axel Moehrenschlager, Pritpal Soorae, and Tammy E. Steeves

Part IV Case Studies — 413

14 Reintroduction of the Endemic Plant *Manglietiastrum sinicum* (Magnoliaceae) to Yunnan Province, China — 415
Weibang Sun, Lei Cai, and Peter M. Hollingsworth

15 Applying Adaptive Management to Reintroductions of Pyne's Ground-Plum *Astragalus bibullatus* — 422
Matthew A. Albrecht

16 Five Reasons to Consider Long-Term Monitoring: Case Studies from Bird Reintroductions on Tiritiri Matangi Island — 429
Doug P. Armstrong, Elizabeth H. Parlato, and John G. Ewen

17 Multiple Reintroductions to Restore Ecological Interactions in a Defaunated Tropical Forest — 436
Marcelo Lopes Rheingantz, Alexandra dos Santos Pires, and Fernando A. S. Fernandez

18 Bringing Jaguars and Their Prey Base Back to the Iberá Wetlands, Argentina — 443
Emiliano Donadio, Talía Zamboni, and Sebastián Di Martino

19 The Return of the Eurasian Beaver to Britain: The Implications of Unplanned Releases and the Human Dimension — 449
Roisin Campbell-Palmer, Andrew Bauer, Simon Jones, Ben Ross, and Martin J. Gaywood

20 The Role of Community Engagement in Conservation Translocations: The South of Scotland Golden Eagle Project (SSGEP) — 456
Catherine Barlow

21 The European Native Oyster and the Challenges for Conservation Translocations: The Scottish Experience — 462
Cass Bromley and David W. Donnan

22 **Slow and Steady Wins the Race: Using Non-native Tortoises to Rewild Islands off Mauritius** 469
Carl G. Jones, Vikash Tatayah, Rosemary Moorhouse-Gann, Christine Griffiths, Nicolas Zuël, and Nik Cole

23 **Assisted Colonisation as a Conservation Tool: Tasmanian Devils and Maria Island** 476
Carolyn Hogg and Phil Wise

Index 484

Colour plates can be found between pages 230 and 231.

Contributors

MATTHEW A. ALBRECHT
Missouri Botanical Garden, St Louis, MO, USA

DOUG P. ARMSTRONG
Wildlife Ecology Group, Massey University, Palmerston North, New Zealand

CATHERINE BARLOW
South of Scotland Golden Eagle Project, Galashiels, UK

ANDREW BAUER
Scotland's Rural College, Edinburgh, UK

JOE M. BELLIS
School of Biological and Environmental Sciences, Liverpool John Moores University, UK

ANDY BLACKMORE
Ezemvelo KwaZulu-Natal Wildlife and the School of Law, University of KwaZulu-Natal, Pietermaritzburg, South Africa

SALLY BLYTH
Beauly, Inverness-shire, UK

MICHELLE BOGARDUS
Pacific Islands Fish and Wildlife Office, United States Fish and Wildlife Service, Honolulu, HI, USA

CASS BROMLEY
NatureScot, Newburgh, Aberdeenshire, UK

LEI CAI
Kunming Institute of Botany, Chinese Academy of Sciences, Yunnan, China

ROISIN CAMPBELL-PALMER
The Beaver Trust, Pitlochry, UK

STEFANO CANESSA
Institute for Ecology and Evolution, Bern University, Bern, Switzerland

CLAUDIA CARRARO
Institute of Zoology, Zoological Society of London, UK

NIK COLE
Durrell Wildlife Conservation Trust, Jersey, UK

SARAH J. CONVERSE
US Geological Survey Washington Cooperative Fish and Wildlife Research Unit, School of Environmental and Forest Sciences & School of Aquatic and Fishery Sciences, University of Washington, Seattle, WA, USA

SARAH E. DALRYMPLE
School of Biological and Environmental Sciences, Liverpool John Moores University, UK

SEBASTIÁN DI MARTINO
Fundación Rewilding Argentina, Acassuso-Buenos Aires, Argentina

EMILIANO DONADIO
Fundación Rewilding Argentina, Acassuso-Buenos Aires, Argentina

DAVID W. DONNAN
NatureScot, Perth, UK

JOHN G. EWEN
Institute of Zoology, Zoological Society of London, UK

FERNANDO A. S. FERNANDEZ
Departamento de Ecologia–Instituto de Biologia, Universidade Federal do Rio de Janeiro, Brazil

FLOOR FLEURKE
Tilburg University, The Netherlands

BEATRICE FRANK
Capital Regional District Regional Parks, Victoria, BC, Canada

MARTIN J. GAYWOOD
University of the Highlands and Islands, Inverness, UK

JENNY A. GLIKMAN
Instituto de Estudios Sociales Avanzados (IESA-CSIC), Cordoba, Spain

SARAH GREEN
Centre for Ecosystems, Society and Biosecurity, Forest Research, Midlothian, UK

CHRISTINE GRIFFITHS
Ebony Forest, Black River, Mauritius

MARIA HÄLLFORS
Biodiversity Centre, Finnish Environment Institute, Helsinki, Finland

LAUREN A. HARRINGTON
WildCRU, University of Oxford, UK

CAROLYN HOGG
School of Life & Environmental Sciences, The University of Sydney, NSW, Australia

PETER M. HOLLINGSWORTH
Royal Botanic Garden Edinburgh, UK

CARL G. JONES
Durrell Wildlife Conservation Trust, Jersey, UK

SIMON JONES
Loch Lomond & The Trossachs National Park Authority, Balloch, UK

NATASHA LLOYD
Calgary Zoological Society, Calgary, AB, Canada

FRANCINE MADDEN
Center for Conservation Peacebuilding (CPeace), Washington, DC, USA

JOYCE MASCHINSKI
Institute for Conservation Research and Center for Plant Conservation, San Diego Zoo Global, Escondido, CA, USA

PHILLIPA MCCORMACK
Faculty of Law, University of Tasmania, Hobart, TAS, Australia

SAMANTHA MEYSOHN
Portland, Oregon, USA

RUTH J. MITCHELL
The James Hutton Institute, Aberdeen, UK

AXEL MOEHRENSCHLAGER
IUCN SSC Conservation Translocation Specialist Group, Calgary, AB, Canada

ROSEMARY MOORHOUSE-GANN
Durrell Wildlife Conservation Trust, Jersey, UK

LINDA E. NEAVES
Fenner School of Environment and Society, The Australian National University, Canberra, ACT, Australia

ROB OGDEN
Royal (Dick) School of Veterinary Studies and the Roslin Institute, University of Edinburgh, UK

KEVIN A. PARKER
Parker Conservation Ltd, Nelson, New Zealand

ELIZABETH H. PARLATO
Wildlife Ecology Group, Massey University, Palmerston North, New Zealand

ALEXANDRA DOS SANTOS PIRES
Departamento de Ciências Ambientais, Universidade Federal Rural do Rio de Janeiro, Brazil

MARCELO LOPES RHEINGANTZ
Departamento de Ecologia–Instituto de Biologia, Universidade Federal do Rio de Janeiro, Brazil

BEN ROSS
NatureScot, Inverness, UK

ANTHONY W. SAINSBURY
Institute of Zoology, Zoological Society of London, UK

CAMILLA SANDSTRÖM
Department of Political Science, Umeå University, Sweden

PHILIP J. SEDDON
Department of Zoology, University of Otago, Dunedin, New Zealand

PRITPAL SOORAE
IUCN Conservation Translocation Specialist Group, Abu Dhabi, UAE

MARK STANLEY-PRICE
WildCRU, University of Oxford, UK

TAMMY E. STEEVES
School of Biological Sciences, University of Canterbury, Christchurch, New Zealand

WEIBANG SUN
Kunming Institute of Botany, Chinese Academy of Sciences, Yunnan, China

VIKASH TATAYAH
Mauritian Wildlife Foundation, Vacoas, Mauritius

ARIE TROUWBORST
Department of Public Law & Governance, Tilburg University, The Netherlands and Faculty of Law, North-West University, South Africa

PHIL WISE
Save the Tasmanian Devil Program, Hobart, TAS, Australia

TALÍA ZAMBONI
Fundación Rewilding Argentina, Acassuso-Buenos Aires, Argentina

ALEXANDRA ZIMMERMANN
WildCRU, University of Oxford, UK

NICOLAS ZUËL
Ebony Forest, Black River, Mauritius

Foreword

We are at a critical point in time where humanity's relationship with nature is at a crossroads. Challenges such as the biodiversity and climate crises are immense, but nature-based solutions developed through innovation and actioned through collaboration can overcome many of the obstacles that lie before us. We must be bold. We must be courageous. And we must act now.

An inspiring approach to avert extinction and enable ecosystem recovery is that of conservation translocations. Returning species to the wild from programmes where they are under human care or where they are moved among wild populations can yield profound outcomes. I can tell you from personal experience that difficulties can be tackled if we combine sound science, planning, and action with unrelenting commitment and tenacity. I have been privileged to support the return of the Scimitar-horned oryx to Chad after the species was Extinct in the Wild for decades. Overcoming such immense challenges with collaborators on multiple continents showed the true power and possibility that conservation translocations can have.

Complex problems seldom have simple solutions. We need to ensure that conservation and sustainable development can go hand in hand – after all, the needs of nature and humanity are forever intertwined. Within the International Union for Conservation of Nature (IUCN) we aim to strive for a world where all people have a quality of life that gives them both dignity and opportunities. Through a diverse global membership and science-based approaches, the IUCN seeks to rise above the polarisation of ideas that often prevents progress. The IUCN Conservation Translocation Specialist Group within the Species Survival Commission evidences this approach by engaging with diverse stakeholders including practitioners, academics, Indigenous Peoples, local communities, conservation organisations, and governments around the world. Such inclusive engagement addresses a myriad of biological, social, cultural, legal, and economic considerations to seek successful outcomes for nature and for society.

Tackling big problems requires a diversity of knowledge and perspectives. As such I am so pleased to present this first authoritative text on conservation translocations. Contributors from all around the world not only showcase lessons learned to date but also set the stage for future actions that will help species large and small, restore ecosystems from oceans to land, and yield benefits for humanity that transcend geography and culture.

Let us now translate such knowledge into action. Let us work together with courage and optimism to create the change that the world needs now.

Razan Al Mubarak, President,
International Union for Conservation of Nature

Preface

This is a book about people moving species to help conserve our planet. In some ways the very fact that we have to use such drastic measures is a sad reflection of how damaged our environment has become. Desperate times call for desperate measures. Natural habitats and ecosystems have become degraded or destroyed, and populations of many animals, plants, and other species have become fragmented, small, and unviable, and their ability to disperse to new areas reduced or not possible. In response, various types of conservation intervention have been used to try and mitigate the damage we have caused, including conservation translocation.

Conservation translocation is no longer just a tool of last resort, but is increasingly being used in more proactive and creative ways, not only to save species but to restore habitats and ecosystems. Such projects also, when done in the right way, can have strong public appeal and help to engage people with nature. They can give people hope by showing that positive action can make a real, visible difference and contribute small but important and cumulative solutions to the global biodiversity crisis. But when they are done in the wrong way, and in particular when local communities, stakeholders, and Indigenous Peoples are not involved, then damage can be done, important support can be lost, and the chances of long-term success can be reduced.

We started work on this book in 2018, and yet in the short time between then and now we have seen significant new and ongoing challenges for our own species, and related developments in societal attitudes and concerns. We had no idea back then that most of the main writing of the book would be done during a global pandemic, which of course affected all our authors in different and sometimes very difficult ways, as it did for so many others. The 'anthropause' resulting from decreased human activity meant some people had an opportunity to reflect more on our complex relationship with nature. During this same period, we also had the Intergovernmental Science-Policy Platform on Biodiversity and Ecosystem Services (IPBES) report that demonstrated powerfully the increasingly desperate state of the biodiversity crisis, and other numerous high level reports emphasising the severity of predicted climatic change. This has been accompanied by rising frustration from people, especially the young, that there is still insufficient action to address these fundamental threats to our future existence. This was reflected by the high levels of public expectation on our political representatives to deliver environmental solutions at the 2021 COP26 conference on climate change in Glasgow, and the planned 2022 COP15 conference on biodiversity in Montréal.

So where does conservation translocation fit in all this? It is, of course, just one specific type of conservation intervention and arguably one we should try and avoid using by first ensuring large areas of habitats and ecosystems are protected and maintained. But, since the latter has often not happened or is insufficient, conservation translocation can have a vital role. Our aim in this book is therefore to provide our readers with some of the latest science and experiences in this field, which will help to maximise the chances of project success and reduce the risks of failure. We have set out the book to introduce some of the concepts and ideas behind conservation translocation, how projects should be planned, the specific issues that need to be considered when considering or running a project, and how to deal with complex decision-making. Issues such as legislation, animal welfare, plant and animal disease, genomics, and engaging with people are covered. So too are the more novel and challenging types of conservation translocation – not just reintroductions and reinforcements, but assisted colonisation, ecological replacement and associated multi-species translocations, ecological restoration, and rewilding. We also look to the future to consider how conservation translocation may develop over time. You will see that the first part of the book covers these main topics, but the last ten, shorter chapters provide case studies from around the world covering a range of animal and plant taxa, places, topics, and challenges. These demonstrate a key aspect of conservation translocations, which is that every one is different! But there is a standard, and best-practice approach (based on the IUCN Guidelines for Reintroductions and Other Conservation Translocations) that can be applied to all.

We are very conscious that conservation translocations, and other types of intervention, have not always been done well and it is therefore important that the conservation community recognises this and finds ways to improve. This especially applies to how people have been involved, or not involved, in some projects. Historically many projects have been run by ecological and biological specialists, meaning that the requirements of the species concerned are often well accommodated, but the lack of specialists in socio-economic fields has meant that the views, concerns, and aspirations of local/Indigenous Peoples most affected have not. And yet for conservation to work into the longer term, in this human-dominated world, we need to bring people with us and widen the 'ownership' of projects, as ultimately it is the local champions and land users who remain to follow through the necessary action on the ground, long after the professional conservationists have moved on. A project that builds trust and ownership not only improves the chances of its own success, but also increases the likelihood of other conservation interventions being successfully run in the same communities.

This book has also tried to demonstrate the diversity of projects involved in conservation translocation, although inevitably there is a bias towards those parts of the world best known by the editors. The book has strong Scottish origins that are reflected in much of the content, but there are also major contributions from editors and authors from Aotearoa New Zealand, Argentina, Australia, Brazil, Canada, China, England, Finland, Mauritius, Netherlands, South Africa, Spain, Sweden, Switzerland, UAE, USA, and Wales. Some of our 61 contributors are academics working in biological, social science, or legal fields, others work for non-governmental organisations, zoos, botanical gardens, consultancies, and public bodies. This reflects the range of skills and expertise often required in conservation translocation projects. We have also tried to demonstrate that the use of conservation translocation can apply to a range of taxa and environments, and were particularly keen to ensure that plant translocations and marine translocations were included as well as the usual terrestrial, animal (especially vertebrate) examples.

It is now widely accepted that ecological restoration and conservation are needed at transformational scales if we are to address our interconnected biodiversity and climate crises. Conservation translocation will be an increasingly significant part of that work, and our aim here is to inform the necessary positive action. We hope you find this volume a useful and inspiring source of information, whether you are a professional or voluntary conservationist, academic, student, land or water manager, or someone simply wanting to know more about this exciting area of work.

Martin J. Gaywood, John G. Ewen,
Peter M. Hollingsworth, and Axel Moehrenschlager

Acknowledgements

The editors would like to thank especially our many contributors, many of whom had to work through particularly challenging situations during the pandemic. MG would like to thank Michael Usher, the Commissioning Editor for the series, for originally inviting him to develop and produce the book, and for his patient guidance throughout the project. We are grateful to Aleksandra Serocka, Matt Lloyd, Dominic Lewis, and Jenny van der Meijden at Cambridge University Press for their support and guidance, as well as to Indra Siddharthan and Ruth Swan. Thanks also to Caitlin Andrews for her invaluable contribution, and the many photographers who provided images for the book. We also wish to express our gratitude to the IUCN, SSC, and CTSG for supporting the use of their logos on our book, to Jon Paul Rodríguez and Chris Mahon at the IUCN, and to Razan Al Mubarak for her inspirational foreword. Finally a particular thanks to our families who have patiently had to wait for this vast project to be completed!

Part I

Conservation Translocations: Getting Started

1 · *Moving Species: Reintroductions and Other Conservation Translocations*

MARTIN J. GAYWOOD AND MARK STANLEY-PRICE

1.1 Background

The increasing threats to our wildlife species have been reported for decades. However, the last few years have seen a dramatic increase in public awareness and concern, with a call for political representatives and decision makers to make 'transformative' changes to improve the prospects for nature. How we respond over the next decade will prove crucial if we wish to maintain and restore our biological diversity and ecosystem services.

In May 2019 the Intergovernmental Science-Policy Platform on Biodiversity and Ecosystem Services (IPBES) published its landmark report. It made a sobering read: around one million species are threatened with extinction, the abundance of native species in most land-based habitats has fallen by 20 per cent, mostly since 1900, and at least 680 vertebrate species have become extinct since the sixteenth century (IPBES, 2019). The five main, modern drivers of these impacts were listed as changes in land and sea use, the direct exploitation of organisms, climate change, pollution, and invasive non-native species – all of which carry the fingerprints of human activity. It is not surprising that many scientists now recognise a new geological time interval, the 'Anthropocene', defined by the conditions and processes on Earth profoundly altered by human impact (Crutzen & Stoermer, 2000) and characterised by the developing sixth mass extinction. Furthermore, a headline message of the 'Dasgupta Review' of the economics of biodiversity is that 'our economies, livelihoods and well-being all depend on our most precious asset: Nature' (Dasgupta, 2021). We ignore this at our peril.

We have been increasingly adept at recognising and measuring changes in nature. But the more difficult work involves identifying solutions and applying them. In fact, good tools already exist, and the IPBES report not only describes the scale of the challenge but also proposes ways forward. It lists methods that have '...been successful in preventing the extinction of some species', including the practice of 'translocation'. The report concludes that 'transformative change' is required to ensure a more sustainable future, and that the biodiversity challenge can be addressed effectively if that change starts now.

The specific tool of 'conservation translocation' has become increasingly used in the battle to save species and restore ecosystems. There are multiple formal definitions, but in short they describe people deliberately moving and releasing organisms where the primary goal is a conservation benefit. 'Reintroductions' are the best known type, and specifically refer to the translocations of organisms to places where they have become extinct, or where they could have been reasonably expected to occur, in order to try to re-establish viable populations. The science and practice surrounding conservation translocation have grown massively in recent decades, the result being that there are now many types and sub-types, with an increasingly confusing array of different terms. The International Union for Conservation of Nature (IUCN) (2013) has therefore come up with helpful, standard definitions (see Figure 1.1 and Box 1.1) that are widely accepted and employed, and indeed are used throughout this book.

Even so, conservation translocation is a tool that needs careful consideration before being used. Such projects are often complex, expensive, and time consuming, with a strong element of risk (not only in biological but also socio-economic terms) and some past failures (e.g. Griffith et al., 1989).

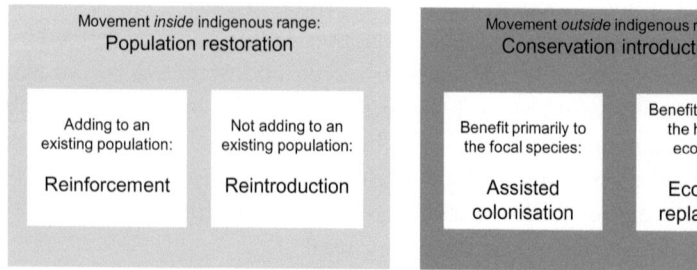

Figure 1.1 Overview of the types of conservation translocations (based on IUCN, 2013, and also applied in National Species Reintroduction Forum, 2014).

Box 1.1 *Definition of terms*

These definitions are based on the 2013 IUCN Guidelines, and are also widely applied in country-based approaches around the world. Also see Figure 1.1.

Conservation translocation – the intentional movement and release of a living organism where the primary objective is a conservation benefit. This usually involves improving the conservation status of the focal species and/or restoring natural habitat or ecosystem functions or processes.

Conservation translocations can entail releases either within or outside the species' indigenous range (the known or inferred distribution generated from historical records or physical evidence of the species' occurrence), and can be subdivided into the following categories:

1. **Population restoration** – a conservation translocation within the indigenous range, including:
   (a) **Reinforcement** – translocation of an organism into an existing population of the same species.

   Reinforcement aims to enhance population viability, for instance by increasing population size, by increasing genetic diversity, or by increasing the representation of specific demographic groups or stages. [Also known as: Augmentation; Supplementation; Re-stocking; Enhancement.]
   (b) **Reintroduction** – translocation of an organism inside its indigenous range from where it has disappeared, to re-establish a viable population of the focal species.
2. **Conservation introduction** – a conservation translocation outside the indigenous range, including:
   (a) **Assisted colonisation** – translocation of an organism outside its indigenous range where the primary purpose is to benefit the focal species.

   This is typically aimed at establishing populations in locations where the current or future conditions are likely to be more suitable than those within the indigenous range. The scale of assisted colonisation can range from local movement to wide-scale international range shifts. [Also known as: Benign introduction; Assisted migration; Managed relocation.]
   (b) **Ecological replacement** – translocation of an organism outside its indigenous range where the primary purpose is

> to perform a specific ecological function that has been lost through extirpation or extinction.
>
> Ecological replacement usually involves replacing the extinct taxon with a related subspecies or species that will perform the same or similar ecological function. [Also known as: Taxon substitution; Ecological substitutes/proxies/surrogates; Subspecific substitution, analogue species.]
>
> In all cases, conservation translocations have the primary goal of achieving a conservation benefit, which is defined as an improvement in the status of the focal species, habitat, or ecosystem.

Some of the more controversial and sometimes poorly executed examples, especially where local people were not involved in the decision-making, have resulted in conflict (Glikman et al., this volume), the result being that some people regard all 'reintroductions' as something that should be resisted, which makes organising new projects more difficult. The effect of translocation on the welfare of the individual animals involved has also been questioned (Harrington et al., this volume). And there are alternatives that should always be considered first (IUCN, 2013): area-based solutions such as wider habitat management; species-based solutions such as targeted control of invasive non-native animals and plants; social/indirect solutions such as setting up protected areas or changing legislation; or no action. Therefore, translocation has sometimes been described as a tool of last resort. But it is also an approach that works and has saved species and populations, and restored ecosystems (Novak et al., 2021). The release and return of a long-lost animal or plant can also be an exciting, inspiring, and engaging event for people, and demonstrates that there are still things we can do that make a positive difference. It is no longer just a tool of last resort – the urgency of the biodiversity crisis is such that we need to look at how we can be more creative and proactive with conservation translocation, for example through using certain influential species or combinations of species to help restore and upgrade ecosystem processes, whilst at the same time applying best practice.

Much of this edited volume was written during the 'anthropause' resulting from the COVID-19 pandemic (Rutz et al., 2020), during which people's minds turned even more to the crises in nature we are all grappling with, and the solutions we urgently need. The book brings together authors from across the world who use the IUCN (2013) guidelines as a way of making sure best practice is used in conservation translocations, thereby increasing the chances of success and decreasing the chances of failure. It looks at the key challenges that face practitioners, decision-makers, and other stakeholders who deal with conservation translocation, and provides the latest science-based theory and practice. Specific, fast-developing, and more radical topics are also covered, and attempts are made to look into the crystal ball and predict what will become most important, especially as we try to learn how to deal with a rapidly changing environment. Finally, a series of case studies is presented in the book that provide a taste of the range of species, ecosystems, places, and issues in which conservation translocation is used. This first chapter attempts to summarise some of this and provide a foundation for the details that follow.

1.2 A Very Short History of Translocations

Conservation is not the only reason people have translocated, or moved, species over the centuries. Seddon et al. (2012) recognised at least seven other motivations:

- Non-lethal management of problem animals.
- Commercial and recreational.
- Biological control.
- Aesthetic.
- Religious.
- Animal rights activism and animal liberation.
- ★Wildlife rehabilitation.

★(Although increasingly such 'welfare translocations' may sometimes be viewed as also having a conservation motivation where there have been short-term, enforced absences of animals from the wild (e.g. for gorillas and orang-utan). See the discussion of temporarily displaced species in Moehrenschlager et al. (this volume).)

Conservation translocation is also not new, although its early practitioners would not have described their actions using this terminology. For example, in Scotland the capercaillie *Tetrao urogallus* became extinct in the eighteenth century and was reintroduced in the 1830s (Stevenson, 2007), and the red squirrel *Sciurus vulgaris* became extinct in parts of the country around the same time, with animals from England and Scandinavia used to reintroduce or reinforce the Scottish population (Tonkin et al., 2016). Kakapo *Strigops habroptilus* in New Zealand and snowy egrets *Egretta thula* in the USA were both subjects of pioneering conservation translocations in the 1890s (Armstrong et al., 2018). The Eurasian beaver *Castor fiber* population was restricted to around 1200 animals scattered across a few isolated populations in Europe and Asia by 1900, but translocation began in 1922 using Norwegian animals to Sweden (Halley & Rosell, 2002), followed by dozens more initiatives across many European countries over the following decades. The motivations for some of the earlier translocations are sometimes unclear, and may not have been purely 'conservation' – for example the capercaillie is a game bird and the beaver has been an important resource for the fur trade.

Reintroduction as a modern conservation tool became progressively more used during the second half of the twentieth century, with high-profile examples including the Arabian oryx *Oryx leucoryx* to Oman (Stanley-Price, 1989) (Figure 1.2) and California condor *Gymnogyps californianus* to the western USA and Mexico (Walters et al., 2010). However, the inherent riskiness of reintroduction meant that, up to thirty years ago, failure rates were relatively high (Griffith et al., 1989). This overall growth in the use of the tool led to the establishment of a dedicated 'Reintroduction Specialist Group' (RSG) by the IUCN

Figure 1.2 Rangers in the Sultanate of Oman protect and monitor the first herd of Arabian oryx released in 1982 (photo: Mark Stanley-Price). (A black and white version of this figure will appear in some formats. For the colour version, please refer to the plate section.)

Species Survival Commission in 1988 to help the development of best practice. Ten years later the RSG published its first 'Guidelines for Reintroductions' (IUCN, 1998), a simple, pragmatic approach that thousands of practitioners around the world have subsequently used to support their decision-making.

1.3 From Reintroductions to All Conservation Translocations: Species Conservation, Ecological Restoration, and Rewilding

The RSG later produced a revised version of their key publication, with the new title 'Guidelines for Reintroductions and Other Conservation Translocations' (IUCN, 2013). In late 2018 the RSG then announced a change in its name to the 'Conservation Translocation Specialist Group' (CTSG). So why has there been this subtle change in scope from just species reintroductions to all forms of conservation translocation? In part it reflects the increasingly complex range of conservation challenges and issues that are being identified, and the fast developing science and practice. Therefore reinforcement, assisted colonisation, and ecological replacement, as well as reintroduction, provide a broad suite of actions that can help address the significant and increasing threats of climate change, disease transmission, habitat loss, and others. At the same time the guidelines continue to provide a simple framework to advise how such work should be done.

The numbers of such projects have also increased dramatically over the last three decades. Seddon et al. (2007) looked at the numbers of papers referring to reintroduction and found very small totals before the early 1990s (no single year reached double figures) but then a rapid increase, with a total of 454 papers for the 15 years up to 2005. The IUCN RSG/CTSG have been publishing case studies in their 'Global Conservation Translocation Perspectives' series since 2008, and by the time of their 2021 volume they had amassed details of 418 projects (Soorae, 2021). This trend is continuing, and it seems likely there have been thousands of projects taking place in recent decades, based on what continues to be published in journals and the grey literature, and the significant proportion that are not formally reported.

In addition, the types of projects are changing and diversifying. The primary goal of any translocation of this type must be a conservation benefit. This has often involved improving the conservation status of a focal species, for example reintroducing a threatened species to part of its indigenous range to help restore the population. There are plenty of examples of this approach, and often habitat restoration (and/or other actions, such as managing invasive non-native species) is required at prospective destination sites to treat the cause of a species' decline before the translocation can take place.

However, there is increasing use of conservation translocation to contribute directly to the restoration of the natural habitat or ecosystem functions and processes, rather than just focussing on the conservation benefits to the translocated focal species. There are a number of imaginative and ambitious examples involving 'keystone species' (those which have a disproportionately large effect on their environment relative to their abundance (Paine, 1995)), including 'ecosystem engineers' (organisms that directly or indirectly control the availability of resources to other organisms by causing physical state changes in biotic or abiotic materials (Jones et al., 1997)). These terms sometimes generate controversy in the scientific community, but they are useful to communicate and highlight the important ecosystem-level role certain individual species can play. Examples include the reintroduction of predators such as wolves *Canis lupus* to Yellowstone National Park in the USA, burrowing and digging species such as black-tailed prairie dog *Cynomys ludovicianus* to North American prairies and eastern bettong *Bettongia gaimardi* to south eastern Australia (Munro et al., 2019), reef-forming species such as corals (Swan et al., 2015), and the wetland-creating Eurasian beaver across many European countries (Figure 1.3).

'Ecological restoration' is a topic that has received considerable and increasing attention in recent years (the science behind it is called 'restoration ecology'). It is defined as '…the process of assisting the recovery of an ecosystem that has been degraded or destroyed' (Society for Ecological Restoration, 2004). However, advocates increasingly recognise that restoration has to be integral to land management in the modern world, and that goals for the future ecosystem should be achievable, rather than based on some arbitrarily selected point in the past (Hobbs & Harris, 2001). The term 'restoration' can therefore be problematic: it can be perceived as too backward looking if the focus is too much on composition, but less so if the focus is the return of ecological processes. The science and practice surrounding species reintroductions have also been developing over the last few decades, and the opportunities for synergy and collaboration between these two fields have started to be more fully recognised and realised. The translocations of key species to degraded systems are now regularly promoted as elements of

Figure 1.3 Beavers are ecosystem engineers and have been used in conservation translocation projects as a means of restoring ecological processes. At this Scottish site a metre-wide stream was dammed by beavers, resulting in an extensive beaver pond and associated wetland habitat (photo: Martin Gaywood). (A black and white version of this figure will appear in some formats. For the colour version, please refer to the plate section.)

wider ecological restoration. For example there are now numerous studies that have shown the importance of apex consumers and their role in 'trophic cascading', and the trophic 'downgrading' that can result when such species are removed by humans (Estes et al., 2011). The process of returning such species, and restoring and upgrading our ecosystems, is conservation translocation.

More recently still the term 'rewilding' has gained prominence and caught the public imagination to such an extent that it is now used to describe all manner of projects at all types of scales. Rewilding projects can also generate controversy and division, as some view them as an attempt to return to previous, natural ecosystems where people's livelihoods are given lower priority. This is a particularly sensitive issue in rural communities with long and complex socio-political histories of land use and ownership. In part, this reflects the range of definitions that different advocates use, although many recognise the complexities of the human dimension and the need to work and engage with those who are best placed to 'steward' the land concerned, including Indigenous Peoples (Moehrenschlager et al., this volume). Some definitions of rewilding include a particular focus on species reintroduction, such as that of Naundrup and Svenning (2015): 'Reintroduction of extirpated species or functional types of high ecological importance to restore self-managing functional, biodiverse systems'. Others have a wider scope, such as that of Carver et al. (2021), who attempted to identify guiding principles for rewilding, noting that '...rewilding sits upon a continuum of scale, connectivity, and level of human influence, and aims to restore ecosystem structure and functions to achieve a self-sustaining autonomous nature'. Both of these descriptions of rewilding focus more on restoring or reorganising ecological functions and processes than on trying to return and recreate places to the wild state of some particular historical moment of time.

Clearly there is overlap between the concepts of rewilding and ecological restoration, although the targets of the latter are generally more pre-defined than those of the former. Either way, these are concepts where the restoration process might involve not only 'passive' colonisation and recolonisation of sites by species but active intervention through conservation translocation. Seddon (this volume) provides a comprehensive assessment of how ecological restoration, rewilding, and conservation translocation compare and contrast, and where the commonalities lie.

Such approaches have to recognise that the starting points for such conservation activities are ecosystems that have been transformed by human activity. Indeed the term 'novel ecosystems' has been used to describe this phenomenon, although the term has courted controversy as some suggest it may predispose people to abandon attempts at restoration since it could be perceived as too difficult and costly (Aronson et al., 2014; Hobbs et al., 2014). However, the fact remains that many ecosystems have been or are being modified substantially; a full return to a system as it appeared before human impact will often not be possible, especially in light of continuing climatic change, but restoration can still achieve significant improvements to biodiversity and wider ecosystem functions. A challenge lies in where and how such restoration efforts should be prioritised, noting that the IPBES report concluded that participatory spatial planning on a landscape approach is vital where multiple land uses coexist, to enable the allocation and management of land to achieve social, economic, and environmental objectives in landscape mosaics (IPBES, 2019).

1.4 Drivers of Conservation Translocation

The IPBES recognises five main, modern drivers that have resulted in the current biodiversity crisis. In order of significance these are: changes in land and sea use, direct exploitation of organisms, climate change, pollution, and invasive non-native species. There are examples of conservation translocations that have tried to respond to all these drivers in different ways. In the following sections we also consider the inter-related issue of pathogens and disease transmission.

1.4.1 Changes in Land and Sea Use

Probably the majority of conservation translocations are done in response to the loss and decrease in quality of land and marine habitats, the challenge often being to find destination sites where the habitat is still suitable or can be adequately restored. Examples include the reintroduction or reinforcement of plant species such as woolly willow *Salix lapponum* (Figure 1.4) and alpine blue sow-thistle *Cicerbita alpina*

Figure 1.4 Woolly willow is a montane species that is vulnerable to grazing pressures. In Scotland it is now restricted and at risk, but an ongoing reinforcement programme has taken place over several years to try to restore this and other subarctic willow scrub species (photo: Lorne Gill/NatureScot). (A black and white version of this figure will appear in some formats. For the colour version, please refer to the plate section.)

(Marriott et al., 2015; Royal Botanic Garden Edinburgh, 2021) to specific sites in the Cairngorms, Scotland, where grazing pressure is less intense.

1.4.2 Direct Exploitation

Species that have been over-exploited include the freshwater pearl mussel *Margaritifera margaritifera* (Figure 1.5), and Scotland remains a country where illegal pearl fishing is still a threat. Freshwater pearl mussels have therefore been reintroduced to secret locations in the Scottish Highlands, and varying levels of success have been recorded through monitoring (Watt et al., 2018).

1.4.3 Climate Change

Climate change will continue to be a cause of species loss, but also a driver of more novel and creative approaches to how conservation translocation might be used to mitigate at least some of the impacts (Hopkins et al., 2007;

Figure 1.5 The freshwater pearl mussel has been subject to a range of pressures including pollution, river engineering, and pearl fishing, to the extent that it is now critically endangered. A range of conservation interventions is being applied and tested in Scotland, including the use of conservation translocations (photo: Sue Scott/NatureScot). (A black and white version of this figure will appear in some formats. For the colour version, please refer to the plate section.)

Scottish Natural Heritage, 2019). There is now increasing interest and debate in this field. Some species are particularly vulnerable where climate change reduces suitable habitat in their current locations, but their ability to respond is hampered by poor dispersal abilities. Climate change also does not act in isolation; it interacts with other potentially problematic factors such as habitat fragmentation and loss, and the presence of any natural or human-made barriers in the landscape. Brooker et al. (2011) highlighted the complexity in identifying the potential vulnerability of individual species to such impacts, especially those whose autecology was less understood, such as many cryptogams, other plants, and invertebrates. Even so, efforts have been made to identify species in Britain that may be particularly vulnerable (Brooker et al., 2011; Pearce-Higgins et al., 2015), the latter study identifying 640 of 3048 plant and animal species as at high risk from climate change under a 2°C warming scenario, and 188 at medium risk. Such measures of risk are helpful in informing decisions on where action may be prioritised, although there will be a level of uncertainty over the adaptability of such species to changing conditions.

In light of the increasing urgency of threats from climate change, how can we better use the potential of conservation translocation? The use of reintroduction and reinforcement will continue to have a role in some situations, focussing on the restoration of populations within the indigenous range of the species concerned. But if climate change, and other interacting factors, make potential destination sites in the indigenous range unsuitable in the short to medium term, the alternative option is to translocate a species outside its indigenous range, in other words to do an 'assisted colonisation', a type of 'conservation introduction'. However, the accidental or reckless introduction of non-native species has been the cause of substantial biological and socio-economic problems around the world, to the extent that the IPBES highlights this as a main cause of biodiversity loss as referred to in Section 1.4. Therefore the use of introduction for conservation purposes is an extreme measure that comes with added risk, although the climate emergency is now such that we need to increasingly consider such approaches, informed by guidelines that help us assess such risks (IUCN, 2013, 2017; Moehrenschlager, this volume). This is explored further in Section 1.5.1.

1.4.4 Pollution

The story of the peregrine falcon *Falco peregrinus* is a well-known example of where the impact of pollution prompted conservation translocation.

Approximately 1173 peregrines were released in the Midwestern USA and adjacent regions of Ontario and Manitoba (Tordoff & Redig, 2002) in a successful attempt to replace the population that had been extirpated by chlorinated hydrocarbon poisoning resulting from pesticide use in the 1950s.

1.4.5 Invasive Non-native Species

The threat posed by invasive non-native species (invasives) has probably been best documented in Australia and New Zealand. Many species recovery projects there have the difficult challenge of tackling the presence of invasives first, often followed by conservation translocation. One approach has been the use of offshore islands as destination sites where invasives are not present or can be more easily managed, or the use of large, fenced exclosures to form 'mainland islands' from which invasives are removed. Although this latter approach has received criticism because of the costs involved and the questions over long-term viability (Scofield et al., 2011), mainland islands have been increasing in numbers and scale. Examples include the Maungatautari Restoration Project, a 3,363 ha forested ecological island in the North Island of New Zealand, enclosed by a 47 km fence and from which most introduced animals, such as feral domestic cats *Felis catus*, brush-tailed possums *Trichosurus vulpecula*, and black rats *Rattus rattus* have been removed. Since it was established in 2001 at least seven indigenous bird species, such as North Island brown kiwi *Apteryx mantelli* and takahe *Porphyrio hochstetteri*, have been reintroduced to the enclosed area (Smuts-Kennedy & Parker, 2013). Australia also has a number of such exclosures that apply the same principles, including the Newhaven wildlife sanctuary that covers 262,000 ha in central Australia and involves a major programme of removing species such as feral domestic cat, fox *Vulpes vulpes*, and dromedary camel *Camelus dromedarius*, that will ultimately be followed by planned reintroductions of mala *Lagorchestes hirsutus*, central rock rat *Zyzomys pedunculatus*, and golden bandicoot *Isoodon auratus* (Australian Wildlife Conservancy, 2022).

1.4.6 Pathogens

The IPBES has also highlighted the heightened risks to people from zoonoses (infectious diseases that can be transmitted from non-human animals to humans) as human activities intensify and increased contact with wildlife results. This is not something that conservation translocations can address directly. However, at the basic level care should always be taken to ensure that any public health risks, and risks to domesticated animal health, are properly assessed and mitigated when conservation translocations are carried out. Furthermore, the risk of disease transmission between wild animal species, and between wild plant species, should always be considered.

The challenges associated with disease transmission in wild species have become more significant, partly as a result of the increased presence of invasives that can act as hosts (e.g. crayfish plague carried by North American signal crayfish *Pacifastacus leniusculus* which has devastated native crayfish *Astacus* spp. populations in Europe) and partly through the complicating effects of climate change. In some cases the impact of disease on native species populations has been so dramatic that conservation translocation has been used to move animals to isolated destination sites that can act as refuges. This has sometimes involved assisted colonisation, one of the best known examples being the 2012 translocation of Tasmanian devils *Sarcophilus harrisii* to Maria Island, which is not within the indigenous range of the species. By mid-2014 over 100 individuals were present, and the population is now used as a source of trial releases back to mainland Tasmania, and to genetically augment the wild diseased populations (Hogg & Wise, this volume). The decision to translocate had to take into account the risks of devil predation to the resident bird colonies on the island, some of which have subsequently been impacted.

The act of conservation translocation of an organism also involves moving a 'biological package' of organisms hosted by the focal species. Consequently the practitioner also needs to be very aware of the disease transmission risks associated with a translocation and how they can be mitigated. Sainsbury and Carraro (this volume) cover this in detail for animal diseases, and Mitchell et al. (this volume) for plant diseases.

1.5 More Radical Approaches to Conservation Translocations

1.5.1 Assisted Colonisation

Assisted colonisation is an emerging tool (Hällfors & Dalrymple, this volume). The risks and uncertainties involved in using it, and the necessary mitigation required, will vary depending on factors such as the biology of the species concerned, the 'reversibility' of any release, and the species' potential impact in biological and socio-economic terms.

Figure 1.6 Assisted colonisation of the fruticose terricolous arctic-alpine lichen *Flavocetraria nivalis* has been tested and monitored within the Cairngorm Mountains, Scotland. Each individual transplant was tagged as shown, to assist future identification (photo: Lorne Gill/NatureScot). (A black and white version of this figure will appear in some formats. For the colour version, please refer to the plate section.)

At one extreme the relatively localised, assisted colonisation of the arctic-alpine crinkled snow lichen *Flavocetraria nivalis* has been tested between mountains in the Cairngorms, Scotland (Brooker et al., 2018) (Figure 1.6). This demonstrated the difficulty in building predictive models of habitat suitability for immobile species that respond strongly to very local conditions, and the need to also involve expert judgement and the use of many individual transplants to increase the likelihood of success. Perhaps unsurprisingly, this project generated little controversy and was generally regarded as an acceptable experimental trial to test whether arctic-alpine crypytogams potentially vulnerable to climate change would benefit from this type of intervention.

However, the acceptability of using assisted colonisation for some other species, especially vertebrates, and over far greater translocation distances, will no doubt prove to be a more contentious topic. Some have highlighted concerns that we have not yet developed sufficient understanding of the impacts of introduced species to be able to make informed decisions about translocating them (Ricciardi & Simberloff, 2009). However, Thomas (2011) argued that consideration could be given to translocating species at risk from climate change from further afield, such as the Provence chalkhill blue butterfly *Polyommatus hispanus* and the de Prunner's ringlet butterfly *Erebia triaria* from southern Europe to southern England. More bold, and inevitably more controversial, suggestions made in the same paper include vertebrates such as the Pyrenean desman *Galemys pyrenaicus*, Spanish imperial eagle *Aquila heliacea adalberti*, and Iberian lynx *Lynx pardinus*, with the argument that Britain could be an ideal assisted regional colonisation area (ARC) and contribute to the conservation of globally threatened species. It will be interesting to see how the acceptability or unacceptability of such dramatic, proposed interventions develops over the years to come. Such proposals will need to assess the importance of different 'values', and apply appropriate structured decision-making tools (Ewen et al., this volume).

1.5.2 Ecological Replacements

Assisted colonisation is a form of conservation translocation where the benefit is primarily to the focal species concerned. However, in another form of conservation introduction called 'ecological replacement', again the translocation involves moving a species outside its historic, indigenous range but the primary purpose is to perform a specific ecological function that has been lost through extinction. The benefit is therefore primarily for the relevant, wider habitat or ecosystem. Again, this is a relatively novel approach with different levels of risks to address, and it has only been used or proposed a few times (see Hällfors & Dalrymple, this volume).

Probably the best known examples have been with giant tortoises. The Giant Tortoise Restoration Initiative led by the Galápagos Conservancy and Galápagos National Park Directorate uses a range of conservation actions, including ecological replacement. Giant tortoises specific to Santa Fe Island and Pinta Island are now extinct ('Lonesome George', the famous Pinta Island tortoise *Chelonoidis abingdonii*, died in 2012), and the tortoises of Floreana went extinct in 1835 although hybrids exist on another island (Galápagos Conservancy, 2022). The giant tortoises of these islands provided important ecosystem roles such as trampling vegetation, opening areas, and dispersing seeds, thereby providing habitat conditions that support other island species. Work is now underway to identify which alternative, extant giant tortoise species could be introduced to these islands to continue these ecosystem roles. On Pinta the decision on the species to be used in the long term is still awaited, but in the meantime sterilised giant tortoises have been introduced to allow their impacts to be monitored. Initial results have demonstrated increased local vegetation patchiness and shown that even moderate density tortoise populations can reverse woody plant encroachment (Hunter & Gibbs, 2013). Meanwhile, on Santa Fe Island, fertile Española giant tortoises *Chelonoidis hoodensis*, endemic to Española Island, were first introduced in 2015, again followed by post-release monitoring.

Similar ecological replacements of giant tortoise species have also been used in the Mascarene Islands, where five native species became extinct by the mid-1800s. On Ile aux Aigrettes, Mauritius, the non-native Aldabra giant tortoise *Aldabrachelys gigantea* was therefore introduced as an ecological replacement, first into enclosures and then into the wild. Its role in creating and maintaining 'tortoise lawns' in open areas, thereby contributing to a more heterogeneous habitat mosaic beneficial to biodiversity, improving seed germination of an endemic ebony species after ingestion, and other factors, is described by Jones et al. (this volume). Animals have now been translocated to Round Island as well, and there are plans for other destination island sites.

The predicted effects of climate change and disease on woodland communities in Britain are also creating debate over how interventionist we should be, including whether we should be starting to replace some dominant tree species that are likely to decline increasingly over time (e.g. ash *Fraxinus excelsior* as a result of ash dieback fungal disease *Hymenoscyphus fraxineus*, Mitchell et al., 2014). One form of what one might arguably class as ecological replacement has been used for some time in Britain and elsewhere in Europe. The aurochs *Bos primigenius* became extinct in the seventeenth century, although much of its gene pool lives on in the domesticated cattle breeds we have today. These large herbivores would have played influential roles in maintaining structural diversity, and consequently biological diversity and wider ecosystem functioning (van Wieren, 1995), although we cannot be certain of their precise ecological effects. More 'traditional' breeds of modern cattle and ponies are likely to perform some of these roles, and they are often used in modern nature conservation management. Perhaps these are not examples of planned, deliberate ecological replacement in the strict sense but, in the absence of aurochs, such traditional breeds may have an increasingly important role in large scale ecological restoration and rewilding projects, and maintaining special grassland and open woodland habitats (Hodder et al., 2005). This might seem to go against the general push to reduce the numbers of cattle because of their contributions to creating carbon emissions, but traditional breeds of cattle used in low intensity situations are small in number compared to conventional farming in industrialised countries, and can have benefits in climate change mitigation (Pyke & Marty, 2005). Recently a small number of European bison *Bison bonasus* have been released within an enclosed reserve in Kent, UK. Some would describe this as an ecological replacement of species such as the steppe bison *Bison priscus* that became extinct in the late Pleistocene.

1.5.3 Multi-species Conservation Translocations

Conservation translocations usually focus on single focal species, but increasingly some projects are attempting to translocate groups of species in order to restore communities and thereby wider ecosystems. For example Foundation Rewilding Argentina's work in the Iberá wetlands resulted in the first jaguars being reintroduced in 2021, with associated ongoing conservation translocations of prey species such as Pampas deer *Ozotoceros bezoarticus*, giant anteaters *Myrmecophaga tridactyla*, and collared peccaries *Pecari tajacu* (Donadio et al., this volume). On Wedge Island, Australia, the release of an ecosystem engineer, the southern hairy-nosed wombat *Lasiorhinus latifrons*, was accompanied by translocations of black-footed wallaby *Petrogale lateralis pearsonii* and brush-tailed bettong *Bettongia penicillata*. The wombat burrows increased habitat complexity and were subsequently used by the wallabies and bettongs as well (Ostendorf et al., 2016), and all three species have increased in numbers. Arguably, even the sowing of diverse, native species seed mixtures on ex-arable soils to produce species-rich swards could be described as a type of multi-species conservation translocation with the aim of restoring grassland communities, although the reinstatement of the associated microbial and faunal communities is more of a challenge (Walker et al., 2004), and requires time to allow natural processes to take effect.

Such multi-species translocations involve an additional level of complexity because of the potential interactions between the species involved. Plein et al. (2016) developed models that highlighted the need to consider the types of interspecific interactions (e.g. consumer-resource, mutualism, and competition), the sequencing of releases, and founder sizes on the type of translocation strategies that should be used. Another modelling exercise looked at options for restoring disturbed plant-pollinator communities and concluded that reintroducing multiple, highly interacting generalist species worked best for restoring the species richness of lightly disturbed communities, whereas the introduction (rather than reintroduction) of generalist species was more effective for more significantly disturbed communities (LaBar et al., 2014).

1.5.4 De-extinction and Genetic Interventions

The IUCN (2016) has produced 'Guiding Principles' on de-extinction, defined as 'the process of creating an organism that resembles an extinct species'. To date most de-extinction proposals have involved mammal and bird species. The methods used have ranged from back breeding (e.g. of domesticated cattle to produce Heck cattle that resemble the extinct aurochs), to CRISPR gene-editing technology. An example of the latter is the work underway to produce birds with the traits of the extinct passenger pigeon *Ectopistes migratorius*, a species once so numerous that it has been considered an ecosystem engineer of North American forests (Revive & Restore, 2021). Seddon (this volume) explores this topic in more detail.

De-extinction therefore raises the future prospect of conservation translocations involving extinct species, or at least functional proxies of such species. Genetic technology is also being developed for interventions on extant species (Neaves et al., this volume). For example, permission is currently being sought to allow the planting in the wild of genetically engineered American chestnut *Castanea dentata*, an ecologically important species of eastern North America devastated by an introduced fungal blight (Newhouse & Powell, 2020). The genetically engineered trees are blight resistant. Meanwhile, in late 2020, a Przewalski's horse *Equus ferus przewalskii* named Kurt was born, and made global news by becoming the first animal cloned for conservation purposes. We can expect the development and use of such technologies to accelerate, presenting powerful new tools for conservation translocations and other conservation interventions, but also complex biological, social, and ethical challenges.

1.5.5 Mitigation Translocations

These involve the '...removal of organisms from habitat due to be lost through anthropogenic land use change and release at an alternative site' (IUCN, 2013). Examples include developments where legal permissions to proceed may be conditional to mitigating specific environmental impacts, including on particular species that may be present. Such reactive, economically motivated uses of translocation contrast with the more proactive, purely conservation motivated translocations described elsewhere in this chapter. Mitigation translocations can clearly serve a conservation purpose where the alternative is leaving the organisms at the development site and risking their destruction, but their use can be controversial. The concern is that translocation is too often put forward as a publicly acceptable 'solution' by developers when protected species, or habitats, are present at a site. Alternative, and potentially less risky solutions, may be more appropriate, including the least risky option from a conservation perspective of not permitting the development at all where the conservation value of a site is high.

Even so, mitigation translocations have become increasingly used. Despite the fact that such translocations can often be well resourced, their long-term effectiveness for the species concerned remains uncertain. Indeed the general pattern has often been that the resourcing goes primarily into the planning and execution of the translocation itself, but far less into the post-release monitoring and reporting that is so essential to allow assessments of effectiveness. Germano et al. (2015) noted that many mitigation translocations fail; there was a failure to document outcomes, and a need for the billion dollar environmental consulting industry to address such shortcomings. There are some species of conservation value that have become regular candidates of mitigation translocations. Britain is the European stronghold for the great crested newt *Triturus cristatus*, a highly protected but fairly widespread species. A review of 460 licensed projects involving the species found that only 22 provided post-development monitoring data, of which 16 reported that at least one small population was sustained (Lewis, 2012). The Conservation Evidence (2021) web site scores the 'effectiveness' of this type of action for great crested newt at 50 per cent.

Guidelines have been published that emphasise the value of properly considering options when assessing development proposals, and ensuring that approved mitigation translocations are properly designed and monitored (IUCN, 2013; Randall et al., 2018). The mitigation translocation of Mojave desert tortoises *Gopherus agassizii* from areas scheduled for solar energy developments was judged to have been well designed and monitored, and to have demonstrated the wider conservation value of reporting the details of properly considered pre- and post-release work (e.g. Dickson et al., 2019).

1.6 The Biological Considerations behind Conservation Translocations: Using Science and Guidance to Make Better Decisions

The challenges of any conservation translocation are often complex and involve a range of risks and uncertainty. The applied science of reintroduction biology has therefore developed since the 1990s, in part prompted by observations that many earlier reintroduction projects had failed. Efforts have been made to encourage more strategic approaches, to address specific issues *a priori* (i.e. beforehand, based on reasoning and experience), and to apply appropriate research such as predictive modelling techniques and experimental studies to improve the outcomes of released captive-bred animals (Seddon et al., 2007). The aim has been to help decision-making become more evidence-based, thereby improving the likelihood of project success.

Taylor et al. (2017) carried out an analysis of how recent advances in reintroduction biology are actually being applied in reintroduction practice. They looked at four broad areas where the science has been developing: population establishment, population persistence, meta-populations, and ecosystems. A total of 361 reintroduction-related papers were examined, and they found that 61% of papers addressed questions at the population establishment level, 32% at the population persistence level, 4% at the meta-population level, and just 3% at the ecosystem level. They also found that 49% of all the papers clearly stated *a priori* questions, although this increased over time to 64% by 2016, which might suggest an improvement in best practice. The authors of this and other studies noted the need for decision-making to be better incorporated into reintroduction biology, to target management uncertainties and to apply adaptive management approaches rather than trial and error (Canessa et al., 2016; Albrecht, this volume; Ewen et al., this volume).

Such rigorous application of scientific approaches should improve the likelihood of success for conservation translocations, and therefore provide benefits beyond improving the conservation status of the species or ecosystem concerned. Such benefits include: ensuring limited financial resources are most efficiently and effectively directed; improving the likelihood of relevant stakeholders remaining enthused and engaged during the work; and ensuring that potentially limited stocks of the relevant donor plants and animal populations (which are often threatened species) are utilised in the best way possible. However, the meta-analyses of such projects reported in the scientific literature, such as those referred to in the paragraph above, will inevitably be skewed by focussing on those written up by practitioners with access to scientific expertise and able to publish in peer-reviewed journals. The details of many (probably most) projects do not get published in these fora, and some do not apply more robust scientific approaches for a variety of reasons. In many cases the project personnel may feel overwhelmed by some of the apparent complexity of the methods advocated by reintroduction biologists. Others are simply not convinced of their value; they may feel that 'perfection can be the enemy of good', and assume such approaches will result in delay.

The aim, therefore, is to find ways by which beneficial projects can be supported with simple, pragmatic guidance leading to action. The developing scientific literature is a vital source of information, but it is not always accessible to some of the key audiences (in a physical or readable sense) who may be involved in conservation translocation practice. Therefore other ways of sharing knowledge and experience are also needed. Increasingly, more 'straightforward' decision support tools are being developed and made available to help practitioners prioritise their efforts and deal with the inevitable uncertainties they can come up against (e.g. Ewen et al., this volume; IUCN, 2017). Further guidance on the types of biological and socio-economic issues to consider is included in the IUCN Guidelines (IUCN 2013), related documents such as the Scottish Code for Conservation Translocations (National Species Reintroduction Forum (NSRF), 2014), and this book (e.g. Dalrymple & Bellis, this volume; Ewen et al., this volume; Maschinski & Albrecht, this volume). The Scottish Code includes a list of the potential biological and socio-economic risks associated with any translocation that should be considered during an initial appraisal (Box 1.2). Similarly, assessments can be made of the potential benefits of the translocation (Box 1.3) and legal requirements. If the initial appraisal process suggests clear benefits, and the legal requirements and risks seem surmountable through a realistic level of mitigation, the next step is to formalise the planning process and address any key issues (especially the higher risk issues) in more detail.

Future options to improve capacity should include the development of more 'plain language' guidance – for example handbooks and simple web-based material that summarise the latest applied research, and the provision of simple online, decision-making tools that can be accessed anywhere in the world. The IUCN CTSG has already organised workshops and training in different countries to help with capacity building,

Box 1.2 *Examples of biological and socio-economic risks associated with conservation translocations (National Species Reintroduction Forum, 2014)*

Biological:
- Distance of the translocation.
- Threat to the source population.
- Establishment following the translocation may cause loss/reduction of important habitat.
- Establishment may cause loss/reduction of important species.
- Translocation may spread pests and diseases.
- Hybridisation threat (intra-specific or inter-specific).
- Species is likely to spread beyond the confines of the destination site (this can be a measure of success, but risks need to be considered).
- Potential for animal welfare concerns to released animals or those they interact with.

Socio-economic:
- The likelihood of strong social resistance by some to translocation.
- Harm to human health and well-being.
- Harm to human livelihoods.
- Insufficient resources may prevent successful implementation of the translocation plan.
- Major financial cost once the translocation has been completed (e.g. control measures if the population has greater impacts than envisaged).

Box 1.3 *Examples of benefits associated with conservation translocations (National Species Reintroduction Forum, 2014)*

Focal species – reducing extinction risk and/or improving the conservation status of a species:

- Increasing the number of individuals, improving population structure, and/or increasing the number of locations at which species occur.
- Improving the genetic health and resilience of a population by directly introducing genetic diversity.
- Establishing 'bridging populations' to facilitate migration and/or gene flow.

- Establishing populations in areas where the species will experience reduced levels of threat (e.g. by moving organisms into more suitable 'climate space', disease-free areas, or localities with suitable management).

Habitats/ecosystems – improving the conservation status of an ecosystem, habitat, and/or other species:

- Increasing the overall species richness of a habitat to enhance its biodiversity value.
- Increasing habitat quality (e.g. translocating species to change grazing regimes).
- Improving ecosystem services and functions (e.g. translocating species to provide pollinator services).

People – socio-economic benefits that may arise as a result of a conservation translocation:

- Enriched human experiences and environmental awareness due to increased contact with biodiversity.
- Increased benefits to humans from ecosystem services (e.g. pollination).
- Increased income (e.g. revenue from ecotourism where the translocated species leads to increased visits or spend).

and produced some initial, web-based training videos (IUCN, 2021). It will take time to further develop and promote these tools but, for the mean time, all practitioners should be encouraged to monitor their projects after release (a failing of many projects as reported in Sutherland et al., 2010) and report the results even, and indeed especially, if they fail. This does not necessarily have to be in scientific journals that require intensive peer-review; it could be as submissions to repositories such as the IUCN CTSG's 'Global Conservation Translocation Perspectives' series of case studies published every few years (most recently Soorae, 2021).

1.7 The People Considerations behind Conservation Translocations

Historically many conservation projects, including translocations, have tended to focus predominantly on biological considerations. This has been in part due to the fact that such projects often involve specialists and practitioners with more biological backgrounds and training, and much of the relevant scientific literature and international guidance still has that strong focus. Biological expertise is a vital element of these types of projects where the primary aim is to improve the conservation status of focal species and wider habitats and ecosystems.

However, as time has moved on there has been increasing recognition that the socio-economic, 'people' side of this work (including cultural and spiritual considerations) needs to improve to ensure more effective, fair, and democratic approaches are used. Furthermore, better engagement with key stakeholders helps to increase the likelihood of project success and therefore biodiversity benefits. The human population and its use of the land and sea continues to increase, and therefore the likelihood of conservation translocations overlapping with people's interests, and sometimes creating conflict or opportunities, will also rise.

1.7.1 Addressing the Risks and Opportunities for People

The revised IUCN Guidelines (2013) therefore included an expanded section on 'social feasibility' compared to the original 1998 version. In Scotland, the socio-economic aspects of such projects have been given a particular focus, starting with the formation of the 'National Species Reintroduction Forum' (NSRF), which includes members of not only the usual conservation and environmental non-governmental organisations (NGOs) and government bodies but also organisations representing land and water management and interests. Its origin was partly grounded on the experiences of two high-profile projects – the reintroductions of the white-tailed eagle *Haliaeetus albicilla* (Figure 1.7) and the Eurasian beaver to Scotland – both of which resulted in intense and passionate debates within the public, political, and media arenas. The formal, licensed reintroduction of Eurasian beaver was complicated by unofficial escapes or illegal releases, with consequent impacts on trust between the conservation and land management communities, and ultimately on the prospects of other, future mammal reintroduction proposals (Campbell-Palmer et al., this volume). The white-tailed eagle release resulted in complaints by farmers and crofters over the lack of consultation before the original releases, and debate over how they should be compensated for the loss of lambs they reported.

There are now fora in Scotland that were specifically set up to enable the management and conservation issues surrounding these species to be

Figure 1.7 Following its extirpation in the early twentieth century, the reintroduction of the white-tailed eagle to the west coast of Scotland began in the 1970s, with a second phase in the 1990s. This was followed by the 'East of Scotland Sea Eagle Project' that ran from 2007 to 2012. There are now estimated to be about 150 breeding birds in the country (photo: Lorne Gill). (A black and white version of this figure will appear in some formats. For the colour version, please refer to the plate section.)

discussed. Disagreements and concerns remain but both species are becoming increasingly established, and some socio-economic benefits have started to be realised (Morling, 2022). The NSRF therefore used these and other experiences to inform the design of its Scottish Code for Conservation Translocations, based around the IUCN Guidelines framework but with a strong Scottish focus and consideration of people issues, as well as the usual biological aspects (Box 1.2). The Code was therefore not just the product of biological specialists but also representatives of land and water managers and users, conservationists, government agency specialists, and others.

Conservation translocation projects therefore should always consider the human dimension to their projects (Dalrymple & Bellis, this volume; Glikman et al., this volume; Moehrenschlager et al., this volume). This will vary tremendously depending on the species concerned. The translocation of a lichen species from one hill top to another may be biologically complex, but the socio-economic implications are likely to be limited to engagement with the relevant land managers, interest groups, and statutory authorities. However, the translocation of a predator (which in some ways might arguably be less biologically complex, or at least better researched, than a lichen translocation) will often be a contentious proposition and involve a far wider range of stakeholders.

Therefore the planning and implementation of conservation translocations, especially those with complex people considerations, need good governance and stakeholder involvement (Martin et al., 2012). Ultimately the communities living in areas where organisms are released or planted will not only be most affected by such projects but also in a key position to facilitate the long-term viability of any restored population. Complex projects will benefit from the use of specific planning tools (e.g. Ewen et al., this volume) and the involvement of neutral facilitators (IUCN, 2017). Key stakeholder individuals and organisations need to be identified and engaged as early on in the process as possible. Such engagement should be a genuine, two-way listening process with the proposers providing information and other stakeholders invited to set out their own values, experiences, and any concerns they have or benefits they wish to access. Trust can be built by providing clarity over who is responsible for making decisions, and transparency over how decisions are made. Engagement with indigenous communities is fundamental (e.g. McMurdo Hamilton et al., 2021; Moehrenschlager et al., this volume).

Conservation translocations tend to be driven by biological specialists, but where there are complex socio-economic aspects to a project then it is necessary to draw on the professional input of experts in public engagement, social aspects, and wildlife conflict. It would be unthinkable not to use experts on the animal or plant concerned during a conservation translocation, so on that basis it seems common sense to involve experts in the social sciences and humanities for projects with significant human dimensions.

The potential real or perceived negative impacts of a translocation on people have to be considered early in project development, but there may be a range of direct and indirect socio-economic benefits too (Box 1.3). The IPBES report, in addition to identifying the five main drivers of biodiversity loss, also identified two other significant, indirect drivers: the lack of connection people have with nature, and the lack of value and importance placed on nature (IPBES 2019). Conservation translocation has a role to play here too, by helping to engage people in positive solutions to challenges that can sometimes feel overwhelming in their scale. It is, by its very nature, a very visible, immediate, and concrete activity that the public can easily relate to, thereby providing a way of promoting wider biodiversity issues. The media, public, and political interest generated when an animal is released or a rare plant put back in the wild can be substantial, and provides opportunities for practitioners to explain the underlying problems that caused the decline of the species in the first place and why such desperate measures are needed to try to restore them.

Such projects can also help to promote the wider ecosystem value that translocations can have as part of ecosystem restoration, creation, or resilience initiatives, whether tagged as 'rewilding' or not (e.g. through reintroducing 'ecosystem engineer' or other 'keystone' species). Beavers, for example, provide a range of ecosystem services (Gaywood, 2015). These include 'provisioning ecosystem services' such as increased ground water storage, 'regulation and maintenance ecosystem services' such as flow stabilisation and flood prevention, and 'cultural ecosystem services' that relate to people's recreational, educational, and spiritual interactions with the environment. All of these contribute to human well-being and have socio-economic impacts.

The South of Scotland Golden Eagle Project has given particular attention to cultural ecosystem services, involving local communities and those further afield, with dedicated project officers covering stakeholder engagement and public outreach (Barlow, this volume). This has included supporting local 'Eagle Schools' that learn about and champion the project and link with schools from areas where the donor animals came from. In New Zealand there has been a large increase in the number of community-led projects that can result in wider engagement and interest in environmental issues (Department of Conservation, 2012). The reintroduction of bison *Bison bison* to North America's Great Plains was found to have immediate positive benefits on visitors and in connecting people to conservation (Wilkins et al., 2019). There can be direct economic benefits too; for example £4.9 to £8 million of annual tourist spend on the Isle of Mull in Scotland results from people wishing to see the white-tailed eagles that resulted from reintroduction (Morling, 2022).

1.7.2 Working with Other Environmental Stakeholders

The IUCN guidelines for conservation translocations (IUCN, 2013) are used by many sectors involved in wildlife conservation. Since they were first developed over three decades ago, subsequent scientific developments and practical experiences have informed their further refinement. They have also provided a framework on which to build guidelines and codes of best practice that are more relevant at domestic levels (e.g. DEFRA, 2021; National Species Reintroduction Forum, 2014) and legal recommendations at regional scales (The Standing Committee of the Convention on the Conservation of European Wildlife and Natural Habitats, 2012). However, the promotion and application of such best practice frameworks designed for conservation translocations tend to be more prevalent within certain sectors of biodiversity conservation, reflecting the unintentional silo approaches that often develop in many aspects of environmental work. Are there opportunities to widen the application of such frameworks, and to learn from the approaches developed by other sectors?

Examples of translocations that have wider environmental and other motivations, and have not always been classed as conservation translocations by practitioners, include:

- Forestry. The translocation of tree species is a widespread forestry practice that historically tended to result from more socio-cultural and commercially driven motivations. However, tree planting is increasingly viewed by governments as a 'low tech' means of capturing carbon and helping to meet international climate change targets. There will be increasing opportunities to build in biodiversity benefits to reforestation and ecosystem restoration and creation (including 'rewilding'), using tree species and genetic strains resilient to future environmental change.
- Freshwater and marine fisheries. Fishery management has traditionally had commercial interests and sustainable harvesting as key motivations behind the translocation of fish species, crustaceans, and molluscs, etc. However, the conservation of specific fish species, and other evolutionary significant units, is recognised to be of increasing importance. Native oyster translocation is being used to develop sustainable fisheries and at the same time restore oyster species and oyster reef habitats (e.g. Bromley & Donnan, this volume).
- Habitat management and restoration. Examples include work involving sea grass beds, species-rich grasslands, native woodlands, and standing freshwaters. These can involve the translocation of a range of plant species, from seed to mature plant life stages, for environmental

benefits. There is increasing use of coral translocations to restore reef habitats (Swan et al., 2015).

These types of translocations have many of the associated risks that also apply to more species-focussed conservation translocations, such as the transmission of pathogens, the potential invasive effects of non-native species, and socio-economic impacts. Even so, such projects are often led by bodies with primarily non-conservation roles who may not necessarily be aware of conservation translocation best practice frameworks. Equally, they will have developed their own protocols and approaches to address their particular challenges, from which conservation translocation practitioners could learn and benefit. The motivations of such projects increasingly overlap, especially as governments require biodiversity, climate change, and other environmental considerations to be addressed holistically as part of wider initiatives associated with food production, timber supply, and other land and water uses. This therefore seems a good time to develop new and wider multidisciplinary links between the conservation and key land and water management sectors to increase opportunities for synergy and coordination, exchange knowledge and identify nature based solutions, and promote conservation translocation best practice more widely, thereby reducing risks and increasing benefits. The same argument applies within the conservation sectors, with species conservation translocation practitioners needing to collaborate more closely with those who focus on wider habitat and ecosystem restoration, to ensure that the big landscape-scale projects we need can better meet their biodiversity potential.

1.7.3 Don't Forget the Law

There will often be associated international and domestic legal considerations for conservation translocations. These may work in a supportive way, or a restrictive way. They may relate to the protective status of the species concerned and their movement between countries. The site of the donor population, or the proposed destination sites, may be designated for nature conservation or other purposes and therefore covered by relevant legislation. Other legislation may apply in terms of non-native species, animal welfare, biosecurity, land access and permissions, and where dangerous species are involved. These types of issues are covered in detail by Trouwborst et al. (this volume).

1.8 Conservation Translocations into the Twenty-First Century

Although conservation translocation has a longer history, the last three decades in particular have seen major advances in the ways many projects are designed, planned, and operated. This chapter has attempted to introduce and summarise these concepts, and they are described in more detail elsewhere in this book.

The urgent call for more transformative approaches to address the biodiversity crisis will result in conservation translocations being used more often and more widely. The good news is that there is now a wealth of knowledge and information to draw on, the result of not only numerous academic studies but also practical experience from *ex situ* and *in situ* work. There will always be levels of risk and uncertainty when planning conservation translocation projects, and sometimes these might be considered high – for example with assisted colonisations that attempt to 'rescue' species and move them to places they have never inhabited before. Therefore the tools of modelling, and the use of field-based trials where time allows, will become more frequently used to inform the project design of more complex conservation translocations. It will continue to be important for practitioners to be open and transparent and to share experiences, including the failures, in a way that is accessible to as wide an audience as possible.

So where will we be in 2030? At the time of writing we are all still dealing with the COVID-19 pandemic, but the lockdowns seem to have heightened people's thoughts about nature and the ways in which we can try to repair the damage we have done. Even before COVID-19, calls were increasing to bring about transformational change and urgent action to address the unfolding and, to a large extent, preventable biodiversity and climate crises. Conservation and other forms of environmental management have to be bolder, more ambitious, and radical. There may be associated biological and socio-economic risks, but these can often be managed through the use of tried and tested best practice, professionalism, and sensitivity. Increasingly these assessments of risks, and benefits, have to take account of the alternative risks of inaction or insufficient action.

Until now, conservation translocation has tended to be used in more extreme situations when the alternatives are limited. However, we are now moving to a point where it will need to become more commonly used as we struggle to respond to increasing threats. It will remain a method that can benefit many individual threatened species, but increasingly we need to find opportunities to move species that have key ecosystem roles and thereby restore and create habitats and ecosystems and make them more resilient. We are now in the UN Decade on Ecosystem Restoration, signifying an expectation that governments and people will act quickly and effectively. Conservation translocation practitioners need to be ready to respond.

1.9 Key Messages

- The May 2019 IPBES report emphasised the scale of the current biodiversity crisis and the need for transformative change, but highlighted that the tools exist to enable this change.
- Conservation translocation is an increasingly used tool that involves people deliberately moving and releasing organisms where the primary goal is conservation; it includes species reintroductions, reinforcements, assisted colonisations, and ecological replacements.
- It can be complex, expensive, time consuming, and sometimes controversial, but when best practice guidelines are followed it can be a very effective conservation method and a way of exciting and engaging people in environmental issues.
- Conservation translocations have an important role to play not only in improving the conservation status of individual species but also in ecological restoration and rewilding by moving keystone and other influential species.
- As the climate continues to change, species with poor dispersal abilities or opportunities will be at particular risk. Assisted colonisation, which involves moving species outside their indigenous range, is likely to become an increasingly used method. It is also a tool that may become increasingly used to avoid threats from the transmission of pathogens.
- Other more radical forms of conservation translocation, such as ecological replacements, multi-species conservation translocations, and the use of de-extinction and genetic interventions, are also likely to be given stronger consideration within the wider framework of ecological restoration.
- There have been significant advances in the science of reintroduction biology over the last three decades. However, new ways of transferring and sharing such information are needed to enable a wider spectrum of practitioners to have easier access to knowledge and guidance.

- In the past the biological considerations of conservation translocations have often heavily outweighed the people considerations. However, it is increasingly important that socio-economic factors are also built into projects and that relevant experts are involved in order to reduce conflict and improve the chances of success.
- Some level of biological and socio-economic risk will be present for most conservation translocations, but risk can often be managed through the use of sensitivity, professionalism, and the application of tried and tested best practice.
- Species reintroduction and other forms of conservation translocations will be an increasingly important tool if we are to restore, and make more resilient, our damaged ecosystems.

References

Albrecht, M. A. (2023) Applying adaptive management to reintroductions of Pyne's ground-plum *Astragalus bibullatus*. In Gaywood, M. J., Ewen, J. G., Hollingsworth, P. M. and Moehrenschlager, A. (eds.) *Conservation Translocations*. Cambridge, Cambridge University Press.

Armstrong, D. P., Seddon, P. J. & Moehrenschlager, A. (2018) Reintroduction. In Fath, B. D. (ed.) *Encyclopedia of Ecology*, 2nd ed., vol. 1. Oxford, Elsevier, pp. 458–466.

Aronson, J., Murcia, C., Kattan, G. H., Moreno-Mateos, D., Dixon, K. & Simberloff, D. (2014) The road to confusion is paved with novel ecosystem labels: a reply to Hobbs et al. *Trends in Ecology & Evolution*. 29, 646–647.

Australian Wildlife Conservancy (2022) *Newhaven*. Available from: www.australianwildlife.org/where-we-work/newhaven/ [Accessed 18 May 2022].

Barlow, C. (2023) The role of community engagement in conservation translocations: the South of Scotland Golden Eagle Project (SSGEP). In Gaywood, M. J., Ewen, J. G., Hollingsworth, P. M. and Moehrenschlager, A. (eds.) *Conservation Translocations*. Cambridge, Cambridge University Press.

Bromley, C. & Donnan, D. W. (2023) The European native oyster and the challenges for conservation translocations: the Scottish experience. In Gaywood, M. J., Ewen, J. G., Hollingsworth, P. M. and Moehrenschlager, A. (eds.) *Conservation Translocations*. Cambridge, Cambridge University Press.

Brooker, R. W., Britton, A., Gimona, A., Lennon, J. & Littlewood, N. (2011) Literature review: species translocations as a tool for biodiversity conservation during climate change. Scottish Natural Heritage Commissioned Report number 440, Inverness.

Brooker, R. W., Brewer, M. J., Britton, A. J., et al. (2018) Tiny niches and translocations: the challenge of identifying suitable recipient sites for small and immobile species. *Journal of Applied Ecology*. 55, 621–630.

Campbell-Palmer, R., Bauer, A., Jones, S., Ross, B. & Gaywood, M. J. (2023) The return of the Eurasian beaver to Britain: the implications of unplanned releases and the human dimension. In Gaywood, M. J., Ewen, J.G., Hollingsworth, P. M. and Moehrenschlager, A. (eds.) *Conservation Translocations*. Cambridge, Cambridge University Press.

Canessa, S., Guillera-Arroita, G., Lahoz-Monfort, J.J., et al. (2016) Adaptive management for improving species conservation across the captive-wild spectrum. *Biological Conservation*. 199, 123–131.

Carver, S., Convery, I., Hawkins, S., et al. (2021) Guiding principles for rewilding. *Conservation Biology*. 2021, 1–12.

Conservation Evidence (2021) *Translocate great crested newts*. Available from: www.conservationevidence.com/actions/858 [Accessed 12 January 2021].

Crutzen, P. J. & Stoermer, E. F. (2000) The Anthropocene. *Global Change Newsletter*. 41, 17–18.

Dalrymple, S. E. & Bellis, J. M. (2023) Conservation translocations: planning and the initial appraisal. In Gaywood, M. J., Ewen, J. G., Hollingsworth, P. M. and Moehrenschlager, A. (eds.) *Conservation Translocations*. Cambridge, Cambridge University Press.

Dasgupta, P. (2021) *The Economics of Biodiversity: The Dasgupta Review*. London, HM Treasury.

DEFRA (2021) *Reintroductions and other conservation translocations: code and guidance for England*. Department of Environment for Rural Affairs, UK.

Department of Conservation (2012) *Translocation guide for community groups*. Department of Conservation, New Zealand.

Dickson, B., Scherer, R., Kissel, A., et al. (2019) Multiyear monitoring of survival following mitigation-driven translocation of a long-lived threatened reptile. *Conservation Biology*. 33, 1094–1105.

Donadio, E., Zamboni, T. & Di Martino, S. (2023) Bringing jaguars and their prey base back to the Iberá Wetlands, Argentina. In Gaywood, M. J., Ewen, J. G., Hollingsworth, P. M. and Moehrenschlager, A. (eds.) *Conservation Translocations*. Cambridge, Cambridge University Press.

Estes, J. A., Terborgh, J., Brashares, J. S., et al. (2011) Trophic downgrading of planet Earth. *Science*. 333, 301–306.

Ewen, J. G., Canessa, S., Converse, S. J. & Parker, K. A. (2023) Decision-making in animal conservation translocations: biological considerations and beyond. In Gaywood, M. J., Ewen, J. G., Hollingsworth, P. M. and Moehrenschlager, A. (eds.) *Conservation Translocations*. Cambridge, Cambridge University Press.

Galápagos Conservancy (2022) *Restoring Existing Tortoise Populations*. Available from: www.galapagos.org/conservation/our-work/tortoise-restoration/restoring-existing-populations/ [Accessed 18 May 2022].

Gaywood, M. J. (ed.) (2015) *Beavers in Scotland: A Report to the Scottish Government*. Inverness, Scottish Natural Heritage.

Germano, J., Field, K., Griffiths, R., et al. (2015) Mitigation-driven translocations: are we moving wildlife in the right direction? *Frontiers in Ecology and the Environment*. 13, 100–105.

Glikman, J. A., Frank, B., Sandström, C., et al. (2023) The human dimensions and the public engagement spectrum of conservation translocation. In Gaywood, M. J., Ewen, J. G., Hollingsworth, P. M. and Moehrenschlager, A. (eds.) *Conservation Translocations*. Cambridge, Cambridge University Press.

Griffith, B., Scott, J. M., Carpenter, J. W. & Reed, C. (1989) Translocation as a species conservation tool: status and strategy. *Science.* 245, 477–480.

Halley, D. J. & Rosell, F. (2002) The beaver's reconquest of Eurasia: status, population development and management of a conservation success. *Mammal Review.* 32, 153–178.

Hällfors, M. & Dalrymple, S. E. (2023) Assisted colonisation and ecological replacement. In Gaywood, M. J., Ewen, J. G., Hollingsworth, P. M. and Moehrenschlager, A. (eds.) *Conservation Translocations.* Cambridge, Cambridge University Press.

Harrington, L. A., Lloyd, N. & Moehrenschlager, A. (2023) Animal welfare, animal rights, and conservation translocations: moving forward in the face of ethical dilemmas. In Gaywood, M. J., Ewen, J. G., Hollingsworth, P. M. and Moehrenschlager, A. (eds.) *Conservation Translocations.* Cambridge, Cambridge University Press.

Hobbs, R. J. & Harris, J. A. (2001) Restoration ecology: repairing the earth's ecosystems in the new millennium. *Restoration Ecology.* 9, 239–246.

Hobbs, R. J., Higgs, E. S. & Harris, J. A. (2014) Novel ecosystems: concept or inconvenient reality? A response to Murcia et al. *Trends in Ecology & Evolution.* 29, 645–646.

Hodder, K., Bullock, J. M., Buckland, P. & Kirby, K. (2005) Large herbivores in the wildwood and modern naturalistic grazing systems. English Nature Research Report number 648, Peterborough.

Hogg, C. & Wise, P. (2023) Assisted colonisation as a conservation tool: Tasmanian devils and Maria Island. In Gaywood, M. J., Ewen, J. G., Hollingsworth, P. M. and Moehrenschlager, A. (eds.) *Conservation Translocations.* Cambridge, Cambridge University Press.

Hopkins, J., Walmsley, C., Gaywood, M., Thurgate, G. & Allison, H. (2007) *Conserving biodiversity in a changing climate: guidance on building capacity to adapt.* London, Defra on behalf of the Biodiversity Partnership.

Hunter, E. A. & Gibbs, J. P. (2013) Densities of ecological replacement herbivores required to restore plant communities: a case study of giant tortoises on Pinta Island, Galápagos: giant tortoise-woody plant interactions. *Restoration Ecology.* 22, 248–256.

IPBES (2019) *Summary for Policymakers of the Global Assessment Report on Biodiversity and Ecosystem Services.* Bonn, Germany, IPBES secretariat.

IUCN (1998) *IUCN Guidelines for Re-introductions.* Gland, Switzerland, IUCN Species Survival Commission.

IUCN (2013) *Guidelines for Reintroductions and Other Conservation Translocations.* Gland, Switzerland, IUCN Species Survival Commission.

IUCN (2016) *IUCN SSC Guiding Principles on Creating Proxies of Extinct Species for Conservation Benefit.* Gland, Switzerland, IUCN Species Survival Commission.

IUCN (2017) *Guidelines for Species Conservation Planning.* Gland, Switzerland, IUCN.

IUCN (2021) *Training.* Available from: https://iucn-ctsg.org/training/ [Accessed 6 September 2021].

Jones, C. G., Lawton, J. H. & Shachak, M. (1997) Positive and negative effects of organisms as physical ecosystem engineers. *Ecology.* 78, 1946–1957.

Jones, C. J., Tatayah, V., Moorhouse-Gann, R., Griffiths, C., Zuël, N. & Cole, N. (2023) Slow and steady wins the race: using non-native tortoises to rewild islands off Mauritius. In Gaywood, M. J., Ewen, J. G., Hollingsworth, P. M. and Moehrenschlager, A. (eds.) *Conservation Translocations*. Cambridge, Cambridge University Press.

LaBar, T., Campbell, C., Yang, S., Albert, R. & Shea, K. (2014) Restoration of plant–pollinator interaction networks via species translocation. *Theoretical Ecology*. 7, 209–220.

Lewis, B. (2012) An evaluation of mitigation actions for great crested newts at development sites. PhD thesis, University of Kent, Canterbury.

Marriott, R. W., McHaffie, H. & Mardon, D. K. (2015) Woolly willow. In Gaywood, M. J., Boon, P. J. & Thompson, D. B. A. (eds.) *The Species Action Framework Handbook*. Perth, Scottish Natural Heritage.

Martin, T. G., Nally, S., Burbidge, A. A., et al. (2012) Acting fast helps avoid extinction. *Conservation Letters*. 5, 274–280.

Maschinski, J. & Albrecht, M. (2023) Conservation translocations for plants. In Gaywood, M. J., Ewen, J. G., Hollingsworth, P. M. and Moehrenschlager, A. (eds.) *Conservation Translocations*. Cambridge, Cambridge University Press.

McMurdo Hamilton, T., Caness, S., Clark, K., et al. (2021) Applying a values-based decision process to facilitate comanagement of threatened species in Aoterea New Zealand. *Conservation Biology*. 35,1162–1173.

Mitchell, R. J., Beaton, J. K., Bellamy, P. E., et al. (2014) Ash dieback in the UK: a review of the ecological and conservation implications and potential management options. *Biological Conservation*. 175, 95–109.

Mitchell, R., Green, S. & Hollingsworth, P. M. (2023) Plant health, biosecurity, and conservation translocations. In Gaywood, M. J., Ewen, J. G., Hollingsworth, P. M. and Moehrenschlager, A. (eds.) *Conservation Translocations*. Cambridge, Cambridge University Press.

Moehrenschlager, A., Soorae, P. & Steeves, T. E. (2023) From genes to ecosystems and beyond: addressing eleven contentious issues to advance the future of conservation translocations. In Gaywood, M. J., Ewen, J. G., Hollingsworth, P. M. and Moehrenschlager, A. (eds.) *Conservation Translocations*. Cambridge, Cambridge University Press.

Morling, P. (2022) *The economic impact of white-tailed eagles on the Isle of Mull*. Sandy, UK, The RSPB.

Munro, N. T., McIntyre, S., Macdonald, B., et al. (2019) Returning a lost process by reintroducing a locally extinct digging marsupial. *PeerJ*. 7, e6622.

National Species Reintroduction Forum (2014) *The Scottish Code for Conservation Translocations*. Inverness, Scottish Natural Heritage.

Naundrup, P. J. & Svenning, J.-C. (2015) A geographic assessment of the global scope for rewilding with wild-living horses (*Equus ferus*). *PLoS ONE*. 10, e0132359.

Neaves, L. E., Ogden, R. & Hollingsworth, P. M. (2023) Genomics and conservation translocations. In Gaywood, M. J., Ewen, J. G., Hollingsworth, P. M. and Moehrenschlager, A. (eds.) *Conservation Translocations*. Cambridge, Cambridge University Press.

Newhouse, A. E. & Powell, W. A. (2020) Intentional introgression of a blight tolerance transgene to rescue the remnant population of American chestnut. *Conservation Science and Practice.* 3, e348.

Novak, B. J., Phelan, R. & Weber, M. (2021) U.S. conservation translocations: over a century of intended consequences. *Conservation Science and Practice.* 3, e394.

Ostendorf, B., Boardman, W. & Taggart, D. (2016) Islands as refuges for threatened species: multispecies translocation and evidence of species interactions four decades on. *Australian Mammalogy.* 38, 204–212.

Paine, R. T. (1995) A conversation on refining the concept of keystone species. *Conservation Biology.* 9, 962–964.

Pearce-Higgins, J. W., Ausden, M. A., Beale, C. M., Oliver, T. H. & Crick, H. Q. P. (eds.) (2015) *Research on the Assessment of Risks & Opportunities for Species in England as a Result of Climate Change.* Natural England Commissioned Report number 175, Peterborough.

Plein, M., Bode, M., Moir, M. & Vesk, P. (2016) Translocation strategies for multiple species depend on interspecific interaction type. *Ecological Applications.* 26, 1186–1197.

Pyke, C. R. & Marty, J. (2005) Cattle grazing mediates climate change impacts on ephemeral wetlands. *Conservation Biology.* 19, 1619–1625.

Randall, L., Lloyd, N. & Moehrenschlager, A. (2018) *Guidelines for Mitigation Translocations of Amphibians: Applications for Canada's Prairie Provinces.* Calgary, Alberta, Canada Centre for Conservation Research, Calgary Zoological Society.

Revive & Restore (2021) *Passenger Pigeon Project.* Available from: https://reviverestore.org/about-the-passenger-pigeon/ [Accessed 3 May 2021].

Ricciardi, A. & Simberloff, D. (2009) Assisted colonization is not a viable conservation strategy. *Trends in Ecology & Evolution.* 24, 248–253.

Royal Botanic Garden Edinburgh (2021) *Conservation Genetics.* Available from: www.rbge.org.uk/science-and-conservation/genetics-and-conservation/molecular-ecology/conservation-genetics/ [Accessed 11 January 2021].

Rutz, C., Loretto, M.-C., Bates, A. E., et al. (2020). COVID-19 lockdown allows researchers to quantify the effects of human activity on wildlife. *Nature Ecology & Evolution.* 4, 1156–1159.

Sainsbury, A. W. & Carraro, C. (2023) Animal disease and conservation translocations. In Gaywood, M. J., Ewen, J. G., Hollingsworth, P. M. and Moehrenschlager, A. (eds.) *Conservation Translocations.* Cambridge, Cambridge University Press.

Scofield, R., Cullen, R. & Wang, M. (2011) Are predator-proof fences the answer to New Zealand's terrestrial faunal biodiversity crisis? *New Zealand Journal of Ecology.* 35, 312–317.

Scottish Natural Heritage (2019) *SNH's Climate Change Commitments - Towards a Nature-Rich Future.* Inverness, Scottish Natural Heritage.

Seddon, P. J. (2023) The role of conservation translocations in rewilding and de-extinction. In Gaywood, M. J., Ewen, J. G., Hollingsworth, P. M. and Moehrenschlager, A. (eds.) *Conservation Translocations.* Cambridge, Cambridge University Press

Seddon, P., Armstrong, D. & Maloney, R. (2007) Developing the science of reintroduction biology. *Conservation Biology.* 21, 303–312.

Seddon, P. J., Strauss, W. M. & Innes, J. (2012) Animal translocations: what are they and why do we do them. In: Ewen, J. G., Armstrong, D. P., Parker, K. A. and Seddon, P. J. (eds.) *Reintroduction Biology: Integrating Science and Management.* West Sussex, Wiley-Blackwell.

Smuts-Kennedy, C. & Parker, K. (2013) Reconstructing avian biodiversity on Maungatautari. *Notornis.* 60, 93–106.

Society for Ecological Restoration (2004) *SER International Primer on Ecological Restoration.* Available from: www.ser-rrc.org/resource/the-ser-international-primer-on/ [Accessed 29 April 2021].

Soorae, P. S. (ed.) (2021) *Global Conservation Translocation Perspectives: 2021. Case Studies from around the Globe.* IUCN, Gland, Switzerland and Environment Agency, Abu Dhabi, UAE.

Stanley-Price, M. R. (1989) *Animal Reintroductions: The Arabian Oryx in Oman.* Cambridge, Cambridge University Press.

Stevenson G. B. (2007) An historical account of the social and ecological causes of capercaillie *Tetrao urogallus* extinction and reintroduction in Scotland. PhD thesis, University of Stirling.

Sutherland, W. J., Armstrong, D., Butchart, S. H. M., et al. (2010) Standards for documenting and monitoring bird reintroduction projects. *Conservation Letters.* 3, 229–235.

Swan, K., McPherson, J., Seddon, P. & Moehrenschlager, A. (2015) Managing marine biodiversity: the rising diversity and prevalence of marine conservation translocations. *Conservation Letters.* 9, 239–251.

Taylor, G., Canessa, S., Clarke, R., et al. (2017) Is reintroduction biology an effective applied science? *Trends in Ecology & Evolution.* 32, 873–880.

The Standing Committee of the Convention on the Conservation of European Wildlife and Natural Habitats (2012) *Recommendation No. 158 (2012) of the Standing Committee, adopted on 30 November 2012 on Conservation translocations under changing climatic conditions.* Strasbourg, France, Council of Europe.

Thomas, C. D. (2011) Translocation of species, climate change, and the end of trying to recreate past ecological communities. *Trends in Ecology & Evolution.* 26, 216–221.

Tonkin, M., Garritt, J., Bryce, J. & Cole, M. (2016) Red and grey squirrels. In Gaywood, M. J., Boon, P. J. and Thompson, D. B. A. (eds.) *The Species Action Framework Handbook.* Perth, Scottish Natural Heritage.

Tordoff, H. & Redig, P. (2002) Role of genetic background in the success of reintroduced peregrine falcons. *Conservation Biology.* 15, 528–532.

Trouwborst, A., Blackmore, A., Blyth, S., Fleurke, F., McCormack, P. & Gaywood, M. J. (2023) Conservation translocations and the law. In Gaywood, M. J., Ewen, J. G., Hollingsworth, P. M. and Moehrenschlager, A. (eds.) *Conservation Translocations.* Cambridge, Cambridge University Press.

Van Wieren, S. (1995) The potential role large herbivores in nature conservation and extensive land use in Europe. *Biological Journal of the Linnean Society.* 56, 11–23.

Walker, K., Stevens, P., Stevens, D., Mountford, J., Manchester, S. & Pywell, R. (2004) The restoration and re-creation of species-rich lowland grassland on land formerly managed for intensive agriculture in the UK. *Biological Conservation.* 119, 1–18.

Walters, J. R., Derrickson, S. R., Michael Fry, D., Haig, S. M., Marzluff, J. M. & Wunderle, J. M. (2010) Status of the California condor (*Gymnogyps californianus*) and efforts to achieve its recovery. *The Auk.* 127, 969–1001.

Watt, J., Hastie, L. C. & Cosgrove, P. J. (2018) Monitoring the success of freshwater pearl mussel reintroductions. Scottish Natural Heritage Research Report number 956, Inverness.

Wilkins, K., Pejchar, L. & Garvoille, R. (2019) Ecological and social consequences of bison reintroduction in Colorado. *Conservation Science and Practice.* 1, e9.

2 · *Conservation Translocations: Planning and the Initial Appraisal*

SARAH E. DALRYMPLE AND JOE M. BELLIS

2.1 Background

The term 'conservation translocation' unites a suite of interventions involving the movement of plants, animals, and fungi for conservation benefit. However, a myriad of starting points, motivations, technical constraints, and biological and socio-economic risks can be experienced by translocation practitioners. The IUCN Guidelines for Reintroductions and Other Conservation Translocations (IUCN, 2013; hereafter referred to as the IUCN Guidelines) were greatly expanded compared to the previous version for good reason: the variety of issues to consider and the expectation for rigorous planning and monitoring have grown in parallel with the diversity of focal species and number of attempted translocations. The IUCN Guidelines present the key points in undertaking a translocation, from considering other conservation options to developing best practice in monitoring. This chapter aims to be the bridge between the IUCN Guidelines and the content presented in the remainder of this volume, and makes the role of the conservation practitioner explicit in this process.

The material does not have to be followed in the sequence presented in this chapter, but we recommend that it is read carefully in its entirety. Practitioners should then choose the most relevant sections from the remainder of the book to read in full before planning and undertaking a translocation.

2.2 Conservation Motivations and Translocation Contexts

Conservation translocations aim to result in 'conservation benefit' but the form of the benefit can vary depending on the standpoint of a conservation practitioner and the motivation for considering conservation translocations (Brichieri-Colombi & Moehrenschlager, 2016). In order to demonstrate this, and to provide a workable way to access conservation translocation guidance, we present three broad conservation benefits and the type of practitioners and activities that may correspond to these. These motivations are not intended to define particular approaches beyond the needs of this chapter, nor are they mutually exclusive, but they may affect the operational considerations of the practitioner. It is also useful to appreciate the diversity of practice that constitutes conservation translocations: the juxtaposition of one approach with others might make it easier to appreciate the range of considerations necessary, and, we hope, will improve the appraisal process for conservation translocations as a result.

2.2.1 Species Recovery

The motivation to employ translocations in species recovery stems from the need to restore species (or lower taxonomic levels such as subspecies or ecotypes) to a status equivalent to some former condition, whether this is the restoration of species to particular sites, recovery to demographic baselines, or the more general reinstatement of species to regions within their indigenous ranges from which they have been extirpated. Species recovery might be a requirement of national or international legislation, be driven by government or government agency policy, or come from a 'grass-roots' or non-governmental body that has recognised a species decline. Typical individuals and roles that might be considering recovery-driven translocations include managers of sites from where the focal species has been extirpated and those with a responsibility for species recovery described in management plans and other statutory and non-statutory mechanisms.

2.2.2 Maintaining Species' Current Status

Although this motivation, to maintain the status of a species, may not be cited as frequently as the scenario described in Section 2.2.1, it is possible that species might be translocated before declines can be evidenced by population losses or range contraction. Pre-emptive translocations might anticipate future threats or, perhaps more likely, current threats that have yet to cross thresholds of a species' survival. Examples might include the potential for rapid transmission of disease in the current range, or climate change when there is a high confidence that the species is sensitive to changing temperature or precipitation. Given that this motivation for translocations might arise from pro-active assessments of risk, it is likely that the proponents will be those with the responsibility for the protection of the species at national (or even international) levels.

2.2.3 Ecosystem Functioning

The motivation to restore ecosystem function is driven by the need to fill a vacant ecological function such as pollination, predation, nutrient turnover, etc.; the identity and status of the species being translocated is of secondary importance to its ecosystem role, and it is the ecosystem that is the principal recipient of anticipated conservation benefits. This motivation might be associated with those who have responsibilities at site level but might also be considered by those who are looking at ecosystem function loss at regional levels when a keystone species has been extirpated across its range, an approach that has been used extensively in marine ecosystems (Swan et al., 2016).

2.2.4 Translocation Contexts

These three motivations feed into five translocation contexts whereby an individual or organisation has taken lead responsibility (through voluntary or legislative mechanisms) for:

1. A focal species that is in decline (this decline may be recognised locally, nationally, and/or internationally).
2. A site that is missing a species.
3. A site that is missing an ecosystem function.
4. A species that may or may not currently be threatened, but there are anticipated threats.
5. A population threatened at site level by development (i.e. mitigation-driven translocation).

Table 2.1 links the motivations and translocation contexts, and expands upon these to highlight practical considerations specific to each of the roles.

2.3 Demonstrating a Need

Although we have described possible motivations in Section 2.2.4, there is also an onus to demonstrate the *need* for a conservation translocation that

Table 2.1. *Translocation contexts and key biological considerations when undertaking translocation programmes. The content emphasises differences between the five 'contexts' and considerations common to all translocations – this is further developed in the main text.*

Translocation context	Potential translocation types	Establishing a need	Identifying a destination site	Identifying a source population	Translocate/ release	Possible scenarios prompting exit strategy initiation
1. Your focal species is in decline and/or threatened	Reintroduction, reinforcement, assisted colonisation	Species is internationally, nationally, or locally declining. Threats should be well understood before selecting translocation instead of other interventions	Depending on threats, may include sites within indigenous species range, or sites outside the range. Choice of site may be dictated by the availability of source individuals – the destination site should therefore aim to provide conditions to which the founders are pre-adapted	The most viable population(s) should be considered as candidate source population(s). This decision should be based on sufficient information on the status of the species in its whole range. These may be *in situ* or *ex situ* populations	May involve 'bulking up' in *ex situ* facilities. Specifics of translocation and release will be unique to every release event	If conducting within-range translocations, the most likely failure scenario is that the translocations do not work and a viable population is not created. However, it is possible that unanticipated negative interactions are initiated, and these are more likely for assisted colonisation beyond the indigenous range

2.	You have a site that is missing a previously present species	Reintroduction	Translocation may form a later stage of site restoration, or the missing species may perform a recognised ecological function that needs reinstating	Site is already known	Source population will need to be matched as closely as possible to the site	All known and inferred causes of original extirpation must be removed/ minimised, as must other potentially harmful changes to the site since extirpation, prior to release	Monitor for unanticipated negative interactions/ impacts. Although this is a site-level reintroduction, practitioners should be aware that the site may have changed in the interim since extirpation of the species
3.	You have a site that is missing a previously present ecosystem function	Reintroduction, ecological replacement	As for 'missing species' scenario but it may be the ecological function that is the focus	Site is already known	Must be viable source population. With ecological replacement, the species to be translocated must first be selected but this might be on the basis of having viable populations across a wider extent than the	As for 'missing species' context above in the case of reintroduction, but anticipate substantially more interest (much of which may be negative) at the point of release when considering ecological	Monitor for unanticipated negative interactions/ impacts. Whilst the translocated species may or may not have been found in this site before, there is always potential for negative impacts. In reintroduction (*cont.*)

Table 2.1. (*cont.*)

Translocation context	Potential translocation types	Establishing a need	Identifying a destination site	Identifying a source population	Translocate/ release	Possible scenarios prompting exit strategy initiation
				extirpated species	replacement. Additional preparation/ assessment required and expected	it may be because the site has changed in the interim since extirpation. In ecological replacement, there is greater potential for negative impacts arising from novel interactions
4. You have recognised future threats to a species and it may or may not currently be threatened	Reintroduction, reinforcement, assisted colonisation	Conservation interventions should aim to maintain resilience against future threats. Translocation may form an optimal intervention	It will need to be demonstrated that candidate destination site(s) are not currently threatened and onset of threats such as disease or climate change can be	Source population needs to be healthy and deemed to be pre-adapted to the destination site	Again, reintroductions (and reinforcements) will probably just need to follow normal procedures but assisted colonisations are likely to gain	Exit strategy initiated at the first signs of negative impacts

				when dispersal limitation is a barrier to escaping the threat	avoided in the long term		more external attention which may be quite negative – additional preparation/assessment required and expected
5.	You have a population threatened at site level by development	Mitigation translocation	Need only arises in the regrettable situation that the conservation status cannot be maintained at the site and translocation is undertaken due to imminent loss of site/population of focal species	Needs to match conditions supporting population to be lost	Source population is known	May be forced by the timeline of development causing anticipated loss	Exit strategy must be in place but given the circumstances of mitigation translocations, there is a larger risk of net negative impacts regarding conservation outcomes – if the translocation fails or must be deliberately ended due to unforeseen events, the source

(*cont.*)

Table 2.1. (*cont.*)

Translocation context	Potential translocation types	Establishing a need	Identifying a destination site	Identifying a source population	Translocate/ release	Possible scenarios prompting exit strategy initiation
						population has already been destroyed and conservation harm rather than conservation benefit will have been achieved

meets the objective of achieving conservation benefit. Conservation translocations are just one of a range of interventions classed as species management by the IUCN (2013), and the vast majority of translocation-based interventions are undertaken as part of a suite of management options to promote the health and resilience of the focal species and ecosystems. The specific management options will be selected according to need and feasibility, based on specific (and often local) contexts (see Ewen et al., this volume). With all the possible conservation options presented, and taking into account the complexity, cost, and perceived risk of conservation translocation, it is unsurprising that conservation translocations are often considered to be an option of 'last resort'. However, this is an unhelpful perspective from which to view translocations: waiting until the 'last resort' might alternatively (and perhaps somewhat confrontationally) be referred to as a 'last ditch' effort to prevent local or global extinctions. By this, we mean that waiting to act until every other option has been exhausted leaves the translocation at risk of failure due to the genetically poor source material, low numbers and associated demographic instability, and short time frame in which to act when a species is being pulled back from the brink of extinction. We recommend that conservation translocations be considered proactively before populations reach critical levels that might jeopardise the translocation and compromise the species' status in the future, and be designed to complement other interventions such as invasive species removal, anti-poaching, prevention of illegal trade, etc.

To do this, conservation translocations might be explored using techniques such as population viability analysis (PVA) to predict the survival probability of translocated individuals and the subsequent demographic contribution resulting from the reproduction, recruitment, and dispersal of surviving individuals. These analyses are routinely undertaken by management groups that are planning releases of individuals from *ex situ* facilities but PVA will also simulate the outcomes of wild-to-wild translocations. For PVAs to be useful, ideally data are needed on the demographic parameters of a threatened population, the carrying capacity of the localities supporting modelled populations, the genetic diversity of a species, and reliable estimates of the frequency and impact of catastrophes such as disease. As a result, PVAs are data intensive, and poor-quality data affects the confidence with which models can be interpreted. However, when used to compare outcomes between different scenarios, PVAs can help focus discussion on whether, for example, habitat restoration would yield better outcomes than translocation, or if the combination of interventions might prolong population survival more effectively than one intervention alone.

The case for undertaking, or opposing, translocation can be made by individuals from a variety of backgrounds and is not the soley preserve of conservationists. The need for a functioning and resilient biosphere is a legal human right in many countries. Non-specialists can express this right by participating in conservation decision-making, especially where it relates to (but is not restricted to) the vicinity in which they live and work. It would be easy to relegate stakeholders to groups to be consulted only once a conservation translocation has been proposed. However, the suggestion of possible translocation projects by any individual, including those from community groups, should be taken seriously; at the very least, an initial appraisal of feasibility should be undertaken with support from bodies that can comment on ecological, socio-economic, and legislative considerations. In doing so, such approaches can make the process of initiating translocation projects more accessible and facilitate greater collaboration with those outside the professional conservation community. As an example, the Scottish Code for Conservation Translocations (National Species Reintroduction Forum, 2014) engenders this approach (also see Glikman et al., this volume, for detailed discussion on human dimensions of translocations).

2.4 Identifying and Evaluating Destination Sites

For those operating within the translocation contexts 1 and 4 (where the context is defined by a species approach to conservation, and the destination site has not already been determined), the selection of a suitable site for release represents a key stage in the project. Even for those who might be operating in the translocation contexts 2, 3, and 5, where the site is the starting point for conservation management, many of the same considerations must be addressed and much of the material presented in this section is still relevant.

In the past, destination sites have often been selected based on expert knowledge and subjective impressions of their similarity with sites of current occurrence. This approach is likely to be responsible for poor destination site selection, ranking as one of the most frequently reported causes of failure in conservation translocations. In recognition of the need for more objective assessment of the suitability of destination sites, techniques such as habitat suitability modelling (Osborne & Seddon, 2012; IUCN, 2013), statistical similarity comparisons (Ferrarini et al., 2016;

Brzosko et al., 2018), and strengths, weaknesses, opportunities, threats (SWOT) analysis (White et al., 2015) are introduced here.

2.4.1 Techniques for Site Selection

Some of the techniques available for rigorous site selection described in this section require a level of statistical expertise and computing capability that may be beyond the capacity of typical conservation practitioners, particularly those whose responsibilities or interests lie with site or species management and who do not have access to the required resources, and those who are not professional conservationists. However, we have included this material in order to alert the conservation translocation community to what is possible, and is currently being done, in the research surrounding translocations. We hope that practitioners may become more familiar with these techniques so that, at the very least, they have the confidence to approach research institutions and discuss their potential role as partners when building a team for planning conservation translocations (Guisan et al., 2013).

2.4.1.1 Habitat Suitability Modelling

Habitat suitability models (HSMs; also frequently referred to as species distribution models, ecological niche models, or bio-climate envelope models when climate variables are the main covariates) are the principal tools used to derive spatially and temporally explicit predictions of environmental suitability. Correlative HSMs identify statistical relationships between observed locations of species occurrence (presence or absence) and environmental variables. The use of HSMs for destination site selection was discussed in detail by Osborne and Seddon (2012). Since then, translocation projects involving insects (Maes et al., 2019), birds (Kalle et al., 2017), freshwater fish (Malone et al., 2018), and lichens (Brooker et al., 2018) have applied HSM techniques to estimate the suitability of candidate destination sites.

For those interested in using HSMs for destination site selection, the maximum entropy approach is one of the most popular methods for predicting habitat suitability (Phillips et al., 2006). This approach can be applied using the software package 'MaxEnt', which has the advantage of a graphical user interface that makes it more user-friendly than other approaches available in the statistical software environment 'R' (e.g. *biomod2*, *dismo*, and *sdm*) (R Core Team, 2018). *biomod2* is the most popular R package for running HSMs and provides the user with a choice of around a dozen statistical methods for making spatial predictions of suitability (Thuiller et al., 2009). A book authored by three of the leading specialists in the field of HSMs delivers a very useful introduction to the key stages of running a model and provides extensive examples using the *biomod2* package in R (Guisan et al., 2017).

While the potential of HSMs to inform destination site selection is significant, care is required when constructing and interpreting models. Decisions regarding data preparation, selection of predictor variables, choice of statistical method, model fitting, and evaluation can all impact HSM predictions, which in turn can affect the resulting management proposals. Poorly constructed and interpreted models that are not fit for such targeted decision-making can lead to erroneous conservation actions. To avoid this, decisions underpinning model construction need to be clearly reported and justified by, for example, following a standardised protocol (e.g. ODMAP https://odmap.wsl.ch/ Zurell et al., 2020).

2.4.1.2 Other Methods

A commonly employed method of reducing subjectivity in destination site selection is to statistically compare the conditions at sites where the focal species survives with those at potential destination site(s). Multivariate approaches such as principal components analysis and Mahalanobis distance can be used to quantify how similar the conditions at the destination site are to those present at extant population sites (Thatcher et al., 2006; Ferrarini et al., 2016; Brzosko et al., 2018). The relative usefulness of this type of approach depends on the factors that have shaped the distribution of the target species. For example, an apparently narrow distribution may be a reflection of dispersal limitations or direct anthropogenic impacts, such as over-hunting or persecution, rather than the specific habitat requirements of a species. Consequently, care is needed when constructing and interpreting outputs from these types of analyses (and HSMs) as a species' recent distribution may not be located in optimal habitat conditions.

Quantitative SWOT analysis has also been proposed for objective evaluation of potential destination sites (White et al., 2015). The method is based on comparing internal, controllable aspects of a destination site (strengths and weaknesses) with external, uncontrollable factors (opportunities and threats). The aggregated numeric scoring can be used to judge the potential management actions a destination site might need to reach specified goals and objectives. SWOT analysis has been widely used for strategic planning in business and industrial settings but remains in its infancy in the field of conservation. It may be effectively utilised to go beyond ecological considerations to address socio-economic opportunities and threats as well.

2.4.2 Biological Risks in Destination Site Selection

All types of conservation translocation can have biological consequences for both the translocated species and the recipient ecological community (IUCN, 2013). Consequences for the translocated individuals may result from inter-specific competition, predation, hybridisation (intra- and inter-specific), unfavourable climates, interactions with invasive species, and the introduction of harmful diseases or parasites to conspecifics. The selection of a suboptimal destination site may lead to the translocation failing to result in a viable population and the translocation project failing to meet its objectives. However, for all the potential to incur biological impacts due to unanticipated interactions, unfortunately, the most likely cause of translocation failure is the original cause of population extirpation. A review of freshwater fish reintroductions found that ensuring that the initial cause of decline had been adequately addressed prior to release at a destination site was the best predictor of successful establishment (Cochran-Biederman et al., 2015).

Many of these biological risk factors are also relevant to species already present at the destination site. Of particular concern is the risk that translocated stock are carriers of diseases that could infect other species (Kock et al., 2010), especially if sourced from *ex situ* conditions such as zoos, botanical gardens, or captive-breeding centres, due to the greater exposure to pathogenic diseases at these facilities (Walker et al., 2008) (see Sainsbury & Carraro, this volume, and Mitchell et al., this volume, for information on animal and plant health in translocations). Translocations to sites beyond the natural range (i.e. assisted colonisation or ecological replacement) carry further risks as the focal species (and associated parasites) could become invasive. Despite most non-native species introductions appearing to have only minor impacts (Mack et al., 2000) – a recent review has found that only 1.4 per cent of biocontrol attempts resulted in deleterious ecosystem impacts (Novak et al., 2021) – some accidental releases and poorly planned biological control attempts have led to substantial alterations of biotic communities and cost millions in damages and control measures (see Hällfors & Dalrymple, this volume, on assisted colonisation and ecological replacement).

2.4.3 Socio-economic Risks

Since so much of the global environment has been modified by humans, it is inevitable that some translocated organisms will affect humans and their livelihoods in positive and negative ways. The strength and direction of this impact will depend on the human population density and economic wealth of the area surrounding the destination site and the degree of local, and sometimes national (for more contentious species), stakeholder involvement in the translocation project. A socio-economic assessment by Scottish Natural Heritage compared the benefits and costs of two relatively recent Eurasian beaver *Castor fiber* reintroductions in Scotland: one carefully planned trial release in Argyll with high stakeholder involvement, and one unlicensed release in Tayside with very little stakeholder involvement (Gaywood, 2015). The key conclusion was that there are far greater socio-economic risks from unplanned releases with no public involvement (see Campbell-Palmer et al., this volume, for more detail). While it is important to note that the two areas have quite different human environments (the Argyll site being more remote than the Tayside sites), they highlight the risks and missed opportunities associated with unlicensed or poorly planned releases.

In addition to potential economic damage, there is also the prospect of translocated species posing health risks to humans within the vicinity of a destination site. Managers working with dangerous species, such as large carnivores and venomous or poisonous species, must be well informed with respect to the potential threats posed by their focal taxa. Projects involving dangerous animals or plants are likely to be met with stakeholder resistance. Transparent public awareness campaigns aiming to educate stakeholders on potential costs, benefits, and threats can help to reduce stakeholder resistance, but these should be undertaken before translocations become politicised and opposition becomes entrenched (Hiroyasu et al., 2019). There are also public health risks associated with disease transmission. Quarantine into the UK, for example, focuses on public health risks such as rabies (perhaps a dramatic one – but beavers from Germany, for example, have to go through quarantine/health screening to take account of this and a range of other public health issues such as *Echinococcus multilocularis*).

Translocations that are accidental, or not planned in a conservation context, have high potential to cause indirect ecological consequences that could impact local livelihoods. Translocated species, especially if beyond the natural range, could threaten food supplies or ecosystem services such as clean water, erosion control, pollination, or nutrient cycling (Charles & Dukes, 2008). For example, the non-native shrub tamarisk *Tamarix* spp., which was planted for horticultural and erosion control purposes in the western USA during the 1800s, is estimated to reduce ecosystem services such as irrigation and municipal water, hydropower, and flood control by 100 million USD annually (Zavaleta, 2000). If a translocated species causes significant unacceptable consequences, the remedial costs could be very high and opportunities for funding future conservation translocations are less likely (although robust evidence suggests these instances are very rare – Novak et al., 2021).

2.4.4 Legislative Constraints

Site designations and species protection under national or international statutes are a necessary part of conservation; however, legislation may also place limits on translocation operations despite the intention that releases result in conservation benefit. Destination sites located inside protected areas offer advantages such as diverse habitat (Gray et al., 2016) and more security against future habitat degradation, but translocations into such an area could significantly affect the recipient ecosystem by altering the qualifying habitat or disrupting other species of conservation significance. The level of protection afforded to areas, the protection status of the target species, and the type of translocation (i.e. reinforcement, reintroduction, assisted colonisation, or ecological replacement) will determine which government licences are required before a translocation is permitted, and legislation in some countries may rule against assisted colonisation and ecological replacement altogether. In Scotland, the Wildlife & Countryside Act 1981 presumes against the translocation of species outside their 'native' range, and thus a 'non-native species licence' is required before a release is deemed to be legal (National Species Reintroduction Forum, 2014; also see Trouwborst et al., this volume, for more details on legislation and permissions).

2.5 Identifying a Source Population

Identifying individuals for translocation starts with an evaluation of the candidate populations and subsequent selection of those populations that can tolerate the removal of adequate numbers of individuals and have ecological similarities with the destination sites. For some species, there may be little choice of source population(s) because the number of wild populations is so few. In mitigation translocations (translocation context 5 above), the translocation is undertaken to avoid mortality of that population due to development so the 'source' population is more properly identified as the population at risk. Much of this section is therefore not relevant to those operating within context 5.

Qualities of suitable source populations will need to be assessed on a species-by-species basis. Basic thresholds for suitability should include viable populations that exhibit typical fluctuations in size for the species, and the lack of expression of genetic problems. Ideally, populations can be shown to have some capacity to 'spare' enough individuals for translocation without any prolonged reduction in population size. This might be evidenced by density-dependent mortality of seedlings in plants or similar markers in animals such as smaller-than-average territories signalling that the removal of individuals will not adversely affect the source population. In other cases, the reproductive buffer of multiple broods or huge seed numbers indicates that species have traits that can safeguard the population against removal of individuals. However, there may be extreme cases where individuals of a species are brought into captivity as a final act to prevent imminent extinction and obviously there is little room for selection on the part of the practitioner.

When several populations can be shown to be suitable sources, selection can then focus on the process of identifying which would be optimal. There is evidence to demonstrate that ecological similarity between source and destination sites can achieve better results in translocations due to the translocated individuals being pre-adapted to the new conditions (Dalrymple & Broome 2010; Lawrence & Kaye 2011). Identification of the destination site will inform this process, and some of the same methods described in Section 2.4.1 can also be used to select the source population based on its similarity to the proposed destination site.

The number of source sites to select will also vary depending on properties of the species that indicate whether it is optimal to use one or several source populations. When species are susceptible to outbreeding depression, the combination of genetic material from divergent populations can interrupt co-evolved traits when mixed with individuals that are genetically different (see Neaves et al., this volume). Where possible, this should be evaluated using molecular techniques, but if this option is not feasible then field observations could also provide indications of potential phenotypic differences in morphology, timing of reproduction, and behaviour. Inbreeding depression is also a potential problem: using sources that are too similarly related may be detrimental to reproduction and/or survival of any resultant offspring (Neaves et al., this volume).

As part of the planning phase, a full evaluation of the status of the focal species is necessary and this includes legislative constraints – for example, licensing requirements for taking individuals from natural populations or permits for removing individual organisms from protected areas (Trouwborst et al., this volume).

2.6 Translocate/Release

Of all the sections in this chapter, the release phase is the most difficult for which to provide overarching advice. This is because the process of capturing/collecting individual animals, plants, or fungi (or their propagules), keeping them in safe conditions in the interim, and releasing them into their new locations will be very much directed by the ecology of the species and the situations particular to the translocation programme. We recommend that subsequent chapters of this book are consulted for those considering plant and fungi translocations (Mitchell et al., this volume; Maschinski & Albrecht, this volume), and animal translocations (Ewen et al., this volume; Harrington et al., this volume; Sainsbury & Carraro, this volume). Other considerations may benefit from referring to chapters on genetics (Neaves et al., this volume), engaging the wider public (Glikman et al., this volume), and legislation (Trouwborst et al., this volume). The remainder of this section briefly covers overarching principles that are relevant to all translocations.

2.6.1 Minimise or Eradicate the Cause of Original Extirpation

For those translocations where the intention is to release organisms in sites within the indigenous range and from which the focal species has been lost, it is crucial (and fairly obvious) that the original causes of extirpation are dealt with to the point where these threats are removed at the site, or at least reduced in severity to be tolerable to the species of concern.

2.6.2 Rule out Other Potential Threats

The cause of a species decline will often be attributed to a particular threat or sometimes a number of threats and, as stated above, these must be (and usually are) dealt with prior to a translocation taking place.

However, there may be a number of additional threats that were previously unidentified or, in the case of reintroductions and reinforcements, may have come to operate at a site in the interim since the species was lost; practitioners must try to anticipate these threats prior to release. Poor climate suitability is probably a much-overlooked cause of translocation failure (Bellis et al., 2020) and a good example of a threat that might not be implicated in the original loss but could cause problems for releases. Climate change can reasonably be expected to become more significant if conservation translocation practitioners continue to ignore its extensive, albeit sometimes subtle, impacts. There are various other threats that are similarly incremental in their changes but can render a translocation site unsuitable. Examples include atmospheric pollution, the encroachment of invasive species, changes in fire frequency, exploitation of water sources, and various others that might change the underlying properties and composition of an ecosystem without any apparent changes to a casual observer. A similar process of threat anticipation must take place when planning conservation introductions into sites outside the species' indigenous range. In all cases, this might be very challenging when little is known about causes of decline in the species' indigenous range but the success of the translocation programme may be undermined without a full assessment.

2.6.3 Multiple Releases at Multiple Sites

Where possible, practitioners should consider whether releases to multiple sites are possible and optimal, and if multiple release events can be accommodated in the translocation plan. This can protect against environmental variability across both time and space to a certain extent, lowering the probability that a release will be thwarted by atypical conditions. Of course, there may be good reasons for not doing this: limited availability of founders, suboptimal destination sites, or logistical constraints that mean supporting infrastructure can only be available for a short time. These may all mean that translocations are limited to one site and/or one release event. In this situation practitioners should be aware that, for some animals, releasing a territorial species into a destination site can result in them taking up the best quality habitat, and sometimes with extended territories. Subsequent releases into the same areas can therefore be difficult if there is not sufficient potential habitat left to set up their own territories, with associated risks of aggression/fighting between individuals, and increased risk of dispersal away from the release areas (this can happen with beavers for example, see Nolet & Baveco, 1996).

2.6.4 Exploring Different Release Strategies

There will often be a number of strategies that can be used to release a group of individuals to the wild. These methods should be explored, and experienced practitioners should not simply rely on the approaches used before (Walsh et al., 2015). Alternatives might be experimentally compared through trials of translocation methods. Population modelling is not without its limitations (see Section 2.3 for a discussion on PVA) but where direct evidence for the efficacy of different release methods is lacking, modelling different numbers of released individuals or strategies such as soft- versus hard-release can be a good way of exploring a range of options before committing resources to one particular release strategy over another (Rout et al., 2007).

Numerous guidelines and studies of conservation translocations emphasise the importance of releasing adequate numbers of individuals (Germano & Bishop 2009; Bellis et al., 2019). Although this has been interpreted to mean that the more individuals are translocated, the higher the probability of success, this is only likely to be the case when the correct habitat conditions have been identified. To avoid unnecessary mortality, potential animal welfare implications (Harrington et al., this volume), and the associated waste of resources and time, we recommend that large releases are only undertaken after destination sites have been trialled with smaller numbers of individuals. This may not be possible or optimal in all circumstances (see note on territoriality in Section 2.6.3), but is something that should be actively considered in the early planning phases.

2.6.5 Stakeholder Involvement

Involving stakeholders in the release of a threatened species can be a very positive act; the release of individual plants, animals, or fungi is the defining moment in what can often be years of work in planning and monitoring translocated populations. By inviting representatives of stakeholder groups to the release, it is likely that participation can improve the sense of ownership and pride in a conservation translocation (Gaywood & Stanley-Price, this volume; Glikman et al., this volume). In exceptional cases it may not be appropriate to communicate widely the locations of destination sites, where the focal species could be particularly sensitive to human disturbance including exploitation and persecution. If the potential risk of persecution is a reason for keeping a destination site secret, this should ring loud warning bells about the appropriateness of releasing individuals at all. Ideally, the preparatory stages will have addressed such controversies prior to the release, and involving stakeholders from local communities can engage potential opponents, identify concerns, and determine mitigating actions that are acceptable to all involved.

2.7 Monitoring

If questioned, most conservation practitioners would recognise the importance of monitoring in assessing the performance of conservation interventions and providing a rationale to undertake similar work in the future. However, monitoring is also crucial to effective management of conservation translocations because there are so many aspects of a proposal that can go awry: in the event of the translocation not progressing as hoped, monitoring can flag the need for alternative or additional management and, in the worst-case scenario, will prompt the initiation of an exit strategy. A rigorous (but not necessarily laborious) monitoring programme is a key part of adaptive management and should be incorporated into a cycle of action, monitoring, evaluation and, where appropriate, revision and implementation of the adjusted plan (Ewen et al., this volume; Albrecht, this volume). For this reason, the existence of a monitoring plan from the very start of the translocation programme is a requirement not just for reporting on performance but also for building trust with stakeholders who might be anxious about the translocation outcomes. Monitoring programmes must include indicators of success and failure, often linked to the targets of translocation programmes, which are feasible to evaluate through monitoring protocols, but might also include indicators of poor performance that can be incorporated into the exit strategy (see Section 2.8).

The following is a summary of considerations around sampling and reporting of monitoring, and possible variables to be recorded with a brief rationale for why each one is useful. These have been adapted and expanded upon from the IUCN Guidelines (IUCN 2013). Some are relatively straightforward and may be undertaken as 'citizen science' initiatives; others are more technical and require expertise and/or specialised equipment and facilities. This section should prompt practitioners to build their own monitoring programmes to be relevant and specific to their species and contexts.

2.7.1 Sampling

The suite of indicators described in the following paragraphs will be highly species dependent; therefore the numbers of individuals sampled, or the size and number of plots and monitoring stations, will vary considerably. However, it is important to make sure that monitoring programmes include enough plots, individuals, or patches to represent a translocated population adequately. Most ecological systems are complex and it is sometimes difficult to find conclusive explanators for outcomes, or even to identify trends in population size. As a starting point, it is often useful to replicate sampling protocols used to monitor wild populations.

2.7.2 Population Size

This is the most basic of metrics to be reported. 'Population size' generally refers to individuals in a defined location that are able to interbreed (be aware that in some contexts, the population is defined as the global number of individuals of that species). After a conservation translocation has taken place, population size at the destination site might be assessed annually, or more often if the number of individuals is to be assessed at different life stages in short-lived organisms (see Section 2.7.3). It might be tempting to record population size less frequently when the focal organisms are long-lived but we would strongly encourage practitioners to record or estimate population size every year for the first 10 years of the programme to account for strong inter-annual variation in weather patterns and/or food sources and other resources (for a case study that demonstrates the value of long-term monitoring, please see Armstrong et al., this volume). Annual monitoring can also guard against falsely optimistic reports early in the post-release phase: a population may appear to be performing well in the years immediately after translocation, but this could indicate a falsely optimistic outcome as a result of unusually favourable conditions that are not normal for the site. Repeated assessments of population size also enable the calculation of long-term population growth rate, which in turn allows important inferences to be made about the conditions in which a population can be found: increasing population size (or a population growth rate of more than 1) indicates that the niche of the species has been met and different growth rates will allow the assessment and comparison of the suitability of sites supporting both natural and translocated populations (see Box 2.1 on defining translocation success).

2.7.3 Demographic Indicators

The key indicator of demographic health is the maintenance or growth of population size, but it is often useful to collect more specific demographic information. In some cases, the maintenance of population size is evidence of reproduction, especially where the species is short-lived, but in many species the reporting of stable population size alone is not enough to indicate reproduction. Long-lived species may maintain the population through survival of the translocated individuals, and where individuals have been released into extant populations (reinforcement) or close to natural extant populations, the metric of population size cannot discriminate between survival of translocated individuals and those already present in the vicinity.

Completion of a species' life cycle is evidenced by the successful transition of offspring through juvenile stages to become reproductive adults. This is often very difficult to monitor for the entirety of the lifecycle and therefore key demographic stages can be identified to provide insight into particularly important or vulnerable points in a lifecycle. Examples include monitoring the number of seedlings in fixed plots, or brood size in nests. In animals, behaviours are well-established indicators of reproductive status – breeding calls, mating, nest-building, and successfully reared young can all be taken as positive signs of the translocated population having found suitable habitat.

2.7.4 Other Indicators of Fitness

In an extension of looking for proxies for reproduction, it can also be useful to record parameters of individual health and fitness prior to reproduction. The size and growth form of plants and fungi is usually straightforward to assess and monitor at individual level using fixed plots or tagged plants that are visited repeatedly. For animals, indicators of body condition, ability to source food and nest sites, and the expression of normal behaviours provide useful information for assessing the potential for translocated populations to become established.

2.7.5 Genetic Indicators

Monitoring the genetic health of translocated individuals and/or populations can be technical and resource intensive and would normally be considered to be outside the normal working practices of conservation organisations. However, molecular analysis is becoming cheaper and therefore more accessible, and the techniques are often undertaken at research centres, such as universities, or in commercial laboratories. Genetic diversity is the key metric and can be measured as percentage heterozygosity, allelic richness, or percentage polymorphic loci. Once a population is established, field-based measures of morphology and growth can infer the extent of genetic adaptation of the released individuals to the conditions at the release site when compared to the same measures at the source population. Translocations might also be used in the context of genetic rescue whereby populations are reinforced with individuals that can improve genetic diversity. In this case, monitoring genetic diversity over time is necessary to track the impacts of bringing in new stock. Reciprocal translocations are a particularly robust way of untangling genetic effects from the phenotypic response (i.e. the environmental response) in taxa such as plants, fungi, and the more sedentary animals (e.g. Hällfors et al., 2020).

2.7.6 Ecological Monitoring

The role of monitoring extends beyond that of simply assessing the status of the translocated population and should also aim to evaluate the impacts on the surrounding ecosystem at the destination site. This will demonstrate if a translocation has resulted in a species being functionally viable within the host community (i.e. fulfilling typical ecological roles such as pollination, herbivory, predation, nutrient cycling, etc. (Akçakaya et al., 2019)). It can also flag adverse effects that might trigger alternative management and possibly even an exit strategy.

2.7.7 Socio-economic Monitoring

Key indicators of positive or negative outcomes can be expressed in terms of social, cultural, or economic measures (Gaywood & Stanley-Price, this volume). These might be very important to monitor when there are concerns about negative impacts on people's livelihoods or well-being resulting from the translocation. However, it is useful also to measure benefits so that balanced evaluations can be made of the legacy of a translocation programme. In addition to potentially being victims or beneficiaries of conservation translocations, people are often the primary threat. This is especially the case in the context of poaching for consumption or illegal wildlife trade of the focal species. Socio-economic monitoring should also incorporate measures that might flag the return of a threat such as increased poaching or another damaging human activity so that mitigating action is appropriate for addressing the cause of population decline.

2.7.8 Range Size

The monitoring and reporting of geographic distribution as Extent of Occurrence and Area of Occupancy is an important component of IUCN Red List assessments and should include translocated populations. For many species, the increase in range size and/or occupied habitat is a key goal providing the motivation for translocations and reporting on this can demonstrate the success of the translocation and impacts on the recovery of the species.

2.7.9 Reporting Monitoring Results

There is huge potential for others to learn from translocation monitoring data and, arguably, the data are more valuable if the translocation is unsuccessful and the cause(s) of failure can be identified. A common message from numerous reviews of conservation translocations is that translocation outcomes are not reported consistently in terms of the metrics and duration of the monitoring period (e.g. Dalrymple et al., 2011; Nason et al., 2021). We strongly recommend that practitioners explore many of the newly available platforms for data sharing and publishing to enable others to benefit from their experience. Grey literature, books, and peer-reviewed literature are valuable but can be subject to organisational restrictions on access, the limits of print-runs, and subscription paywalls or pay-to-publish fees, respectively. Platforms such as the British Ecological Society's Applied Ecology Resources and the IUCN's Conservation Translocation Specialist Group's online database are free to use and provide search functionality to limit returns to relevant studies.

2.8 Building an Exit Strategy

An exit strategy is a plan of how to end the programme of release and monitoring, and is implemented once predefined conditions are met. In the context of conservation translocations, these are generally associated with a negative outcome arising from the release, such as problems with the health or fitness of translocated individuals, or if the translocated individuals become problematic themselves. However, some translocation programmes might also employ an exit strategy when the objectives of a conservation translocation have been met, thus ensuring a smooth end to a project with provision for further monitoring and management where necessary. Different possible exit strategy scenarios are outlined in Table 2.1, with the emphasis on exit strategies initiated when a translocation project fails to meet its objectives.

The exit strategy is in itself a mechanism by which effective stakeholder engagement can be achieved. The writing of the strategy allows all stakeholders to voice their concerns and have those concerns formally acknowledged with a contingency plan should those situations arise. An effective exit strategy needs a representative range of stakeholders to be involved; ideally, this would happen through several mechanisms, including meetings at locations that are accessible to stakeholders and communications in widely accessible formats (Reed et al., 2018). The aim of this engagement is to identify a set of scenarios that would represent the range of possible outcomes from a translocation. These might be positive and negative, and all stakeholders should be asked to elucidate both extremes in order to identify what might be common ground for a positive scenario (i.e. translocated individuals forming a self-sustaining population whilst problematic behaviours are managed or mitigated). Ruiz-Miranda and Consorte-McCrea (2019) promote the idea of exit strategies that also plan for positive outcomes so that there is a planned handover period in which the organisation that instigated the translocation ends its formal involvement and management is passed on to other groups. Once these scenarios have been identified, a number of indicators can be set to flag the onset of future scenarios as early in the progression as possible, and incorporate these indicators into the monitoring plan.

2.9 Seeking Advice, Getting Help, and Building a Team

A starting point for planning translocations is to seek out relevant guidelines, of which the IUCN Guidelines are the key resource. However, there are now many similar documents that are specific to different taxonomic groups (see the selection presented by the IUCN CTSG at https://iucn-ctsg.org/policy-guidelines/taxon-specific-guidelines/) and countries (e.g. Scottish Code for Conservation Translocations (National

Species Reintroduction Forum, 2014)) that can provide more specialised advice and contacts. Successful translocation teams are most likely to have a mix of expertise, interests, roles, and experience as indicated by the variety of topics covered in this chapter. Representation will be needed from landowners, conservation organisations – both *in situ* and *ex situ*, local community groups, local authorities, and government agencies (national agencies usually supply advice and application forms for licences and permits and should be consulted even if these are not required). In addition, there are often many opportunities to involve the science community based in research institutes and universities. For example, student research projects can generate useful monitoring data and explore causal factors behind differential success rates; they can also offer access to research literature and analytical tools and techniques that are unavailable to most non-academic bodies.

When building a team, representation should be addressed sensitively. For example, it helps to avoid branding that conveys ownership or dominance of one organisation over others, and working practices are made more constructive by acknowledging each party's agendas and motivations for participating. The National Species Reintroduction Forum in Scotland is an example of a widely representative group that has very effectively formed a platform by which conservation translocations are considered to maximise conservation benefit whilst minimising socio-economic and biological risk (Gaywood & Stanley Price, this volume).

2.10 Conclusion

Although this chapter provides only a cursory overview of the many stages and considerations in carrying out conservation translocations, we hope that it has provided a starting point for a thorough appraisal of potential translocation projects. There are many benefits of effective planning, including being more efficient with resources and increasing the chance of biological success. A thorough plan is also more likely to result in the effective buy-in of stakeholders. There is now a huge body of evidence on conservation translocations that should inform detailed planning and appraisal ahead of undertaking these ambitious projects. We hope that this chapter, and the rest of this volume, prompts practitioners to exploit this experience and undertake translocations that achieve biological success and build positive alliances beyond the conservation sector to increase environmental benefits to the wider society.

> Box 2.1 *Defining success*
>
> In a recent assessment of the causes of failure in conservation translocations (Bellis et al., 2020) we developed a common definition of success that compared the outcomes of translocations across a range of species. Typically, the definition of a successful translocation varies and is dependent on the aims, motivations, and post-release time limitations of the translocation project (Robert et al., 2015). The most commonly used definition of 'success' in translocation programmes is whether the programme results in a self-sustaining population of the target species (e.g. Griffith et al., 1989; Fischer & Lindenmayer, 2000). However, monitoring frequency and duration can affect the reported outcomes even within this relatively straightforward definition. By formulating guidelines that take into consideration the generation time of a species, the time between translocation and monitoring, and the demographic trends of the population, it should be possible to standardise the definition of a translocation success.
>
> We considered a translocation successful if it met three criteria: (i) the time elapsed between the most recent release and most recent post-release monitoring exceeded the lifecycle duration of the species, (ii) the same period also exceeded 10 years to account for interannual variability in environmental conditions that might artificially indicate optimal conditions at shorter timescales, and (iii) the most recent monitoring results indicated population persistence at the release site. In the case of the last criterion, evidence of population maintenance or growth is the true measure of success, but using population persistence as the minimum requirement for success does at least allow robust comparisons when other data are missing. Our definition does not exclude populations that are subject to some form of management *in situ*.

2.11 Key Messages

- Although conservation translocations always aim to result in 'conservation benefit' for a species or ecosystem, there are many motivations and contexts in which people undertake conservation translocations.
- Establishing a need for a translocation involves considering other ways to deliver the 'conservation benefit' to a species or ecosystem. Typically, translocations are initiated alongside other conservation measures.

- Destination sites must be carefully selected, with attention paid to climate, biotic interactions, and other species' requirements for survival. Socio-economic risks and benefits and legislative constraints should also be considered when evaluating destination sites.
- Translocated plants, animals, or fungi may be sourced from *in situ* or *ex situ* populations, or a combination of the two. Ideally, source populations are those that can support the removal of individuals for translocation without negative consequences for individuals that remain in the wild. It also helps to have ecological similarity between source and destination sites when using wild individuals.
- Releases should only occur where the original causes of extirpation/decline have been minimised. The choice of one or several sites and the possibility of several releases over multiple timepoints should be considered, perhaps using pilot releases to evaluate suitability.
- Monitoring is crucial to being able to report the outcome of the conservation translocation but, perhaps more importantly, allows the translocation team to adapt to changes in status and may prompt the implementation of an exit strategy. Sharing monitoring results is important so that others can benefit from lessons learnt.
- Seek as much advice as you can in the planning phases and build a team that can deliver the various components of the project.

References

Akçakaya, H. R., Rodrigues, A. S. L., Keith, D. A., et al. (2019) Assessing ecological function in the context of species recovery. *Conservation Biology*. 34, 561–571.

Albrecht, M. A. (2023) Applying adaptive management to reintroductions of Pyne's ground-plum *Astragalus bibullatus*. In Gaywood, M. J., Ewen, J. G., Hollingsworth, P. M. and Moehrenschlager, A. (eds.) *Conservation Translocations*. Cambridge, Cambridge University Press.

Armstrong, D. P., Parlato, E. H. & Ewen, J. G. (2023) Five reasons to consider long-term monitoring: case studies from bird reintroductions on Tiritiri Matangi Island. In Gaywood, M. J., Ewen, J. G., Hollingsworth, P. M. and Moehrenschlager, A. (eds.) *Conservation Translocations*. Cambridge, Cambridge University Press.

Bellis, J., Bourke, D., Maschinski, J., Heineman, K. & Dalrymple, S. (2020) Climate suitability as a predictor of conservation translocation failure. *Conservation Biology*. 34, 1473–1481.

Bellis, J., Bourke, D., Williams, C. & Dalrymple, S. (2019) Identifying factors associated with the success and failure of terrestrial insect translocations. *Biological Conservation*. 236, 29–36.

Brichieri-Colombi, T. A. & Moehrenschlager, A. (2016) Alignment of threat, effort, and perceived success in North American conservation translocations. *Conservation Biology.* 30, 1159–1172.

Brooker, R. W., Brewer, M. J., Britton, A. J., et al. (2018) Tiny niches and translocations: the challenge of identifying suitable recipient sites for small and immobile species. *Journal of Applied Ecology.* 55, 621–630.

Brzosko, E., Jermakowicz, E., Ostrowiecka, B., et al. (2018) Rare plant translocation between mineral islands in Biebrza Valley (northeastern Poland): effectiveness and recipient site selection. *Restoration Ecology.* 26, 56–62.

Campbell-Palmer, R., Bauer, A., Jones, S., Ross, B. & Gaywood, M. J. (2023) The return of the Eurasian beaver to Britain: the implications of unplanned releases and the human dimension. In Gaywood, M. J., Ewen, J. G., Hollingsworth, P. M. and Moehrenschlager, A. (eds.) *Conservation Translocations.* Cambridge, Cambridge University Press.

Charles, H. & Dukes, J. S. (2008) Impacts of invasive species on ecosystem services. In *Biological Invasions.* Berlin Heidelberg, Springer, pp. 217–237.

Cochran-Biederman, J. L., Wyman, K. E., French, W. E. & Loppnow, G. L. (2015) Identifying correlates of success and failure of native freshwater fish reintroductions. *Conservation Biology.* 29, 175–186.

Dalrymple, S. E. & Broome, A. (2010) The importance of donor population identity and habitat type when creating new populations of small *Melampyrum sylvaticum* from seed in Perthshire, Scotland. *Conservation Evidence.* 7, 1–8.

Dalrymple, S. E., Stewart, G. B. & Pullin, A. S. (2011) *Are (re-) introductions an effective way of mitigating against plant extinctions? CEE review 07-008 (SR32).* Collaboration for Environmental Evidence. Available from: www.environmentalevidence.org/SR32.html [Accessed 6 June 2020].

Ewen, J. G., Canessa, S., Converse, S. J. & Parker, K. A. (2023) Decision-making in animal conservation translocations: biological considerations and beyond. In Gaywood, M. J., Ewen, J. G., Hollingsworth, P. M. and Moehrenschlager, A. (eds.) *Conservation Translocations.* Cambridge, Cambridge University Press.

Ferrarini, A., Selvaggi, A., Abeli, T., et al. (2016) Planning for assisted colonization of plants in a warming world. *Scientific Reports.* 6, 6–11.

Fischer, J. & Lindenmayer, D. B. (2000) An assessment of the published results of animal relocations. *Biological Conservation.* 96, 1–11.

Gaywood, M. J. (ed.) (2015) *Beavers in Scotland: A Report to the Scottish Government.* Inverness.

Gaywood, M. J. & Stanley-Price, M. (2023) Moving species: reintroductions and other conservation translocations. In Gaywood, M. J., Ewen, J. G., Hollingsworth, P. M. and Moehrenschlager, A. (eds.) *Conservation Translocations.* Cambridge, Cambridge University Press.

Germano, J. M. & Bishop, P. J. (2009) Suitability of amphibians and reptiles for translocation. *Conservation Biology.* 23, 7–15.

Glikman, J. A., Frank, B., Sandström, C., et al. (2023) The human dimensions and the public engagement spectrum of conservation translocation. In Gaywood, M. J., Ewen, J. G., Hollingsworth, P. M. and Moehrenschlager, A. (eds.) *Conservation Translocations.* Cambridge, Cambridge University Press.

Gray, C. L., Hill, S. L. L., Newbold, T., et al. (2016) Local biodiversity is higher inside than outside terrestrial protected areas worldwide. *Nature Communications*. 7, 12306.

Griffith, B., Scott, J. M., Carpenter, J. W. & Reed, C. (1989) Translocation as a species conservation tool: status and strategy. *Science*, 245, 477–480.

Guisan, A., Thuiller, W. & Zimmermann, N. E. (2017). *Habitat Suitability and Distribution Models: With Applications in R*. Cambridge, Cambridge University Press.

Guisan, A., Tingley, R., Baumgartner, J. B., et al. (2013) Predicting species distributions for conservation decisions. *Ecology Letters*. 16, 1424–1435.

Hällfors, M. & Dalrymple, S. E. (2023) Assisted colonisation and ecological replacement. In Gaywood, M. J., Ewen, J. G., Hollingsworth, P. M. and Moehrenschlager, A. (eds.) *Conservation Translocations*. Cambridge, Cambridge University Press.

Hällfors, M., Lehvävirta, S., Aandahl, T., et al. (2020) Translocation of an arctic seashore plant reveals signs of maladaptation to altered climatic conditions. *PeerJ*. 8, e10357.

Harrington, L. A., Lloyd, N. & Moehrenschlager, A. (2023) Animal welfare, animal rights, and conservation translocations: moving forward in the face of ethical dilemmas. In Gaywood, M. J., Ewen, J. G., Hollingsworth, P. M. and Moehrenschlager, A. (eds.) *Conservation Translocations*. Cambridge, Cambridge University Press.

Hiroyasu, E. H. T., Miljanich, C. P. & Anderson, S. E. (2019) Drivers of support: the case of species reintroductions with an ill-informed public. *Human Dimensions of Wildlife*. 24, 401–417.

IUCN (2013) *Guidelines for Reintroductions and Other Conservation Translocations. Version 1*. Gland, Switzerland, IUCN Species Survival Commission.

Kalle, R., Combrink, L., Ramesh, T. & Downs, C. T. (2017) Re-establishing the pecking order: niche models reliably predict suitable habitats for the reintroduction of red-billed oxpeckers. *Ecology and Evolution*. 7, 1974–1983.

Kock, R. A., Woodford, M. H. & Rossiter, P. B. (2010) Disease risks associated with the translocation of wildlife. *OIE Revue Scientifique et Technique*. 29, 329–350.

Lawrence, B. A. & Kaye, T. N. (2011) Reintroduction of *Castilleja levisecta*: effects of ecological similarity, source population genetics, and habitat quality. *Restoration Ecology*. 19, 166–176.

Mack, R. N., Simberloff, D., Lonsdale, W. M., et al. (2000) Biotic invasions: causes, epidemiology, global consequences, and control. *Ecological Applications*. 10, 689–710.

Maes, D., Ellis, S., Goffart, P., et al. (2019) The potential of species distribution modelling for reintroduction projects: the case study of the Chequered Skipper in England. *Journal of Insect Conservation*. 23, 419–431.

Malone, E. W., Perkin, J. S., Leckie, B. M., et al. (2018) Which species, how many, and from where: integrating habitat suitability, population genomics, and abundance estimates into species reintroduction planning. *Global Change Biology*. 24, 3729–3748.

Maschinski, J. & Albrecht, M. (2023) Conservation translocations for plants. In Gaywood, M. J., Ewen, J. G., Hollingsworth, P. M. and Moehrenschlager, A. (eds.) *Conservation Translocations*. Cambridge, Cambridge University Press.

Mitchell, R., Green, S. & Hollingsworth, P. M. (2023) Plant health, biosecurity and conservation translocations. In Gaywood, M. J., Ewen, J. G., Hollingsworth, P. M. and Moehrenschlager, A. (eds.) *Conservation Translocations*. Cambridge, Cambridge University Press.

Nason, S. E., Lloyd, N., Kelly, C. D., et al. (2021) Maximizing the effectiveness of qualitative systematic reviews: a case study on terrestrial arthropod conservation translocations. *Biological Conservation*. 254, 108948.

National Species Reintroduction Forum (2014) *The Scottish Code for Conservation Translocations*. Inverness, Scottish Natural Heritage.

Neaves, L. E., Ogden, R. & Hollingsworth, P. M. (2023) Genomics and conservation translocations. In Gaywood, M. J., Ewen, J. G., Hollingsworth, P. M. and Moehrenschlager, A. (eds.) *Conservation Translocations*. Cambridge, Cambridge University Press.

Nolet, B. A. & Badeco, J. M. (1996) Development and viability of a translocated beaver *Castor fiber* population in the Netherlands. *Biological Conservation*. 75, 125–137.

Novak, B. J., Phelan, R. & Weber, M. (2021) U.S. conservation translocations: Over a century of intended consequences. *Conservation Science and Practice*. 3, e394.

Osborne, P. E. & Seddon, P. J. (2012) Selecting suitable habitats for reintroductions: variation, change and the role of species distribution modelling. In Ewen, J. G., Armstrong, D. P., Parker, K. A. and Seddon P. J. (eds.) *Reintroduction Biology: Integrating Science and Management*. Chichester, Wiley-Blackwell, pp. 73–104.

Phillips, S. B., Aneja, V. P., Kang, D. & Arya, S. P. (2006) Modelling and analysis of the atmospheric nitrogen deposition in North Carolina. *Ecological Modelling*. 190, 231–259.

R Core Team (2018) R: A language and environment for statistical computing. R Foundation for Statistical Computing, Vienna, Austria. www.R-project.org/

Reed, M. S., Vella, S., Challies, E., et al. (2018) A theory of participation: what makes stakeholder and public engagement in environmental management work? *Restoration Ecology*. 26, S7–S17.

Robert, A., Colas, B., Guigon, I., et al. (2015) Defining reintroduction success using IUCN criteria for threatened species: a demographic assessment. *Animal Conservation*. 18, 397–406.

Rout, T. M., Hauser, C. E. & Possingham, H. P. (2007) Minimise long-term loss or maximise short-term gain?: Optimal translocation strategies for threatened species. *Ecological Modelling*. 201, 67–74.

Ruiz-Miranda, C. R. & Consorte-McCrea, A. (2019) *Exit Strategies and animal reintroductions (and wildlife conservation in general)*. IUCN-SSC/CTSG Human-Wildlife Interactions Working Group report. doi:10.13140/RG.2.2.11350.40002

Sainsbury, A. W. & Carraro, C. (2023) Animal disease and conservation translocations. In Gaywood, M. J., Ewen, J. G., Hollingsworth, P. M. and

Moehrenschlager, A. (eds.) *Conservation Translocations*. Cambridge, Cambridge University Press.

Swan, K. D., McPherson, J. M., Seddon, P. J. & Moehrenschlager, A. (2016) Managing marine biodiversity: the rising diversity and prevalence of marine conservation translocations. *Conservation Letters*. 9, 239–251.

Thatcher, C. A., Manen, F. T. Van & Clark, J. D. (2006) Identifying suitable sites for Florida Panther reintroduction. *Journal of Wildlife Management*. 70, 752–763.

Thuiller, W., Lafourcade, B., Engler, R. & Araújo, M. B. (2009) BIOMOD - A platform for ensemble forecasting of species distributions. *Ecography*. 32, 369–373.

Trouwborst, A., Blackmore, A., Blyth, S., Fleurke, F., McCormack, P. & Gaywood, M. J. (2023) Conservation translocations and the law. In Gaywood, M. J., Ewen, J. G., Hollingsworth, P. M. and Moehrenschlager, A. (eds.) *Conservation Translocations*. Cambridge, Cambridge University Press.

Walker, S. F., Bosch, J., James, T. Y., et al. (2008) Invasive pathogens threaten species recovery programs. *Current Biology*. 18, 853–854.

Walsh, J. C., Dicks, L. V. & Sutherland, W. J. (2015) The effect of scientific evidence on conservation practitioners' management decisions. *Conservation Biology*. 29, 88–98.

White, T. H., de Melo Barros, Y., Develey, P. F., et al. (2015) Improving reintroduction planning and implementation through quantitative SWOT analysis. *Journal for Nature Conservation*. 28, 149–159.

Zavaleta, E. (2000) The economic value of controlling an invasive shrub. *AMBIO: A Journal of the Human Environment*. 29, 462–467.

Zurell, D., Franklin, J., König, C., et al. (2020) A standard protocol for reporting species distribution models. *Ecography*. 43, 1–17.

Part II
Conservation Translocations: The Key Issues

3 · *Conservation Translocations and the Law*

ARIE TROUWBORST, ANDY BLACKMORE,
SALLY BLYTH, FLOOR FLEURKE, PHILLIPA
MCCORMACK, AND MARTIN J. GAYWOOD

3.1 Introduction

This chapter approaches conservation translocations from a legal perspective, by introducing and discussing relevant legal instruments. Standard legal research methodology is used, involving the identification and interpretation of relevant legislation.

Law can influence translocations in two basic ways, by playing a supportive and/or a restrictive role. In the first situation, legislation can provide support, a mandate, and/or an obligation to consider or perform a translocation. In the second situation, legislation can impose limitations on translocation possibilities and/or make translocations conditional on meeting certain requirements. Such limitations and conditions may follow from legislation on native species protection, area protection, invasive non-native species (INNS, also known as invasive alien species), disease, trade, animal welfare, property, public safety, etc.

Requirements of the first and second types of law may flow from domestic (national) law and from international law – more specifically, global and regional treaties and other binding international legal instruments. Given the unfeasibility of discussing the domestic legislation of all countries in the world, this chapter will focus predominantly on international instruments, which are in force for many countries simultaneously and the content of which is accordingly reflected in many domestic legal systems. In addition, concise summaries of the legal framework of several selected countries or jurisdictions are provided for illustrative purposes (Boxes 3.1–3.4). Examples are provided for Australia, South Africa, Scotland, and the European Union. The precise extent of applicable legislation in a given instance will depend on the country (or countries), species, areas, and type of translocation involved.

3.2 Law, Its Context and Interpretation

The principal source of law is legislation, that is, binding rules issued by governments and other public entities at international, national, and lower levels. Other sources include contracts between private entities, customary (unwritten) law, and court decisions. On the international plane, most law is created through binding agreements concluded between national governments. These come by different names, including treaty, convention, agreement, and protocol. A treaty is binding only for those countries (and regional organisations) that have become contracting parties to it, which normally requires a formal act of ratification. Some treaty-based international organisations can also adopt binding decisions or legislation. A particularly advanced example is the European Union (EU), whose organs have issued a plethora of regulations and directives with binding force in the 27 EU member states (see Box 3.4). Its binding nature distinguishes law from other, non-binding instruments in the broader domain often referred to as 'policy'. Political declarations, statements of intent, strategies, action plans, guidelines, recommendations, codes of conduct, and memoranda of understanding are all examples of policy instruments that are *not* legally binding, though such instruments may nevertheless play important roles in the functional use of legal instruments that govern translocations.

A crucial step in the actual application of law is its *interpretation*. The precise implications of a legal obligation in a specific set of circumstances may not always be immediately apparent from the obligation's formulation, especially in the case of rules with a general scope. Differing opinions regarding the correct interpretation of rules are at the heart of many legal disputes. According to the basic rules recorded in the 1969 (Vienna) Convention on the Law of Treaties, a treaty must be interpreted 'in accordance with the ordinary meaning to be given to the terms of the treaty in their context and in the light of its object and purpose' (Article 31(1)). In addition to such textual, contextual, and teleological (objective-driven) interpretation, account must be taken of any 'subsequent practice' or 'subsequent agreement' indicating the views of the parties regarding the treaty's interpretation, and any other 'relevant rules of international law' (Article 31(3)). Notably, this entails a potentially significant role for *non*-binding decisions adopted by Conferences of the Parties (COPs) and other treaty bodies, and for other non-binding instruments, in the interpretation of binding treaty provisions. Other potential interpretive aids include general principles of environmental law (such as the precautionary principle, see Section 3.5) and relevant court decisions.

The non-binding IUCN Guidelines for Reintroductions and Other Conservation Translocations (IUCN, 2013, hereinafter referred to as the IUCN Guidelines) are of particular importance in the interpretation and application of relevant binding international and national law. The 1979 Convention on the Conservation of European Wildlife and Natural Habitats (Bern Convention) provides a good example. Article 11(2)(a) of the Convention commits each party to 'encourage the reintroduction of native species of wild flora and fauna when this would contribute to the conservation of an endangered species, provided that a study is first made in the light of the experiences of other Contracting Parties to establish that such reintroduction would be effective and acceptable'. The interpretation of this binding provision is informed by various Recommendations – non-binding by themselves – that have been adopted over the years by the Convention's Standing Committee, the main decision-making body, in which all parties are represented. One of these, Recommendation No. 158 (2012) on conservation translocations under changing climatic conditions, calls on parties to '[f]ollow the revised IUCN guidelines for Reintroductions and Other Conservation Translocations . . . when conducting translocations'.

Also beyond Europe, the IUCN Guidelines typically play an important part in the application of international and national legislation on translocations to individual cases (see, e.g., https://iucn-ctsg.org/policy/). The role of both the IUCN Guidelines and courts in the current context can be illustrated using the topical example of the proposed conservation translocation of Southern African cheetah *Acinonyx jubatus jubatus* from Namibia to India (Aggarwal, 2020). The Asiatic cheetah *Acinonyx jubatus venaticus* had been extirpated in India in the last century, and the legal arguments involved the appropriateness of using the Southern African subspecies for the translocation. In 2012, the Supreme Court of India outlawed such a translocation, in part because of incompatibility with a previous version of the IUCN Guidelines (Supreme Court of India, Centre for Environmental Law, WWF. v. Union of India et al., 15 April 2013). In 2020, the same court declared the legality of a renewed proposal to translocate the species on an experimental basis (Supreme Court of India, Centre for Environmental Law, WWF. v. Union of India et al., 28 January 2020).

In a legal context, as elsewhere, due attention needs to be paid to the distinctions between the various types of conservation translocation – reinforcement, reintroduction, assisted colonisation, ecological replacement – as these distinctions can be significant from a legal perspective. Similar care needs to be taken with regard to key terms such as 'native', 'historic' and 'natural' or 'indigenous range', and 'alien', 'non-native', or 'invasive' species. The meaning of such terms can vary from one legal regime to the next, and can be defined by ecologists in slightly different ways to the definitions used in law. Moreover, they may not be used by governmental stakeholders in an entirely clear or consistent manner, whether it be an authority granting or refusing a translocation permit, or a court reviewing such a decision.

Finally, legal frameworks are dynamic: new legal instruments are adopted, or old ones amended. Changes are particularly frequent at national levels. International treaty regimes are more constant by comparison, but they are not immutable either. Whereas amendments to binding international treaty provisions are rare, the body of strategic and technical documents informing the application of those binding commitments changes regularly. For instance, the next Conference of the Parties (COP) of the 1992 Convention on Biological Diversity (CBD), due to be held in 2022, is expected to set new strategic goalposts for biodiversity conservation for the post-2020 period.

3.3 Legislation Promoting or Requiring Translocations

Law can provide support, a mandate, and/or an obligation to consider or conduct a conservation translocation (Rees, 2001). This may involve both general obligations to conserve or restore certain species, sites, or habitats, and specific obligations concerning conservation translocations.

3.3.1 Restoration Entailing Translocation

General obligations to conserve or restore biodiversity abound in international legal instruments (Gardner, 2003; Bastmeijer, 2016; Telesetsky et al, 2017). The precise relevance of such obligations for present purposes will depend on their formulation and the circumstances involved. An example with a global scope is Article 8(f) of the CBD, requiring contracting parties (195 countries and the EU), 'as far as possible and as appropriate', to '[r]ehabilitate and restore degraded ecosystems and promote the recovery of threatened species, inter alia, through the development and implementation of plans or other management strategies' (Article 8(f)). A similarly phrased provision with a regional scope is Article 24 of the 1992 Convention for the Conservation of the Biodiversity and the Protection of Priority Wilderness Areas in Central America.

Further examples can be found in the 2003 Protocol to the Carpathian Convention on Conservation and Sustainable Use of Biological and Landscape Diversity (Carpathian Biodiversity Protocol). Each of its seven parties 'shall develop and implement policies and strategies in its national territory aiming at the conservation, *restoration* and sustainable use of biological ... diversity of the Carpathians' (Article 4, emphasis added), and 'take measures in its national territory with the objective to ensure the long-term conservation, *restoration* and sustainable use of natural habitats in the Carpathians' (Article 8(2), emphasis added). The Protocol defines the term 'restoration' as the 'return of an ecosystem or habitat to its original structure, natural composition of species, and natural functions' (Article 3(r)). Clearly, therefore, depending on the circumstances, the carrying out of particular conservation translocations (particularly reinforcements, reintroductions, and ecological replacement) may be necessary to comply with the two cited obligations.

3.3.2 Reintroduction and Reinforcement

In addition to such general conservation and restoration provisions, many legal instruments contain provisions expressly calling for conservation translocations, particularly reintroductions. For instance, the CBD sets out a general obligation to adopt, 'as far as possible and as appropriate, ... measures for the recovery and rehabilitation of threatened species and for their reintroduction into their natural habitats under appropriate conditions' (Article 9(c)). The aforementioned Carpathian Biodiversity Protocol requires its parties to 'cooperate on activities aiming at reintroduction of native species of fauna and flora' (Article 12(3)). A legal instrument covering another European mountain range, the 1994 Protocol to the Alpine Convention Relating to Nature Conservation and Landscape Conservation (Alpine Nature Protocol), contains an obligation to 'undertake to promote the reintroduction and distribution of wild, indigenous animal and plant species and also subspecies, breeds and ecotypes' (Article 16(1)).

According to Article 11(2)(a) of the Bern Convention, cited previously, each of the Convention's 51 parties – including all EU member states and the EU itself – 'undertakes' to 'encourage the reintroduction of native species'. The EU's 1992 Directive 92/43/EC on the Conservation of Natural Habitats and of Wild Fauna and Flora (Habitats Directive) sets out a requirement for EU member states to 'study the desirability of re-introducing' native species listed in the Directive's Annex IV

(Article 22(a)). Incidentally, as Rees (2001) has rightly noted, the latter duty to 'study the desirability' of reintroductions appears to set a lower standard than the Bern Convention's requirement to 'encourage' them, raising questions regarding the Directive's compatibility with the Convention in this regard.

All of the above are binding provisions. Their meaning and the way they ought to be applied have been elaborated in various non-binding documents. Recommendations adopted over several decades by the Bern Convention's Standing Committee provide a range of interesting examples. Two of these are of general application, whereas others are species specific. Especially significant is Recommendation No. 58 (1997), providing general guidance regarding reintroduction and reinforcement translocations. It was adopted to 'improve the implementation of' Article 11(2)(a) and to 'take account of the particular case of population reinforcements'. The aforementioned Recommendation No. 158 (2012) provides guidance regarding conservation translocations as a response to climate change. Both of these will be revisited later in Sections 3.3.3, 3.4.1, and 3.5.

Species that have received targeted attention from the Standing Committee in connection with translocations include the Eurasian lynx *Lynx lynx* (Recommendations No. 20 (1991) and No. 204 (2019)), Iberian lynx *Lynx pardinus* (Recommendation No. 107 (2003)), and Eurasian otter *Lutra lutra* (Recommendation No. 53 (1996)). For instance, Recommendation No. 20 called on parties to 'consider the possibility' of carrying out reintroduction programmes for Eurasian lynx 'in areas where the species has become extinct or is endangered', and to coordinate such projects between neighbouring states. Regarding the Iberian lynx, Recommendation No. 107 on the Odelouca Dam specifically urged Portugal to 'co-operate with the Spanish authorities ... with a view to reintroduction of the species', and to safeguard potential habitat from the adverse impacts of dam-building.

A duty to reintroduce or reinforce may also arise from a failure to observe certain obligations to conserve. The protected area regime of the Habitats Directive is a case in point. When, pursuant to the Directive, a protected area has been designated for a certain species, and the species subsequently disappears from the area due to non-natural causes, this entails not only a violation of Article 6 of the Directive by the EU member state involved, but also a duty to restore the species to the site. If this is unlikely to happen through natural recolonisation, reintroduction will be obligatory. A topical example is the extinction of the wolf

Canis lupus population in the Sierra Morena region in southern Spain (Trouwborst, 2014; López-Bao et al., 2018).

3.3.3 Assisted Colonisation and Ecological Replacement

Assisted colonisation and ecological replacement, which are both classed in the IUCN Guidelines as 'conservation introductions', are not nearly as firmly anchored in the law as reinforcement and reintroduction are (Camacho, 2010; McCormack, 2018; Brodie et al., 2021). All legal instruments addressed so far lack binding provisions explicitly addressing assisted colonisation or ecological replacement. In particular sets of circumstances, however, it may be that a duty to perform an assisted colonisation or ecological replacement can be inferred from general obligations to conserve or restore specific species or ecosystems (Trouwborst, 2015).

At any rate, assisted colonisation is considered in various non-binding documents, including Bern Convention Recommendation No. 158 (2012), and climate change-related decisions by the parties to the CBD (COP Decision X/33 (2010)) and the 1979 Convention on the Conservation of Migratory Species of Wild Animals (CMS) (COP Resolution 12.21 (2017)). Recommendations to consider assisted colonisation in these decisions are couched in careful terms. CBD COP Decision X/33 (par. 8(e)) reads as follows, where the alternative term 'assisted migration' is used instead of 'assisted colonisation':

Bearing in mind that under climate change, natural adaptation will be difficult and recognizing that *in situ* conservation actions are more effective, also consider *ex situ* measures, such as relocation, assisted migration and captive breeding, among others, that could contribute to maintaining the adaptive capacity and securing the survival of species at risk, taking into account the precautionary approach in order to avoid unintended ecological consequences.

In a similar vein, the programme of measures on climate change appended to CMS COP Resolution 12.21 recommends parties to consider 'assisted colonization, including translocation, as appropriate, for those migratory species most severely threatened by climate change while bearing in mind the need to minimise the potential for unintended ecological consequences'. However, overall, the existing measures and guidance provided by the CBD, CMS and Bern Convention are limited. Brodie et al. (2021) recommended that the CBD establish a committee to create protocols for assisted colonisation that all countries could implement.

3.4 Legislation Creating Conditions or Restrictions for Translocations

The second major role that law can play in the present context is to restrict the options for conservation translocations, principally in order to ensure the application of best practice and to minimise the risks of adverse impacts. As the IUCN Guidelines put it, a conservation translocation 'may need to meet regulatory requirements at any or all of international, national, regional or sub-regional levels' (IUCN 2013). For example, if a translocation involves the cross-border movement of plants or animals belonging to species listed in the appendices to CITES, it will require prior application for an export permit, import permit, and/or other CITES documentation. It must also comply with plant and animal health standards set by, respectively, the 1951 International Plant Protection Convention's Commission on Phytosanitary Measures, and the World Organisation for Animal Health (OIE).

Several further licences or other legal permissions may need to be obtained for the capture, keeping, transport, and/or release of an organism, pursuant to national and international legislation. A variety of requirements may need to be met in order to acquire such permissions, ranging from scientific studies and public participation to disease prevention and animal welfare standards. As these examples indicate, relevant areas of law extend well beyond nature conservation law and may also include, for example, rules concerning access to land, the keeping of dangerous animals, and other public safety issues. For want of space, we have selected only some of these to examine in further detail: rules on translocations themselves, INNS, and protected species. These areas have been selected because of the combination of their practical importance and the significant legal questions to which they give rise. Various other areas of law are addressed in Boxes 3.1–3.4 illustrating national and regional jurisdictions.

3.4.1 Rules on Conservation Translocations

The very provisions in international and national legislation that call for translocations, particularly reintroductions, also tend to specify certain conditions under which translocations are to take place. These conditions

Box 3.1 *Australia*

The state and territory (sub-national) governments have primary responsibility for legal frameworks for translocations in Australia. As a result, specific legal requirements differ across the country, although in all cases translocations are facilitated under general conservation legislation and translocation-specific policies, and are constrained by state and national biosecurity laws that prohibit the introduction of invasive or 'pest' species and pathogens.

Conservation translocations are typically conducted as a recovery action implemented pursuant to statutory recovery plans. Recovery plans must (or in some cases 'may') be developed for native species listed on statutory threatened species lists, and plans are taken forward by recovery teams, which typically include representatives from government, research institutions, and conservation organisations. For example, the 'Save the Tasmanian Devil Program' is carrying out the recovery plan for the Tasmanian Devil *Sarcophilus harrisii* that was developed under the Tasmanian Nature Conservation Act 2002 and adopted under the national Environment Protection and Biodiversity Conservation Act 1999. The Devil is endangered as a result of an aggressive and contagious facial tumour disease (Tasmanian Government, 2020) and its recovery plan includes an international genetic research collaboration and captive breeding programme, as well as assisted colonisation of healthy individuals to an offshore island as an 'insurance population' (Hogg & Wise, this volume). Similarly, as climate warming and drying trends have reduced the viability of habitat in its historical distribution, the recovery plan for the critically endangered western swamp tortoise *Pseudemydura umbrina* includes assisted colonisation of tortoises to wetlands further south, making it Australia's first climate-driven translocation (Threatened Species Recovery Hub, 2019). The tortoise translocation involved multiple phases of modelling, consultation and expert reporting, and trial reintroductions within its native range before two assisted colonisations were carried out in 2016 and 2018.

While the terms of individual recovery plans vary, there are some consistencies in translocation governance across Australia. For example, all translocation policies and procedures require a Translocation Proposal to be submitted to the relevant conservation agency as part of statutory assessment and approval processes. Translocation Proposals must include information about: the target species' biological and ecological requirements (including threats to its conservation status); justifications for the proposed translocation (i.e. demonstrating that benefits outweigh risks); detailed risk assessments including an 'exit plan' in case of failure; and stakeholder engagement planning.

Until recently, the state of Tasmania had the only policy in Australia that explicitly referenced climate change as a driver for translocations. The limited provision for future climate-driven translocations is notable given Australia's high levels of biodiversity, species endemism, and exposure to the effects of climate change. However, new policies in the state of New South Wales (Office of Environment and Heritage, 2019, clauses 21–23)('NSW Operational Policy') and the Australian Capital Territory (ACT Government, 2017) explicitly acknowledge climate as a driver for translocation and a crucial consideration in assessing the suitability of proposals.

Species proposed for translocation often have great spiritual, economic, and social significance to Indigenous People. In some jurisdictions, Translocation Proposals explicitly require engagement with Indigenous stakeholders, over and above other stakeholder engagement. The new NSW Operational Policy (uniquely) enables translocation of *non-threatened*, protected fauna if 'there is a strong case that the translocation would be of high . . . Aboriginal cultural significance' (clause 28c). Recent research indicates that Indigenous People in Australia have translocated species around the continent for millennia, both to reinforce existing populations and to introduce new ones (Silcock, 2018). Indigenous People have also been instrumental in supporting historical and more recent translocations by government agencies and conservation groups (Lundie-Jenkins, 1999).

However, while there is some evidence that translocation projects are increasingly involving Indigenous communities directly in planning, implementation, and monitoring phases, important challenges remain. Perhaps the most important is the terminology used in legal instruments – particularly 'assisted colonisation' – which can cause serious offence and reinforce the trauma of Aboriginal dispossession and European colonisation of the continent, its peoples, and its environments (Van Dooren, 2019). Another important challenge is the emerging reality of the impact of climate change on some species, increasing the need to translocate some species beyond their natural ranges (as was the case for the western swamp tortoise), which removes them from Indigenous land where they may have been part of deep and complex cultural relationships. Existing laws and policies for translocation provide little guidance for managing these challenges (McCormack, 2018).

Australia's laws, policies, and procedures for governing translocations are being tested as the rate and scale of climate change and its effects exceed scientific projections. As a result, Australian policymakers and translocation practitioners will likely be at the forefront of a steep, global learning curve, deploying translocations and other conservation tools in novel ways to conserve as much of the continent's rich biodiversity as possible, for generations to come.

---

primarily concern scientific research, for example, by requiring evidence about the expected effectiveness and impacts of conservation translocations, and obligations to inform and/or consult relevant stakeholders and the public at large. The obligation imposed by Article 11(2)(a) of the Bern Convention, for example, only requires parties to encourage reintroductions 'when this would contribute to the conservation of an endangered species', and furthermore makes any reintroduction conditional upon the outcome of a prior study showing 'that such reintroduction would be effective and acceptable'. According to Recommendation No. 58 (1997), for a release to qualify as a 'reintroduction', it must concern 'a species or lower taxon which has previously been observed as a naturally occurring and self-sustaining population in historical times, but which has declined or disappeared as a result of human intervention or a natural disaster' (Annex, par. 1(a)). Under the Habitats Directive, the requirement to 'study the desirability' of reintroducing species also applies only 'where this might contribute to their conservation, provided that an investigation, also taking into account experience in other Member States or elsewhere, has established that such re-introduction contributes effectively to re-establishing these species at a favourable conservation status and that it takes place only after proper consultation of the public concerned' (Article 22(a)).

Two treaties on migratory birds, both of which are part of the broader CMS regime, are also instructive. The 2001 Agreement on the Conservation of Albatrosses and Petrels (ACAP) obliges its thirteen parties to 'take a precautionary approach when re-establishing albatrosses

Box 3.2 *South Africa*

The translocation of wildlife in South Africa is primarily regulated by the 2004 National Environmental Management: Biodiversity Act 10 (NEMBA) and is complemented by various provisions in provincial statutes and policy. Chapter 4 of NEMBA provides for, among other things, the protection of species that are threatened or in need of protection to ensure their survival in the wild, and of species subject to international obligations concerning the regulation of international trade in endangered species. The use of wildlife that is deemed to be endangered, threatened, or requiring special protection is regulated, by way of a permit, principally through the Threatened or Protected Species Regulations (ToPS). These ToPS species are supplemented by additional lists of species administered under regulations in relevant provincial statutes and ordinances. Legal provisions in respect of wildlife translocation also extend to agricultural legislation that regulates the occurrence and movement of animal diseases, and animal welfare. Continued ownership of, and hence responsibility for, wildlife is conferred by compliance with the 1991 Game Theft Act 105. Non-compliance with this Act may render *res nullius* (i.e. owned by no-one) the translocated wildlife and hence negate the protection conferred by 'ownership'. The principal purpose of the Game Theft Act is to override the common law precept that all game is unowned until a person takes physical possession through hunting, seizing, capturing, or uniquely marking (e.g. branding) an animal with a predetermined intention to take possession of it (Blackmore, 2020). By adequately enclosing the property (i.e. by encircling animals with a game-proof fence), this Act confers ownership thereof on the landowner even though the animals are free-roaming within the (sometimes large) area involved (Blackmore, 2020).

Translocation of wildlife in the ToPS is managed by assessing the risk to the receptor sites in terms of habitat degradation and fragmentation, hybridisation, disease, and human safety. The ToPS further prohibits the translocation of listed species into a formal protected area that occurs outside their respective natural ranges (i.e. a conservation introduction). This provision is to be extended, in the forthcoming amendments to these Regulations, to other free-roaming or extensive wildlife systems that are not declared protected areas. In these instances, the prohibition may be (and in certain circumstances, be) relaxed to a permit requirement where a comprehensive risk assessment has been undertaken and adopted by the relevant conservation authority. This amendment will bring uniformity to the current legislative disparity between formal protected areas, and unregistered game farms and extensive wildlife areas.

In addition, the translocation and release of certain wildlife species may be regulated under the Alien and Invasive Species Regulations provided for in Chapter 5 of NEMBA. Invasive species are those listed that may, through rapid colonisation, cause damage outside their natural range, whereas alien species are those species that are not indigenous to South Africa or species that, prior to the promulgation of these Regulations, have been purposefully translocated outside their natural distribution, for instance, non-indigenous nyala *Tragelaphus angasii* displacing the indigenous Cape bushbuck *Tragelaphus sylvaticus* (Ehlers Smith et al., 2020). Translocation, as a form of 'use' of both listed alien and invasive species, is either regulated through a conditional permit or may be prohibited generally or specifically within certain circumstances; a species may also be required to be eradicated after individuals have been translocated where they are found to cause significant adverse effects on biodiversity and other impacts. For non-listed species, the NEMBA and its Regulations require a duty of care to be exercised.

The 1996 Constitution of the Republic of South Africa provides for norms and standards to bring uniformity to the application of the country's law. Consequently, the NEMBA provides for norms and standards for the use and management of biodiversity, which may apply nationally, to a specific area or areas, or to a specific category or categories of biodiversity. One set of norms and standards that is relevant to the translocation of wildlife is the National Norms and Standards for the Management of Elephants in South Africa (Elephant Norms and Standards). These Norms and Standards set in place, among other things, the minimum requirements for the physical transport of all species of elephants, and the circumstances when an elephant or group of elephants may or may not be transported.

The requirement for public consultation is prescribed in the NEMBA for administrative decisions that may be appealed or taken on review by the courts. This requirement does not, however, extend to the issuing of permits, including those required for the translocation of wildlife. Additional public consultation for the translocation of wildlife, and in particular potentially damage-causing wildlife, may be included in the regulations or norms and standards set in place by this Act. Interestingly, South Africa is yet to adopt national guidelines, policy, norms and standards, or regulations that provide for consultation prior to the translocation of wildlife, save for elephants, that may cause human–wildlife conflict. As a consequence, other than when registering or amending the registration of a captive facility, there are no mandatory requirements in ToPS or any other Regulation for a person wishing to translocate wildlife to consult with neighbouring communities and landowners. The exception is the consultation required when a landowner voluntarily applies for registration of their own game farm or extensive wildlife area. In this instance, the landowner would need to declare at least the species to be translocated. As mentioned, the Elephant Norms and Standards require the landowner or the person wanting to establish an elephant population on an extensive wildlife area to notify adjacent landowners and communities and any other person who may be directly affected by the intended translocation, and request them to provide written comment within a specified time. This comment is considered by the conservation authority when considering the necessary permit application. During this process, the conservation authority is required to evaluate whether the property is adequately enclosed (i.e. the game-proof perimeter fence meets the necessary standards) to contain elephants or any other potentially damage-causing wildlife.

Outside of this legal framework, neighbouring individuals or local communities have a common law right and in certain circumstances have access to various provisions in general administrative law to seek compensation from the owner of an extensive wildlife area for damage caused by escaped wildlife. The granting and extent of the compensation would be influenced by, among other things, the consultation that had taken place with the affected people. Furthermore, local communities are increasingly becoming owners of protected areas and extensive wildlife systems through, in particular, the country's land restoration programme in terms of the 1994 Restitution of Land Rights Act 22. In these circumstances, the communities in their role as owners or shareholders would naturally need to be consulted by their protected area or extensive wildlife area

> manager, at least on the translocation into their areas of wildlife having a propensity to escape and cause damage (i.e. African elephant *Loxodonta africana africana*, lion *Panthera leo*, wild dog *Lycaon pictus*, hippopotamus *Hippopotamus amphibius*) or cause disease in livestock (e.g. buffalo *Syncerus caffer* and wildebeest *Connochaetes* spp.).

and petrels into parts of their traditional breeding range', whereby they shall 'develop and follow a detailed re-establishment scheme', which 'shall be based on best scientific evidence and should be publicly available' (Annex 1, par. 1.3). The 1995 Agreement on the Conservation of African-Eurasian Migratory Waterbirds (AEWA) sets out similar requirements for its 80 parties, adding that re-establishment plans 'should include assessment of the impact on the environment' (Annex 3, par. 2.4). By way of a final illustration, Article 16 of the Alpine Nature Protocol states:

(1) The Contracting Parties shall undertake to promote the reintroducing and distribution of wild, indigenous animal and plant species and also subspecies, breeds and ecotypes, on condition that there are the necessary prerequisites and, by doing this, there is a contribution to the preservation and strengthening of those species and that no effects unsustainable to nature and the landscape, or to human activities, are caused.
(2) Scientific knowledge is to be applied for reintroducing and distributing these species. ... Following the reintroduction, it will be necessary to control and, if required, regulate the development of these animal and plant species.

This provision aptly illustrates one of law's defining features, namely that a few words can make a great practical difference. In particular, the condition that 'no effects unsustainable to ... human activities' may be caused, appears to raise the (regulatory) bar for having a proposed reintroduction approved, to considerable heights. The regulatory bodies would need to be content that, for example, appropriate monitoring, mitigation and management are in place.

Bern Convention Recommendation No. 58 (1997) paints a more representative picture of the types of requirements that may apply under national legislation regulating reintroductions and reinforcements. It urges parties to make all reintroduction operations contingent on the prior granting of a permit, in accordance with the following conditions, which are modeled closely on a previous version of IUCN guidance (IUCN 1998) on reintroductions (Appendix, par. 1(b)(i)):

- 'a permit should be granted only if the original causes of extinction of the species in question have been eliminated and the habitat requirements of the species are satisfied;
- the organisms reintroduced should belong to a subspecies or type as close as possible to the original stock, and preferably to the subspecies previously occurring in the area;
- the reintroduction envisaged should not cause substantial damage to agriculture or to forestry, to fishing and aquaculture, either marine or inland;
- the procedure for dealing with applications for permits should include:
  o an assessment of the possible effects of the reintroduction on the environment, on other species and on social and economic interests;
  o consultation of a scientific body designated for this purpose;
  o public hearings, where it is shown that the reintroduction may have a social and economic impact or, at least, consultation of the persons concerned, especially local authorities and landowners;
  o consultation [sic] neighbouring states where reintroduced organisms are liable to cross borders;'

Furthermore, the Recommendation calls, among other things, for penalties to be imposed for any translocations carried out without a permit or in breach of permit conditions, and for the recognition of 'civil liability of those responsible for unlawful reintroductions for any resulting damage and for the cost of any necessary eradication measures' (Appendix, par. 1(b)(ii)-(iii)). This relates to several further legal questions, including the uncertainty that may arise regarding the legal status of animals which belong to an internationally protected species but have been reintroduced illegally (as occurred with Eurasian beavers *Castor fiber* in Tayside, Scotland, the legal status of which was contentious for years (Pillai & Heptinstall, 2013), but became European Protected Species in 2019).

The importance of complying with authoritative expert guidelines is a common feature of international and national law and policy on conservation translocations. Under the national legislation of many countries, compliance with the 2013 IUCN Guidelines tends to significantly increase the chances of obtaining a translocation permit, if it is not an outright condition. Other expert guidelines may also be involved. To illustrate, the aforementioned Bern Convention Recommendation No. 53 (1996) on otters requests parties to make sure that 'any possible reintroduction programmes are designed and implemented following the guidelines laid down by IUCN's otter specialist group' (par. 3.7). To provide a national example, in Scotland there is a mandatory requirement to address the Scottish Code for Conservation Translocations (National Species Reintroduction Forum, 2014) when applying for a licence. The Scottish Code is based on the IUCN Guidelines (see also Box 3.3).

3.4.2 Rules on Invasive Non-native Species (INNS)

Rules on the prevention and control of INNS (or 'invasive alien species', as they are often referred to in law and policy documents) can, self-evidently, pose hurdles for certain conservation translocations – particularly assisted colonisation and ecological replacement. One illustration is the reference to 'unintended ecological consequences' in the CBD and CMS COP decisions on assisted colonisation, as cited previously. Generally, to meet such safeguards, which reflect the additional level of risk involved in conservation introductions, proposals need to be particularly thorough with regard to risk identification, mitigation, and justification of conservation benefit.

Each party to the CBD is under an obligation, 'as far as possible and as appropriate', to '[p]revent the introduction of, control or eradicate those alien species which threaten ecosystems, habitats or species' (Article 8(h)). The CBD COP defined an 'alien species' as a species 'introduced outside its natural past or present distribution', which is considered 'invasive' when it threatens native biodiversity, and adopted a set of guiding principles on how to implement Article 8(h) (CBD COP Decision VI/23 (2002), Annex). According to Guiding Principle 10:

An appropriate risk analysis, which may include an environmental impact assessment, should be carried out as part of the evaluation process before coming to a decision on whether or not to authorize a proposed introduction to the country or to new ecological regions within a country. States should make all efforts to permit only those species that are unlikely to threaten biological diversity. The burden of proof that a proposed introduction is unlikely to threaten biological diversity should be with the proposer of the introduction or be assigned as appropriate by the recipient State. Authorization of an introduction may, where appropriate, be accompanied by conditions (e.g., preparation of a mitigation plan, monitoring procedures, payment for assessment and management, or containment requirements).

Box 3.3 *Scotland*

In Scotland many species of conservation importance are protected by Scots law. Key legal instruments are the 1981 Wildlife & Countryside Act and the 1994 Conservation (Natural Habitats, &c.) Regulations (both as amended). How a species is protected varies: for some animal species it is illegal to intentionally/deliberately or recklessly capture, injure, kill, or disturb them, or to damage or destroy their breeding site or resting place. Some plants and fungi, including their seeds/spores, are protected from intentional/deliberate or reckless picking, uprooting, or destruction. In addition, it can be illegal to keep, possess, transport, or sell some protected species.

The conservation translocation of protected species generally requires a licence issued by the statutory nature conservation organisation, NatureScot. As part of the licensing process the proposal is assessed against certain considerations, typically:

- Is there a legal purpose or reason for the translocation?
- What alternative actions have been considered and why have these been discounted?
- What is the likely impact of the translocation on the conservation status of the population/species concerned?

Sometimes the donor or release sites are protected areas too. If the capture or release of a protected species affects a 'European' site (the most heavily protected type of site in Scotland, which may be a Special Protection Area classified for birds, or a Special Area of Conservation designated for a range of habitats and non-bird species), a 'Habitats Regulations Appraisal' must be completed (see later in this box) to make sure there is no adverse impact. Also, a species translocation may result in an offence against another protected species

In addition to such pre-release considerations, protected species legislation can play a role after release. A translocated protected species is usually legally protected at the release site. If its population eventually requires control then a further species licence is required from NatureScot. An additional licence may also be required for research or monitoring methods that are likely to disturb the translocated species.

Scots law protects areas of land as well as certain species. There are many different types of protected places but European sites and Sites of Special Scientific Interest (SSSIs) are particularly important (although the UK has left the European Union, the Scottish Parliament has passed legislation that continues to protect European sites). How a place is protected varies over time and so online resources with up to date information are available to practitioners to identify if donor or release sites are protected (and to check to what extent a species is protected). The protection afforded to places varies according to the reason(s) for designation. There is a consenting process for translocation action on certain protected places. For SSSIs, consent from NatureScot is required if the translocation actions are 'Operations Requiring Consent' (ORCs) on the SSSI. For European sites, NatureScot must carry out a 'Habitats Regulations Appraisal' before deciding to give consent or issue a licence.

Under the 1981 Act there is a presumption against the release or planting of species beyond their 'native range'. This legislation is particularly relevant to conservation translocations. For animals, the following actions are illegal without a licence: releasing, allowing to escape from captivity, or causing to be at a place outside the control of any person, outside its native range. For plants and fungi, likewise illegal are the actions of planting, or causing to grow, in the wild at a place outside the native range. Thus, conservation translocations involving the introduction of a species outside its native range must be carried out under a 'non-native species licence' from NatureScot. There are exceptions to this. For example: animals kept within secure enclosures are not considered to be released; some planting is not considered to be in the wild in certain locations; and some commonly planted and 'low risk' plants are exempt by an exception order.

The 1981 Act defines native range as 'the locality to which the animal or plant of that type is indigenous, and does not refer to any locality to which that type of animal or plant has been imported (whether intentionally or otherwise) by any person'.

Plants, fungi, and animals introduced to Scotland, even a long time ago (e.g. rabbit *Oryctolagus cuniculus*), are considered to be outside their native range. The law furthermore entails that a non-native species licence is required to introduce a species to an area where no records of it exist. Former natives (species once native to a location but now extinct there and unable to recolonise naturally) are outside their native range, as defined under the law. All reintroductions to sites require a non-native species licence (e.g. vendace *Coregonus albula*,

> white-tailed eagle *Haliaeetus albicilla*, and red kite *Milvus milvus*). This is useful, as it means that NatureScot can ensure that the best practice approach set out in the Scottish Code for Conservation Translocations (National Species Reintroduction Forum, 2014) is applied before permission is given.
>
> Conservation translocations may also require movement between countries, and therefore the spread of harmful pests and diseases, as well as stress and harm to sentient species, must be avoided. Additional licensing requirements might be needed too (e.g. freshwater fish species translocations require a licence from Marine Scotland). Given this complexity, NatureScot recommends completion of a 'Translocation Project Form' as a checklist for practitioners. The Form serves as a licence application and details the entire translocation from start to finish, promoting understanding and learning for future projects.

Many regional treaties reflect similar requirements, though their formulations differ, which may entail different consequences for conservation translocations. A few examples are offered here. Parties to the 2003 (Revised) African Convention on the Conservation of Nature and Natural Resources shall 'strictly control the intentional ... introduction, in any area, of species which are not native to that area' (Article IX(2)(h)). The Bern Convention likewise commits its parties to 'strictly control the introduction of non-native species' (Article 11(2)(b)). AEWA parties shall 'prohibit the deliberate introduction of non-native waterbird species into the environment ... if this introduction ... would prejudice the conservation status of wild flora and fauna' (Article III(2)(g)). Parties to the 1985 Protocol to the Nairobi Convention Concerning Protected Areas and Wild Fauna and Flora in the Eastern African Region (East Africa Protocol) 'shall take all appropriate measures to prohibit the intentional or accidental introduction of alien or new species which may cause significant or harmful changes to the Eastern African region' (Article 7). Parties to the Alpine Protocol are required to ensure that 'no wild animal or plant species are introduced into a region that were not previously present naturally for a verifiable historic period', whereas exceptions can be made for introductions which are 'necessary for certain uses and will not lead to negative effects on nature and the landscape' (Article 17). In line with such international obligations, many national laws have made it a criminal offence to introduce a non-native species without a permit.

Another, related issue that is likely to become increasingly significant is the release of genetically modified organisms (GMOs). The release of genetically modified American chestnut *Castanea dentata* resistant to a fungus that has decimated this ecologically important species is already being given consideration in the USA. Legislation already exists that covers the use of GMOs, such as the EU Directive 2001/18/EC on the Deliberate Release into the Environment of Genetically Modified Organisms. The release of such organisms for conservation purposes therefore needs to address such legislation, as well as the more conservation-focused legislation relating to 'non-native' species described above.

3.4.3 Rules on Species Protection

Further restrictions follow from international and national species protection legislation, particularly when such laws include prohibitions to 'take' specimens of protected species. A typical example is the CMS. For the endangered mammal, bird, fish, and reptile species listed in its Appendix I, parties 'shall prohibit the taking of animals belonging to such species' (Article III(5)). The CMS defines 'taking' broadly as 'taking, hunting, fishing, capturing, harassing, deliberate killing, or attempting to engage in any such conduct' (Article 1(i)). Exceptions to this prohibition may be made for a limited set of reasons only, including 'for the purpose of enhancing the propagation or survival of the affected species' or if 'extraordinary circumstances so require', and on the conditions that the exception is 'precise as to content and limited in space and time', and that the taking does 'not operate to the disadvantage of the species' (Article III(5)). Therefore, if an envisaged translocation involves the taking from the wild of an Appendix I organism, the proponent must demonstrate, convincingly and in advance, that the aforementioned conditions are met.

A representative regional treaty is the 1990 Protocol Concerning Specially Protected Areas and Wildlife to the Convention for the Protection and Development of the Marine Environment of the Wider Caribbean Region (Caribbean SPAW Protocol). It requires prohibitions on, among other things, the 'collecting' and 'possession' of plants (and their seeds) listed in its Annex I, and the 'taking', 'possession', and 'disturbance' of animals (and their eggs) listed in Annex II (Article 11(1)). Exemptions from these prohibitions may be granted by national authorities 'for scientific, educational or management purposes necessary to ensure the survival of the species' (Article 11(2)). The authorities involved must report any exemptions to the UN Environment Programme (UNEP) in order for the Protocol's Scientific and Technical Advisory Committee to 'assess the pertinence of the exemptions granted' (Article 11(2)).

Similar taking prohibitions occur in many other legal instruments, typically accompanied by an exemption clause setting out particular conditions. Below is a non-exhaustive inventory of such instruments, and the (lists of) species for which taking restrictions apply.

- 1986 Australia - China Agreement for the Protection of Migratory Birds and their Environment – *birds in Annex.*
- ACAP – *birds in Annex 1.*
- AEWA – *birds in Annex 3, Table 1, Column A.*
- 1990 Agreement on the Conservation of Seals in the Wadden Sea (WSSA) – *all seals.*
- 1991 Agreement on the Conservation of European Bats (EUROBATS) – *bats in Annex 1.*
- 1992 Agreement on the Conservation of Small Cetaceans of the Baltic, North East Atlantic, Irish and North Seas (ASCOBANS) – *all 'small cetaceans'.*
- 1996 Agreement on the Conservation of Cetaceans of the Black Sea, Mediterranean Sea and contiguous Atlantic Area (ACCOBAMS) – *all cetaceans.*
- 2007 Agreement on the Conservation of Gorillas and their Habitats (Gorilla Agreement) – *all gorillas.*
- 1940 Convention on Nature Protection and Wild-Life Preservation in the Western Hemisphere – *flora and fauna in Annex.*
- 1968 African Convention on the Conservation of Nature and Natural Resources – *flora and fauna in Annex.*
- East Africa Protocol – *flora in Annex I, fauna in Annex II.*
- Bern Convention – *flora in Appendix I, fauna in Appendix II.*
- 1995 Protocol to the Barcelona Convention Concerning Specially Protected Areas and Biological Diversity in the Mediterranean – *flora and fauna in Annex II.*
- Habitats Directive – *flora and fauna in Annex IV.*
- EU Directive 2009/147/EC on the Conservation of Wild Birds (Birds Directive) – *all native EU birds.*

Most of the exemption clauses in these instruments seem to offer sufficient leeway for authorities to permit conservation translocations, as long as certain conditions can be met (see, for example, Box 3.4 on EU law).

Box 3.4 *European Union*

At EU level the principal instruments fostering conservation translocations are the Birds Directive and the Habitats Directive. Measures taken pursuant to the latter 'shall be designed to maintain or *restore*, at favourable conservation status, natural habitats and species of wild fauna and flora of Community interest' (Article 2(2), emphasis added). Conservation translocations are part of the toolbox to achieve such restoration. This is confirmed by Article 22(a), requiring member states to 'study the desirability of re-introducing' native species listed in the Directive's Annex IV. Various translocation projects have received financial aid through the EU 'LIFE' funding scheme.

Restrictions for translocations also follow from the Habitats Directive, which prohibits the taking of protected species listed in its Annex IV (Article 12). Article 16(1)(d), however, offers a derogation possibility for 'the purpose of research and education, of repopulating and re-introducing these species and for the breeding operations necessary for these purposes, including the artificial propagation of plants'. In a similar way, the Birds Directive prohibits the taking of all native birds but enables derogations for the purposes of research, re-population, reintroduction, and breeding necessary for these purposes (Article 9(b)).

In addition to these nature conservation instruments, there are several other EU instruments that create conditions or restrictions for translocations. Regarding trade, the EU has implemented the 1973 Convention on International Trade in Endangered Species of Wild Fauna and Flora (CITES) through a set of regulations that are directly applicable in all 27 EU member states. Most important is 1997 Regulation No. 338/97 on the protection of species of wild fauna and flora by regulating trade therein (Basic Regulation), which is accompanied by Regulation No. 865/2006 (Implementing Regulation) and Regulation No. 792/2012 (Permit Regulation). Together, these regulations detail the obligatory procedures and documents for the movement and possession of listed species, which may require import permits, export permits, re-export certificates, import notifications, and/or internal trade certificates.

Furthermore, Regulation No. 1143/2014 on the Prevention and Management of the Introduction and Spread of Invasive Alien Species should be taken into account, for instance when assisted colonisation is considered in the face of climate change or other threats. Restrictions related to INNS also flow from the Habitats Directive (Article 22(b)) and the Birds Directive (Article 11). The latter, for instance, requires member states to ensure that 'any introduction of species of bird which do not occur naturally in the wild state in the European territory of the Member States does not prejudice the local flora and fauna', and requires consultation of the European Commission regarding such introductions.

Finally, the EU has adopted an array of legislation on animal and plant health, veterinary checks, and animal welfare requirements. Examples include:

- 1991 Directive 91/496/EEC Laying Down Principles Governing the Organisation of Veterinary Checks on Animals Entering the Community from Third Countries.
- 1992 Directive 92/65/EEC Laying Down Animal Health Requirements Governing Trade in and Imports into the Community of Animals, Semen, Ova and Embryos.
- 2004 Directive 2004/68/EC Laying Down Animal Health Rules for the Importation into and Transit through the Community of Certain Live Ungulate Animals.
- 2004 Regulation No. 1/2005 on the Protection of Animals during Transport and Related Operations.

Some of the subsidiary CMS treaties may be problematic, however. Translocating cetaceans from or within the sea areas to which ASCOBANS and ACCOBAMS apply is likely to pose legal problems, as these treaties only allow exceptions to their taking prohibitions for research purposes and, in the case of ACCOBAMS, emergency situations (ASCOBANS, Annex, par. 2 & 4; ACCOBAMS, Article II(1)-(2); see also Trouwborst et al., 2013). The same applies regarding seals in the Wadden Sea, which may be exceptionally taken only for research, or to aid 'diseased or weakened seals or evidently abandoned suckling seals' (WSSA, Article VI(2)). Curiously, under the Gorilla Agreement, it appears that no exceptions whatsoever can be granted for the taking of gorillas (Article III(2)(a)).

As noted in the introduction to this chapter, and acknowledged in the IUCN Guidelines, many countries have 'formal legislation restricting the capture and/or collection of species within their jurisdiction' (IUCN,

2013), with clear implications for conservation translocations. Many of the legal instruments operating at national and sub-national scales implement, and enable countries to demonstrate their compliance with, the international obligations described in this chapter.

3.5 Questions and Conundrums

In this section we highlight, in non-exhaustive fashion, several additional legal questions and conundrums concerning conservation translocations.

The first issue concerns the legal consequences of taxonomic blurred lines, namely, instances where there is a taxonomic debate over the splitting of certain species or the recognition of certain subspecies, or where broadly agreed taxonomic changes occurred in the recent past. In such cases, due attention must be paid to the (sub)species' names as they are listed under applicable legislation, as the latter are likely to determine the legal position. The time lags involved in the updating of legal instruments' annexes in accordance with advancing phylogenetic and other scientific insights can cause potentially significant incongruences (Zhou et al., 2015, 2016). In practice, whether the source population for an envisaged conservation translocation is recognised as a different subspecies from the extinct population in the target area – specifically, whether the translocation is a reintroduction or a conservation introduction (i.e. an assisted colonisation or ecological replacement as defined in the IUCN Guidelines) – can determine legal success or failure. For instance, the Supreme Court of India in its aforementioned 2020 ruling determined that:

> the word 're-introduce' has been erroneously applied since it is an admitted fact that the African Cheetahs never inhabited in [*sic*] India. Therefore, if an attempt is made to relocate the African Cheetah within the territory of India, it will amount to an 'introduction' . . . and not a 're-introduction'. (Supreme Court of India, Centre for Environmental Law, WWF. v. Union of India et al., 28 January 2020)

In this case, however, this did not stop the Court from ultimately giving the green light to the operation. Similarly, Bern Convention Recommendation No. 58 (1997) expressly declares that 'the release of a non-native sub-species into a given territory should be considered an introduction', subject to rules on non-native species, instead of a reintroduction or reinforcement. To add to the confusion, however, this categorical statement is then directly contradicted in the guidelines on reintroductions appended to the very same Recommendation, which stipulate that 'the organisms reintroduced should belong to a subspecies or type *as close as possible* to the original stock, and *preferably* to the subspecies previously occurring in the area' (Annex, par. 1(b)(i), emphasis added). These types of legal issue are likely to become more common, and potentially complex, as the expanding use of modern genetic tools allows us to learn more about the detail of biological diversity and inform definitions of species, subspecies, and other evolutionary significant units.

The second issue concerns baselines. Big differences can exist between national and international laws, policies, and guidelines regarding the appropriate past baseline for determining whether a species is native to a given area, and thus its eligibility for reintroduction or conservation introduction. To provide just one illustration, Bern Convention Recommendation No. 58 (1997) defines a 'species native to a given territory' as a 'species that has been observed there in the form of naturally-occurring and self-sustaining population in historical times'. It is not hard to imagine disagreement arising over what constitutes sufficient proof of a species having been 'observed' as such and, indeed, how to define the time span of 'historical times'.

The third and final issue concerns the role of the precautionary principle (also known as the precautionary approach), which has been mentioned a few times in the preceding analysis. This central principle of international environmental law and policy has been devised to aid decision-making in uncertain circumstances. As the CBD formulates the principle, 'where there is a threat of significant reduction or loss of biological diversity, lack of full scientific certainty should not be used as a reason for postponing measures to avoid or minimize such a threat' (Preamble). Precaution entails giving nature the benefit of any doubt, by doing whatever is expected, on the basis of the best information that *is* available, to achieve the best outcome, and then monitoring the effects of the chosen approach, and adjusting it in light of any new information.

Precaution is a prominent feature of law and policy on INNS, and of environmental law in general. The CBD's guiding principles on INNS, for instance, declare that '[d]ecisions concerning intentional introductions should be based on the precautionary approach' (CBD COP Decision VI/23 (2002), Annex, Guiding Principle 10). Despite the precautionary principle's apparent simplicity, its precise implications remain prone to confusion and misunderstanding, particularly regarding the placing and weight of the burden of proof – that is, the answer to the question: Who needs to prove what? (Cooney & Dickson 2005; Trouwborst et al., 2020).

The CBD guidance, reproduced above, calls for a risk analysis, whereby the 'burden of proof that a proposed introduction is unlikely to threaten biological diversity should be with the proposer of the introduction or be assigned as appropriate by the recipient State' (CBD COP Decision VI/23 (2002), Annex, Guiding Principle 10).

Such evidentiary constructions are typical in fields where threats to biodiversity are, in principle, more clear-cut and one-sided. Examples include pesticide use and road-building, the environmental effects of which are predominantly (if not universally) negative, thus justifying an approach whereby new projects are in one way or another considered 'guilty until proven innocent'. It would be a mistake to think, however, that treating *any* proposed project or change in this way is an automatic consequence of applying the precautionary principle (Cooney, 2004; Cooney & Dickson, 2005; Trouwborst et al., 2020). As reflected in the CBD's preambulatory statement cited above, the purpose of the principle is not to restrict new activities but to avert *threats* to biodiversity or the environment at large. What that requires will vary according to the circumstances. In some cases, the most precautionary course of action will be to restrict human activities, whereas in others it will precisely involve allowing them. To pick a controversial example, the prohibition or restriction of trophy hunting 'may be the precautionary response in some circumstances, but may not be in others', as curbing this activity may improve or worsen the prospects of wildlife populations, depending on the circumstances (Rosser et al., 2005). The idea that precaution does not equal a presumption against change or human activities *as such* is clearly reflected in the neutral formulation of CITES and CMS COP Resolutions adopted to guide parties in deciding what level of legal restrictions to apply to each species (CITES Resolution Conf. 9.24 (Rev. COP13); CMS COP Resolution 11.33 (2017)):

> by virtue of the precautionary approach and in case of uncertainty regarding the status of a species [or the impact of trade on the conservation of a species], the Parties shall act in the best interest of the conservation of the species concerned and ... adopt measures that are proportionate to the anticipated risks to the species.

Thus, determining the implications of the precautionary principle for conservation translocations, particularly assisted colonisation and ecological replacement, is a nuanced affair. It is also clear that complications may arise from the default application of the principle in INNS law and policy, which is predicated – understandably, given the global track record of disastrous introductions – on a 'guilty until proven innocent' approach. The anticipated increased future need for assisted colonisation and other forms of conservation translocations calls for a more balanced evidentiary approach, geared towards securing the best overall conservation outcome (Gaywood & Stanley-Price, this volume; Hällfors & Dalrymple, this volume). International biodiversity law and policy has begun to grapple with this issue, as illustrated for instance in a CBD COP Decision quoted earlier, which calls on parties to 'consider . . . assisted migration . . . that could contribute to . . . securing the survival of species at risk, taking into account the precautionary approach in order to avoid unintended ecological consequences' (CBD COP Decision X/33 (2010), par. 8(e)). Another statement on assisted colonisation, adopted by the parties to the Bern Convention, clearly exemplifies the pitfalls of an overly simplistic application of the precautionary principle in this context (Recommendation No. 158 (2012)):

> where there is inadequate information to assess that a translocation outside indigenous range bears low risks, the Precautionary Principle should be applied and such a translocation should not be carried out.

Requiring 'adequate information' demonstrating 'low risk' as a precondition for all assisted colonisation operations would appear to reflect a misreading of the precautionary principle. To grasp this, it helps to imagine a case where the choice is between averting an uncertain risk to biodiversity in the target area on the one hand, and averting the global extinction of a species that has no future in its remaining, indigenous range, on the other.

3.6 Conclusion

The legal requirements that will apply to any contemplated conservation translocation depend, among other things, on the countries, areas, and species involved, and there may be significant differences between one project and the next. Sometimes these requirements will be relatively easy to meet, whereas at the other end of the spectrum the law can sometimes raise insurmountable obstacles. As the foregoing makes clear, it is highly recommended that a proponent carry out a careful analysis of applicable legislation in the early stages of considering a particular conservation translocation. Depending on the circumstances, such an analysis can make the difference between a project's failure or success.

3.7 Key Messages

- Law can influence conservation translocations in two basic ways; it can play a supportive and/or a restrictive role.
- Regarding the first role, legislation can provide support, a mandate, and/or an obligation to consider or perform a translocation. This may involve general obligations to conserve or restore certain species or ecosystems, as well as specific obligations concerning species reintroductions or other types of translocation.
- Regarding the second role, legislation can impose limitations on translocation possibilities and/or make translocations conditional on meeting certain requirements. Such limitations and conditions may follow from legislation on native species protection, area protection, invasive non-native species, disease, trade, animal welfare, and several other areas of law.
- Requirements of both types may flow from international and national law.
- The legal requirements that will apply to any projected conservation translocation depend on the countries, areas, and species involved, and there may be significant differences between one project and the next.
- The applicable requirements may be easy to meet for some projects, and prohibitive for others.
- It is crucial that a careful analysis of applicable domestic legislation is carried out in the early stages of any contemplated conservation translocation. Practitioners should also make themselves aware of the relevant international legislation to provide them with the wider legal context of their work and to help them influence the development of future legislation.

References

ACT Government (2017) *Conservator Guidelines for the Translocation of Native Flora and Fauna in the ACT*. Canberra, Australia, ACT Government.

Aggarwal, M. (2020) *India considers bringing back the cheetah, even as lions and other species wait for conservation attention*. Available from: https://india.mongabay.com/2020/01/india-considers-bringing-back-the-cheetah-even-as-lions-and-other-species-wait-for-conservation-attention/ [Accessed 31 January 2020].

Bastmeijer, K. (2016) Ecological restoration in international biodiversity law: a promising strategy to address our failure to prevent. In Bowman, M., Davies, P. and Goodwin, E. (eds.) *Research Handbook on Biodiversity and Law*. Abingdon, Edward Elgar, pp. 387–413.

Blackmore, A. (2020) Climate change and the ownership of game: a concern for fenced wildlife areas. *Koedoe*. 62, a1594.

Brodie, J. F., Lieberman, S., Moehrenschlager, A., et al. (2021) Global policy for assisted colonization of species. *Science*. 372, 456–458.

Camacho, A. E. (2010) Assisted migration: redefining nature and natural resource law. *Yale Journal on Regulation*. 27, 171–255.

Cooney, R. (2004) *The Precautionary Principle in Biodiversity Conservation and Natural Resource Management: An Issues Paper for Policy-Makers, Researchers and Practitioners*. Gland, Switzerland, IUCN.

Cooney, R. & Dickson, B. (eds.) (2005) *Biodiversity & the Precautionary Principle*. London, Earthscan.

Ehlers Smith, Y. C., Ehlers Smith, D. A., Ramesh, T. & Downs, C. T. (2020) Co-occurrence modelling highlights conservation implications for two competing spiral-horned antelope. *Austral Ecology*. 45, 305–318.

Gardner, R.C. (2003) Rehabilitating nature: a comparative review of legal mechanisms that encourage wetland restoration efforts. *Catholic University Law Review*. 52, 573–20.

Gaywood, M. J. & Stanley-Price, M. (2023) Moving species: reintroductions and other conservation translocations. In Gaywood, M. J., Ewen, J. G., Hollingsworth, P. M. and Moehrenschlager, A. (eds.) *Conservation Translocations*. Cambridge, Cambridge University Press.

Hällfors, M. & Dalrymple, S. E. (2023) Assisted colonisation and ecological replacement. In Gaywood, M. J., Ewen, J. G., Hollingsworth, P. M. and Moehrenschlager, A. (eds.) *Conservation Translocations*. Cambridge, Cambridge University Press.

Hogg, C. & Wise, P. (2023) Assisted colonisation as a conservation tool: Tasmanian devils and Maria Island. In Gaywood, M. J., Ewen, J. G., Hollingsworth, P. M. and Moehrenschlager, A. (eds.) *Conservation Translocations*. Cambridge, Cambridge University Press.

IUCN (1998) *Guidelines for Re-introductions*. Gland, Switzerland, IUCN.

IUCN (2013) *Guidelines for Reintroductions and Other Conservation Translocations. Version 1.0*. Gland, Switzerland, IUCN Species Survival Commission.

López-Bao, J. V., Fleurke, F., Chapron, G. & Trouwborst, A. (2018) Legal obligations regarding populations on the verge of extinction in Europe: conservation, restoration, recolonization, reintroduction. *Biological Conservation*. 227, 319–325.

Lundie-Jenkins, G. (1999) Reintroduction of the mala to Aboriginal land in the Tanami Desert, Northern Territory. PhD thesis, University of New England.

McCormack, P. C. (2018) Conservation introductions for biodiversity adaptation under climate change. *Transnational Environmental Law*. 7, 323–345.

National Species Reintroduction Forum (2014) *Scottish Code for Conservation Translocations*. Inverness, Scottish Natural Heritage.

Office of Environment and Heritage (2019) *Translocation Operational Policy*. Available from: www.environment.nsw.gov.au/research-and-publications/publications-search/translocation-operational-policy [Accessed 28 December 2020].

Pillai, A. & Heptinstall, D. (2013) Twenty years of the Habitats Directive: a case study on species reintroduction, protection and management. *Environmental Law Review*. 15, 27–46.

Rees, P. A. (2001) Is there a legal obligation to reintroduce animal species into their former habitats? *Oryx*. 35, 216–223.

Rosser, A. M., Tareen, N. & Leader-Williams, N. (2005) The precautionary principle, uncertainty and trophy-hunting. In Cooney, R. and Dickson, B. (eds.) *Biodiversity & the Precautionary Principle*. London, Earthscan, pp. 55–72.

Silcock, J. L. (2018) Aboriginal translocations: the intentional propagation and dispersal of plants in Aboriginal Australia. *Journal of Enthnobiology*. 38, 372–405.

Tasmanian Government (2020) *Sarcophilus harrisii (Tasmanian Devil): species management profile*. Available from: www.threatenedspecieslink.tas.gov.au/Pages/Tasmanian-Devil.aspx [Accessed 28 December 2020].

Telesetsky, A., Cliquet, A. & Akhtar-Khavari, A. (2017) *Ecological Restoration in International Environmental Law*. London & New York, Routledge.

Threatened Species Recovery Hub (2019) *Assisted colonisation of Australia's rarest reptile: the western swamp turtle*. Available from: www.nespthreatenedspecies.edu.au/projects/assisted-colonisation-of-australia-s-rarest-reptile-the-western-swamp-turtle [Accessed 28 December 2020].

Trouwborst, A. (2014) The EU Habitats Directive and wolf conservation and management on the Iberian Peninsula: a legal perspective. *Galemys: Spanish Journal of Mammalogy*. 26, 15–30.

Trouwborst, A. (2015) The Habitats Directive and climate change: is the law climate-proof? In Born, C., Cliquet, A., Schoukens, H., Misonne, D. and Van Hoorick, G. (eds.) *The Habitats Directive in Its EU Environmental Law Context*. Abingdon & New York, Routledge, pp. 303–324.

Trouwborst, A., Caddell, R. & Couzens, E. (2013) To free or not to free? State obligations and the rescue and release of marine mammals: a case study of 'Morgan the orca'. *Transnational Environmental Law*. 2, 117–144.

Trouwborst, A., Loveridge, A. J. & Macdonald, D. W. (2020) Spotty data: managing international leopard (*Panthera pardus*) trophy hunting quotas amidst uncertainty. *Journal of Environmental Law*. 32, 253–278.

Van Dooren, T. (2019) Moving birds in Hawai'i: assisted colonisation in a colonised land. *Cultural Studies Review*. 25, 41–64.

Zhou, Z., Newman, C., Buesching, C. D., Meng, X., Macdonald, D. W. & Zhou, Y. (2015) Outdated listing puts species at risk. *Nature*. 525, 187.

Zhou, Z., Newman, C., Buesching, C. D., Meng, X., Macdonald, D. W. & Zhou, Y. (2016) Revised taxonomic binomials jeopardize protective wildlife legislation. *Conservation Letters*. 9, 313–315.

4 · Decision-Making in Animal Conservation Translocations: Biological Considerations and Beyond

JOHN G. EWEN, STEFANO CANESSA,
SARAH J. CONVERSE, AND
KEVIN A. PARKER

4.1 Introduction

Conservation translocations are defined by the IUCN (2013) as 'the intentional movement and release of a living organism where the primary objective is a conservation benefit: this will usually comprise improving the conservation status of the focal species locally or globally, and/or restoring natural ecosystem functions or processes'. Thus, conservation translocations may be undertaken to achieve a wide variety of objectives (Ewen et al., 2014), but by definition they all have at least one conservation objective. As a result, biological considerations are often a major focus in translocation planning efforts.

Biological considerations are complex and diverse. They include the physiological, behavioural, demographic, and ecological considerations relevant to the selection of appropriate individuals for holding, breeding, transfer, and release; the treatment and conditioning of individuals during these phases; the selection of source and release locations; post-release management of released individuals, their progeny, and their habitat; and more. Ignoring these considerations when making decisions can lead to failure. For example, North Island miromiro, or tomtits, *Petroica macrocephala toitoi* were released on Tiritiri Matangi, New Zealand, in 2004 at a time when their dispersal abilities were unclear. At least one bird returned to the source population and the translocation failed, likely due to dispersal (Parker et al., 2004). The IUCN Guidelines for Reintroductions and Other Conservation Translocations (IUCN, 2013; hereafter 'the IUCN Guidelines') note a variety of biological

Table 4.1. *Most relevant sections of the IUCN Guidelines for Reintroductions and Other Conservation Translocations (IUCN, 2013) that give pointers to the key biological considerations of conservation translocations. Italic text shows components of Section 6 that are not directly relevant for biological considerations.*

Section 5 Feasibility and design	Section 6 Risk assessment
5.1 Biological feasibility	**6.6 Main risk factors**
5.1.1 Basic biological knowledge	• Risk to source population
5.1.2 Habitat	• Ecological risk
5.1.3 Climate requirements	• Disease risk
5.1.4 Founders	• Associated invasion risk
5.1.5 Animal welfare	• Gene escape
	• *Socio-economic risk*
	• *Financial risk*

considerations relevant for the different phases of a conservation translocation and a set of generic biological risks that should be considered (Table 4.1). Others have elaborated on biological considerations, with recent literature providing methods on how to approach these considerations (e.g. Parker et al., 2012; Converse et al., 2013a; Parker et al., 2015; Moehrenschlager & Lloyd, 2016) and reviews of how well they have been integrated in reported conservation translocation programmes (Berger-Tal et al., 2020).

However, it is probably unrealistic for practitioners to directly evaluate and address all possible biological considerations when planning a conservation translocation effort. In any given situation, some considerations are almost certainly more important in determining outcomes than others: the challenge is thus deciding which biological considerations to emphasise. Unfortunately, practitioners faced with the challenge of identifying the key biological considerations often fall victim to cognitive biases, leading to poor choices. For example, availability bias occurs when an event that comes readily to mind is considered more common than it actually is, such that we might overestimate the danger of flying when there has been recent news of a plane crash (Tversky & Kahneman, 1973). If management agencies and their 'go to' research partners dealt with a particular issue in a previous conservation translocation effort, they may be more likely to consider it important in all subsequent conservation translocations. As a result, we often see plans with fantastically detailed genetic management plans, for example, or with a meticulous disease risk analysis, seemingly with lesser regard to other biological considerations. This puts translocations at increased risk for poor outcomes: imagine extended quarantine in highly artificial environments prior to release to prevent co-introduction of possibly novel parasites, resulting in reduced post-release survival due to stress. It also risks overlooking creative solutions: imagine dogmatic adherence to a threshold habitat patch size deemed capable of supporting a genetically viable population when smaller patches may be of superior quality, and more readily available, but will need to be managed via ongoing translocations between patches. On the other hand, trying to capture all considerations within a single framework is immensely challenging. Finding the right balance creates a conundrum for practitioners. At worst this can create a fear to act, thereby stalling any conservation progress, or a constant call for yet more 'research' before action (Converse & Grant, 2019).

We cannot avoid complexity; it is a reality of conservation translocations. Instead, we need to develop an approach for dealing with it in a thoughtful and deliberative manner. Such an approach is critical to giving decision makers the confidence to act while ensuring the best chance of success. While such a decision-making approach cannot guarantee a good outcome, it can make it more likely. In our work (also see IUCN guidance on the application of the IUCN Guidelines; Canessa et al., 2019a), we draw on structured decision-making (hereafter SDM, also known as decision analysis; Gregory et al., 2012; Runge et al., 2020) to help develop frameworks needed to guide conservation translocation decisions. A decision framework provides us with a kind of cognitive map, allowing us to think more clearly in spite of complexity. We point out that this is relevant for all aspects of conservation translocations, scalable from single aspects, such as deciding what pre-release behavioural conditioning to use, through to overarching decisions on whether and how to undertake a conservation translocation considering all biological and socio-economic objectives.

In this chapter we describe common biological considerations in animal translocations and introduce SDM as a way of dealing with the complexity inherent in conservation translocation decision-making. We focus on sorting through complexity by creating a clear decision framework that makes biological (and other) considerations explicit. In particular, we begin by (i) describing some of the more common biological considerations and the time frames over which they are relevant. We then (ii) introduce the general principles of SDM and (iii) show how to treat biological considerations and related information within a decision framework. Finally, (iv), we provide some thoughts on the treatment of uncertainty within these frameworks.

4.2 Common Biological Considerations and When They Matter

Many of the biological considerations in animal translocations that are mentioned in Table 4.1 (and hence the IUCN Guidelines) are the focus of chapters in this volume, including disease risks (Sainsbury & Carraro, this volume), genetics (Neaves et al., this volume) and animal welfare (Harrington et al., this volume), and we do not cover these aspects in detail here. Rather, we have selected a few other biological considerations that we frequently encounter. We present these along a translocation pathway from sourcing individuals, their movement and release, and the establishment and growth phases of the programme. Using the translocation pathway can help us to recognise where the opportunities for management are and the possible limitations of this management, and to consider priorities for which biological considerations to focus on at which point.

We think two useful questions to start with are: where are the animals coming from, and where are they going? The answers to these questions occupy a continuum. At one end of the continuum wild animals are going from source sites to release sites that are virtually identical. In our experience these translocations can be relatively straightforward, especially if the source and release site are close together. This is because the biological variables that the animals must cope with are not radically different from those they are used to, and the translocation process, especially holding and transport, can be kept to a bare minimum. As an example, wild North Island tīeke or saddleback *Philesturnus rufusater* were successfully translocated from New Zealand's Whatupuke Island to Lady Alice Island in 1972 (Lovegrove, 1996). The two islands are approximately 400 m apart with similar climate and vegetation associations. Thus, translocation logistics were simple, with birds captured and released on the same day. At the other end of the continuum, the source and release sites can be radically different and far apart. An extreme example is the sihek or Guam kingfisher *Todiramphus cinnamominus*. It was declared extinct in the wild in 1988 and now exists as a small meta-population across several zoos on the US mainland and at a government facility on Guam (Trask et al., 2021). Reintroducing sihek to the wild requires finding high-quality release sites where animals that have spent multiple generations in captivity can adapt to living in the wild, specifically finding food, shelter, and mates, and avoiding predators and competitors. Furthermore, sihek are currently held in mixed species captive collections, thereby requiring careful health screening and quarantine protocols. They will also need to be transported across large distances (thousands of kilometres), thereby potentially exposing them to a wide range of stressors during quarantine and transport, and after release (for further information on disease risks see Sainsbury & Carraro, this volume).

Biological considerations are clearly important in the examples cited above, but in radically different ways. Furthermore, while uncertainty is relatively low when translocating North Island tīeke, it will be high when translocating sihek. Therefore, while translocating North Island tīeke between islands is almost routine, translocating sihek will require years of careful thought and planning. So how do we select the best individuals for translocation, what are the best release sites, and how does the selection of individuals and release sites influence post-release stress in translocated animals?

4.2.1 Selecting Appropriate Source Individuals

In an ideal world, we would have abundant source populations comprised of wild, genetically diverse individuals in habitats identical to our release sites. However, reality is often quite different: there might not be many animals alive anywhere that are available for a conservation translocation.

A common decision is whether to use wild-sourced or captive-reared individuals (Parker et al., 2012; Parker et al., 2015). One consideration is that the translocation process (capture, holding, moving, and release) is likely to be more stressful for wild animals relative to captive-reared animals (Parker et al., 2012). However, wild animals typically (but not always) do better after release (Griffith et al., 1989; Fischer & Lindenmayer, 2000; Parker et al., 2012). This is backed by extensive literature showing short- and long-term adaptation to captivity (Frankham et al., 1986; Allendorf, 1993; Snyder et al., 1996; McPhee, 2004; Hakånsson & Jensen, 2005; Hakånsson & Jensen, 2008). It is entirely logical that a wild animal is almost always going to be better adapted to life in the wild than a captive-bred animal. So, if wild animals are available from a population that can sustain harvest (or that might otherwise be lost for some reason), this will frequently be a better choice than using captive-reared animals. Where alternatives exist, we recommend careful comparison of predicted consequences across all objectives as described in Section 4.3. For example, it might seem tempting to use captive source populations because they provide a larger number of easily obtained individuals, sometimes at a lower cost. However, they may also suffer higher post-release mortality and as a result there may be a lower chance of population establishment and greater welfare concerns.

While there are reasons to prefer wild founders for conservation translocations, they will not always be available or appropriate. Wild populations are often too small or too poorly understood to risk a harvest for translocation. Alternatively, the distance between source and release sites might be so great that the translocation process of capture, holding, movement, and release might be considered too stressful for wild animals to endure.

If captive animals must be used as a source population, the primary uncertainty is their ability to adapt to life in the wild (Swaisgood, 2010; Parker et al., 2012, 2015). A lack of adaptability to conditions in the wild can be overcome to some extent if large numbers of captive-reared animals are available for release (i.e. at least some will survive and persist). However, there are significant welfare concerns in releasing large numbers of animals with the expectation that many will die (see more discussion in Harrington et al., this volume). Alternative approaches to improving post-release survival could be considered, including rearing animals in as natural a setting as possible, especially with respect to physical (e.g. climate, topography, seasonality) and biological (e.g. vegetation associations, roosting, denning and nesting sites, food, group size and composition) aspects of their habitat (Swaisgood, 2010; Parker et al., 2012, 2015). This also may entail consideration of some ethically challenging (because of stress inflicted on captive animals) aspects of captive rearing, especially behavioural conditioning so that captive-reared animals learn to avoid predation (Van Heezik et al., 1999; Edwards et al., 2021) and/or become proficient predators themselves by learning to capture live prey.

The use of captive animals will also influence release strategies. For instance, an immediate release strategy is generally most appropriate for wild-caught animals, thereby minimising the novelty, lack of control, and unpredictability of prolonged and stressful periods of captivity (Parker et al., 2012) (Figure 4.1). In contrast, a delayed release might be more appropriate for animals that are habituated to captivity and for whom wild habitats will be novel, unpredictable, and stressful

Figure 4.1 Immediate release of wild-caught kākāriki in Aotearoa New Zealand that also showcases local community involvement (photo: Darren Markin). (A black and white version of this figure will appear in some formats. For the colour version, please refer to the plate section.)

(Parker et al., 2012). Therefore, along with delayed release strategies there might be extended post-release support that is especially important for, but not exclusive to, captive-reared animals, for example supplementary feeding (Ewen et al., 2015), or even more radical strategies, such as teaching animals to migrate (Ellis et al., 2003; Nichols et al., 2010).

Decisions must also be made on when to release animals and the composition of the release group, specifically how many to release and the age and sex ratio of release groups. These decisions may be relevant to recovery plans and therefore need to be solved long before they are implemented: for example, the number of releases needed can influence the choice of source population. The default is often to release as many animals as possible, an approach supported by meta-analyses suggesting that the release of more animals is correlated with a higher chance of establishment (e.g. Griffith et al., 1989; Fischer & Lindenmayer, 2000). However, analysis of the relationship between success and the number of individuals released is complex for at least two reasons. First, there may exist an asymptotic relationship between the number released and establishment success (Armstrong & Wittmer, 2011) although the shape of this relationship has rarely been explored and likely varies by species. Second, managers may tend to release larger numbers of animals when they have high confidence in a site. The key point is that while the numbers of animals released is clearly important (Griffith et al., 1989; Wolf et al., 1996; Fischer & Lindenmayer, 2000; Cassey et al., 2008; Germano & Bishop, 2008), it may have less bearing on success than the habitat quality of the release site. For instance, some translocations have been successful following the release of a small number of animals into high-quality habitats (e.g. Chatham Island black robins *Petroica traversi* to Mangere and Rangatira Islands, New Zealand; Butler & Merton, 1992). Furthermore, releases of large numbers of animals do not compensate for low-quality habitat (e.g. the failed translocation of 656 popokatea *Mohoua albicilla* over 14 years to the Waitākere Ranges, New Zealand (K. A. Parker, unpublished data)). Managers must consider how many individuals they are willing to risk at an uncertain site and in light of biological considerations such as Allee effects (where individual survival or reproductive success is reduced at low abundance; Armstrong & Wittmer, 2011), individual differences in behaviour (Richardson et al., 2017, 2019), changes in sociality (Franks et al., 2020), and post-release effects (which reduce survival relative to 'normal' background levels; Armstrong et al., 2017).

Selecting an appropriate sex ratio in the release cohort will largely be driven by biological knowledge of social behaviour and mating systems (Parker et al., 2012). However, we note that when uncertainty about a destination site is high, less valuable individuals, usually males, might be released first as a means of testing the site's suitability. If survival is acceptable, further releases might be carried out with a view to establishing a breeding population (a caveat here is that some translocated animals might disperse in search of conspecifics and mates). The age ratio should also be driven by biological knowledge, trial, and experimentation, because the optimal ratio varies between taxa. For example, in our experience with wild New Zealand forest birds, older birds are often more productive breeders relative to younger birds. However, younger birds seem to adapt more readily to holding during translocation by maintaining or gaining weight, whereas adult birds often lose weight (K. A. Parker, unpublished data). Unfortunately, the extent to which this pattern influences post-release success is currently unclear, because monitoring small passerines can be expensive and difficult. But for some species there is robust information on age-specific survival (Pinter-Wollman et al., 2009; Imlay et al., 2010).

4.2.2 Selecting Appropriate Destination Sites

Conservation translocations are typically, but not always, carried out within the indigenous range of a species, for example following local extirpation (i.e. reintroduction (IUCN, 2013)), and where natural recolonisation is unlikely to occur on a time scale acceptable to managers. To have a chance of success, the conditions that animals need to persist must be present in the destination area, although these might also be provided through supportive management. Unfortunately, the concept of habitat is often poorly used and poorly defined in conservation translocation planning (Stadtmann & Seddon, 2018). Hall et al. (1997) describe habitat as '...the resources and conditions in an area that produce occupancy – including survival and reproduction – by a given organism', and the aim of translocations and conservation in general is usually to maintain that occupancy over time. This includes all physical (e.g. climate, aspect, altitude, soil type) and biological (e.g. predators, competitors, vegetation associations, prey species, parasites, landscape connectivity) aspects of an area where a species lives. Habitat quality refers to '...the ability of the environment to provide conditions appropriate for individual and population persistence' (Hall et al., 1997), specifically survival, reproduction, and population growth. Habitat quality is continuous, ranging from low to high, and can be very difficult to define explicitly, although there are useful proxies (Hall et al., 1997). Annual population growth rate is a useful proxy for measuring translocation success as it needs to be greater than one for population growth to occur, until density dependence, or other limiting factors, regulate population growth. It is often assumed that places currently occupied by a species represent high-quality habitat. However, patterns of decline are such that remnant populations do not necessarily occupy the highest-quality habitat. In addition, habitat conditions need not replicate past states so long as they provide the critical habitat characteristics that a translocated species requires. This includes moving animals out of indigenous range (i.e. conservation introduction (Chauvenet et al., 2013; Hällfors & Dalrymple, this volume)), and supportive management, such as the provisioning of supplemental food and nest boxes (Ewen et al., 2015).

Habitat connectivity, and the concomitant ability for species to disperse between habitat patches, is typically seen as a positive landscape feature. However, habitat connectivity that facilitates dispersal from managed destination areas into adjacent unmanaged areas can have problematic effects, for example exposure to predation from non-native predators outside areas where these predators are controlled (Richardson et al., 2015b). Dispersal generally affects population growth at two levels. First, post-release dispersal, which is dispersal from the site of release to where individuals settle and breed following release, can cause the loss of individuals from the founding population, thereby reducing the probability of establishment and persistence. For example, in an analysis of 14 reintroduced toutouwai/North Island robin populations, Parlato & Armstrong (2013) showed that habitat connectivity was a key factor in determining individual establishment following translocation, with individuals released at highly connected sites having a lower establishment probability than those at less connected sites, such as islands or isolated mainland reserves. Second, natal dispersal, the distance moved from site of birth to where individuals settle and breed, and breeding dispersal, the distance moved between breeding attempts, can also reduce persistence if individuals move from managed to unmanaged sites (Richardson et al., 2015b). Critically, all types of dispersal can limit population growth, erode genetic diversity, and reduce the likelihood of long-term persistence of a translocated population. Of course, in some cases a lack of connectivity is desirable, for example when establishing distinct subpopulations to reduce the risks of a single threat harming them all simultaneously (an example is provided by Hogg & Wise, this volume). The opposite is also true: connectivity is desirable in particular where the conservation translocation aims to strengthen connections between isolated populations.

The tendency for translocated species to disperse after release is highly variable, sometimes difficult to predict, and, importantly, may not match predictions based on natal or breeding dispersal (Richardson et al., 2015b). For instance, some translocated species show very strong post-release dispersal regardless of habitat connectivity (Parker et al., 2004; Brunton et al., 2008; Ortiz-Catedral, 2010) whereas others are less likely to disperse (Newman, 1980; Richard & Armstrong, 2010). The inherent post-release and natal/breeding dispersal abilities of a translocated species directly interact with the landscape features of the release site, specifically the degree to which it is connected to surrounding habitats, although the nature of this relationship is imperfectly understood and connectivity is sometimes difficult to characterise. Many species, including some with relatively strong dispersal abilities, rarely leave isolated sites (e.g. islands and forest remnants). In contrast, species with poor dispersal abilities can move out of protected areas if connected to habitat that the species will willingly move through (Richard & Armstrong, 2010).

A variety of alternative approaches have been used to try to reduce post-release dispersal, albeit with variable results. Holding animals in captivity at the release site (delayed release) has been tried with many taxa, but the results have been extremely variable and it seems most useful when releasing captive-reared animals (Parker et al., 2012; Smuts-Kennedy & Parker, 2013; Parker et al., 2015; Richardson et al., 2015a, 2015b). Supplementary feeding has also been used with success for some species (Rickett et al., 2013), but has been less useful for others (Richardson et al., 2015b). Acoustic anchoring (playback of pre-recorded calls) has also been used for some forest birds in New Zealand, but does not appear to be effective in reducing dispersal (Leuschner, 2007; Molles et al., 2008; Bradley et al., 2011). Another option for mitigating the impact of dispersal in the early stages of establishment is the release of large numbers of individuals, either in one big release or over several years, but see the caveats noted above.

4.2.3 Stress

Stress is clearly a critical aspect of any translocation, influencing both the successful establishment of translocated populations and the welfare of the individual animals themselves. We decline to comment on stress management for humans. However, we do note that many animals are attuned to the emotional state of human handlers and that highly stressed, unsure, or poorly skilled handlers will often have to cope with stressed animals, whereas calm, competent handlers will have a lesser impact on animals. Therefore, the emotional state of the translocation team is important.

Detailed work has been published on understanding stress and its role in translocations, especially by Dickens et al. (2009, 2010) and Parker et al. (2012, 2015), and we direct the reader to these works (also see Harrington et al., this volume for links between stress and welfare; Sainsbury & Carraro, this volume for links between stress and disease). However, at a practical level we think it is essential that translocation managers understand the two primary types of stress (acute stress and chronic stress) and what they can do about them during a translocation.

Acute stress is a fast response to a potentially threatening, but short-lived, situation. This results in behavioural and physiological responses that are considered beneficial and are likely adaptive (Romero, 2004). However, the problem with the translocation process is that animals are exposed to multiple, continuous, unpredictable, novel stressors, and repeated acute stress responses (Dickens et al., 2009, 2010). This can overwhelm an animal's physiological system, moving from the beneficial acute stress response to a state of chronic stress (Romero et al., 2009). Chronic stress can lead to a temporary suspension of physiological and behavioural responses that are important for animals to cope with the translocation process (Dickens et al., 2009, 2010). This includes suppression of the immune and reproductive systems, an altered metabolism, a decrease in the fight or flight response, and altered behavioural coping ability (Dickens et al., 2009, 2010). This in turn can lead to increased susceptibility to disease-causing parasites and pathogens, starvation, vulnerability to predation, and dispersal, all of which decrease the probability of individual survival (Dickens et al., 2009, 2010) and recruitment and, ultimately, population persistence. Unfortunately, some level of stress is unavoidable during a translocation. However, we can minimise the effects of stress by: (i) maximising an animal's perception of control; (ii) minimising unpredictability; and (iii) minimising novelty (Dickens et al., 2009, 2010; Parker et al., 2012, 2015).

If an animal has control over a stressful situation, the acute stress response might be moderated or even avoided. This has been most clearly demonstrated in an experimental setting with rats exposed to an electric shock – rats that could control the duration of the shock showed a lower stress response relative to those that could not (Weiss, 1968). In some translocated prey species, post-release survival has been higher when they are provisioned with food and shelter (Cabezas & Moreno, 2007) (i.e. providing a sense of control over their essential needs). Similarly, when we translocate forest birds, their holding aviaries are heavily furnished with vegetation to provide multiple refuges and resting places along with multiple opportunities to feed, drink, and bathe. This maximises opportunities for meeting essential behavioural and physiological needs, including avoiding dominant or aggressive conspecifics (Parker et al., 2015; Ewen et al., 2018).

Predictability often enables animals to cope better with stress (Levine & Ursin, 1991). For example, captive animals should be fed on a regular schedule, ideally by the same staff. Even staff wearing the same clothing likely increases the predictability of this event. In addition, novelty (i.e. exposure to an unfamiliar environment) is also an important stressor, including for animals translocated to a new environment

(Marin et al., 2007; Dickens et al., 2009). Novel environments, whether during holding or after release, are unpredictable and difficult to control. Unfortunately, the translocation process is characterised by a series of novel events. However, by carefully scrutinising all aspects of the translocation pathway, from capture to release, novelty can be reduced, for example by minimising handling events, enclosure changes, staff changes, and modes of transport, and by providing post-release support (Parker et al., 2012, 2015).

4.3 How to Approach Conservation Translocation Decisions

SDM comprises a set of principles and methods that facilitate decision-making (Gregory et al., 2012). SDM has been successfully used in a wide range of conservation problems (Runge et al., 2020) and is increasingly utilised in conservation translocations (e.g. Converse et al., 2013b; Canessa et al., 2014; Brignon et al., 2018; Gerber et al. 2018; Ferrière et al., 2021; Panfylova et al., 2019, see Box 4.1). In 2016, the IUCN Conservation Translocation Specialist Group (IUCN CTSG) developed training in application of the IUCN Guidelines (2013) based on SDM methods, a course that has run annually and delivered training to practitioners across 33 countries. Much of the source material for the IUCN CTSG training is derived from Gregory et al. (2012) along with SDM-focused training modules offered by the US Fish and Wildlife Service (Runge et al., 2011a). IUCN CTSG training continues, and additional online opportunities and accompanying texts are under development.

SDM conceives all decisions as being made of a consistent set of components. These components include a clear statement of the decision problem, identification of values relevant to the decision (management objectives), the various approaches that might be taken to realise the objectives (alternative actions), predicted effects of alternative actions in terms of the management objectives (arising from models of the system), and a method for selecting the 'best' of the alternatives. Problem-solving using SDM focuses on breaking down complex problems into their component parts. Good decisions can be defined as those that develop a clear and transparent framework for using available information to evaluate a range of alternative solutions for achieving a set of specified objectives, while taking into account uncertainty in knowledge and the decision maker's tolerance of risk. In our experience, most conservation translocations will require decisions, often difficult ones, involving biological considerations.

4.3.1 The SDM Cycle

The IUCN CTSG training is built around a somewhat standard SDM cycle (Figure 4.2). First, a **conservation goal statement** is specified that includes a short statement of what is in scope, what decisions are needed in the conservation translocation, who is making these decisions, and when the decisions must be made. Setting the bounds is critical to keep a team focussed on solving the same problem, which can help avoid wasted arguments. Second, the **fundamental and means objectives are specified**. Objectives are likely to include a mix of desired conservation outcomes, for example the successful establishment of a new population,

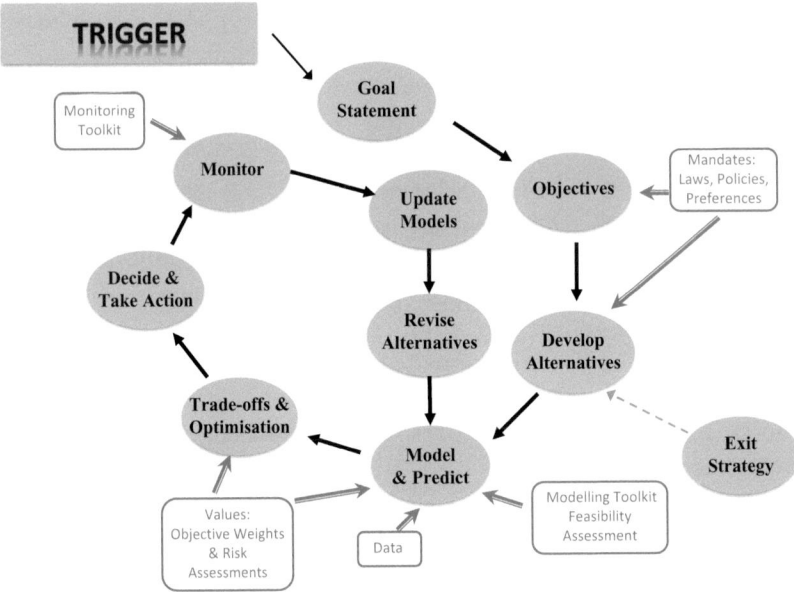

Figure 4.2 Structured decision-making cycle for conservation translocations including monitoring and updating models to revise decisions under adaptive management. This process is valid for all aspects of conservation translocation decision-making, scalable from single aspects such as deciding what pre-release behavioural conditioning to use through to overarching decisions on whether and how to undertake a conservation translocation considering all biological and socio-economic objectives.

and other objectives, for example, avoiding compromising the welfare of released individuals as well as important socio-economic factors. Fundamental objectives are those that are of essential importance to the decision maker(s). In contrast, means objectives can help us achieve fundamental objectives but are not themselves of essential importance. In our experience, failing to distinguish these two types of objective can lead decision makers in the wrong direction.

For example, a management focus on reducing exposure to beak and feather disease virus in Mauritius parakeet nests through chemical treatment was deemed important, and monitoring showed success in reducing infections (Fogell et al., 2019). However, the fundamental objective in treating nests was not to reduce viral infections per se but rather to improve reproductive success in this recovering population, something the chemical treatment was in fact reducing (Fogell et al., 2019). In this example, the focus on a means objective (reduced infection) moved the programme further from its fundamental objective (improved demographic rates; Fogell et al., 2019). A focus on means objectives, or more broadly failing to specify objectives at the outset of conservation translocations, is common, and authors have discussed this problem at length elsewhere (e.g. Ewen et al., 2015; Chauvenet et al., 2016).

In the third step, **alternative strategies** are specified, involving differing approaches to potentially achieve the fundamental objectives. These can include both no-translocation and translocation options, as well as specifying exit strategies. The strategies should be sufficiently detailed that the fourth step, **predicting consequences**, can be done effectively. Consequences are predicted for all objectives, and therefore it is necessary to develop models of a variety of forms, such that predictions can be made for all the objectives. Expert knowledge, experience, and published evidence can be used to help develop alternatives and to predict their consequences. Finally, in the fifth step, a **choice is made** for a best alternative based on the decision makers' value across objectives and their tolerance for the risk that may arise due to uncertainty in the predicted consequences. Uncertainty is expected to be high in most conservation translocations. Ignoring uncertainty or using it as a reason to delay action indefinitely can have negative effects on outcomes; however, it is possible to employ a variety of methods to account for uncertainty in the decision process and to identify alternatives that are robust to uncertainty. The SDM approach is scalable and can be used for any component of a conservation translocation.

An important recognition of SDM is that there are a few common impediments that can reliably make decisions difficult and that there are related tools available to assist with each of these impediments (Runge et al., 2020). Two of the most prominent of these are the challenge of making trade-offs between competing objectives (e.g. decreasing costs versus improving conservation outcomes) and the need to deal with uncertainty. While all conservation translocations, by definition, include one or more conservation objectives, it is also critical to account for other fundamental objectives that virtually always influence decisions about conservation translocations, for example, costs, social values, and the like. Much has been written about dealing with trade-offs in conservation translocations (e.g. Converse et al., 2013b; Chauvenet et al., 2016) and a variety of other conservation settings (Converse, 2020a). We direct readers to these sources and encourage them to reach out to the authors for information on the IUCN CTSG training previously mentioned.

In this chapter, we focus attention on the challenge presented by uncertainty. Complex biological considerations suggest a complex model of the system. A model of some type – whether conceptual or quantitative – is required to make predictions about how our alternative actions will influence attainment of our objectives. In fact, we all carry models in our heads for making a huge variety of decisions in our work and personal lives on a daily basis, though we often are not explicitly aware of these models. In the face of the complexity inherent in conservation translocations, making our models explicit – and ideally quantitative – is valuable because it helps us and others critically evaluate our assumptions and it provides a mechanism for integrating new information into our models.

4.3.2 Influence Diagram

Given the complexity of conservation translocations, a useful place to start is with an influence diagram (Figure 4.3). An influence diagram makes an explicit link between management objectives – in the case of the hypothetical situation represented in Figure 4.3, the conservation objective of 'species persistence' for a translocated turtle, shown at the bottom of the influence diagram – and management alternatives – in this case captured by a variety of boxes at the highest level in the influence diagram (e.g. 'Lethal Predator Control', 'Release Cohort Demography', and 'Law Enforcement'). Our hypothetical influence diagram demonstrates several important concepts.

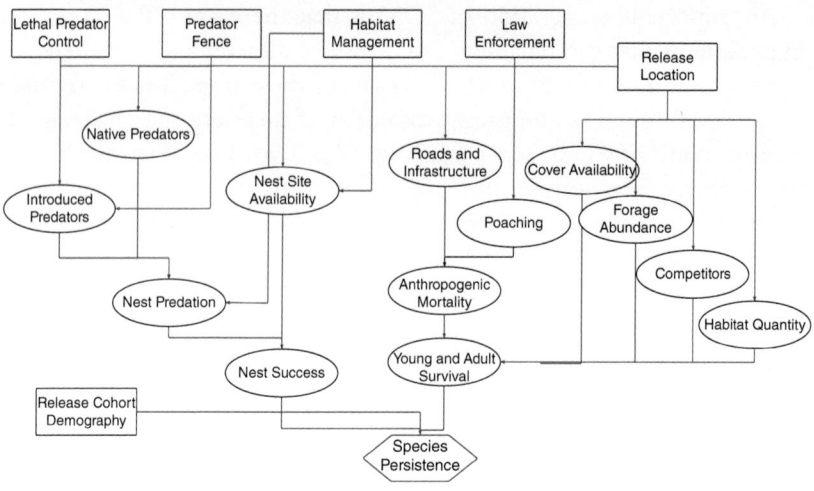

Figure 4.3 A hypothetical influence diagram including biological considerations and action elements for a reintroduced turtle species. Threats include predator impacts on nest success, anthropogenic mortality due to roads and poaching, and degraded habitat quality. It can be helpful to distinguish management actions we apply (squares) to achieve our objectives (hexagon) and the biological nature of the system we are working in (ovals).

First, models should always be designed to link our alternative actions to our objectives, and thus it is counterproductive to start focussing on the science of the situation until we understand what we are trying to achieve and how we might achieve it; note how 'Model and Predict' is the fourth step in Figure 4.2. We find that people tend to jump to science too early in the process (Converse & Grant, 2019) instead of first considering the previous steps. It is the role of applied science to help provide information to make informed decisions about what considerations are most critical (Taylor et al., 2017), but that question can only be answered in the context of clear objectives and alternatives.

Second, the influence diagram demonstrates that the types of alternatives we consider in conservation translocations are themselves often quite complex. In this case, as is typical, they are what we call 'strategy-based alternatives'. Strategy-based alternatives are those that are composed of a set of unlike components. We lay out the options for each component and then pick one option for each component to build alternatives. Table 4.2 is a strategy table for the hypothetical situation represented in Figure 4.3. Note that each component in the table addresses what we have referred to as a biological consideration: how

Table 4.2. *Strategy table related to the hypothetical conservation translocation project illustrated in Figure 4.3. Two alternative strategies are shown (Alternative 1 – 'Full Throttle' and Alternative 2 – 'Cost Savings'). Each strategy is composed of one option from each of six components. For example, one component is lethal predator control, and the options include no predator control, moderate predator control, or intensive predator control. Alternative 1 utilises intensive predator control while Alternative 2 utilises moderate predator control.*

Component	Options	'Full Throttle'	'Cost Savings'
Lethal predator control	None, moderate, intensive	Intensive	Moderate
Predator fence	No fence, exclusion of large mammals, exclusion of all mammals	Exclusion of all mammals	No fence
Release cohort	Number adult females/number adult males	80/80	40/40
Habitat management	Prescribed fire, chemical control of invasive weeds, both fire and chemical weed control, no habitat management	Both fire and chemical weed control	No habitat management
Law enforcement	None, moderate, intensive	Intensive	Moderate
Release location	Site A, Site B, Site A and B	Site A and B	Site A

many and which individuals to release, where to release them, what post-release management to undertake, and so on. Thus, a useful approach to getting started with integrating necessary biological considerations into our predictive models, after developing clear objectives for the conservation translocation, is to gather a group of experts, identify potential biological considerations, and list decision 'components' to deal with them. For example, beginning with the IUCN Guidelines, experts might identify considerations such as 'disease risk', which would in turn prompt consideration of options to deal with this disease risk. Evaluating the large literature on conservation translocations (e.g. Converse et al., 2013a) can also help identify considerations and suggest potential options for dealing with the considerations that are identified. The key point is to not default to a set of 'standard considerations' but instead to query the particular situation and address the needs of that situation.

This step could then lead into initial efforts to build an influence diagram, which will require careful thinking about how biological considerations might act in the system. This can be a highly iterative process: by thinking about the system using the influence diagram, more biological considerations might occur to experts and be added into the diagram, leading to clearer thinking about alternative components and options. We have focussed on biological considerations here, but we advocate for this approach for creating alternatives considering all fundamental objectives and drawing on relevant experts for both biological and non-biological considerations as required. This can be a highly productive part of the planning process and can allow for a great deal of creativity without judgement (Aslan et al., 2014, Game et al., 2014, Converse & Sipe, 2021). Evaluation of proposed alternatives can come later.

Once we have our influence diagram, the next step is to turn this conceptual model into a (quantitative) predictive model. A variety of tools exist for doing this (e.g. Converse & Armstrong, 2016) although providing details on these tools is beyond the scope of this chapter. However, one thing all models should do is in some way represent uncertainty. Uncertainty can be captured in models in a variety of ways. First, we might have multiple models of the system representing different competing hypotheses. For example, Converse et al. (2019) detailed a conservation translocation programme in which no fewer than 13 hypotheses were developed to explain post-translocation breeding failure. Second, we might encompass uncertainty in a continuous distribution for one or more parameters governing how the system behaves, such as the effect of management on post-release population growth (e.g. Ferrière et al., 2021). We say more in Section 4.4 about approaches for dealing with uncertainty.

4.4 Learning and the Challenge of Complexity

Throughout this chapter we have introduced a range of biological considerations for translocations, that is, biological dynamics that can interact with our translocation management and influence its outcomes. Most biological considerations are surrounded by uncertainty, first because our understanding of those dynamics and interactions is typically limited, and second (and more relevant for this chapter) because it is not clear whether and how our translocation plans should change in response to those biological considerations. We typically turn to science to reduce uncertainty and make decisions 'better' (more informed). The role of science in conservation translocations is therefore closely linked to the ultimate objectives: not to achieve scientific knowledge per se, but an improved management outcome, be it a greater population growth rate or a more cost-effective release method (Taylor et al., 2017). However, in our experience we have found this link between biological considerations, learning, and management to be a weak spot in many translocation plans.

For example, a review of published literature for one species we work closely with, the hihi, reveals that scientific examination has highlighted several potential biological considerations (Figure 4.4). However, few of those studies provide direct, explicit indications for management (e.g. 'release N animals at age Y to obtain a population growth rate of X'); they mostly highlight biological considerations, in response to which management *might* be modified, in a way that *might* ultimately improve outcomes.

We commonly see a double gap here. First, expectations and uncertainty about biological considerations are rarely made explicit using, for example, evidence synthesis or formal expert elicitation. Where explicit estimates are missing or remain verbal and not quantitative, it is harder to objectively evaluate monitoring data, or to understand how expert advice changes in response to new evidence. Second, although translocation managers usually express a desire to gather new information, they are typically much less clear on how to respond to that information. For example, we encounter many plans in which monitoring is applied to estimate post-release survival as accurately as possible, but very few of those plans indicate clearly what to do if survival is below a certain threshold, or even what that threshold is.

Both issues (missing or implicit prior estimates and lack of clear response plans) leave decision makers vulnerable to trial-and-error approaches and knee-jerk reactions. The main danger of a trial-and-error approach is that it is mostly implicit: management changes when something new happens, but it is unclear how reliable the new information is, how much it really changes previous knowledge, and even if the management response to the new information is the correct one (Runge, 2011). These problems are especially relevant for translocations of endangered species, where sample sizes are small, opportunities to learn limited, and mistakes can have severe consequences (Canessa et al., 2016a). In such situations, there is a concrete risk of overestimating the strength of inference that can be drawn from small trials or imperfect

Habitat

- Habitat complexity may reduce need for management (Makan et al., 2013) but see (Doerr et al., 2017)
- Tourism disturbance may not influence hihi breeding success (Lindsay et al., 2008)
- Environmental load of *Aspergillus fumigatus* spores may limit hihi viability (Perrot and Armstrong, 2012)
- Pollination efficiency/ecological function is important (Anderson et al., 2011)

Health

- Disease risks for translocating hihi need to be considered (Ewen et al., 2012; Dalziel et al., 2017)
- Nest mites limit fecundity (Armstrong et al., 2007; Ewen et al., 2009)
- Many disease hazards are known to affect hihi (Cork et al., 1999; Alley et al., 1999; Ewen et al., 2007)

Genetics/genomics

- Inbreeding depression can reduce fitness (Brekke et al., 2010)
- Bottlenecks cause loss of genetic diversity (Brekke et al., 2011)
- Lack of adaptive potential a problem for hihi (de Villemereuil et al., 2019a & b; Duntsch et al., 2020)
- Breeding system influences Ne/N and inbreeding (Wang et al., 2010; Brekke et al., 2015)
- Inbreeding may effect fertility (Hemmings et al., 2012)

Competition & Predation

- Interspecific competition may limit translocation success (Rasch 1985)
- Non-native predators can limit population viability (Angehr, 1984)
- Native predators may limit population viability (Low, 2010) but see (Armstrong et al., 2010)

Translocation method

- Delayed release may be worse than immediate release for survival (Castro et al., 1994; Richardson et al., 2014).
- Familiarity may not influence post release survival (Franks et al., 2020)

Climate

- Changes in precipitation and temperature can reduce population viability (Chauvenet et al., 2013)
- In site management may mitigate climate change (Correia et al., 2015)

Demography

- Post-release effects can limit translocation success (Armstrong et al., 2017)
- Natal dispersal can shape establishment success (Richardson et al., 2010)
- Harvest impacts on both source and destination sites (Panfylova et al., 2019)
- Adult sex ratio fluctuations may not influence fitness (Ewen et al., 2011)
- Age of birds translocated is important (Low et al., 2007)

Behaviour

- Personality can influence survival (Richardson et al., 2019)
- Personality can influence natal dispersal behaviour (Richardson et al., 2017)
- Sociality influences natal and post-release dispersal (Richardson and Ewen, 2016)
- Sociality can influence post-release survival (Franks et al., 2020)
- Cognitive ability may influence success (Franks and Thorogood, 2018)

Supplementary Feeding

- Feeding can promote survival and fecundity (Armstrong and Ewen, 2001; Armstrong et al., 2007)
- What you feed can benefit one sex more than other (Walker et al., 2013)
- What you feed can boost capacity to cope with parasites (Ewen et al., 2009)
- What you feed influences personality (Richardson et al., 2019)
- Micronutrients may be important for hihi health (Ewen et al., 2006)

Figure 4.4 A review of the main biological considerations suggested as important to consider in hihi conservation translocations as expressed by the authors of those studies.

monitoring data (Christie et al., 2019), or of falling prey to confirmation bias (Canessa et al., 2020).

To counter these risks, where biological considerations create uncertainty and learning is envisaged to assist management, we recommend framing learning in terms of a management decision: a choice between different alternatives with the aim of achieving some objective. In particular, it is useful to ask three questions:

1. How would a specific biological consideration affect our decisions *in practice*?
2. How much do we know about it and how much could we learn *in practice*?
3. If we had more information about it, how would we respond *in practice*?

These questions can be addressed by value of information analysis, an SDM tool that is increasingly used to guide translocation trials and monitoring (Runge et al., 2011b; Canessa et al., 2015a; Converse et al., 2019; Canessa et al., 2020; Converse, 2020b). Value of information analysis requires us to express uncertainty in terms of management objectives, link those objectives to practical decisions, and choose the best monitoring approach. For example, Runge et al. (2011) used a combination of expert elicitation and multi-criteria decision analysis to identify a preferred management alternative for recovery of the eastern migratory population of whooping cranes under current biological uncertainty and calculated the expected value of partial information to identify where learning would be most useful.

In another example, a reintroduction project for European pond turtles in Italy, Canessa et al. (2015b) had several biological considerations about release age: older, larger turtles might better avoid predation but might also be more habituated to captivity or more likely to disperse away from the site after release (and be effectively lost to the population). Conversely, releasing younger turtles would be cheaper and allow numbers to increase. Value of information was used to determine how uncertainty in age-specific survival would influence the decision to release three-, four-, or five-year-old turtles, calculating the optimal age and cohort size for a trial release (Canessa et al., 2015b). Model estimates were then updated using mark-recapture analysis of monitoring data, which confirmed initial priors about the survival of younger turtles: releasing these younger animals provides the same chance of success as well as improving cost-effectiveness (Canessa et al., 2016b).

Because most conservation translocations consist of repeated decisions, monitoring the results of management actions provides additional information about biological considerations, which can then feed back into decision-making through an adaptive management process (for example see Albrecht, this volume). Again, to maximise the effectiveness and efficiency of monitoring, it is necessary to frame it clearly within the translocation decision process, not as a separate process. For example, Canessa et al. (2019b) framed the choice of translocation methods for yellow-bellied toads under uncertainty about survival considerations using a decision tree. Before starting an experimental comparison of methods, clearly defined decision rules were agreed among managers: for example, wild egg translocations would stop if recruitment in the source population after harvest was estimated to be worse than the 2.5 per cent confidence limit for recruitment in the absence of harvest. As actions were implemented and experimental results became available in the following years, decisions rules were applied accordingly, progressively directing management towards the safest and most efficient translocation method (Canessa et al., 2019b).

4.5 Conclusion

The biological considerations for animal conservation translocations are as varied as the animal taxa that are translocated. In this chapter we have used professional experience to introduce some biological considerations we frequently encounter, although we have largely avoided genetics (Neaves et al., this volume), disease (Sainsbury & Carraro, this volume), and welfare (Harrington et al., this volume) given the focus on these elsewhere in this volume. Rather than attempt to be exhaustive, we have chosen to focus on how the IUCN Guidelines (2013) introduce biological considerations and how we recommend practitioners consider them in their own translocation projects. Although it is not our focus, we recommend that the same SDM process be used to incorporate biological considerations in plants and fungi as well as non-biological considerations in conservation translocations (Converse et al., 2013b). Conservation translocations are made up of a series of management decisions along a pathway, from selecting translocation over non-translocation alternatives through to choice of individuals to translocate, from where, and what support to provide them at the release site. Managers, or management teams, must make these decisions in the face of biological considerations, and we hope they can draw on science to inform them. Biological scientists involved in conservation translocations can best support good outcomes by ensuring the work they do is directly embedded within the management framework rather than independent of it with the belief that information will be useful. This classic model of scientists and management working independently has achieved success, but we view it as inefficient for species conservation. Regardless of the approaches used, it remains challenging to know which combination of biological considerations will provide the best outcomes.

Our focus on the translocation pathway and the biological considerations that will affect establishment reflects where most evaluation is done for conservation translocations. The IUCN Guidelines (2013), however, also suggest another central biological consideration focussed on restoring ecological function. The ecosystem level, rather than the species level, has rarely been the impetus behind conservation translocations to date (but see Donadio et al., this volume; Gaywood & Stanley-Price, this volume; Seddon, this volume; Rheingantz et al., this volume). For example, only 6 per cent of recently reviewed translocations explicitly state ecosystem restoration as an objective (Seddon & Armstrong, 2019). Ecological function may be a more common motivation in plants and algae (Swan et al., 2016). Often the emphasis is placed on risks that translocated species may bring to an ecosystem rather than the functional benefits they impart (Armstrong & Seddon, 2008; Polak & Saltz, 2011). We believe this is an area that will receive much more attention in the future. Ecological function can be a valid objective driving reinforcement and reintroduction, and must be the basis for ecological replacement. Furthermore, functional objectives are the cornerstone of rewilding initiatives, where translocations frequently play a prominent role (Gaywood & Stanley-Price, this volume; Seddon, this volume). Ecological function requires successful species establishment but is not guaranteed by it (function may already be achieved by alternative species, or established populations may not be sufficient or able to achieve desired functions). We encourage, and look forward to, much more practitioner and research activity in this space.

In summary, this chapter has three main take-home messages for approaching the biological considerations of conservation translocation. The first message is that drawing upon science (biological knowledge) will inform management decisions. This is best achieved by using decision-support frameworks like SDM. All decisions can be informed in this way, even for rare and/or declining species where data are scarce; in fact, uncertainty increases the need for formal decision-making. SDM processes and tools, like value of information, allow an objective evaluation of whether it is better to learn before acting, to act before learning (under uncertainty), or to do both and learn from the outcomes of the action itself (i.e. adaptive management). The second message is that scientists supporting conservation translocations should provide predictions of management outcomes that are as accurate as possible. Scientists may benefit by: recognising that various sources of data are valid, including direct study of the species involved in a translocation, surrogates for these species, and expert knowledge; knowing the right methods to obtain and interpret data from those sources; and being explicit about the uncertainty around predictions. Our third and final message is that both managers and scientists would benefit from understanding that decisions are being made despite uncertainty and will represent a mix of objective scientific prediction and subjective attitudes to the uncertainty represented in these. Keeping these principles in mind can help navigate often extremely challenging conservation translocation planning and provide the best chances of incorporating the relevant biological considerations to maximise success. Boxes 4.1 and 4.2 provide examples of how SDM has been used to assist decision-making in two conservation translocation programmes.

Box 4.1 *Reinforcement of a hihi population in New Zealand*

Reinforcement is one common form of conservation translocation (IUCN, 2013). The ecision whether to do one can be made challenging by uncertainty over the current state of the source and destination populations and the expected effects of translocation on them. These factors are captured in Section 5.1.4 (3) and Section 6.6 *Risk to source populations* of the IUCN Guidelines (IUCN, 2013).

Here we summarise the work of Panfylova et al. (2019), who assisted the New Zealand Hihi Recovery Group (HRG) in making a rational reinforcement decision in the face of uncertainty. Hihi *Notiomystis cincta* are a small (ca. 40 g) endemic and threatened passerine of New Zealand (Figure 4.5). Although once found throughout the northern half of New Zealand, hihi are now restricted to one remnant population (on the offshore island of Hauturu-o-Toi/Little Barrier) and six reintroduced populations spread throughout their former range.

Figure 4.5 Male hihi, *Notiomystis cincta* (photo: John Sibley). (A black and white version of this figure will appear in some formats. For the colour version, please refer to the plate section.)

Conservation Goal Statement

In March 2013, the HRG approved reintroduction of hihi to Bushy Park Tarapuruhi, an 87 ha lowland rainforest remnant surrounded by a fence (Xcluder™ 'kiwi') designed to exclude introduced mammals. The 44 birds released were from 220 ha Tiritiri Matangi Island. The Bushy Park Tarapuruhi hihi population declined over the first two years after release due to poor initial survival of the reintroduced females (only four of 21 females survived the first months). Armstrong et al. (2017) showed that this poor female survival was due to transient effects of the translocation (cost of release effects). If this was accounted for in future projections, Panfylova et al. (2016) predicted that the vital rates after the immediate post-release period would allow the hihi population to persist. However, uncertainty in these projections and an underlying belief that more birds would improve reintroduction outcomes led the Bushy Park Tarapuruhi Sanctuary Trust to propose reinforcing the population with additional birds from the Tiritiri Matangi Island population. The HRG therefore needed to make a recommendation in early 2015 on whether to support a reinforcement or not.

Objectives

Four fundamental objectives were identified by the HRG that captured the context of the Bushy Park Tarapuruhi request. These were:

1. Maximise the size of the source population (measured as number of females in September 2016).
2. Maximise persistence of the destination population (measured as whether there would be >0 females at Bushy Park Tarapuruhi in 10 years).
3. Maximise size of the destination population (to increase public appreciation by seeing birds, measured as number of females in 10 years).
4. Minimise cost (measured as the amount of money spent on translocation).

Alternatives

Four management alternatives were considered:

1. No reinforcement for at least 2 years from 2015.
2. Translocation of an additional 15 juvenile females from Tiritiri Matangi island in March 2015.
3. Delaying the reinforcement to March 2016.
4. Relocating the hihi from Bushy Park Tarapuruhi to another reserve.

Discussion of these alternatives resulted in the fourth alternative being immediately rejected as strongly unappealing by the HRG. The two remaining translocation alternatives were set at 15 females given the group's aversion to the risk associated with larger harvests. The non-translocation alternative similarly represented a more cautious approach that was attractive to some in the group who wanted to determine the outcome of the original translocations before committing more birds. While 'wait-and-see' is sometimes used as an alternative to making a decision, here the HRG explicitly identified waiting as an alternative (alternative 1), and evaluated it alongside the other alternatives.

Predicted consequences
Panfylova et al. (2019) estimated the consequences of each alternative on each objective using the agreed-upon metrics. The first three objectives were based on demographic performance: existing resighting and breeding data for both the source and destination population were used to obtain probability distributions for performance measures (including 95% credible intervals; modelled in the OpenBUGS programme, Spiegelhalter et al., 2010; for details see Panfylova et al., 2019). Cost of a translocation was known and set as the same for the translocation alternatives and as zero for the 'no reinforcement' alternative.

Decision
Panfylova et al. (2019) used Simple Multi-Attribute Rating Technique (SMART; Barron and Barrett, 1996), a decision-analytic method, to evaluate the trade-offs between the three remaining alternatives. This required HRG representatives (including those from both the source and destination sites) to assign weights representing how much they valued each objective given the differences among alternatives. The weights were multiplied by the standardised estimated consequences for each objective and then summed across objectives to obtain scores for each alternative. Given uncertainty in consequence estimation, Panfylova et al. (2019) resampled the outcomes across the range of uncertainty for each objective. Although the distributions of final scores overlapped greatly, the 'no reinforcement' alternative was largely dominant over other management options, that is, it was clearly the best choice in most biological and cost projections, and the choice was ambiguous in the remaining projections. The decision was also unaffected by variation in stakeholder values, as repeating the analysis using the objective weights of each representative did not change the outcome.

Box 4.2 *Optimising release strategies through learning in the Vancouver Island marmot*

Conditioning individuals for translocation can help to overcome strong post-release effects whereby survival is reduced during an acclimation period. Post-release effects can be particularly strong when using captive-bred individuals for release, as they are behaviourally naïve to wild environments, including awareness and appropriate responses to predators. Pre-release behavioural training has received some attention (see references in Section 4.2.1) but there are uncertainties about how effective learnt responses are post release. These factors are captured in the IUCN Guidelines (IUCN, 2013), specifically Annex 7, which covers release and implementation (see Point 14 on behavioural conditioning). In this example, Lloyd et al. (2019) showed how behavioural conditioning can improve post-release survival in the critically endangered Vancouver Island marmot *Marmota vancouverensis*. The Vancouver Island marmot is an herbivorous large rodent species endemic to Vancouver Island, British Columbia, Canada. It has recovered from fewer than 30 individuals in the wild in 2003, primarily because of captive breeding and release.

Conservation goal statement

The current recovery efforts for Vancouver Island marmots include continued captive breeding and the release of animals to reinforce the extant meta-populations. One destination site is Strathcona Provincial Park on Vancouver Island, where there is a small population of marmots and a potentially higher abundance of predators, which could explain the lower survival of animals that have been released directly from captivity into this site. In addition, a critical period of post-release mortality occurs during a captive-bred marmot's first wild hibernation; survival is lower compared to wild-born individuals, as hibernation behaviour may also have a learnt component (Lloyd et al., 2019). Animals that survived their first hibernation had similar survival rates to wild-born individuals (Jackson et al., 2016). The Vancouver Island Marmot Recovery Team wanted to test how to improve post-release survival at the Strathcona site to speed recovery of this meta-population.

Objectives

A single objective was considered for this decision problem:

1. Maximise population growth at the Strathcona site. The measurable criterion was survival to prime breeding age for captive-bred

and released marmots (survival to ≥4 years of age and the second spring post release at Strathcona).

Alternatives

Two alternatives were compared:

1. Translocating captive-bred marmots directly to the Strathcona site.
2. Translocating captive-bred marmots using a stepping-stone approach, whereby animals were first released into a population at Mt Washington, Vancouver Island, for one year and were then translocated to the Strathcona site. Marmots could be translocated as yearlings, two-year-olds, or adults (≥ three years old). The Mt Washington population is characterised by high survival to breeding age, likely due to a lower abundance of predators and an established population that provides the opportunity for social learning. It was hypothesised that time spent in Mt Washington would provide sufficient exposure to predators and other threats to encourage learning without substantial loss due to predation.

Although the management decision focussed on captive-bred individuals, wild-born individuals were also tracked to determine how behaviourally compromised captive-bred marmots were.

Predicted consequences

All translocated marmots were fitted with radio transmitters and survival was modelled from monthly monitoring data as a function of release type, age, and a range of other factors in a multi-event mark-recapture recovery model (see Lloyd et al., 2019, for details). The highest probability of survival to breeding age of released captive-bred marmots was achieved via a stepping-stone approach using yearling animals (mode = 0.13, 95% credible intervals = 0.05–0.30), followed by releasing yearling captive-bred animals directly to the Strathcona site (mode = 0.04, 95% credible intervals = 0.01–0.24). Despite the improvement in survival to breeding age of captive-bred marmots using a stepping-stone approach, survival remained lower than for wild-born marmots translocated directly from Mt Washington. Moving older age classes resulted in substantially lower survival to prime breeding age than moving yearlings.

Decision

In this case an event tree summarised the survival consequences of combinations of translocation pathway and marmot age (Lloyd et al.,

2019). Based on the single criterion of maximising survival to breeding age of captive-bred marmots released to the Strathcona site, it is optimal to release yearlings via the stepping-stone approach. The stepping-stone approach incorporated species-specific biological considerations by considering the need and ability of released marmots to learn ways of surviving in the wild. However, Lloyd et al. (2019) recognised that a broader recovery plan for this species would also need to consider the Mt Washington site and its capacity to be host as a stepping-stone, along with cost considerations and other objectives. The information provided in this experiment will help managers in future recovery decisions by explicitly comparing alternatives based on measured consequences.

4.6 Key Messages

- Biological considerations are often a major focus in conservation translocations. They include the physiological, behavioural, demographic, and ecological considerations relevant to management decisions about translocation planning, implementation, and evaluation.
- The vast array of biological, and other, considerations that managers must wrestle with renders conservation translocations exceedingly complex, and a framework that supports thinking through this complexity to inform decisions in a transparent and deliberative fashion is indispensable.
- Structured decision-making (SDM) is a framework that is well suited to help managers deal with the complexity of their decisions, and SDM facilitates the integration of science (biological knowledge) to inform decisions.
- Scientists supporting conservation translocations have many tools at their disposal to help them provide predictions of management outcomes that are as accurate as possible, recognising that various sources of data are valid and that there is substantial guidance available on the appropriate methods to obtain, analyse, and interpret available data.
- Decisions will represent a mix of objective scientific prediction and subjective attitudes regarding trade-offs between objectives and the uncertainty surrounding predictions.
- All conservation translocation decisions can be informed using SDM irrespective of their focus being biological, non-biological, or, perhaps most realistically, a mix across these concerns.

References

Albrecht, M. A. (2023) Applying adaptive management to reintroductions of Pyne's ground-plum Astragalus bibullatus. In Gaywood, M. J., Ewen, J. G., Hollingsworth, P. M. and Moehrenschlager, A. (eds.) *Conservation Translocations*. Cambridge, Cambridge University Press.

Allendorf, F. W. (1993) Delay of adaptation to captive breeding by equalizing family size. *Conservation Biology*. 7, 416–419.

Alley, M. R., Castro, I. & Hunter, J. E. B. (1999) Aspergillosis in hihi (*Notiomystis cincta*) on Mokoia Island. *New Zealand Veterinary Journal*. 47, 88–91.

Anderson S. H., Kelly, D., Ladley, J. J., Molloy, S. & Terry, J. (2011) Cascading effects of bird functional extinction reduce pollination and plant density. *Science*. 331, 1068–1071.

Angehr, G. R. (1984) *Ecology and behaviour of the stitchbird with recommendations for management*. Wellington, New Zealand Wildlife Service, Department of Internal Affairs. Unpublished report.

Armstrong, D. P. & Ewen, J. G. (2001) Testing food limitation in reintroduced hihi populations: contrasting results for two islands. *Pacific Conservation Biology*. 7, 87–92.

Armstrong, D. P. & Seddon, P. J. (2008) Directions in reintroduction biology. *Trends in Ecology and Evolution*. 23, 20–25.

Armstrong, D. P. & Wittmer, H. U. (2011) Incorporating Allee effects into reintroduction strategies. *Ecological Research*. 26, 687–695.

Armstrong, D. P., Castro, I. & Griffiths, R. (2007) Using adaptive management to determine requirements of re-introduced populations: the case of the New Zealand hihi. *Journal of Applied Ecology*. 44, 953–962.

Armstrong, D. P., Castro, I., Perrott, J. K., Ewen, J. G. & Thorogood, R. (2010) Impacts of pathogenic disease and native predators on threatened native species. *New Zealand Journal of Ecology*. 34, 272–273.

Armstrong, D. P., Le Coeur, C., Thorne, J. M., et al. (2017) Using Bayesian mark-recapture modelling to quantify the strength and duration of post-release effects in reintroduced populations. *Biological Conservation*. 215, 39–45.

Aslan, C. E., Pinsky, M. L., Ryan, M. E., Souther, S. & Terrell, K. A. (2014) Cultivating creativity in conservation science. *Conservation Biology*. 28, 345–353.

Barron, F. H. & Barrett, B. E. (1996) The efficacy of SMARTER – simple multi-attribute rating technique extended to ranking. *Acta Psychologica*. 93, 23–36.

Berger-Tal, O., Blumstein, D. T. & Swaisgood, R. R. (2020) Conservation translocations: a review of common difficulties and promising directions. *Animal Conservation*. 23, 121–131.

Bradley, D. W., Ninnes, C. E., Valderrama, S. V. & Waas, J. R. (2011) Does 'acoustic anchoring' reduce post-translocation dispersal of North Island robins? *Wildlife Research*. 38, 69–76.

Brekke, P., Bennett, P. M., Wang, J., Pettorelli, N. & Ewen, J. G. (2010) Sensitive males: inbreeding depression in an endangered bird. *Proceedings of the Royal Society of London B*. 277, 3677–3684.

Brekke, P., Bennett, P. M., Santure, A. W. & Ewen, J. G. (2011) High genetic diversity in the remnant island population of hihi and the genetic consequences of re-introduction. *Molecular Ecology.* 20, 29–45.

Brekke, P., Ewen, J. G., Clucas, G. & Santure, A. (2015) Determinants of male floating behavior and floater reproduction in a threatened population of the hihi (*Notiomystis cincta*). *Evolutionary Applications.* 8, 796–806.

Brignon, W. R., Peterson, J. T., Dunham, J. B., Schaller, H. A. & Schreck, C. B. (2018) Evaluating trade-offs in bull trout reintroduction strategies using structured decision making. *Canadian Journal of Fisheries and Aquatic Sciences.* 75, 293–307.

Brunton, D. H., Evans, B. A. & Ji, W. (2008) Assessing natural dispersal of New Zealand bellbirds using song type and song playbacks. *New Zealand Journal of Ecology.* 32, 147–154.

Butler, D. & Merton, D. V. (1992) *The Black Robin. Saving the World's Most Endangered Bird*. Oxford, Oxford University Press, p. 294.

Cabezas, S. & Moreno, S. (2007) An experimental study of translocation success and habitat improvement in wild rabbits. *Animal Conservation.* 10, 387–391.

Canessa, S., Hunter, D., McFadden, M., Marantelli, G. & McCarthy, M. A. (2014) Optimal release strategies for cost-effective reintroductions. *Journal of Applied Ecology.* 51, 1107–1115.

Canessa, S., Guillera-Arroita, G., Lahoz-Monfort, J. J., et al. (2015a) When do we need more data? A primer on calculating the value of information for applied ecologists. *Methods in Ecology and Evolution.* 6, 1219–1228.

Canessa, S., Ottonello, D. & Salvidio, S. (2015b) Population modelling to assess supplementation strategies for the European pond terrapin Emys orbicularis in Liguria. In Doria, G., Poggi, R., Salvidio, S. and Tavano, M. (eds.) *Atti X Congresso Nazionale della Societas Herpetologica Italica* (Genova, 15–18 October 2014). Pescara, Italy, Ianieri Edizioni, pp. 385–391.

Canessa, S., Guillera-Arroita, G., Lahoz-Monfort, J. J., et al. (2016a) Adaptive management for improving species conservation across the captive-wild spectrum. *Biological Conservation.* 199, 123–131.

Canessa, S., Genta, P., Jesu, R., et al. (2016b) Challenges of monitoring reintroduction outcomes: insights from the conservation breeding program of an endangered turtle in Italy. *Biological Conservation.* 204, 128–133.

Canessa, S., Armstrong, D. P., Converse, S. J., et al. (2019a) *IUCN/SSC Conservation Translocation Specialist Group Training for Effective Conservation Translocations*. Available from: https://iucn-ctsg.org/training/ [Accessed 5 July 2021].

Canessa, S., Ottonello, D., Rosa, G., Salvidio, S., Grasselli, E. & Oneto, F. (2019b) Adaptive management of amphibian recovery programs: a real-world application for an endangered amphibian. *Biological Conservation.* 236, 202–210.

Canessa, S., Taylor, G., Clarke, R. H., Ingwersen, D., Vandersteen, J. & Ewen, J. G. (2020) Risk aversion and uncertainty create a conundrum for planning recovery of a critically endangered species. *Conservation Science and Practice.* 2, e138.

Cassey, P., Blackburn, T. M., Duncan, R. P. & Lockwood, J. L. (2008) Lessons from introductions of exotic species as a possible information source for managing translocations of birds. *Wildlife Research.* 35, 193–201.

Castro, I., Alley, J. C., Empson, R. A. & Minot, E. O. (1994) Translocation of hihi or stitchbird *Notiomystis cincta* to Kapiti Island, New Zealand: transfer techniques and comparison of release strategies. In Serena, M. (ed.) *Reintroduction Biology of Australian and New Zealand Fauna*. Chipping Norton, Australia, Surrey Neatty & Sons, pp. 113–120.

Chauvenet, A. L. M., Ewen, J. G., Armstrong, D. P. & Pettorelli, N. (2013) Saving the hihi under climate change: a case for assisted colonization. *Journal of Applied Ecology*. 50, 1330–1340.

Chauvenet, A. L. M., Canessa, S. & Ewen, J. G. (2016) Setting objectives and defining the success of reintroductions. In Jachowski, D. S., Millspaugh, J. J., Angermeier, P. L. and Slotow, R. (eds.) *Reintroduction of Fish and Wildlife Populations*. Oakland, CA, University of California Press, pp. 105–122.

Christie, A. P., Amano, T., Martin, P. A., Shackelford, G. E., Simmons, B. I. & Sutherland, W. J. (2019) Simple study designs in ecology produce inaccurate estimates of biodiversity responses. *Journal of Applied Ecology*. 56, 2742–2754.

Converse, S. J. (2020a) Introduction to multi-criteria decision analysis. In Runge, M. C., Converse, S. J., Lyons, J. E. and Smith, D. R. (eds.) *Structured Decision Making: Case Studies in Natural Resource Management*. Baltimore, MD, Johns Hopkins University Press, pp. 51–61.

Converse, S. J. (2020b) Prioritizing uncertainties to improve management of a reintroduction program. In Runge, M. C., Converse, S. J., Lyons, J. E. and Smith, D. R. (eds.) *Structured Decision Making: Case Studies in Natural Resource Management*. Baltimore, MD, Johns Hopkins University Press, pp. 214–224.

Converse, S. J. & Armstrong, D. P. (2016) Demographic modelling for reintroduction decision making. In Jachowski, D. S., Millspaugh, J. J., Angermeier, P. L. and Slotow, R. (eds.) *Reintroduction of Fish and Wildlife Populations*., Oakland, CA, University of California Press, pp. 123–146.

Converse, S. J. & Grant, E. H. C. (2019) A three-pipe problem: dealing with complexity to halt amphibian declines. *Biological Conservation*. 236, 107–114.

Converse, S. J. & Sipe, H. A. (2021) Finding the win-win strategies in endangered species conservation. *Animal Conservation*. 24, 161–162.

Converse, S. J., Moore, C. T. & Armstrong, D. P. (2013a) Demographics of reintroduced populations: estimation, modelling, and decision analysis. *Journal of Wildlife Management*. 77, 1081–1093.

Converse, S. J., Moore, C., Folk, M. J. & Runge, M. C. (2013b) A matter of tradeoffs: reintroduction as a multiple objective decision. *Journal of Wildlife Management*. 77, 1145–1156.

Converse, S. J., Strobel, B. N. & Barzen J. A. (2019) Reproductive failure in the eastern migratory population: the interaction of research and management. In French, J. B. Jr., Converse, S. J. and Austin, J. E. (eds.) *Whooping Cranes: Biology and Conservation. Biodiversity of the World: Conservation from Genes to Landscapes*. London, Academic Press, pp. 161–178.

Cork, S. C., Alley, M. R., Johnstone, A. C. & Stockdale, P. H. G. (1999) Aspergillosis and other causes of mortality in the stitchbird in New Zealand. *Journal of Wildlife Diseases*. 35, 481–486.

Correia, D. L. P., Chauvenet, A. L. M., Rowcliffe, J. M. & Ewen, J. G. (2015) Targeted management buffers negative impacts of climate change on the hihi, a threatened New Zealand passerine. *Biological Conservation*. 192, 145–153.

Dalziel, A. E., Sainsbury, A. W., McInnes, K., Jakob-Hoff, R. & Ewen, J. G. (2017) A comparison of disease risk analysis tools for conservation translocations. *Ecohealth*. 14, 30–41.

de Villemereuil, P., Rutschmann, A., Ewen, J. G., Santure, A. W. & Brekke, P. (2018) Can threatened species adapt in restored habitat? No expected evolutionary response in lay date for the New Zealand hihi. *Evolutionary Applications*. 12, 482–497.

de Villemereuil, P., Rutschmann, A., Lee, K., Ewen, J. G., Brekke, P. & Santure, A. (2019) Little adaptive potential in a threatened passerine bird. *Current Biology*. 29, 889–894.

Dickens, M. J., Delehanty, D. J. & Romero, L. M. (2009) Stress and translocation: alterations in the stress physiology of translocated birds. *Proceedings of the Royal Society B*. 276, 2051–2056.

Dickens, M. J., Delehanty, D. J. & Romero, L. M. (2010) Stress: an inevitable component of animal translocation. *Biological Conservation*. 143, 1329–1341.

Doerr, L. R., Richardson, K. M., Ewen, J. G. & Armstrong, D. P. (2017) Effect of supplementary feeding on reproductive success of hihi (stitchbird, *Notiomystis cincta*) at a mature forest reintroduction site. *New Zealand Journal of Ecology*. 41, 34–40.

Donadio, E., Zamboni, T. & Di Martino, S. (2023) Bringing jaguars and their prey base back to the Iberá Wetlands, Argentina. In Gaywood, M. J., Ewen, J. G., Hollingsworth, P. M. and Moehrenschlager, A. (eds.) *Conservation Translocations*. Cambridge, Cambridge University Press.

Duntsch, L., Tomotani, B. M., Villemereuil, P. D., et al. (2020) Polygenic basis for adaptive morphological variation in a threatened Aoteroa New Zealand bird, the hihi (*Notiomystis cincta*). *Proceedings of the Royal Society B*. 287, 20200948.

Edwards, M. C., Ford, C., Hoy, J. M., Fitzgibbon, S. & Murray, P. J. (2021) How to train your wildlife: a review of predator avoidance training. *Applied Animal Behaviour Science*. 234, 105170.

Ellis, D. H., Sladen, W. J. L. & Lishman, W. A. (2003) Motorized migrations: the future or mere fantasy? *Bioscience*. 53, 260–264.

Ewen, J. G., Thorogood, R., Karadas, F., Pappas, A. C. & Surai, P. F. (2006) Influences of carotenoid supplementation on the integrated antioxidant system of a free living endangered passerine, the hihi (Notiomystis cincta). *Comparative Biochemistry and Physiology Part A: Molecular & Integrative Physiology*. 143, 149–154.

Ewen, J. G., Thorogood, R., Nicol, C., Armstrong, D. P. & Alley, M. (2007) *Salmonella typhimurium* in hihi, New Zealand. *Journal of Emerging Infectious Diseases*. 13, 788–790.

Ewen, J. G., Thorogood, R., Brekke, P., Cassey, P., Karadas, F. & Armstrong, D. P. (2009) Maternally invested carotenoids compensate costly ectoparasitism in the hihi. *Proceedings of the National Academy of Sciences USA*. 106, 12798–12802.

Ewen, J. G., Thorogood, R. & Armstrong, D. P. (2011) Demographic consequences of adult sex ratio in a reintroduced hihi population. *Journal of Animal Ecology.* 80, 448–455.

Ewen, J. G., Armstrong, D. P., Empson, R., et al. (2012) Parasite management in translocations: lessons from an endangered New Zealand bird. *Oryx.* 46, 446–456.

Ewen, J. G., Soorae, P. S. & Canessa, S. (2014) Reintroduction objectives, decisions and outcomes: global perspectives from the herpetofauna. *Animal Conservation.* 17, 74–81.

Ewen, J. G., Walker, L., Canessa, S. & Groombridge, J. J. (2015) Improving supplementary feeding in species conservation. *Conservation Biology.* 29, 341–349.

Ewen, J. G., Armstrong, D. P., McInnes, K., et al. (2018) *Hihi: Best Practise Guide.* Wellington, New Zealand, Department of Conservation, 133 pp.

Ferrière, C., Zuël, N., Ewen, J. G., Jones, C. G., Tatayah, V. & Canessa, S. (2021) Assessing the risks of changing ongoing management of endangered species. *Animal Conservation.* 24, 153–160.

Fischer, J. & Lindenmayer, D. B. (2000) An assessment of the published results of animal relocations. *Biological Conservation.* 96, 1–11.

Fogell, D., Groombridge, J. J., Tollington, S., et al. (2019) Hygiene and biosecurity protocols reduce infection prevalence but do not improve fledging success in an endangered parrot. *Scientific Reports.* 9, 4779.

Frankham, R., Loebel, D. A., Ryder, O. A., et al. (1986) Modelling problems in conservation genetics using captive Drosophila populations: rapid genetic adaptation to captivity. *Zoo Biology.* 11, 333–342.

Franks, V. R. & Thorogood, R. (2018) Older and wiser? Age differences in foraging and learning by an endangered passerine. *Behavioural Processes.* 148, 1–9.

Franks, V. R., Andrews, C. E., Ewen, J. G., et al. (2020) Changes in social groups across reintroductions and effects on post-release survival. *Animal Conservation.* 23, 443–454.

Game, E. T., Meijaard, E., Sheil, D., & McDonald-Madden, E. (2014) Conservation in a wicked complex world; challenges and solutions. *Conservation Letters.* 7, 271–277.

Gaywood, M. J. & Stanley-Price, M. (2023) Moving species: reintroductions and other conservation translocations. In Gaywood, M. J., Ewen, J. G., Hollingsworth, P. M. and Moehrenschlager, A. (eds.) *Conservation Translocations.* Cambridge, Cambridge University Press.

Gerber, B. D., Converse, S. J., Muths, E., Crockett, H. J., Mosher, B. A. & Bailey, L. L. (2018) Identifying species conservation strategies to reduce disease-associated declines. *Conservation Letters.* 11, 1–10.

Germano, J. M. & Bishop, P. J. (2008) Suitability of amphibians and reptiles for translocation. *Conservation Biology.* 23, 7–15.

Gregory, R., Failing, L., Harstone, M., Long, G., McDaniels, T. & Ohlson, D. (2012) *Structured Decision Making: A Practical Guide to Environmental Management Choices.* Chichester, Wiley-Blackwell.

Griffith, B., Scott, J. M., Carpenter, J. W. & Reed, C. (1989) Translocation as a species conservation tool: status and strategy. *Science.* 245, 477–480.

Hakånsson, J. & Jensen, P. (2005) Behavioural and morphological variation between captive populations of red junglefowl (*Gallus gallus*) - possible implications for conservation. *Biological Conservation*. 122, 431–439.

Hakånsson, J. & Jensen, P. (2008) A longitudinal study of antipredator behaviour in four successive generations of two populations of captive red junglefowl. *Applied Animal Behaviour Science*. 114, 409–418.

Hall, L. S., Krausman, P. R. & Morrison, M. L. (1997) The habitat concept and a plea for standard terminology. *Wildlife Society Bulletin*. 25, 173–182.

Hällfors, M. & Dalrymple, S. E. (2023) Assisted colonisation and ecological replacement. In Gaywood, M. J., Ewen, J. G., Hollingsworth, P. M. and Moehrenschlager, A. (eds.) *Conservation Translocations*. Cambridge, Cambridge University Press.

Harrington, L. A., Lloyd, N. & Moehrenschlager, A. (2023) Animal welfare, animal rights, and conservation translocations: moving forward in the face of ethical dilemmas. In Gaywood, M. J., Ewen, J. G., Hollingsworth, P. M. and Moehrenschlager, A. (eds.) *Conservation Translocations*. Cambridge, Cambridge University Press.

Hemmings, N., West, M. & Birkhead, T. R. (2012) Causes of hatching failure in endangered birds. *Biology Letters*. 8, 964–967.

Hogg, C. & Wise, P. (2023) Assisted colonisation as a conservation tool: Tasmanian devils and Maria Island. In Gaywood, M. J., Ewen, J. G., Hollingsworth, P. M. and Moehrenschlager, A. (eds.) *Conservation Translocations*. Cambridge, Cambridge University Press.

Imlay, T. I., Crowley, J. F., Argue, A. M., et al. (2010) Survival, dispersal and early migration movements of captive-bred juvenile eastern loggerhead shrikes (*Lanius ludovicianus migrans*). *Biological Conservation*. 143, 2578–2582.

IUCN (2013) *Guidelines for Reintroductions and Other Conservation Translocations. Version 1.0*. Gland, Switzerland, IUCN Species Survival Commission.

Jackson, C. L., Schuster, R. & Arcese, P. (2016) Release date influences first-year site fidelity and survival in captive-bred Vancouver Island marmots. *Ecosphere*. 7, 1–16.

Leuschner, N. (2007) Ecology and behaviour of the whitehead (*Mohoua albicilla*) in its translocated ranges in New Zealand. Unpublished thesis, University of Auckland, Auckland, 109 pp.

Levine, S. & Ursin, H. (1991) What is stress? In Brown, M. R., Koob G. F. and Rivier, C. (eds.) *Stress: Neurobiology and Neuroendocrinology*. New York, Dekker.

Lindsay, K., Craig, J. & Low, M. (2008) Tourism and conservation: the effects of track proximity on avian reproductive success and nest selection in an open sanctuary. *Tourism Management*. 29, 730–739.

Lloyd, N. A., Hostetter, N. J., Jackson, C. L., Converse, S. J. & Moehrenschlager, A. (2019) Optimizing release strategies: a stepping-stone approach to reintroduction. *Animal Conservation*. 22, 105–115.

Lovegrove, T. G. (1996) Island releases of saddlebacks *Philesturnus carunculatus* in New Zealand. *Biological Conservation*. 77, 151–157.

Low, M. (2010) Which factors limited stitchbird population growth on Mokoia Island? *New Zealand Journal of Ecology*. 34, 269–271.

Low, M., Pärt, T. & Forslund, P. (2007) Age-specific variation in reproduction is largely explained by the timing of territory establishment in the New Zealand stitchbird *Notiomystis cincta*. *Journal of Animal Ecology*. 76, 459–470.

Makan, T., Castro, I., Robertson, A. W., Joy, M. K. & Low, M. (2014) Habitat complexity and management intensity positively influence fledging success in the endangered hihi (*Notiomystis cincta*). *New Zealand Journal of Ecology*. 38, 53–63.

Marin, M. T., Cruz, F. C. & Planeta, C. S. (2007) Chronic restraint or variable stresses differently affect the behaviour, corticosterone secretion and body weight in rats. *Physiology and Behaviour*. 90, 29–35.

McPhee, M. E. (2004) Generations in captivity increases behavioural variance: considerations for captive breeding and reintroduction programmes. *Biological Conservation*. 115, 71–77.

Moehrenschlager, A. & Lloyd, N. A. (2016) Release considerations and techniques to improve conservation translocation success. In Jachowski, D. S., Millspaugh, J. J., Angermeier, P. L. & Slotow, R. (eds.) *Reintroduction of Fish and Wildlife Populations*. Oakland, CA, University of California Press, pp. 245–280.

Molles, L. E., Calcott, A., Peters, D., et al. (2008) 'Acoustic anchoring' and the successful translocation of North Island kokako (*Callaeas cinerea wilsoni*) to a New Zealand mainland site within continuous forest. *Notornis*. 55, 57–68.

Neaves, L. E., Ogden, R. & Hollingsworth, P. M. (2023) Genomics and conservation translocations. In Gaywood, M. J., Ewen, J. G., Hollingsworth, P. M. and Moehrenschlager, A. (eds.) *Conservation Translocations*. Cambridge, Cambridge University Press.

Newman, D. G. (1980) Colonisation of Copermine Island by the North Island saddleback. *Notornis*. 27, 146–147.

Nichols, R. K., Steiner, J. & Woolaver, L. G. (2010) Conservation initiatives for an endangered migratory passerine: field propagation and release. *Oryx*. 44, 171–177.

Ortiz-Catedral, L. (2010) Homing of a red-crowned parakeet (*Cyanoramphus novaezelandiae*) from Motuihe Island to Little Barrier Island, New Zealand. *Notornis*. 57, 48–49.

Panfylova, J., Bemelmans, E., Devine, C., Frost, P. & Armstrong, D. (2016) Post-release effects on reintroduced populations of hihi. *Journal of Wildlife Management*. 80, 970–977.

Panfylova, J., Ewen, J. G. & Armstrong, D. P. (2019) Making structured decisions for reintroduction populations in the face of uncertainty. *Conservation Science and Practice*. 1, e.90.

Parker, K. A., Hughes, B., Thorogood, R. & Griffiths, R. (2004) Homing over 56 km by a North Island tomtit (*Petroica macrocephala toitoi*). *Notornis*. 51, 238–239.

Parker, K. A., Dickens, M. J., Clarke, R. H. & Lovegrove, T. G. (2012) The theory and practise of catching, holding, moving and releasing animals. In Ewen, J. G., Armstrong, D. P., Parker, K. A. and Seddon, P. J. (eds.) *Reintroduction Biology: Integrating Science and Management*. West Sussex, Wiley-Blackwell, pp. 105–137.

Parker, K. A., Adams, L., Baling, M., et al. (2015) Practical guidelines for planning and implementing fauna translocations. In Armstrong, D. P., Hayward, M. W.,

Moro, D. & Seddon, P. J. (eds.) *Advances in Reintroduction Biology of Australian and New Zealand Fauna*. Clayton South, VIC, Australia, CSIRO Publishing, pp. 255–272.

Parlato, E. H. & Armstrong, D. P. (2013) Predicting post-release establishment using data from multiple introductions. *Biological Conservation*. 160, 97–104.

Perrott, J. K. & Armstrong, D. P. (2012) *Aspergillus fumigatus* densities in relation to forest succession and edge effects: implications for wildlife health in modified environments. *Ecohealth*. 8, 290–300.

Pinter-Wollman, N., Isbell, L. A. & Hart, L. A. (2009) Assessing translocation outcome: comparing behavioural and physiological aspects of translocated and resident African elephants (*Loxodonta africana*). *Biological Conservation*. 142, 1116–1124.

Polak, T. & Saltz, D. (2011) Reintroduction as an ecosystem restoration technique. *Conservation Biology*. 25, 424–427.

Rasch, G. (1985) *The behavioural ecology of the stitchbird*. Unpublished MSc thesis, University of Auckland, New Zealand.

Rheingantz, M. L., dos Santos Pires, A. & Fernandez, F. A. S. (2023) Multiple reintroductions to restore ecological interactions in a defaunated tropical forest. In Gaywood, M. J., Ewen, J. G., Hollingsworth, P. M. and Moehrenschlager, A. (eds.) *Conservation Translocations*. Cambridge, Cambridge University Press.

Richard, Y. & Armstrong, D. P. (2010) Cost distance modelling of landscape connectivity and gap-crossing ability using radio-tracking data. *Journal of Applied Ecology*. 47, 603–610.

Richardson, K. & Ewen, J. G. (2016) Habitat selection in a reintroduced population: social effects differ between natal and post-release dispersal. *Animal Conservation*. 19, 413–421.

Richardson, K., Ewen, J. G., Armstrong, D. P. & Hauber, M. E. (2010) Sex-specific shifts in natal dispersal dynamics in a reintroduced hihi population. *Behaviour*. 147, 1517–1532.

Richardson, K., Castro, I. C., Brunton, D. H. & Armstrong, D. P. (2014) Not so soft? Delayed release reduces long-term survival in a passerine reintroduction. *Oryx*. 49, 535–541.

Richardson, K., Castro, I. C., Brunton, D. H. & Armstrong, D. P. (2015a) Not so soft? Delayed release reduces long-term survival in a passerine reintroduction. *Oryx*. 49, 535–541.

Richardson, K., Doerr, V., Ebrahimi, M. & Parker, K. A. (2015b) Considering dispersal in reintroduction and restoration planning. In Armstrong, D. P., Hayward, M. W., Moro, D. and Seddon, P. J. (eds.) *Advances in Reintroduction Biology of Australian and New Zealand Fauna*. Clayton South, VIC, Australia, CSIRO Publishing, pp. 59–72.

Richardson, K. M., Ewen, J. G., Brekke, P., Doerr, L. R., Parker, K. A. & Armstrong, D. P. (2017) Behaviour in the hand predicts male natal dispersal distances in an establishing reintroduced hihi (*Notiomystis cincta*) population. *Animal Conservation*. 20, 135–143.

Richardson, K. M., Parlato, E. H., Walker, L. K., Parker, K. A., Ewen, J. G. & Armstrong, D. P. (2019) Links between personality, early natal nutrition and survival of a threatened bird. *Philosophical Transactions of the Royal Society B.* 374, 1–7.

Rickett, J., Dey, C. J., Stothart, J., O'Connor, C. M., Quinn, J. S. & Ji, W. (2013) The influence of supplemental feeding on survival, dispersal and competition in translocated brown teal, or pateke (*Anas chlorotis*). *Emu.* 113, 62–68.

Romero, L. M. (2004) Physiological stress in ecology: lessons from biomedical research. *Trends in Ecology & Evolution.* 19, 249–255.

Romero, L. M., Dickens, M. J. & Cyr, N. E. (2009) The reactive scope model - a new model integrating homeostasis, allostasis, and stress. *Hormones and Behavior.* 55, 375–389.

Runge, M. C. (2011) An introduction to adaptive management for threatened and endangered species. *Journal of Fish and Wildlife Management.* 2, 220–233.

Runge, M. C., Cochrane, J. F., Converse, S. J., et al. (2011a) *An Overview of Structured Decision Making: A Two-Day Course for Managers of Natural Resources*, Revised edition., Shepherdstown, WV, National Conservation Training Center.

Runge, M. C., Converse, S. J. & Lyons, J. E. (2011b) Which uncertainty? Using expert elicitation and expected value of information to design an adaptive program. *Biological Conservation.* 144, 1214–1223.

Runge, M. C., Converse, S. J., Lyons, J. E. & Smith, D. R. (2020) *Structured Decision Making: Case Studies in Natural Resource Management.* Baltimore, MD, Johns Hopkins University Press.

Sainsbury, A. W. & Carraro, C. (2023) Animal disease and conservation translocations. In Gaywood, M. J., Ewen, J. G., Hollingsworth, P. M. and Moehrenschlager, A. (eds.) *Conservation Translocations.* Cambridge, Cambridge University Press.

Seddon, P. J. (2023) The role of conservation translocations in rewilding and de-extinction. In Gaywood, M. J., Ewen, J. G., Hollingsworth, P. M. and Moehrenschlager, A. (eds.) *Conservation Translocations.* Cambridge, Cambridge University Press.

Seddon, P. J. & Armstrong, D. P. (2019) The role of translocation in rewilding. In Pettorelli, N., Durant, S. M. and du Toit, J. T. (eds.) *Rewilding.* Cambridge, Cambridge University Press.

Smuts-Kennedy, C. & Parker, K. A. (2013) Reconstructing avian biodiversity on Maungatautari. *Notornis.* 60, 93–106.

Snyder, N. F. R., Derrickson, S. R., Bessinger, S. R., et al. (1996) Limitations of captive breeding in endangered species recovery. *Conservation Biology.* 10, 338–348.

Spiegelhalter, D., Thomas, A., Best, N. & Lunn, D. (2010) *OpenBUGS User Manual, version 3.1.1.* Cambridge, MRC Biostatistics Unit.

Stadtmann, S. & Seddon, P. J. (2018) Release site selection: reintroductions and the habitat concept. *Oryx.* 54, 687–695.

Swaisgood, R. R. (2010) The conservation-welfare nexus in reintroduction programmes: a role for sensory ecology. *Animal Welfare.* 19, 125–137.

Swan, K.D., McPherson, J.M., Seddon, P.J. & Moehrenschlager, A. (2016) Managing marine biodiversity: the rising diversity and prevalence of marine conservation translocations. *Conservation Letters* 9, 239-251.

Taylor, G., Canessa, S., Clarke, R. H., et al. (2017) Is reintroduction biology an effective applied science? *Trends in Ecology and Evolution*. 11, 873–880.

Trask, A. E., Ferrie, G. M., Wang, J., et al. (2021) Multiple life-stage inbreeding depression impacts demography and extinction risk in an extinct-in-the-wild species. *Scientific Reports*. 11, 682.

Tversky, A. & Kahneman, D. (1973) Availability: a heuristic for judging frequency and probability. *Cognitive Psychology*. 5, 207–232.

Van Heezik, Y., Seddon, P. J. & Maloney, R. F. (1999) Helping reintroduced bustards avoid predation: effective anti-predator training and the predictive value of pre-release behaviour. *Animal Conservation*. 2, 155–163.

Walker, L. K., Armstrong, D. P., Brekke, P., Chauvenet, A. L. M., Kilner, R. M. & Ewen, J. G. (2013) Giving hihi a helping hand: assessment of alternative rearing diets in food supplemented populations of an endangered bird. *Animal Conservation*. 16, 538–545.

Wang, J., Brekke, P., Huchard, E., Knapp, L. A. & Cowlishaw, G. (2010) Estimation of parameters of inbreeding and genetic drift in populations with overlapping generations. *Evolution*. 64, 1704–1718.

Weiss, J. M. (1968) Effects of coping responses on stress. *Journal of Comparative and Physiological Psychology*. 65, 251–260.

Wolf, C. M., Griffith, B., Reed, C. & Temple, S. A. (1996) Avian and mammalian translocations: update and reanalysis of 1987 survey data. *Conservation Biology*. 10, 1142–1154.

5 · *Animal Disease and Conservation Translocations*

ANTHONY W. SAINSBURY AND
CLAUDIA CARRARO

5.1 Introduction

Wildlife translocations for conservation purposes (reintroduction, reinforcement, ecological replacement, and assisted colonisation) have become a key conservation tool to help restore species and ecosystem functions (IUCN, 2013). Wild animals face many threats from disease during and after conservation translocation (Griffith et al., 1993; Woodford & Rossiter, 1993; Kirkwood & Sainsbury, 1997) (for similar threats in plant translocations see Mitchell et al., this volume). Both infectious and non-infectious agents have led to disease outbreaks, and examples are outlined later in this chapter. Concomitant with a conservation translocation is the movement of a host and its parasites, a biological package (Davidson & Nettles, 1992), which may influence the health of the host following translocation, other individuals of the same species at the destination, or even other species at the destination including domestic animals and humans. Consequently, a disease risk analysis (DRA) is an important evaluation preceding a conservation translocation and can be used to inform any disease risk management, possible pre- and post-release monitoring, and ultimately whether a translocation should proceed. Disease resulting from conservation translocation also raises a welfare concern, and this is covered by Harrington et al. (this volume). Brief definitions of some common epidemiological terms used in this chapter are provided in Box 5.1.

5.2 Parasite Encounters and Infectious Disease in Conservation Translocations

A number of host/parasite changes occur during and after a wild animal translocation (Davidson & Nettles, 1992; Sainsbury & Vaughan-Higgins, 2012). For example, a non-native parasite might be introduced to the

Box 5.1 *Basic definitions of relevance to this chapter*

Disease is used in its broadest sense, defined as 'any alteration in structure or function of an organism'. Therefore, this includes not only infectious diseases but also trauma, toxicities, and nutritional diseases (non-infectious diseases).

Disease risk analysis (DRA) is a formal assessment of the risks from disease (in the case under discussion here, of a conservation translocation) by identifying the probability of occurrence and the magnitude of any negative consequences with a view to reducing the risks from disease by altering translocation protocols. Disease risk analysis is the overarching term.

Disease risk assessment is a component of disease risk analysis in which the risk of disease is estimated in four stages: release assessment, exposure assessment, consequence assessment, and risk estimation.

Disease risk management is a reasoned explanation of mitigation measures that can reduce risk and are supported by the disease risk assessment.

Disease surveillance is the systematic (continuous or repeated) measurement, collection, collation, analysis, interpretation, and timely dissemination of animal health and welfare data from defined populations. These data are essential for describing health hazard occurrence and contributing to the planning, implementation, and evaluation of risk mitigation actions (Hoinville et al., 2013).

Exposure refers to the actual contact of a living organism with an infectious agent or parasite. It is not equivalent to infection and in fact an organism can be exposed without being infected.

Hazard is defined as a biological, chemical, or physical agent, or a condition of an animal with the potential to cause disease. Hazards include infectious agents and non-infectious agents (e.g. toxins and trauma) and can be categorised as:

- **Source hazards:** those parasites (or strains of those parasites) present at the source that would be novel at the destination site. Translocated animals are a potential vehicle for introduction of these parasites to the destination site.
- **Destination hazards:** those parasites (or strains of those parasites) absent at the source but present at the destination site, and thus novel to the translocated animals.

- **Carrier hazards:** those commensal parasites to which the source population has co-adapted and co-evolved and are effectively carried by translocated animals. They may cause disease, in transit or at destination, when the host is under stress associated with translocation or is subjected to factors that affect the host–parasite relationship, such as alterations in host density.
- **Host-immunodeficiency hazards:** those parasites which, when the host population is subjected to stressors, such as those associated with conservation translocation, or factors that affect parasite dynamics, such as alterations in host density, cause disease in translocated animals in the transit or destination environment. The hosts are infected by such hazards only during transit or at the destination environment. Host-immunodeficiency hazards include parasites moved with materials such as crates, equipment, food, and water, and parasites capable of causing disease when precipitated by transit stressors.
- **Transport hazards:** those parasites that may be encountered during transport (between source and destination) which may be novel to the translocated animals and/or the release environment. Translocated animals can be a potential vehicle for these parasites.
- **Population hazards:** those infectious and non-infectious agents present during transit or at destination, not novel to the translocated animals, which could potentially have a negative impact at the population level.

Infection is defined as the 'acquisition of a potentially harmful organism by another living organism (the host)'. Infection does not necessarily indicate disease: we distinguish between a *commensal infection* which occurs without harm to the organism (without disease), and a *pathogenic infection* which occurs with harm.

Parasites are here considered in the ecological sense to include both *microparasites* (viruses, bacteria, fungi, protozoa) and *macroparasites* (helminths, ectoparasites). A parasite is an organism that lives on, or in, another living organism (the host), derives food, shelter, and other requirements, and does not benefit the host. The host may or may not suffer disease from this relationship. A *commensal parasite* is usually acquired early in life and infection can be sustained without damage although disease may be precipitated in situations of impaired

> immunity and altered host–parasite dynamics. *Infectious agent* is used synonymously with 'parasite'.
> **Pathogens** are described as any disease-causing organism.
> **Pathogenicity** represents the ability of an organism to cause disease.
> **Risk** is the likelihood, or probability, of the occurrence and the likely magnitude of the consequences of an adverse event to health.
> **Wild animal** is used to include both captive wild animals, for example in zoological collections, and free-living wild animals.

destination with the translocated host, the translocated animal might contract a non-native parasite during transport, or the translocated animals might be exposed to a novel parasite at the destination. In addition, commensal parasites might become pathogenic because the stress of translocation has affected host immune competence. Parasite species might also be 'lost' during translocation because intermediate hosts might be absent in the new habitat, transmission might not occur because low numbers of animals are translocated, or parasites might be spread unequally across a source population and un-infected animals chosen for translocation (Blackburn & Ewen, 2017). In addition, an increased number of hosts within a release area (host aggregation) may (i) affect the ecology of the host/parasite interaction and (ii) increase the probability of transmission of infectious agents between hosts (through an increased number of susceptible animals and contact rates, plus possibly increased stress), therefore increasing the basic reproduction number (R_0), the transmission potential, and thus the probability of spread of an infectious agent through a population (R_0 here is defined as the number of secondary infections caused by the introduction of a single infected animal into a population of susceptible individuals).

The best-known examples of non-native parasite introduction arise from human-mediated animal translocations not conducted for conservation reasons, for example chytrid fungus, *Batrachochytrium dendrobatidis* (Daszak et al., 2000), squirrelpox virus (Sainsbury et al., 2008), and rinderpest virus (Dobson & Hudson, 1995). In each case, the animals were translocated across geographical and ecological barriers to an ecosystem where the parasite had not previously occurred. There is substantial evidence of multiple extinctions of amphibians as a consequence of the chytrid fungal epidemic and therefore of implications for biodiversity (Scheele et al., 2019). Squirrelpox viral disease has led to population declines and local extinctions of red squirrels *Sciurus vulgaris* in the United Kingdom (Rushton et al., 2006), and has compromised subsequent attempts at red squirrel conservation translocation (Box 5.2). Rinderpest virus induced a pandemic of rinderpest disease that swept from northern to southern Africa leading to disease and the death of many millions of ungulates with knock-on effects on the ecosystem (Dobson & Hudson, 1995). The ecosystem effects were dramatically illustrated when cattle vaccination for rinderpest virus commenced in East Africa in the 1950s, and over a period of thirty years western white-bearded wildebeest *Connochaetes taurinus mearnsi* numbers recovered from approximately 500,000 to 1,500,000 in the Serengeti ecosystem with concomitant increases in spotted hyaena *Crocuta crocuta* and African lion *Panthera leo*. The population effects of non-native parasite introduction may take decades to develop, as seen with the squirrelpox virus in the United Kingdom. Occurrences of epidemic disease following non-native parasite introduction in the above non-conservation translocation examples, suggest that disease outbreaks might be more likely where conservation translocations cross geographical or ecological barriers (Bobadilla-Suarez et al., 2017).

Alternatively, translocated individuals can encounter parasites associated with diseases at the destination. Grey wolves *Canis lupus* reintroduced to the Yellowstone National Park provide a revealing example of the exposure of translocated animals to novel parasites at the destination. The wolves, sourced from British Columbia and Alberta, were exposed to five novel parasites at the destination, including canine parvovirus, canine herpesvirus, canine hepatitis virus, canine distemper virus, and *Sarcoptes scabiei* (Almberg et al., 2012). Canine distemper virus was shown to trigger epidemic mortality of wolf cubs, and sarcoptic mange was associated with population declines in some packs (Almberg et al., 2012). While the reintroduced adult wolves had been vaccinated against all four viral agents, their offspring, unvaccinated, were immunologically naïve and susceptible to disease. In some cases, the translocated animals may be naïve, and lack immunity, to these resident parasites. For example, eastern woodrats *Neotoma floridana* became infected with *Baylisascaris procyonis*, a neurotropic roundworm of raccoons *Procyon lotor*, and developed neurological disease when they were released in a former range in New York where *B. procyonis* occurs (Davidson & Nettles, 1992). Finally, there can be stressor-associated parasitic disease in translocated wild animals. For example, the known commensal parasite *Isospora normanlevinei* was associated with disease in cirl buntings *Emberiza cirlus*

> Box 5.2 *Squirrelpox in translocated red squirrels*
>
> In 1992, a 1700 ha dedicated Red Squirrel Reserve was established in Thetford Forest, East Anglia, UK. In 1996, 24 red squirrels were translocated to Thetford Forest from either Cumbria or Northumberland, initially into a 1 ha pre-release pen, and after four weeks into the surrounding forest where they were monitored through radiotelemetry. Initially, the animals seemed to have adapted well to the release environment judging by behavioural observations (Venning et al., 1997); however, approximately two months after translocation, an epidemic of squirrelpox viral disease occurred. Disease was confirmed in seven red squirrels by gross pathological description of lesions and electron microscopy of skin from lesions, over the following seven weeks (Carroll et al., 2009). The remaining reintroduced population disappeared at the same time, probably due to the same disease (Carroll et al., 2009). Squirrelpox virus causes a fatal disease in red squirrels recognisable through the exudative erythematous dermatitis surrounding the eyes and other facial areas (Figure 5.1), in axillary and inguinal areas, on the external genitalia, and on the toes of all feet, and has been reproduced experimentally (Tompkins et al., 2002). Following subsequent detailed epidemiological studies, it has become evident that the red squirrels were exposed to squirrelpox virus harboured by non-native, introduced grey squirrels *Sciurus carolinensis* that were present in the release environment. Given that the reintroduced red squirrels had already settled at the release site, stress was not believed to have been a significant contributory factor in the disease outbreak, thus suggesting the epidemic resulted from the exposure of an immunologically naïve population to a novel parasite.
>
> The example of squirrelpox virus and its effect on red squirrels is a salutary warning to those intending to undertake conservation translocations, including those described as 'rewilding' (Seddon, this volume). Conditions at destination locations may make them unsuitable for species conservation translocations. Uncontrolled release of wild animals from source populations that require translocation across geographical or ecological barriers risks the release of a non-native parasite which could lead to catastrophic mortality. Recent uncontrolled introductions of Eurasian beaver *Castor fiber* from continental Europe to the UK are an example

(Campbell-Palmer et al., this volume). We recommend that conservation translocations, especially those that cross geographical and ecological barriers, include a DRA.

Figure 5.1 A healthy red squirrel *Sciurus vulgaris* in an English woodland (a) (photo: Simon Fraser) and exudative erythematous dermatitis of the eyelids and face in a red squirrel with squirrelpox viral disease (b) (photo: Julian Chantrey). (A black and white version of this figure will appear in some formats. For the colour version, please refer to the plate section.)

during a conservation translocation, likely as a result of stress (McGill et al., 2010).

In other examples of diseases occurring after translocation the parasite encounter remains uncertain. For example, approximately 60 per cent of 200 South Island saddlebacks *Philesturnus carunculatus carunculatus* probably died of a combination of avian malaria and avian pox following translocation to two offshore islands in New Zealand (Alley et al., 2010). The South Island saddlebacks were believed to have been previously exposed to *Plasmodium elongatum* associated with avian malaria, but their susceptibility was possibly increased by concomitant avian pox. Both the avian poxviruses involved and *Plasmodium elongatum* may have been carried to the islands with the South Island saddlebacks during translocation.

5.3 Non-infectious Diseases Associated with Conservation Translocations

In addition to the risks arising from infectious diseases, translocated animals also face threats from non-infectious disease agents, which can manifest at any phase of the translocation pathway (see Section 5.4.1). For example, disease might arise from traumatic injury or stress during translocation procedures (e.g. capture, handling, transport), or through exposure to environmental pollutants and toxic compounds, or other anthropogenic factors (e.g. electrocution or persecution) at the destination (Gorman, 1999; Kock et al., 2007). Examples of the impact of lead poisoning can be seen in the reintroduced populations of the California condor *Gymnogyps californianus* (Finkelstein et al., 2012) and the red kite *Milvus milvus* (Pain et al., 2007). The effects of exposure to anticoagulant rodenticides have also been shown in the red kite (Molenaar et al., 2017).

5.4 Assessing the Risk from Disease in Translocation

Assessing the risk from disease in wild animal translocation has been reliant almost exclusively on qualitative DRA. The first published DRA method for wild animals appeared in the early 1990s described by Davidson and Nettles (1992). All succeeding methods have been based on the risk analysis framework provided by Covello and Merkhofer (1993), which has the capability to assess the risk of any human activity from, for example, climbing a ladder to generating nuclear power. Leighton (2002) published a version of this method for wild animals. At the same time, the World Organization for Animal Health (OIE) drew on the work of Covello and Merkhofer (1993) to devise and publish DRA for the importation of domestic animals, and this method has been through two evolutions (OIE, 1999; Murray et al., 2004; Bruckner et al., 2010). Sainsbury et al. (2011) provided more background on the development of DRA for wild animal translocations.

The International Union for Conservation of Nature (IUCN) published options for undertaking DRA for wild animal translocations in 2003 (Armstrong et al., 2003) and joined forces with the OIE in 2014 to publish a more advanced summary of methods, both quantitative and qualitative (OIE & IUCN, 2014). One of these, the ZSL method published by Sainsbury and Vaughan-Higgins (2012) and later modified by Bobadilla-Suarez et al. (2017) and Rideout et al. (2017), is a qualitative DRA tool that adapts the OIE framework (Murray et al., 2004) for use in wild animals undergoing conservation translocations (Sainsbury et al., 2011) by (i) using a holistic approach to hazard inclusion; (ii) defining hazards according to the interactions between the translocated host's immunity and the parasites with which the host interacts; (iii) defining hazards according to geographical and ecological barriers crossed; and (iv) considering both infectious and non-infectious hazards.

Dalziel et al. (2017) evaluated the alternative DRA methods described by Armstrong et al. (2003), McInnes (2011), and Sainsbury and Vaughan-Higgins (2012), and identified the key differences between them. Their analysis showed that the methods described by Armstrong et al. (2003) and McInnes (2011) produced results more quickly but without the transparency offered by the ZSL method, given that the latter is founded on the OIE version and provides in-text citation and an explanation and justification for decisions. The disease risk assessment component of the ZSL method utilises the tried and tested OIE method (Bruckner et al., 2010) and allows for results to be promptly revisited, and revised, as new evidence becomes available. Similarly, other components of the ZSL method are described in a way to provide transparency.

In this chapter we will provide guidance on how to complete a DRA using the ZSL method. The method involves a six-step process as follows:

1. Description of the translocation pathway.
2. Consideration of geographical and ecological barriers.
3. Hazard identification including justification of hazard status.
4. Disease risk assessment.
5. Disease risk management.
6. Disease risk communication.

The ZSL method has been applied to 27 species-specific conservation translocations and examples are published by Sainsbury and Vaughan-Higgins (2012), Jakob-Hoff et al. (2014a), Vaughan-Higgins et al. (2016), Sainsbury et al. (2017), and Peters et al. (2020). In addition, a further 23 reports are freely available from the authors (Table 5.1).

5.4.1 Translocation Pathway

It is important to understand the translocation pathway to identify where differing types of hazards (see Box 5.1 for hazard categories)

Table 5.1. *DRAs undertaken for conservation translocations using the ZSL method and available to readers by contacting the Disease Risk Analysis and Health Surveillance (DRAHS) group at ZSL.*

Species translocated	Source	Destination	Year
Short-haired bumblebee *Bombus subterraneus*	Sweden	England	2016
Wart-biter cricket *Decticus verrucivorus*	Castle Hill, England	Deep Dene, England	2016
Barberry carpet moth *Pareulype berberata*	Captive collections	England	2007
British field cricket *Gryllus campestris*	England	England	2008
Partula snail *Partulidae*	Captive collections	French Polynesia	2013
Red barbed ant *Formica rufibarbis*	Captive collections	England	2008
Chequered skipper *Carterocephalus palaemon*	Belgium	England	2017
Pool frog *Pelophylax lessonae*	Sweden	England	2005
Pool frog *Pelophylax lessonae*	England	England	2018
Mountain chicken *Leptodactylus fallax*	Captive collections	Montserrat	2020
Natterjack toad *Epidalea calamita*	England	England	2020
Sand lizard *Lacerta agilis agilis*	Captive collections	England	2003
Smooth snake *Coronella austriaca*	England	England	2011
Common European adder *Vipera berus*	Captive collections	England	2016
Big-headed turtle *Platysternon megacephalum*	Rehabilitation centres	Vietnam	2019
Golden coin turtle *Cuora trifasciata*	Rehabilitation centre	Hong Kong	2020
Cirl bunting *Emberiza cirlus*	Captive collection	England	2005
Eurasian crane *Grus grus*	Germany	England	2010
Hen harrier *Circus cyaneus*	France/Spain	England	2019
Regent honeyeater *Anthochaera phrygia*	Captive collections	Australia	2014

Table 5.1. (cont.)

Species translocated	Source	Destination	Year
White-tailed eagle *Haliaeetus leucocephalus*	Poland	England	2010
Sihek – Guam kingfisher *Todiramphus cinnamominus*	Captive collections	Palmyra Atoll (US Minor Outlying Islands)	2021
Eurasian beaver *Castor fiber*	Great Britain/ Norway	England	2020
Greater one horned rhino *Rhinoceros unicornis*	Nepal	Nepal	2020
Pine marten *Martes martes*	Scotland	England	2019
European water vole *Arvicola amphibius*	Captive collections	England	2021
Hazel dormouse *Muscardinus avellanarius*	Captive collections	England	2019

might occur. The translocation pathway should represent all locations that the animals will visit or occupy during transit, including all modes of transport, plus the source and destination sites. The translocation pathway can be overlaid with representations of the hazards, and arrows can be used to indicate at which points these hazards may potentially harm wild animals. Different types of hazards act at different stages of the translocation pathway; an early understanding of which hazards might affect any given translocation allows for risk mitigation and/or consideration of alternative translocation routes. An example can be seen in Figure 5.2, where a proposed translocation pathway for the conservation translocation of the currently extinct-in-the-wild sihek (Guam kingfisher; *Todiramphus cinnamominus*) has been visually represented. Briefly, the proposed translocation pathway for sihek involves the translocation of family-sized groups of young birds from Association of Zoos and Aquariums institutions on the North American mainland to the proposed destination, Palmyra Atoll, a US Minor Outlying Island. Eggs will be harvested from breeding pairs and artificially incubated and chicks hand-reared within a specialist quarantine environment, adhering to biosecurity measures to manage the risk from infectious agent transmission.

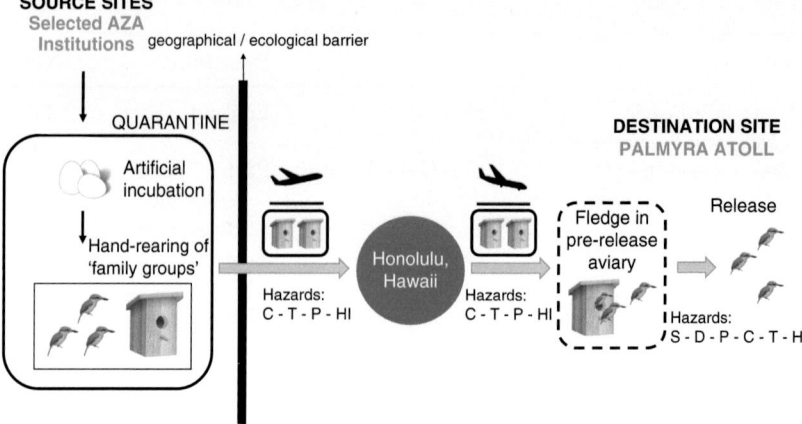

Figure 5.2 Proposed translocation pathway for sihek (Guam kingfisher *Todiramphus cinnamominus*) first trial releases to the wild. The solid line indicates quarantine (including prevention of blood-sucking vector transfer) whereas the dashed line indicates captivity without quarantine (exposure to blood-sucking vectors and external environment through aviary mesh). Hazards are indicated as C, carrier; T, transport; P, population; S, source; D, destination; HI, host immunodeficiency (see Box 5.1 for definitions of hazard categories).

5.4.2 Geographical and Ecological Barriers

'Geographical barriers' are defined as natural, environmental barriers, for example rivers and mountain ranges, whereas 'ecological barriers' may include physical, behavioural, and reproductive factors intrinsic to each animal species (Bobadilla-Suarez et al., 2017). By preventing natural movements and interactions between wild animal populations, barriers that occur between source and destination locations mean that parasites (either species or strains) may differ between the two sites. If either geographical or ecological barriers are crossed during translocation, the risk of exposure to novel parasites is increased. Animal populations that may be naïve and therefore lack immunity might consequently suffer disease. Although some wild animal species may inhabit the same geographical range, they might occupy a different ecological niche, as seen in common European adders *Vipera berus*, and might therefore not be in contact, thus harbouring different parasite complements. Barriers may also be broken when two species that inhabit separate ecological niches or geographical areas are placed in the same wild animal captive collection. Geographical and ecological barriers will be crossed during the proposed sihek translocation pathway (Figure 5.2), therefore all types of hazards may affect translocated birds and other species at the release site.

5.4.3 Hazard Identification

In undertaking hazard identification, two possible approaches can be taken; these are not mutually exclusive. In the first, the possible parasite and non-infectious agent hazards are identified from the literature, from survey, and from expert opinion. Searches can be made of both published and unpublished scientific literature: we have previously used the search engines of Web of Science, Science Direct, Zoological Record, Google Scholar, and PubMed. Surveys can be conducted of those institutions holding the species in captivity. In the second hazard identification method, the screening of source and destination populations is undertaken. Screening is a considerably more costly method, but it can be attractive in: (i) high-risk translocations in which geographical and ecological barriers will be broken, and (ii) where an understanding of parasite complement in the affected populations is weak. This screening approach, which should complement the first approach, was for example taken in the case of pool frog *Pelophylax lessonae*, as described by Sainsbury et al. (2017). Care needs to be taken when designing screening-based approaches because: (i) we do not yet have a clear understanding of the parasites that are likely to induce disease outbreaks in novel environments; (ii) there may not be validated tests available for the detection of infected or carrier animals; (iii) sensitive and specific tests may not be available; (iv) we may not have sufficient knowledge to identify parasite species; and (v), given the continuing discovery of emerging infectious diseases (Daszak et al., 2000), it is likely that many species of wild animals carry parasites of which we are, as yet, unaware.

Whichever method of hazard identification is chosen, a list of hazards is created including: (i) parasites (micro- and macro-parasites) known to be present in the species or closely related taxa, as well as multi-host parasites; and (ii) non-infectious agents. Hazards may have the potential to affect the translocated species, or the wider ecosystem at the destination. An example hazard list is shown in Table 5.2.

Through consideration of (i) geographic distribution, (ii) occurrence, (iii) pathogenesis, (iv) diseases associated with each parasite, and (v) evidence for a negative impact on population numbers, each hazard

Table 5.2. *Hazard identification list (a portion of) gathered for the disease risk analysis for common European adder* Vipera berus *reintroduction to a part of England.*

Potential hazard	Type of parasite	Hazard category	Presence in free-living wild reptiles in the UK	Presence in captive reptiles in the UK (or elsewhere)
Reptilian paramyxovirus (PMV)	Virus	Source	Unknown	Yes
Adenoviruses	Virus	Carrier	Yes	Yes
Reoviruses	Virus	Source	Unknown	Unknown (yes)
Iridoviruses	Virus	Source	Unknown	Yes
Gram negative bacteria	Bacterium	Carrier	Yes	Yes
Mycobacteria other than *Mycobacterium tuberculosis* complex	Bacterium	Population	Unknown	Yes
Chlamydia spp.	Bacterium	Source	Unknown	Unknown (yes)
Aspergillus spp. *Penicillium* spp. *Paecilomyces* spp. *Fusarium* spp.	Fungus	Carrier	Yes	Yes
Candida spp.	Fungus	Carrier	Yes	Yes
Entamoeba invadens	Protozoan	Source	Unknown	Yes

is assigned, when possible, to an appropriate hazard category as defined in Box 5.1. The category is justified using a logical, reasoned approach based on evidence from the literature. Evidence for susceptibility of the translocated species and closely related species to each potential hazard, or similar agents of disease, is considered in carrying out the evaluation. Apparent commensal parasites are also assessed because they can become pathogenic when animals are influenced by stressors. It is not considered necessary to find evidence that parasites are associated with harm to identify them as source, transport, and destination hazards because parasites can precipitate disease in novel wild animal hosts in new environments. For example, turtle bunyavirus was considered a source hazard for the translocation of golden coin turtles *Cuora trifasciata* (family Geoemydidae) from a rehabilitation centre to a release site in Hong Kong on the basis that it was associated with disease in turtles of the family Emydidae in the USA and not golden coin turtles or native turtle species in Hong Kong. Non-infectious agents or events and their association with disease are also evaluated and similarly assigned to their respective hazard category. Box 5.3 provides an example of how a hazard was justified for one conservation translocation.

The number of parasites and non-infectious hazards identified can be considerable: for example, Neimanis and Leighton (2004) identified 122 parasites of potential risk associated with a translocation of Eastern wild turkeys *Meleagris gallopavo silvestris*. Given the enormous task of undertaking qualitative DRA on many hazards, their prioritisation has been repeatedly addressed by authors tackling DRA in wild animals, for example Jakob-Hoff et al. (2014b) and OIE and IUCN (2014). Rideout et al. (2017) addressed this difficulty, concluding that source and destination hazards, particularly multi-host (generalist) microparasites with a reservoir or vector, should be prioritised for disease risk assessment in any DRA because the probability that they are associated with epidemic disease is increased. Microparasites with long infectious and incubation periods should also be high on the priority list (Rideout et al., 2017). Once Rideout et al.'s (2017) suggested priorities have been analysed, we recommend that the following are also analysed:

1. Parasites known to have been associated with disease in the species if it has been translocated in the past.
2. Parasites known to have been associated with outbreaks of disease in the species.
3. Other parasites that are source and destination hazards whether they are of known pathogenicity or not.
4. Zoonotic parasites.
5. Non-infectious agents identified.

If the geographical distribution of a parasite is unclear, it may be difficult to categorise it as a source or destination hazard. Given our limited understanding of parasite epidemiology in free-living wild animal populations it is not surprising that this problem is regularly faced when undertaking a DRA.

If a screening approach is taken for hazard identification, it may be difficult to assess the risk of disease from parasites that have not previously been identified, and the pathogenicity of the same parasites may not be evident. These difficulties were seen in the pool frog conservation

Box 5.3 *Justification of* Francisella tularensis *as a source hazard for the conservation translocation of Eurasian beaver from either Great Britain or Norway to England*

Francisella tularensis is a small, gram-negative coccobacillus that is one of five species within the *Francisella* genus, family *Francisellaceae*. It is the causative agent of tularaemia, an infectious and zoonotic septicaemic disease. Tularaemia was first described in 1911 in rodents exhibiting plague-like clinical signs (McCoy, 1911) and the bacteria were later identified after isolation from Californian ground squirrels *Otospermophilus beecheyi* (McCoy & Chapin, 1912). *F. tularensis* has since been isolated from over 250 species and is considered to have the broadest host range of all zoonotic agents (Mörner, 1992; Gyuranecz, 2012). Eurasian beavers have been implicated as reservoir hosts of *F. tularensis* and one case of human disease has been reported (Mörner & Sandstedt, 1983; Mörner, et al., 1988; Schulze et al., 2016). Tularaemia is a complex disease, and many aspects of its epidemiology are poorly understood, including transmission cycles and reservoir hosts (Hestvik et al., 2015). Mammals within the orders Lagomorpha and Rodentia are thought to be particularly important within the parasite's lifecycles (Gyuranecz, 2012).

Four subspecies of *F. tularensis* are currently recognised: *F. tularensis* subsp. *tularensis*, *F. tularensis* subsp. *holarctica*, *F. tularensis* subsp. *Mediasiatica*, and *F. tularensis* subsp. *novicida*. The moderately virulent *F. tularensis* subsp. *holarctica* is the causative agent of disease in Europe (Gyuranecz, 2012). *F. tularensis* subsp. *holarctica* is associated with aquatic ecosystems. Aquatic mammals, including Eurasian beavers, have been implicated as reservoirs of the bacterium in countries where the disease is endemic (Mörner & Sandstedt, 1983). *F. tularensis* subsp. *holarctica* can also be transmitted by blood-feeding arthropods, including mosquitoes *Aedes aegypti* and ticks *Ixodae* spp. (Výrosteková, 1993; Petersen et al., 2009; Akimana & Kwaik, 2011; Gyuranecz, 2012; Thelaus et al., 2014; Maurin & Gyuranecz, 2016). Mosquitoes become infected through the aquatic cycle during their larval stages, but are not considered to be true reservoirs as transovarial transmission has not been shown, suggesting that the infection will die with the mosquito (Petersen et al., 2009). The tick *Dermacentor reticulatus* is thought to be a true reservoir of *F. tularensis* subsp. *holarctica* and transmits the parasite between mammals in central Europe through a separate terrestrial cycle (Keim et al., 2007).

F. tularensis is widespread across continental Europe and its current geographical range encompasses the Czech Republic, Finland, France, Germany, Liechtenstein, Netherlands, Norway (personal communication, Turid Vikøren, 11th February 2020), Sweden, and Switzerland. It is also suspected to be present in Italy, Denmark, and Russia, and has previously been reported in Austria, Belgium, Bulgaria, Hungary, and Poland, although it is currently absent in these areas. The bacterium is currently considered to be absent from the UK (Donald et al., 2020) and therefore is considered a source hazard for beavers translocated from geographical areas where *F. tularensis* has been detected.

translocation from Sweden to England, and in this case the project's steering committee decided to accept the risk from disease while putting best practice mitigation measures in place (Sainsbury et al., 2017) and at the same time putting resources into post-release health surveillance to detect diseases that the DRA could not effectively evaluate.

5.4.4 Disease Risk Assessment

Disease risk assessment follows four steps: release assessment, exposure assessment, consequence assessment, and risk estimation. Across these steps we suggest using the key statements set out in Table 5.3. Each of the key statements is answered using detailed prose in a reasoned, logical approach using evidence from the literature.

Probability categories assigned to events in the release, exposure, and consequence assessments are chosen using the justifications outlined in Table 5.4. An example disease risk assessment is shown in Table 5.5.

5.4.5 Disease Risk Management

In disease risk management, measures to reduce risk are chosen where they are supported by the disease risk assessment using a reasoned explanation. There are two components to disease risk management. The first, Risk Evaluation, is a statement to verify whether the risk from disease is negligible or higher; if it is non-negligible, disease risk management

Table 5.3. *Description of the key statements in disease risk assessment for wild animal conservation translocations (adapted from Bruckner et al., 2010).*

	Key statements
Release assessment	(i) Describe the biological pathway necessary for the translocated wild animal to become exposed and infected at the source site (ii) Estimate the likelihood of the translocated wild animal being exposed and infected when translocated, thus potentially 'releasing' the hazard at destination
Exposure assessment	(i) Describe the biological pathway necessary for exposure of wild animals of the same and similar species at the destination (ii) Estimate the likelihood of that exposure occurring (iii) Estimate the likelihood of dissemination of the hazard at the destination and the population exposed
Consequence assessment	(i) Estimate the likelihood of at least one translocated animal being infected (ii) Identify the biological, environmental, and economic consequences of the release, establishment, and dissemination of the hazard and their magnitude (iii) Estimate the likelihood of the occurrence of those consequences
Risk estimation	Summarise the results and conclusions of the release, exposure, and consequence assessments and describe the overall risk estimation

Table 5.4. *Interpretation of probability categories used in disease risk assessment (table from European Food Safety Authority (EFSA) Panel on Animal Health and Welfare, 2006, adapted from Murray et al., 2004).*

Probability category	Interpretation
Negligible	Event is so rare that it does not merit being considered
Very low	Event is very rare but cannot be excluded
Low	Event is rare but does occur
Medium	Event occurs regularly
High	Event occurs very often
Very high	Event occurs almost certainly

Table 5.5. *Disease risk assessment for the carrier hazard,* Cyathostoma *spp. (nematodes), for the reintroduction of hen harriers* Circus cyaneus *to England. The disease risk analysis, of which this disease risk assessment was a part, was carried out in 2018/19 for a proposed translocation of wild chicks from either France or Spain to southern England.*

Release assessment

Adult nematodes have a tropism for air sacs, lungs, and trachea where they produce eggs in which infective third-stage larvae develop. The ova are then coughed up, swallowed, and ultimately excreted in the faeces (Lavoie et al., 1999). The parasite has either a direct lifecycle, as experimentally documented by Cole (1999), in which birds ingest fully embryonated eggs from the environment, or an indirect lifecycle in which ova are ingested by arthropods and earthworms that serve as paratenic hosts (Lavoie et al., 1999). Hen harriers, mainly juvenile individuals, are known to feed on invertebrates, especially beetles (*Geotrupes* spp., *Carabus* spp., *Corymbites* spp., Watson (1977)). In some raptor species, the consumption of infected invertebrates present in the gastrointestinal tract of prey species has been suggested as an alternative route of infection (Simpson & Harris, 1992; Lavoie et al., 1999). Usually, *Cyathostoma* spp. infection in wild birds is believed to be of low prevalence and low intensity (Fernando & Barta, 2008).

There is a low likelihood that hen harrier chicks to be reintroduced will be infected with *Cyathostoma* spp. from prey they are fed by their parents while in the nests at the source site.

Exposure assessment

Hen harriers that become infected with *Cyathostoma* spp. will effectively carry the parasite to the destination site where they will excrete the ova in their faeces, thus contributing to the maintenance of the parasite in the environment. Paratenic hosts are likely to be present at the destination site and there is a high likelihood that they will become part of the indirect lifecycle of the parasite. *Cyathostoma* spp. are capable of infecting other definitive hosts and therefore other species may be exposed and become infected at the destination. The hen harrier reintroduction itself is however predicted to have a low impact on the host–parasite dynamic at the destination, considering the ubiquity of *Cyathostoma* spp. and the low population density of the translocated hen harriers. Therefore, the likelihood of exposure and infection of other wild species at the destination because of hen harrier reintroduction is estimated to be low. The likelihood of dissemination of the parasite amongst reintroduced birds and other free-living wild species is also estimated to be low.

Consequence assessment

Cyathostoma spp. infestations in raptors have been associated with several pathological findings: diffuse pyogranulomatous air sacculitis, pneumonia, and bronchitis (Lavoie et al., 1999); severe epicarditis and pleuritis, and granulomatous bronchitis in a female subadult hen harrier (Vaughan-Higgins et al., 2013).

(cont.)

Table 5.5. (*cont.*)

As far as it is understood in raptors, *Cyathostoma* spp. infestations are unlikely, alone, to have a detrimental effect on the health of an individual but in concert with other factors (i.e. secondary bacterial infection and/or stress and immunosuppression) may contribute to fatal disease (Lacina & Bird, 2000; Krone & Cooper, 2002).

The reintroduction procedure itself could act as a stressor to the translocated hen harriers and their susceptibility to disease could increase. There is a low likelihood, however, that stress may precipitate disease in a large proportion of released hen harriers thereby leading to failure of the reintroduction. The likelihood of significant biological, environmental, and economic consequences is predicted to be low.

Risk estimation

There is a low likelihood of reintroduced hen harriers becoming infected with *Cyathostoma* spp. at the source, a low likelihood of exposure and infection of other free-living wild animals at the destination, and a low likelihood of parasite dissemination. The likelihood of significant biological, environmental, and economic consequences is predicted to be low. The overall risk is estimated as LOW.

---

methods are advisable to mitigate the risk from disease. In the second, Option Evaluation, detailed consideration is given to methods available to manage risk, drawing on the literature and consulting with wildlife health experts. The aim of disease risk management is to improve the likelihood of success of the project by ensuring translocated species are in good health through (i) good husbandry practices and health monitoring, and (ii) mitigating the potential negative effects of the hazards identified by the DRA on the translocated species and other species at the destination. When mitigating against disease outbreaks, it is important to successfully conserve commensal parasites that may be translocated with host species (Gompper & Williams, 1998; Pizzi, 2009). In fact, within parasite communities the loss of one parasite species may alter competitive interactions among the remaining species (Sousa, 1990), with potentially detrimental effects on the host. On the other hand, conservation of those commensal parasites of translocated animals will allow the development of appropriate host immune responses prior to the potential exposure to similar parasites in the wild thus hopefully reducing the susceptibility to novel infectious agents and the probability of epidemic disease.

Drawing on the disease risk assessment results, disease risk management outlines recommendations on quarantine and biosecurity protocols that should be built in and followed throughout the translocation pathway as well as recommendations on health examination, therapeutics, and preventive medicine. For example, the DRA carried out for a proposed hen harrier *Circus cyaneus* reintroduction to southern England identified intestinal coccidia as a medium-risk carrier hazard. Coccidia are commensal parasites that are frequently present in relatively low numbers in healthy animals but they may become pathogenic under stressful circumstances such as translocation. We recommended administering a coccidiostat (an anticoccidial drug used to delay the lifecycle or reduce the population of coccidia), alongside good husbandry, to minimise coccidian burdens during translocation whilst conserving the parasite, and developing immunity, in the translocated hen harriers. Developing a detailed health examination protocol is an important part of any disease risk management to ensure released wild animals are in good health.

The limitations of pre-translocation screening need to be clear to all stakeholders. Most diagnostic tests are developed to work with a specific species and are not validated for use on wild animals of the species being tested, thus possibly lacking accuracy and precision. Most tests are developed to detect infectious agents when they are associated with disease. If healthy animals are tested, as would be the case in undertaking pre-release screening, test performance may differ. If, for example, a test had 95% sensitivity and 90% specificity, and the infectious agent had a prevalence of 2% in the population, the positive predictive value (the probability of a positive test being true) would only be 16%. The difficulty is compounded if multiple infectious agents are screened: for example, Greiner and Gardner (2000) showed that if ten infectious agents were screened in a population of 100 animals, using tests of 90% sensitivity in each case, then 65% of the animals would show at least one false positive. Therefore, there is the potential for a large number of test-negative animals to be suspected of being positive for infectious agents of concern and to be unnecessarily removed from the release programme. Consequently, for some translocations we advise investing resources into ensuring that the population for release is healthy and focussing diagnostic capabilities into detecting disease in the population. Pathological examinations are particularly valuable because the focus is on the diseased animals in the population and scarce resources can be used more effectively. Over a period of years, a comprehensive pathological examination programme can develop a detailed understanding of the health of the population, providing more reliable and relevant data for decision-making. In Box 5.4 we provide an example of mitigation measures,

Box 5.4 *Disease risk management produced for the conservation translocation of the short-haired bumblebee* Bombus subterraneus

A programme to reintroduce the short-haired bumblebee started in 2012 with the first 89 individuals captured from Skane County, Sweden, arriving at Royal Holloway University, Surrey, England. Here, bumblebees entered quarantine to be subsequently released at the Royal Society for the Protection of Birds reserve in Dungeness, Kent, England. Further releases occurred in 2013, 2014, 2015, and 2016. Management options, combined with quarantine and hygiene practices, were suggested to effectively mitigate the risk associated with the hazards previously identified through DRA (Vaughan-Higgins et al., 2016). Faecal screening of the queens for four identified source hazards (the trypanosome *Crithidia bombi*, the nematode *Sphaerularia bombi*, the protozoan *Apicystis bombi*, and the microsporidian *Nosema bombi*) was recommended: (i) pre-translocation (so that positive bees could be returned to the source site); (ii) to prevent introduction of non-native parasites; and (iii) as part of the post-release health surveillance. Faecal screening in Sweden allowed early identification of the parasites, thus preventing the importation of infected queens. Biosecurity measures were set out for different stages of the translocation pathway and mainly focused on hygiene and disinfection of all equipment, clothes, and boots. Quarantine guidelines were designed to contain any significant alien infectious agent that might be harboured by a bumblebee queen while deciding on the best action to take. These guidelines are here summarised:

A dedicated, climate-controlled (20–24°C and approximately 50 per cent humidity) room at Royal Holloway will serve as quarantine facility and a minimum of 15 days quarantine is required.

A minimum number of dedicated staff will access quarantine. Staff should not have contact with non-native invertebrates in the two weeks preceding the quarantine period, or during the period when they service quarantine. No visits to zoos or other non-native invertebrate collections are permitted. Avoid contact with native invertebrates during this period. These precautions will reduce the possibility of non-native pathogens from other insects infecting short-haired bumblebees during the reintroduction programme, and non-native parasites imported with the bumblebees infecting native invertebrates.

Transport boxes and all material that accompanies the queens must be destroyed on their arrival at Royal Holloway.

A quarantine barrier should be ensured through the use of (i) dedicated personnel; (ii) dedicated clothing (overalls, boots, and gloves); (iii) dedicated labelled tools which remain in the quarantine facility; (iv) a disinfectant footbath at the barrier; (v) disinfection of all tools used with a bactericidal and virucidal disinfectant used according to manufacturer's recommendations.

All containers/boxes containing queens should be kept solely in the quarantine facility, locked from the rest of the building. On the quarantine door there should be a sign, e.g. 'Quarantine – Do Not Enter', and a coloured tape strip should be placed on the floor around the entrance to the quarantine facility to demarcate this as a 'no go' area.

Staff should change into dedicated overalls, boots, and gloves, and step into the disinfectant footbath, on entering and leaving the quarantine facility. Equipment and clothing should be stored securely at the entrance to the facility. The footbath should be a tub of 300 mm depth (200 mm disinfectant depth), located at the doorway to the quarantine facility and the disinfectant should be changed weekly, or more often if contaminated with organic matter.

Virkon® (DuPont Animal Health Solutions, Sudbury, Suffolk, UK) is the disinfectant of choice for use in footbaths, routine disinfection of cages, and outside cleaning. F10® wipes and spray (F10Biocare, Loughborough) should be used when cleaning indoors. Chlorhexidine (Hibiscrub®) is recommended for routine disinfection of hands.

All quarantine waste should be treated as clinical waste, sealed securely, and sent for incineration.

Bumblebees should be held in containers/boxes which do not allow access to any other insect genera. The quarantine facility has a sealed door to prevent the ingress of other insects. In addition, a strip of insect trap tape (chemical- and attractant-free) will be placed around the internal aspect of the door.

A thorough disinfection of all surfaces and tools must be done at completion of quarantine.

---

including biosecurity and quarantine, provided for the reintroduction of the short-haired bumblebee *Bombus subterraneus* to the UK.

5.4.6 Disease Risk Communication

Based on our work on 27 conservation translocations we recommend keeping the stakeholders in the conservation translocation regularly informed of the objectives and progress at each stage of the DRA.

The stakeholders and the steering committee for the conservation translocation will need to consider other aspects in the conservation planning process, for example an analysis of the factors contributing to species decline, whether the reasons for decline have been sufficiently managed, and whether legal requirements have been met. Therefore, only the steering committee, or a regulatory body, is in a position to make a decision on whether and how the conservation translocation should go ahead. For further information on planning and decision-making see Dalrymple and Bellis, this volume, and Ewen et al., this volume, respectively.

5.5 Post-release Health Surveillance

Given our incomplete understanding of the number and pathogenicity of parasites harboured by wild animals, post-release health surveillance can be important. A parasite of potential threat may be unknown at the time of translocation, and monitoring the health and detection of disease in the translocated population, and wild animals from similar taxonomic groups, may allow us to detect disease problems (Figure 5.3). If a disease outbreak is detected at an early stage it might not be too late to take remedial action and, for example, prevent the establishment of a pathogenic alien parasite. In addition, post-release health surveillance can give an early warning of disease induced by a destination hazard. For example, swabbing for chytrid fungus and sampling for ranavirus were included in the protocol for reintroduced pool frogs based on the high risk posed by the two agents as destination hazards. Likewise, for non-infectious hazards, remedial measures may be possible. For example, as noted earlier, the reintroduced red kite population in England has been susceptible to lead poisoning through the ingestion of carcases shot with lead ammunition during scavenging (Pain et al., 2007). Therefore, public information communications have been distributed to specific stakeholder groups to encourage the collection of carcases after shooting and reduce exposure (Rose et al., 2007). Post-release health surveillance provides evidence on which the DRA for the conservation translocation can be reassessed, hopefully to improve the outcome of subsequent translocations of the same or closely related species. We note, however, that careful consideration should be given to how monitoring is done to ensure it provides useful information over appropriate time scales. In many cases substantial resources are devoted to monitoring that provides

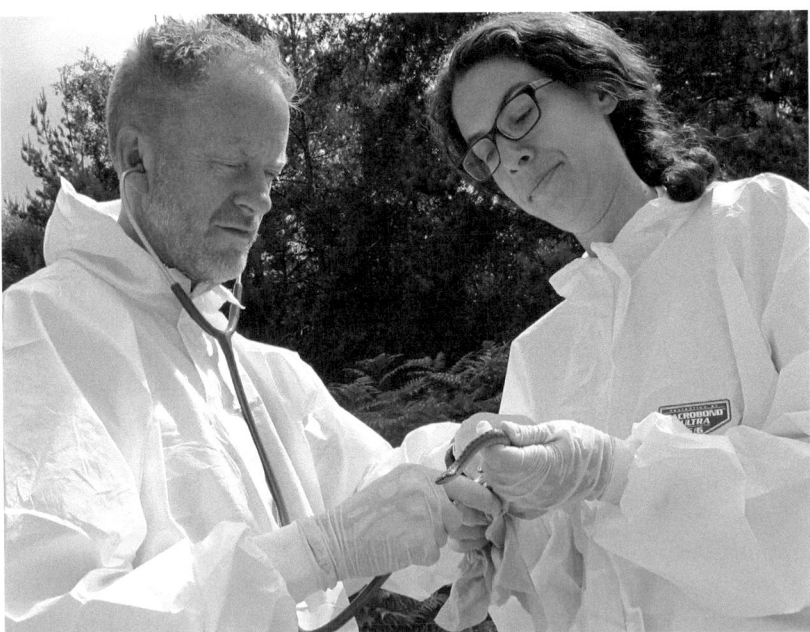

Figure 5.3 A health examination of a smooth snake *Coronella austriaca* as a component of post-release health surveillance. Note the use of personal protective equipment as a component of a biosecurity barrier to reduce the probability of transfer of parasites on to and off the release site (photo: ZSL). (A black and white version of this figure will appear in some formats. For the colour version, please refer to the plate section.)

little useful information, and this can burden conservation translocation projects. For more information on the assessment of value of information in monitoring, see Ewen et al. (this volume).

Changes to ecosystems as a consequence of translocations can take many years to develop. For example, we know that the rinderpest epidemic in Africa lasted many years and, following the initial epidemic, persistent infection influenced population sizes of ungulates with knock-on effects on carnivores (Dobson & Hudson, 1995). Similarly, the effects of squirrelpox virus on the UK ecosystem are still being enacted today, over 140 years since the probable date of the virus's introduction (Sainsbury et al., 2008). While neither of these cases are conservation translocations, they illustrate the dilemmas in choosing appropriate time scales and resources for post-release health surveillance.

5.6 Conclusion

Given that every wild animal represents a biological package of the host and its parasites, conservation translocations inevitably result in disturbance in host–parasite relationships and some of these may precipitate disease outbreaks in translocated species and/or species at the destination. Small populations of translocated species are also at risk from non-infectious diseases such as toxicities. Given these threats, methods have been developed to assess the risk from disease in conservation translocations. Our approach involves a transparent method suitable for the analysis of the risks from disease in the conservation translocation of wild animals. DRAs have been carried out for 27 conservation translocations and the associated reports are available for conservation scientists, practitioners, and policymakers to inform further work. A crucial problem is limited knowledge of the full complement of parasites harboured by wild animals prior to translocation, and their pathogenicity in free-living populations, which restricts full assessment of the risks. Until we have a better understanding of the parasite complement of wild animals, and parasite pathogenicity, carefully planned post-release health surveillance will improve the knowledge base from which DRAs can draw.

5.7 Key Messages

- Disease outbreaks may be a threat to the outcome of conservation translocations and disease risk analysis is best completed before translocation.
- Disease risk analysis is hampered by knowledge of the full complement and pathogenicity of parasites harboured by wild animals.

References

Akimana, C. & Kwaik, Y. A. (2011) Francisella-arthropod vector interaction and its role in patho-adaptation to infect mammals. *Frontiers in Microbiology*. 2, 1–15.

Alley, M. R., Hale, K. A., Cash, W., Ha, H. J. & Howe, L. (2010) Concurrent avian malaria and avipox virus infection in translocated South Island saddlebacks *Philesturnus carunculatus carunculatus*. *New Zealand Veterinary Journal*. 58, 218–223.

Almberg, E. S., Cross, P. C., Dobson, A. P., Smith, D. W. & Hudson, P. J. (2012) Parasite invasion following host reintroduction: a case study of Yellowstone's wolves. *Philosophical Transactions of The Royal Society B-Biological Sciences*. 367, 2840–2851.

Armstrong, D., Jakob-Hoff, R. & Seal, U. S. (eds.) (2003) *Animal Movements and Disease Risk: A Workbook*. Apple Valley, CA, Conservation Breeding Specialist Group (SSC/IUCN).

Blackburn, T. G. & Ewen, J. G. (2017) Parasites as drivers and passengers of human-mediated biological invasions. *Ecohealth*. 14, 61–73.

Bobadilla-Suarez, M., Ewen, J. G., Groombridge, J. J., et al. (2017) Using qualitative disease risk analysis for herpetofauna conservation translocations transgressing ecological and geographical barriers. *Ecohealth*. 14, S47–S60.

Bruckner, G., MacDiarmid, S., Murray, N., et al. (2010) *Handbook on Import Risk Analysis for Animals and Animal Products*. Paris, Office International des Epizooties.

Campbell-Palmer, R., Bauer, A., Jones, S., Ross, B. & Gaywood, M. J. (2023) The return of the Eurasian beaver to Britain: the implications of unplanned releases and the human dimension. In Gaywood, M. J., Ewen, J. G., Hollingsworth, P. M. and Moehrenschlager, A. (eds.) *Conservation Translocations*. Cambridge, Cambridge University Press.

Carroll, B., Russell, P., Gurnell, J., Nettleton, P. & Sainsbury, A. W. (2009) Epidemics of squirrelpox virus disease in red squirrels *Sciurus vulgaris*: temporal and serological findings. *Epidemiology and Infection*. 137, 257–265.

Cole, R. (1999) Tracheal worms. In FrIend, M. & Franson, J. C. (eds.) *Field Manual of Wildlife Diseases: General Field Procedures and Diseases of Birds*. Madison, WI, United States Geological Survey, pp. 229–232.

Covello, V. T. & Merkhofer, M. W. (1993) *Risk Assessment Methods: Approaches for Assessing Health and Environmental Risks*. New York, Plenum Press.

Dalrymple, S. E. & Bellis, J. M. (2023) Conservation translocations: planning and the initial appraisal. In Gaywood, M. J., Ewen, J. G., Hollingsworth, P. M. & Moehrenschlager, A. (eds.) *Conservation Translocations*. Cambridge, Cambridge University Press.

Dalziel, A. E., Sainsbury, A. W., McInnes, K., Jakob-Hoff, R. & Ewen, J. G. (2017). A comparison of disease risk tools for conservation translocations. *EcoHealth*. 14, S30–S41.

Daszak, P., Cunningham, A. A. & Hyatt, A. D. (2000) Emerging infectious diseases of wildlife – threats to biodiversity and human health. *Science*. 287, 443–449.

Davidson, W. R. & Nettles, V. F. (1992) Relocation of wildlife: identifying and evaluating disease risks. *Transactions of the North American Wildlife and Natural Resources Conference*. 57, 466–473.

Dobson, A. P. & Hudson, P. J. (1995) Microparasites: observed patterns in wild animal populations. In Grenfell, B. T. and Dobson, A. P. (eds.) *Ecology of Infectious Diseases in Natural Populations*. Cambridge, Cambridge University Press, pp. 52–89.

Donald, H., Common, S. & Sainsbury, A. W. (2020) Disease risk analysis for the conservation translocation of the Eurasian beaver (*Castor fiber*) to England. Natural England Commissioned Report NECR345, Peterborough.

European Food Safety Authority (EFSA) (2006) Opinion of the Scientific Panel Animal Health and Welfare (AHAW) related with the Migratory Birds and their Possible Role in the Spread of Highly Pathogenic Avian Influenza. *The EFSA Journal*. 357, 1–46.

Ewen, J. G., Canessa, S., Converse, S. J. & Parker, K. A. (2023) Decision-making in animal conservation translocations: biological considerations and beyond. In Gaywood, M. J., Ewen, J. G., Hollingsworth, P. M. and Moehrenschlager, A. (eds.) *Conservation Translocations*. Cambridge, Cambridge University Press.

Fernando, M. A. & Barta, J. R. (2008) Tracheal worms. In Atkinson, C. T., Thomas, N. J. and Hunter, D. B. (eds.) *Parasitic Diseases of Wild Birds*. Ames, IA, Wiley Blackwell, pp. 343–354.

Finkelstein, M. E., Doak, D. F., George, D., et al. (2012) Lead poisoning and the deceptive recovery of the critically endangered California condor. *PNAS*. 109, 11449–11454.

Gompper, M. E. & Williams, E. S. (1998) Parasite conservation and the black-footed ferret recovery program. *Conservation Biology*. 12, 730–732.

Gorman, M. (1999) Oryx go back to the brink. *Nature*. 398, 190.

Greiner, M. & Gardner, I. A. (2000) Epidemiologic issues in the validation of veterinary diagnostic tests. *Preventive Veterinary Medicine*. 45, 3–22.

Griffith, B., Scott, J. M., Carpenter, J. W. & Reed, C. (1993) Animal translocations and potential disease transmission. *Journal of Zoo and Wildlife Medicine*. 24, 231–236.

Gyuranecz, M. (2012) Tularaemia. In Gavier-Widén, D., Duff, J. P. and Meredith, A. (eds.) *Infectious Diseases of Wild Mammals and Birds in Europe*. Oxford, Wiley-Blackwell, pp. 303–309.

Harrington, L. A., Lloyd, N. & Moehrenschlager, A. (2023) Animal welfare, animal rights, and conservation translocations: moving forward in the face of ethical dilemmas. In Gaywood, M. J., Ewen, J. G., Hollingsworth, P. M. and Moehrenschlager, A. (eds.) *Conservation Translocations*. Cambridge, Cambridge University Press.

Hestvik, G., Warns-petit, E., Smith, L. A., et al. (2015) The status of tularemia in Europe in a one-health context: a review. *Epidemiology and Infection*. 143, 2137–2160.

Hoinville, L. J, Alban, L., Drewe, J. A., et al. (2013) Proposed terms and concepts for describing and evaluating animal-health surveillance systems. *Preventive Veterinary Medicine*. 112, 1–12.

IUCN (2013) *Guidelines for Reintroductions and Other Conservation Translocations. Version 1.0*. Gland, Switzerland, IUCN Species Survival Commission.

Jakob-Hoff, R., Carraro, C., Sainsbury, A. W., Ewen, J. & Canessa, S. (eds.) (2014a) *Regent Honeyeater Disease Risk Analysis*. Apple Valley, MN, IUCN SSC Conservation Breeding Specialist Group.

Jakob-Hoff, R. M., MacDiarmid, S. C., Lees, C., et al. (2014b) *Manual of Procedures for Wildlife Disease Risk Analysis*. Paris, World Organisation for Animal Health, published in association with the International Union for Conservation of Nature and the Species Survival Commission, 160 pp.

Keim, P., Johansson, A., & Wagner, D. M. (2007) Molecular epidemiology, evolution, and ecology of *Francisella*. *Annals of the New York Academy of Sciences*. 1105, 30–66.

Kirkwood, J. K. & Sainsbury, A. W. (1997) Diseases and other considerations in wildlife translocations and releases. Proceedings of the World Association of Wildlife Veterinarians. Symposium on Veterinary Involvement with Wildlife Reintroduction and Rehabilitation. Ballygawley, WAWV, pp. 12–16.

Kock, R. A., Soorae, P. S., & Mohammed, O. B. (2007) Role of veterinarians in re-introductions. *International Zoo Yearbook*. 41, 24–37.

Krone, O. & Cooper, J. E. (2002) Parasitic diseases. In Cooper, J. E. (ed.) *Birds of Prey. Health and Diseases*. Oxford, Blackwell Science Ltd., pp. 105–120.

Lacina, D. & Bird, D. M. (2000) Endoparasites of raptors. A review and an update. In Lumeij, J. T., Remple, J. D., Redig, P. T., Lierz, M. and Cooper, J. E. (eds.) *Raptor Biomedicine III*. Lake Worth, FL, Zoological Education Network, Inc, pp. 65–99.

Lavoie, M., Mikaelian, I., Sterner, M., et al. (1999) Respiratory nematodiases in raptors in Quebec. *Journal of Wildlife Diseases*. 35, 375–380.

Leighton, F. A. (2002) Health risk assessment of the translocation of wild animals. *Revue Scientifique et Technique*. 21, 187–195.

Maurin, M. & Gyuranecz, M. (2016) Tularaemia: clinical aspects in Europe. *The Lancet Infectious Diseases*. 16, 113–124.

Mccoy, G. W. (1911) A plague-like disease of rodents. *Public Health Bulletin* 43, 53–71.

Mccoy, G. W. & Chapin, C. W. (1912) Bacterium tularense, the cause of a plague-like disease of rodents. *Public Health Bulletin*. 53, 17–23.

McGill, I. S., Feltrer, Y., Jeffs, C., et al. (2010) Isosporoid coccidiosis in translocated cirl buntings *Emberiza cirlus*. *Veterinary Record*. 167, 656–660.

McInnes, K. (2011) *Translocation Disease Risk Management Process*. Wellington, New Zealand, New Zealand Department of Conservation.

Mitchell, R., Green, S. & Hollingsworth, P. M. (2023) Plant health, biosecurity and conservation translocations. In Gaywood, M. J., Ewen, J. G., Hollingsworth, P. M. and Moehrenschlager, A. (eds.) *Conservation Translocations*. Cambridge, Cambridge University Press.

Molenaar, F. M., Jaffe, J. E., Carter, I., et al. (2017) Poisoning of reintroduced red kites (*Milvus milvus*) in England. *European Journal of Wildlife Research*. 63, 94.

Mörner, T. (1992) The ecology of tularaemia. *Revue Scientifique et Technique*. 11, 1123–1130.

Mörner, T. & Sandstedt, K. (1983) A serological survey of antibodies against *Francisella tularensis* in some Swedish mammals. *Nordisk Veterinaermedicin*. 35, 82–85.

Mörner, T., Sandstrom, G. & Mattsson, R. (1988) Comparison of serum and lung extracts for surveys of wild animals for antibodies to *Francisella tularensis* biovar *palaearctica*. *Journal of Wildlife Diseases*. 24, 10–14.

Murray, N., MacDiarmid, S. C., Wooldridge, M., et al. (2004) Handbook on Import Risk Analysis for Animals and Animal Products. Volume 1. Introduction and Qualitative Risk Analysis. Paris, OIE, World Organisation for Animal Health.

Neimanis, A. P. & Leighton, F. (2004) Health Risk Assessment for the Introduction of Eastern Wild Turkeys *Meleagris gallopavo silvestris* into Nova Scotia. Saskatoon, Canadian Cooperative Wildlife Health Center, 62 pp.

OIE (1999) *International Animal Health Code*. Paris, Office International des Epizooties.

OIE (World Organisation for Animal Health) & IUCN (International Union for Conservation of Nature) (2014) Guidelines for Wildlife Disease Risk Analysis. Paris, OIE, published in association with the IUCN and the Species Survival Commission, 24 pp.

Pain, D. J., Carter, I., Sainsbury, A. W., et al. (2007) Lead contamination in captive and free-living red kites *Milvus milvus* in England. *Science of the Total Environment*. 376, 116–127.

Peters, H., Sadaula, A., Masters, N. & Sainsbury, A. W. (2020) Risks from disease caused by *Mycobacterium orygis* as a consequence of Greater one-horned Rhinoceros (*Rhinoceros unicornis*) translocation in Nepal. *Transboundary and Emerging Diseases*. 67, 711–723.

Petersen, J. M., Mead, P. S. & Schriefer, M. E. (2009) *Francisella tularensis*: An arthropod-borne pathogen. *Veterinary Research*. 40, 7.

Pizzi, R. (2009) Veterinarians and taxonomic chauvinism: the dilemma of parasite conservation. *Journal of Exotic Pet Medicine*. 18, 279–282.

Rideout, B., Sainsbury, A. W. & Hudson, P. J. (2017) Which parasites should we be most concerned about in wildlife translocations? *Ecohealth*. 14, S42–S46.

Rose, C., Carter, I., Newbery, P., et al. (2007) The effects of feeding red kites in England and guidance on best practice. Information Note. Chilterns Conservation Board, Natural England, RSPB, Southern England Kite Group, and ZSL.

Rushton, S. P., Lurz, P. W. W., Gurnell, J., et al. (2006) Disease threats posed by alien species: the role of a poxvirus in the decline of the native red squirrel in Britain. *Epidemiology and Infection*. 134, 521–533.

Sainsbury, A. W. & Vaughan-Higgins, R. J. (2012) Analyzing disease risks associated with translocations. *Conservation Biology*. 26, 442–452.

Sainsbury, A. W., Deaville, R., Lawson, B., et al. (2008) Poxviral disease in red squirrels *Sciurus vulgaris* in the UK: spatial and temporal trends of an emerging threat. *Ecohealth*. 5, 305–316.

Sainsbury, A. W., Armstrong, D. P. & Ewen, J. G. (2011) Methods of disease risk analysis for reintroduction programmes. In Ewen, J. G., Armstrong, D. P., Parker, K. A. and Seddon, P. J. (eds.) *Reintroduction Biology: Integrating Science and Management*. Oxford, Wiley-Blackwell.

Sainsbury, A. W., Yu-Mei, R., Ågren, E., et al. (2017) Disease risk analysis and post-release health surveillance for a reintroduction programme: the pool frog *Pelophylax lessonae*. *Transboundary and Emerging Diseases*. 64, 1530–1548.

Scheele, B. C., Pasmans, F., Skerratt, L. F., et al. (2019) Amphibian fungal panzootic causes catastrophic and ongoing loss of biodiversity. *Science*. 363, 1459–1463.

Schulze, C., Heuner, K., Myrtennäs, K., et al. (2016) High and novel genetic diversity of *Francisella tularensis* in Germany and indication of environmental persistence. *Epidemiology and Infection*. 144, 3025–3036.

Seddon, P. J. (2023) The role of conservation translocations in rewilding and de-extinction. In Gaywood, M. J., Ewen, J. G., Hollingsworth, P. M. and Moehrenschlager, A. (eds.) *Conservation Translocations*. Cambridge, Cambridge University Press.

Simpson, V. R. & Harris, E. A. (1992) *Cyathostoma lari* (Nematoda) infection in birds of prey. *Journal of the Zoological Society of London*. 227, 655–659.

Sousa, W. P. (1990) Spatial scale and the processes structuring a guild of large trematode parasites. In Esch, G. W., Bush, A. O. and Aho, J. M. (eds.) *Parasite Communities: Patterns and Processes*. London, Chapman and Hall.

Thelaus, J., Andersson, A., Broman, T., et al. (2014) *Francisella tularensis* subspecies *holarctica* occurs in Swedish mosquitoes, persists through the developmental stages of laboratory-infected mosquitoes and is transmissible during blood feeding. *Microbial Ecology.* 67, 96–107.

Tompkins, D. M., Sainsbury, A. W., Nettleton, P., Buxton, D. & Gurnell, J. (2002) Parapoxvirus causes a deleterious disease in red squirrels associated with UK population declines. *Proceedings of the Royal Society, London Series B.* 269, 529–533.

Vaughan-Higgins, R., Murphy, S., Carter, I., et al. (2013) Fatal epicarditis in a hen harrier (*Circus cyaneus*) a red-listed bird of high conservation concern in Britain associated with *Cyathostoma species and Escherichia coli* infection. *Veterinary Record.* 173, 477.

Vaughan-Higgins, R. J., Sainsbury, A. W., Beckmann, K. & Brown, M. J. F. (2016) Disease risk analysis for the reintroduction of the short-haired bumblebee *Bombus subterraneus* to the UK. Natural England Commissioned Reports, number 216.

Venning, T., Sainsbury, A. W. & Gurnell, J. (1997) Red squirrel translocation and population reinforcement as a conservation tactic. In Gurnell, J. and Lurz, P. W. W. (eds.) *The Conservation of Red Squirrels, Sciurus vulgaris.* London, Peoples Trust for Endangered Species, pp. 124–144.

Výrosteková, V. (1993) Transstadial transmission of *Francisella tularensis* in the tick, *Ixodes ricinus*, infected during the larval stage. *Ceskoslovenska Epidemiologie, Mikrobiologie, Imunologie.* 42, 71–75.

Watson, D. (1977) *The Hen Harrier.*, Berkhamsted, T. & A. D. Poyser.

Woodford, M. W. & Rossiter, P. B. (1993) Disease risks associated with wildlife translocation projects. *Revue Scientifique et Technique.* 12, 115–135.

6 · *Animal Welfare, Animal Rights, and Conservation Translocations: Moving Forward in the Face of Ethical Dilemmas*

LAUREN A. HARRINGTON, NATASHA LLOYD, AND AXEL MOEHRENSCHLAGER

6.1 Contrasts and Conflicts

Animal welfare, animal rights, and animal conservation are each centred around concern for animals. Yet differences in the primary focus of these distinct, but related, disciplines can lead to conflict. Conflict arises because animal welfare and animal rights focus on individuals, while conservation is concerned with collectives: principally populations or species.

Animal welfare emphasises duties to individual non-human animals (Minteer & Collins, 2005): it requires that we treat animals in a humane manner, avoiding unnecessary suffering (Singer, 2003). Animal welfare proponents believe that beings with 'interests' (those who are capable of enjoyment or suffering) deserve to have those interests (i.e. an interest in avoiding painful experiences) taken into account.

Animal rights differs from animal welfare in that it is concerned with the 'rights' of individual non-human animals (Minteer & Collins, 2005). Animal rights proponents believe that violation of those rights (i.e. the right to life) is wrong, regardless of whether or not the animal actually suffers, and regardless of the broader consequences (Regan, 2003).

Conservation is motivated by an environmental ethic that values, and emphasises duties to, the natural environment including, but not limited to, the individual animals within it (Rolston, 1988; Minteer & Collins, 2005). Conservation biologists are concerned with the long-term viability of populations, and the continued existence of species, biotic communities, and ecosystems as wholes, with the aim of protecting and preserving biological diversity (Soulé, 1985).

Species and populations are, of course, made up of individuals, and therefore species well-being is not entirely separate from individual well-being (Rolston, 1988). It may sometimes, or even often, be the case that animal ethics and environmental ethics are natural allies (Callicott, 1998, 2011). In these cases, there is no conflict. However, there are also situations in conservation where, for example, the killing of animals is required (e.g. to reduce population size or to remove invasive species). Such actions violate animal rights, but not necessarily animal welfare if the killing is carried out humanely (Webster, 1994; although there is some debate on this position, see Yeates, 2010).

Conservation translocation (which includes reintroductions and other movements of animals for conservation purposes; IUCN, 2013) requires direct manipulation of and sometimes considerable risk (and thus potential suffering) to the individual animals involved. Mortality rates are often high, especially in reintroductions of captive-bred animals (Jule et al., 2008). Conservation translocation therefore exemplifies some of the tensions between animal welfare, animal rights, and conservation. In recent years there has been an exponential increase in the number of conservation translocations carried out globally (Seddon & Armstrong, 2016). In parallel, animal welfare has been recognised as an emerging issue in conservation (Sutherland et al., 2015; see also Sekar & Shiller, 2020). Also, animal rights, in the context of potential conflict between individual- and population-level concerns in conservation practice (although not a new concept; Callicott, 1980, 1988, 1998, 2011), has received considerable attention in the conservation literature under the newly coined term 'compassionate conservation' (e.g. Wallach et al., 2018).

Compassionate conservation incorporates animal welfare and animal rights concerns. It has four guiding principles: do no harm, individuals matter, value all wildlife, and peaceful coexistence (Bekoff, 2013a). Some of the proponents of compassionate conservation appear to take a pragmatic view, suggesting that the aim is to 'merge the protection of animals and nature for improved conservation outcomes' (Baker, 2017). With respect to conservation translocation specifically, Baker (2017) states that 'for reasons moral and material, translocation biology offers a clear case for compassionate conservation'. The 'material' case is simply that 'greater consideration of animal welfare, although important in its own right, also has considerable potential to contribute to conservation success' (Harrington et al., 2013). In other words, an animal in a good state of welfare is more likely to survive, to reproduce, and thus to contribute to the re-establishment of the population (Harrington et al., 2013).

However, beyond legal requirements and obligations (which will vary among countries, species, and other factors, such as whether animals require transport by air) (Trouwborst et al., this volume), deciding on what is (ethically) the 'right' action is not always straightforward. Gray (2018), for example, suggests that we consider whether [conservation] actions are 'justified, humane and effective', and advises that when one of these is negative we should consider whether the action is 'worth' taking. The difficulty here lies in the judgement of 'worth', which can depend on the balance between two potentially competing moral obligations, at an extreme: to prevent the extinction of species (or other ecological entity), and to avoid harm to individual animals. Sometimes, intervening and doing 'no' harm may simply be impossible (Gray, 2018), and the decision as to whether a particular intervention is justified or not depends on whether one values species or individuals most highly. Conversely, 'doing nothing' (e.g. not killing an invasive predator) can also result in significant harm to individuals and species (e.g. of the prey species), and the wider ecosystem. These types of dilemmas can be acute in conservation translocations.

Bekoff (2013b), in particular, is critical of reintroductions: he highlights the cases of lynx *Lynx canadensis* reintroduced into Colorado where there was not enough food for them, and the deaths of 160 naïve golden lion tamarins *Leontopithecus rosalia* in Brazil. Where a project is poorly designed or executed, the majority of conservationists, animal welfarists, and compassionate conservationists would presumably agree that the action was 'wrong', although perceptions of what constitutes success or failure of a project can differ among stakeholder groups. The real conflict between welfare and conservation ethics occurs where harm to individuals cannot be avoided to achieve a conservation goal. Bekoff (2013b), for example, is particularly critical of prey species being bred purely to provide food, or experience of killing, for captive-bred predators even though such experience may be critical for the survival of these predators once released into the wild (e.g. Vargas & Anderson, 1998). In these 'hard cases', strict adherence to the principles of compassionate conservation as recently articulated has the potential to be seriously damaging to conservation (e.g. Hayward et al., 2019; Johnson et al., 2019). Callen et al. (2020), for example, point out that neither captive breeding nor 'forced movements' (i.e. translocations) would be permitted under a compassionate conservation approach. Mainstream conservationists are concerned that the compassionate conservation movement holds considerable sway over the general public (where conservation and welfare messages are easily confused), and that, by failing to acknowledge the conservation disbenefits (or the welfare harms to 'other' individual animals), it can lead to public misunderstanding of complex issues and lack of support for conservation interventions.

Throughout this chapter, we consider: First, what can and should be done to improve animal welfare in translocations? A systematic review of conservation translocations of captive-bred and wild-caught animals (Harrington et al., 2013) found that only six per cent of 199 project reports reviewed explicitly referred to animal welfare, despite common circumstances (e.g. high mortality rates, loss of animals, or disease) that may have led to negative welfare states for at least some of the individuals involved. That is not to say that practitioners neglect animal welfare, but a lack of explicit reporting of animal welfare issues encountered during projects thwarts learning (Berger-Tal et al., 2020). Second, under what circumstances do animal ethics (animal welfare and animal rights) truly conflict with conservation ethics? Moral dilemmas in translocations are common, but they are rarely discussed in the conservation translocation literature. On a practical level, our aim is to encourage open communication among practitioners to facilitate improvement of methods and to reduce the risk of repeating the same mistakes, and between practitioners and the wider public to improve understanding of the processes involved. On an ethical level, we do not suggest that we have the answers, nor do we offer a full analysis of these competing interests; rather we hope to encourage broad discussion of these difficult issues with a view to identifying a way forward.

We start by presenting a survey of conservation translocation practitioners, exploring the animal welfare issues actually experienced (the term 'animal welfare' in this case is interpreted broadly), and practitioners' views on these wider concerns. We go on to consider how practitioners can address animal welfare issues in conservation translocations at all stages of the process. Finally, we discuss approaches for dealing with some of the issues that might be encountered, focusing specifically on the more difficult ethical dilemmas that practitioners are confronted with.

6.2 Practitioner Experiences and Views

Many of the issues raised in this chapter occur at the intersection of scientific research (e.g. animal physiology, behaviour, survival, and the multifarious factors they are affected by), practical implementation (e.g. what type of animal marking methods to use, how best to monitor animal movements, when and where to release animals), and ethical considerations (i.e. 'what *should* be done' to best comply with both conservation and animal ethics). As such, the types and range of issues that affect or involve animal welfare, or more broadly, animal ethics, are diverse, and – importantly – whilst deliberations on these issues may be found in scientific journals they are more likely to be found in internal reports or documents produced for ethical and/or animal welfare review boards, and even more likely to be confined to undocumented internal verbal discussions amongst team members. To gain insight into the animal welfare and animal ethics issues encountered in modern-day translocation projects, informed by on-the-ground experience, we carried out an online survey of conservation translocation practitioners. We sought specifically to identify the types of issues encountered, to assess whether or not practitioners believed that they were successful in mitigating them, and to ascertain if they perceived an increased expectation to consider animal welfare in their work. Finally, by posing a number of different scenarios representing a combination of conservation, animal welfare, and other ethical (i.e. animals rights) issues, we sought to elucidate whether and to what extent practitioners perceive, or are concerned by, issues beyond a straightforward obligation to achieve conservation objectives.

The questionnaire was distributed to members of the Conservation Translocation Specialist Group (CTSG) of the IUCN Species Survival Commission (SSC) as well as most primary authors from the Global Reintroduction Perspectives Case studies (Soorae, 2011, 2013, 2016, 2018), via an online survey; 41 practitioners responded, 38 of whom had been involved with translocations of animals. The survey response rate was low (<30 per cent), in part because of only a single send-out, and so the following results should not be considered to be representative of all conservation translocation projects. However they do provide qualitative insights into some of the core issues that might warrant further attention as this field develops. Collectively, questionnaire respondents had conducted translocations on all continents, in projects run by non-governmental organisations (NGOs), government bodies, universities, zoos, and industry, involving both captive-bred and wild-sourced animals. Half of the respondents had 10 or more years' experience in a variety of project roles, 45 per cent as project manager. Thirty-one of the respondents said that they had experienced animal welfare issues in their conservation translocation work.

6.2.1 What Animal Welfare Issues Did Practitioners Encounter?

We initially presented respondents with a number of possible animal welfare issues that might be encountered at each stage of a translocation project and asked if they had ever experienced the issue (Table 6.1), rather than asking open-ended questions. While this imposed our own view of potential animal welfare issues on questionnaire respondents, it allowed us to identify some of the actual issues encountered unbiased by differences in the personal views of the respondents as to precisely what constitutes an animal welfare issue. For example, we included mortality at various translocation stages; this in itself is not usually considered a 'welfare' issue but should be in cases where animal suffering precedes death. We also included 'unknown fate' post release, which does not necessarily equate to poor welfare but does imply uncertainty about the well-being of the individuals involved, which may have suffered during the process. Our specified issues were not intended to be exhaustive but rather to represent a range of potential animal welfare issues that could be encountered.

Bearing in mind the limitations of this brief survey, and that we might have failed to include other frequently experienced issues, almost all issues we asked about were experienced to some extent before, during, or after release. The most commonly experienced potential animal welfare issues involved mortality at the release site due to previously unknown (or inadequately mitigated) threats, distress caused to wild-sourced individuals during capture for translocation, and distress caused by methods of transport (Table 6.1). With respect to the latter two, distress (which occurs when stress is severe, or prolonged, or both) can have serious impacts on health and function (Teixeira et al., 2007; Linklater & Gedir, 2011; Ewen et al., this volume) even in the absence of physical injury.

Post release, the most commonly reported potential animal welfare issues were declining physical health of the released animals (i.e. reduced body condition, physical injuries), mortality in the first few weeks post release, and/or distress, injury, or death of individuals as a result of re-capture, restraint, or collection of samples (e.g. blood samples) for post-release monitoring (Table 6.1).

The most commonly experienced animal welfare issue reported for animals that were not the subject of the translocation, but that were affected by it, was suffering of predators controlled at the release site (Table 6.1). The use of prey species to train captive-bred predators to

Table 6.1. *Potential animal welfare issues presented to survey respondents. Data shown are the percentages of respondents that reported having experienced each issue, and shading from light to dark reflects categories of relatively low (<15%), medium (>15%, <30%), and high (>30%) occurrence (n = 38 respondents who had experience of animal translocations).*

Specified issue★	% response
Issues experienced prior to release (wild-caught individuals)	
Suffering due to confinement or crowding in captivity	34%
Death caused by trapping or capture	34%
Injury caused by trapping or capture	32%
Distress caused by trapping or capture	42%
Negative welfare impacts on animals at source due to removal of other individuals for translocation	5%
Issues experienced prior to release (captive-bred individuals)	
Negative welfare impacts due to unsuitable diet	21%
Mortality (post release) due to lack of training or preparation	8%
Distress, injury, or death experienced by target species during anti-predator training	11%
Distress, injury, or death experienced by prey species used to train predators (i.e. live prey provisioning)	8%
Negative welfare impacts on animals at source due to removal of other individuals for captive breeding	3%
Issues associated directly with release	
Death caused by marking or tracking methods	11%
Injury caused by marking or tracking methods	24%
Distress caused by marking or tracking methods	18%
Social species distressed due to being released without their social group	16%
Solitary species distressed due to being released in a group	0%
Mortality caused by transport method	21%
Injury caused by transport method	24%
Distress caused by transport method	42%
Issues experienced post release	
Unknown fate, due to lack of monitoring	13%
Distress or stress associated with post-release support provided	5%
Poor physical health	26%
Distress, injury, or death caused by trapping and handling for monitoring	21%
Negative welfare impacts on resident populations due to effects of target species	8%

Table 6.1. (*cont.*)

Specified issue*	% response
Disease in released or resident animals	5%
Poor reproduction	18%
Mortality in the first few weeks post release	21%
Issues associated with the release site	
Mortality due to unknown or inadequately mitigated threats at the release site	45%
Mortality due to reduced dispersal or migration in fragmented habitats	18%
Suffering of target or non-target species due to lethal control	18%
Negative welfare impacts on other species due to artificial habitat enhancement	3%

*Where we refer to the death or mortality of animals, we consider this a 'potential' animal welfare issue due to the suffering that may precede death; otherwise, death in the absence of suffering may be considered an animal rights concern (see text).

hunt was reported as an issue by only three respondents, but insufficient pre-release training for captive-bred animals, more generally, was also considered a problem (below). Predator control at release sites and the use of live prey for training purposes are both potentially controversial practices (e.g. Bekoff, 2013b; Wallach et al., 2018) but can be beneficial for the success of conservation outcomes (Shier & Owens, 2006).

We also asked: What are the top three animal welfare issues that you have experienced and consider the most difficult to mitigate? Respondents offered little consensus on this open-ended question, but issues given by more than one respondent were associated with mortality, capture, transport, and marking or tagging methods for monitoring animals post release. More specifically, respondents referred to unmitigated threats, habitat or site-related issues (such as insufficient prey, weather events, and vehicle collisions), difficulties during transportation (stress associated with long journeys and logistical limitations), and unforeseen issues with marking methods (whilst ensuring animals were adequately monitored). There were issues relating to extended periods in captivity (and associated declines in body condition), and respondents referred specifically to quarantine requirements and the need to wait for permission to release animals. Post-release dispersal was referred to, as was the unknown fate of animals lost during projects (and more effective tracking devices being too expensive). For captive-bred animals, respondents suggested that they struggled with ensuring that effective pre-release training was provided for the animals, providing pre-release training that was not stressful for the animals, providing natural parental behaviours to young animals (in the absence of parent animals), and ensuring that animals had natural resistance to disease likely to be encountered in the wild, and were able to adapt to life in the wild.

6.2.2 Were Practitioners Able to Mitigate Animal Welfare Issues Encountered?

Among the 38 respondents who had experience of animal translocations, 61 per cent stated that they were able to mitigate potential animal welfare issues by: carrying out detailed habitat analysis or habitat enhancement (including increased predator control, itself a potential animal welfare issue); reducing animal holding time; changing the source of animals or the age group released; changing the type, timing, or duration of transport (which may involve added cost) or the release methodology; and altering the marking or tagging devices used (using less invasive methods, advanced technology, and/or ensuring the fitting of such devices was carried out only by experienced staff members). To better prepare animals for release and for life in the wild, respondents reported unspecified improvements to pre-release facilities, pre-release training (including predator awareness), holding animals in release pens at the release site, and, post release, the provision of supplemental feeding. Respondents also referred to rescuing released animals and carrying out medical interventions where necessary, as well as monitoring stress.

6.2.3 Do Practitioners Perceive an Increased Expectation to Consider Animal Welfare within Their Projects?

Forty-nine per cent of all survey respondents (n = 41) perceived an increased expectation by others to consider animal welfare in conservation translocations. Overall, 59 per cent felt that an increased expectation for consideration of animal welfare in conservation translocations was helpful towards achieving conservation objectives, while 15 per cent felt that that it hindered their ability to achieve conservation objectives. However, opinion was split on whether animal welfare issues and/or concerns might in the future become restrictive of conservation translocation actions, in terms of regulatory oversight, licensing, and/or public acceptability: 37 per cent were 'somewhat worried', whilst another 37 per cent were either 'hardly worried' or 'not worried at all' (the remainder did not answer the question or replied 'unknown').

Forty-six per cent of respondents stated that they felt a need for greater guidance around animal welfare. Suggestions as to how such guidance could be provided fell into two broad categories: provision of standardised protocols or guidelines, and greater communication among practitioners. Respondents stressed the need for international, standardised, and taxon-specific guidelines that explicitly outline the risks to animal welfare and how they might be dealt with, including taxon- and activity-specific 'best practice' methods (outlining for example the 'best' trapping, handling, and transport methods), methods and approaches for assessing animal welfare status, and guidance on how to balance animal welfare and conservation needs. These types of methods-based guidelines could be made available online as a library or database (suggestions were that they could be made available through the IUCN Conservation Translocation Specialist Group) and accessible to permit-granting institutions. In terms of communication, the emphasis was on raising awareness of animal welfare issues, as well as training, sharing experiences, and learning from others. Respondents specifically suggested contact with others facing the same dilemmas, the sharing of information on evidence-based improved practices, and more open reporting of problems encountered. These would inform a community where challenges are accepted as part of an experimental approach so that practitioners feel comfortable discussing the issues openly and others can learn from their mistakes and avoid them.

6.2.4 Links between Conservation, Welfare, and Ethical Issues

While conservation translocation practitioners identified issues of varying concern (Table 6.1), understanding why such issues are seen as relevant is often difficult to explain or delineate – indeed, it might depend upon personal fundamental values. Individuals might feel that an issue makes them uncomfortable or believe that it should be addressed but may find it difficult to delineate to what extent such concerns are founded upon conservation, welfare, or ethical grounds. Consequently, we posed ten scenarios (Table 6.2) of varying complexity that might arise in any animal conservation translocation project. We asked whether practitioners considered each scenario a conservation concern, an animal welfare concern

Table 6.2. *Possible scenarios presented to survey respondents. Data shown are (a) the percentage of respondents that believed the scenario 'mattered', and (b) the percentage of respondents that believed the scenario raised conservation, animal welfare, or 'other' ethical (i.e. animal rights) concerns (n = 30). Note that the term 'mattered' was not defined a priori (see text). Shading indicates relatively low (<33%), medium (>33%, <66%), and high (>66%) occurrence.*

(a)

	Scenario	Does it matter?
1	A carnivore species in captivity is provided with live prey to learn predatory behaviours before being released into the wild	53%
2	A reintroduction plan is being developed for a species but before the release can take place, random releases are carried out by others not involved in the planning	60%
3	After release, animals are live trapped for post-release monitoring but some individuals are injured in the process	53%
4	Supplementary feeding of the released species unexpectedly leads to disease transfer in the population	63%
5	A reintroduction is not pursued despite favourable risk vs. benefit assessment and sufficient resources to potentially save the species from extinction	69%
6	A reintroduction is pursued to re-establish a species regionally but there is no tangible benefit to the species' global status	47%
7	Abundant captive-bred animals are released instead of wild animals to protect small source populations although captive-bred animals have far lower survival probability	44%
8	Soft releases increase individuals' chance of survival, but animals are hard released because funds are limited and there are large numbers of captive-bred animals available	60%
9	Abundant species at the release site will face predation or competition from reintroduced species	53%
10	Released animals are killed humanely (e.g. through culling to reduce population size)	53%

(b)

	Scenario	Conservation	Welfare	Ethical
1	A carnivore species in captivity is provided with live prey to learn predatory behaviours before being released into the wild	30%	40%	60%

Table 6.2. (*cont.*)

(b)

	Scenario	Conservation	Welfare	Ethical
2	A reintroduction plan is being developed for a species but before the release can take place, random releases are carried out by others not involved in the planning	80%	60%	60%
3	After release, animals are live trapped for post-release monitoring but some individuals are injured in the process	43%	97%	30%
4	Supplementary feeding of the released species unexpectedly leads to disease transfer in the population	90%	63%	17%
5	A reintroduction is not pursued despite favourable risk vs. benefit assessment and sufficient resources to potentially save the species from extinction	100%	3%	47%
6	A reintroduction is pursued to re-establish a species regionally but there is no tangible benefit to the species' global status	63%	0%	17%
7	Abundant captive-bred animals are released instead of wild animals to protect small source populations although captive-bred animals have far lower survival probability	70%	57%	37%
8	Soft releases increase individuals' chance of survival, but animals are hard released because funds are limited and there are large numbers of captive-bred animals available	57%	70%	60%
9	Abundant species at the release site will face predation or competition from reintroduced species	60%	27%	30%
10	Released animals are killed humanely (e.g. through culling to reduce population size)	97%	30%	30%

(hereafter 'welfare concern'), or an 'ethical' concern (i.e. an ethical concern beyond conservation or animal welfare, such as 'animal rights'; we do not imply that conservation and animal welfare do not also have an ethical basis). We also asked whether practitioners thought the issue 'mattered', although we did not attempt to define this term and left interpretation to the respondent.

There was relatively little variation in terms of perceived importance of scenarios. While 69 per cent of respondents (of 30 who had answered this set of questions) believed the issue of not pursuing feasible reintroductions mattered (Scenario 5, Table 6.2a), only 44 per cent saw the idea of releasing abundant, short-lived captive-born animals instead of impacting vulnerable, wild-source populations as important (Scenario 7, Table 6.2a); other percentages reflecting the extent to which practitioners believed the scenario to matter ranged from 47 per cent to 63 per cent (Table 6.2a).

There was, however, considerable variation within and among scenarios in terms of the extent to which each reflected conservation, welfare, and/or ethical concerns (Table 6.2b). Interestingly, the overall take-away message from the responses was that welfare, ethical, and conservation concerns were often interlinked:

- No scenario was seen to be of only conservation, only welfare, or only ethical concern.
- Scenarios most commonly perceived to be of conservation concern were not necessarily also considered to be of welfare concern or vice versa.
- Scenarios most commonly perceived to be of ethical concern were also highly likely to be considered a conservation concern, welfare concern, or both.

These overall conclusions illustrate that conservation translocation proponents, or critics, should avoid labelling different challenges as being 'just a welfare issue' or 'just a conservation issue'; rather, the results show that complex challenges are not only of singular concern. Moreover, insofar as our results are representative of conservation translocation practitioners beyond our survey sample, they suggest that, in contrast to the apparent perceptions of critics, conservation translocation practitioners are fully cognisant of and sympathetic to the potential welfare concerns within possible conservation translocation scenarios (e.g. Scenario 3, Table 6.2a, b).

Interestingly, the scenario most commonly believed to be of conservation concern, and also relatively commonly considered to be of ethical concern, was not about taking a potential action, but about an action not being taken (Scenario 5, Table 6.2b). Scenario 4, unsurprisingly, was commonly considered to be of conservation and relatively commonly of welfare concern (due presumably to the likely impacts on both the target population and the individuals that succumb to the disease); Scenario 2 was commonly considered to be of conservation, and relatively commonly of both welfare and ethical concern; and Scenario 8 was slightly biased towards welfare and ethical concern (raising questions as to whether or not it is acceptable to risk harm or death to an individual when a safer but more expensive approach is available, Table 6.2b). However, similar risks to captive-bred animals were (slightly) less commonly perceived to be of welfare concern (Scenario 7) when the reasoning behind the action was to protect small, extant populations of wild individuals. In short, practitioners appreciate the risks to individual, captive-bred animals associated with particular actions but consider that sometimes those risks are worth taking, depending on the reasoning behind them and the availability of alternatives.

6.3 Practical Considerations

6.3.1 Best Practice: 'Caring for Individuals'

Individual animals may suffer poor welfare at all stages of translocation, and this was evident from our practitioner survey – during initial trapping (to obtain animals for release), captive breeding (when wild animals are not available), holding (e.g. during quarantine periods), or eventual release (Table 6.1). Indeed, stress is inevitable when animals are moved or are released into unfamiliar environments (Dickens et al., 2010; Ewen et al., this volume).

Harrington et al. (2013) outline a number of animal welfare considerations relevant to each stage of the translocation process that should be thoroughly assessed before proceeding with a translocation (summarised in Table 6.3). The implications for individual animal welfare of not adhering to these requirements are straightforward: for example, an animal released into an area where the original cause of decline (e.g. human persecution or disease) has not been resolved (or effectively mitigated) is likely to suffer post release due to factors (e.g. trapping, ill-health, or disease-induced mortality) associated with that cause. The overall process can be summarised simplistically as ensuring that the basic physiological, ecological, and behavioural needs of the animals are met, and that, as stated in the IUCN Guidelines for Reintroductions and Other Conservation Translocations (IUCN, 2013), 'every effort [is] made to reduce stress or suffering'. A deep understanding of the biology of the animals (including their specific dietary, ecological, social, and

Table 6.3. *Questions relevant to animal welfare considerations to ask before proceeding with a conservation translocation. Adapted from Harrington et al. (2013).*

Questions
Initial considerations • Is the cause of decline known? • Has the cause of decline been resolved? • Is the biology of the animal understood? **Pre release (for wild-sourced animals)** • Will removing animals for release impact the wild population? • Are the trapping methods to be used humane? **Pre release (for captive-bred animals)** • Can the species be bred in captivity successfully? • Is long-term care available for individuals that are not released? • Are training needs for the species known and available? • Are the best age classes for release known and available? • Is there disease risk at the release site? • Is there sufficient habitat and space available for release? • Have potential interactions with other native species at the release site been identified (and dealt with)? • Have 'other' threats at the release site been ameliorated? **Release** • Has the best release strategy been identified? • Has the best time of year been identified? • Is there a suitable method of transport available to minimise stress? • What post-release support is available? **Post release** • Is there a suitable monitoring strategy in place? • Are safe marking/identification/tracking methods available? (Is there a need for invasive monitoring?) • Will it be possible to detect animal behaviour, survival, stress, reproduction? • Is there an intervention plan in place?

behavioural requirements), and, when captive-bred animals are involved, ensuring that they are 'fit' for life in the wild (Figures 6.1 and 6.2), is key.

At all stages where animal handling is required, the methods used for restraint, transport, handling, biological sampling, and animal identification should be humane (Figure 6.3) and carried out as quickly as possible with minimal human contact; wherever possible, non-invasive methods should be considered. Practitioners should be aware that the level of stress experienced by captive-bred and wild-sourced animals in response to

Figure 6.1 Care is taken to provide conditions as close to 'natural' as is possible for northern leopard frogs *Lithobates pipiens* in a captive conservation breeding programme supplying animals for release (photo: Jill Hockaday, Calgary Zoo). (A black and white version of this figure will appear in some formats. For the colour version, please refer to the plate section.)

conditions (e.g. variations in temperature, humidity, light, vibration, and noise) to which they are not accustomed, such as during transport, may differ (Moehrenschlager & Lloyd, 2016) and that some species are more susceptible to stress effects than others (Fischer & Romero, 2019). Ungulates and some birds, for example, appear to be particularly susceptible to capture myopathy (Blumstein et al., 2015; Breed et al., 2019). Use of tranquillisers to reduce stress during transport may be an option for some species (Teixeira et al., 2007; Dickens et al., 2010; see also Parker et al., 2012).

For the release strategy, the most suitable age class and optimal time of year should be identified to avoid unnecessary suffering. General recommendations are to release animals at a time of year when resources are plentiful to allow them time to establish territories and to become familiar with their new environment before finding resources (food or shelter) becomes more challenging (Moehrenschlager & Lloyd, 2016). Delayed release, whereby animals are kept in on-site enclosures or pens prior to release (Figure 6.2), together with provision of post-release

Figure 6.2 (a) Young captive-bred European mink *Mustela lutreola* are placed or are born in (b) enclosures at the release site prior to release to provide them with experience of their natural riparian habitat. Hiiumaa Island, Estonia (photos: (a) Andrew Harrington (b) Tiit Maran). (A black and white version of this figure will appear in some formats. For the colour version, please refer to the plate section.)

Figure 6.3 Live trapping and mark-recapture techniques are used to monitor a translocated population of black-tailed prairie dogs *Cynomys ludovicianus*. White plastic covers over the traps provide protection from the sun and therefore aid individual animal welfare during monitoring (photo: Fiona Le Taro, Calgary Zoo). (A black and white version of this figure will appear in some formats. For the colour version, please refer to the plate section.)

support in the form of supplementary food and artificial shelter, is often used to increase site fidelity (e.g. Ruffell & Parsons, 2009) but may also provide animal welfare benefits by allowing animals to locate natural resources in their own time and by alleviating the suffering that would be experienced by individuals unable to immediately locate such resources. Indeed, such support may be beneficial, or even necessary, for some time after release (see Section 6.3.2).

Adhering to these types of animal welfare considerations is conceptually straightforward (although often practically challenging) and can be thought of as caring for individuals while fostering population survival. Post-release support, in particular, is a common approach (Harrington et al., 2013). In this context, animal welfare is complementary to conservation. Problems usually stem from uncertainty (associated with a lack of knowledge and/or unexpected outcomes) and inherent variation among individuals (in various aspects of their behaviour, personality, and/or response to new situations). As regards the latter, however, various authors have suggested that personality testing could be used to select animals that are better able to cope with stressors, or to identify individuals that might require more care or support to survive (e.g. Baker et al., 2016; Baker, 2017). Recent evidence also suggests that personality may be influenced by early developmental conditions (and thus could potentially be manipulated; Richardson et al., 2019). The personality approach has promise (Bremner-Harrison et al., 2004; Baker, 2017) but is complex (e.g. Smith & Blumstein, 2008; Haage et al., 2013, 2017); although it is not a new idea, there are currently few experimental studies and no clear guidelines as to how it may be implemented on a practical level.

6.3.2 Trade-Offs

Caring for individuals clearly has benefits for them but can also have adverse impacts on those same individuals. In using delayed-release methods, providing post-release support, monitoring, and/or intervening (e.g. to treat disease), there are trade-offs in terms of the welfare benefits gained versus the risks or long-term welfare costs. Animals provided with post-release support, for example, may become dependent on artificial resources (Bright & Morris, 1994) and unable to learn and adapt to a novel natural environment. Supplementary food may attract predators and increase predation risk (e.g. Hanmer et al., 2017). For wild-caught animals, delayed release may be more, rather than less, stressful (Batson et al., 2017), and may have negative impacts on long-term survival (Richardson et al., 2013). Shepherding or body guarding Vancouver Island marmots *Marmota vancouverensis* to provide physical protection against predators risks habituation to humans and may increase stress on shepherded individuals (Jachowski et al., 2016) (although the impacts of human habituation are not always negative, Coleman et al., 2008).

Even where post-release support clearly has benefits for animal welfare, there are decisions to be made regarding the long-term outcome and what is and what is not appropriate in terms of supportive management actions. Supplemental food was provided for reintroduced white storks *Ciconia ciconia* for over 30 years (Schaub et al., 2004) but whether or not this population can be considered 'wild' (a fundamental objective of many conservation translocation projects) has been questioned (Chauvenet et al., 2012). One might also ask: when is it, and when is it not, appropriate to rescue translocated individuals that are not doing well (by some welfare metric) in the wild? If released animals are suffering, removing them from the wild might be the only humane option (Moehrenschlager & Lloyd, 2016).

Post-release monitoring also involves trade-offs. Herein lie some key reflections for conservation translocations. First, welfare challenges that arise are often not unique to conservation translocations; instead they are associated with conservation biology techniques in general. Second, monitoring, which may entail welfare issues, may yield information that leads to the refinement of techniques that are not only of value for conservation but also in terms of improving welfare outcomes for individual animals. Monitoring is necessary to assess, for example, how animals respond to different pre-release treatments and post-release strategies, to better understand project success, and to enable adaptive management that will benefit the survival of individuals and thereby conservation outcomes. Monitoring also allows insight into how individual animals fare in the wild, can detect if an animal is suffering, and can alert managers to the possible need for intervention. Yet because monitoring often involves the use of physical devices attached directly to the animals (Figure 6.4), the methods used can themselves cause suffering.

Figure 6.4 A satellite transmitter carefully attached to a burrowing owl *Athene cunicularia* by trained individuals for monitoring survival on migration routes (photo: Calgary Zoo). (A black and white version of this figure will appear in some formats. For the colour version, please refer to the plate section.)

Radio tracking, for example, has helped determine the fate of released individuals and, in experimental situations, revealed optimal release strategies that have informed further releases (e.g. Biggins et al., 1998), but radio transmitters can sometimes have negative impacts on animals (e.g. Moorhouse & Macdonald, 2005). The welfare 'costs' of monitoring can sometimes be reduced by technological improvements (such as reducing the size of tracking devices), but there are limitations: smaller devices may have reduced power and so produce lower quality data. Radio tracking is just one example, but what all tracking devices and direct monitoring methods have in common is the need for repeated capture and handling of animals. This may cause stress and even affect, for example, the animals' ability to adapt and their susceptibility to disease or ability to reproduce.

Assessment of the trade-offs associated with the costs and benefits (not just economic) of direct and/or invasive animal monitoring techniques post release should be an iterative process that incorporates measures of animal welfare (using direct and non-direct indicators, such as physiology, stress levels, and behaviour, Tarszisz et al., 2014; Moehrenschlager & Lloyd 2016) as well as assessment of the quality of the monitoring data obtained and consideration of the questions that can realistically be answered (Ewen et al., this volume). Methods are continually improved as both knowledge and technology progress, and what is best for one species will not necessarily be best for another (Moseby et al., 2014). Practitioners should seek appropriate guidelines, keep up to date with new methodology, consult experts and others in the field working on similar species, and report and publish their own results. That is not to say that answers exist for all possible situations; the reality is that problems are often impossible to predict before release (Moehrenschlager & Lloyd, 2016).

6.3.3 Ethical Dilemmas

A more difficult and conceptually challenging aspect of considering animal welfare, and the interests of individual animals, in conservation translocations, are the true 'ethical dilemmas' where animal ethics potentially clash with broader conservation ethics (Harrington et al., 2013). The most frequently articulated question is: should individual animals be subjected to an uncertain and potentially horrible fate in an attempt to establish a population, save a species, or restore an ecosystem – in other words, should individuals have to suffer for the good of the population or the wider ecosystem within which they live? With compassionate conservation in mind, one could also ask the opposite: should a species be permitted to go extinct, or restoration of an ecosystem not be pursued, to prevent the individuals involved from suffering? These ethical dilemmas are not limited to those individuals to be released; they also include those that may be impacted by the translocation, either in captivity or at the release site. In this case, one might also ask whether individuals of one species (e.g. predators killed at the release site to reduce predation on the released species) should suffer for the good of the population of another species. These questions are highly contentious and even among conservation biologists there can be considerable disagreement, heightened by the often urgent need for conservation action.

One example of an ethical dilemma sometimes encountered in conservation translocations involves the training of captive-bred animals before release. Sometimes only captive-bred animals are available for reintroduction (e.g. when the species is extinct in the wild), but captive-bred animals can lack the skills required to locate suitable food, avoid predators, find refuge/shelter, effectively locomote, and interact appropriately with conspecifics (e.g. Kleiman, 1989; Reading et al., 2013). Training, or more general behavioural enrichment (sometimes termed 'pre-conditioning') has been shown to increase post-release survival for several species (e.g. van Heezik et al., 1999; Shier & Owings, 2006), thereby improving the welfare of individuals released (Reading et al., 2013) and the chance of achieving conservation goals. However, providing exposure to natural situations becomes contentious and potentially problematic when 'training' predators that require prey, particularly when animals destined for release are fed live prey to allow them to 'learn' how to capture and kill their prey. In some countries feeding live prey to captive animals is illegal; where it is permitted, the practice has attracted serious criticism (e.g. Bekoff, 2013b), particularly when the prey in question are mammals (Cottle et al., 2010). However, consider also the ethics of releasing into the wild a naïve captive-bred predator that is unable to recognise and/or kill its prey, and the implications for the welfare of that individual. An animal that does not know how to feed itself will likely starve and suffer; or, if the animal retains some predatory instinct, it might opt for 'easy' domestic prey with the risk that it is trapped, poisoned, or shot by people in retaliation, probably resulting in an inhumane death. Importantly, the ability of a captive-bred predatory species to locate, capture, and kill their natural prey in the wild is usually unknown, and questions regarding the necessity of, and optimal exposure to, training with live prey are rarely subject to scientific research.

There are no easy answers in these conflict situations but articulating the most difficult dilemmas and beginning an open and honest discussion is a good first step.

6.4 Moving Forward

One particular difficulty in dilemmas regarding welfare, conservation, and linked ethical concerns is the lack of approaches to deal with some of these issues. There are practical issues that could benefit from innovative solutions (e.g. how best to track animal movements or monitor their behaviour) and new technologies (e.g. animal tracking devices) but there are also, as we discussed earlier, a number of trade-offs to be considered, and sometimes difficult dilemmas. McMahon et al. (2012) suggested Bateson's Cube (Bateson, 2005) as a simple tool to assess whether a particular piece of research on wild animals (which could, conceptually, be a translocation or other conservation action) should proceed, whereby an explicit assessment is made between the degree of suffering that the animals involved are likely to endure, the importance of the research, and the likely benefit the research will bring. However, assessing conservation importance is in itself not simple, and 'animal suffering' may over-simplify the consideration of individual animals that compassionate conservation demands, which seems to reject any animal mortality at all, regardless of suffering (Draper & Bekoff, 2013; McMahon et al., 2013). Although there are methods for assessing physiological stress and other physical indicators of 'suffering' (albeit infrequently used in conservation translocation, Harrington et al., 2013), quantifying the degree to which animal rights might be violated is more challenging beyond a simple count of the number of individuals that might die. Even if each of the axes of consideration can be appropriately quantified (using, for example, agreed metrics for animal welfare, conservation goals, and the probability of project success), balancing them is still difficult.

One approach that may be helpful is structured decision-making (SDM) (Ewen et al., this volume). SDM is a framework that facilitates a collaborative and iterative process involving value-based discussions, with the aim of outlining, comparing, and evaluating alternative management strategies under various clearly stated objectives (Gregory et al., 2012). SDM has been increasingly used within conservation translocation programmes (Runge et al., 2011; Converse et al., 2013; Ewen et al., 2014; Hayek et al., 2016; Edwards et al., 2019) and could be particularly useful for balancing conservation and welfare considerations (Figure 6.5). One

Animal Welfare and Animal Rights · 203

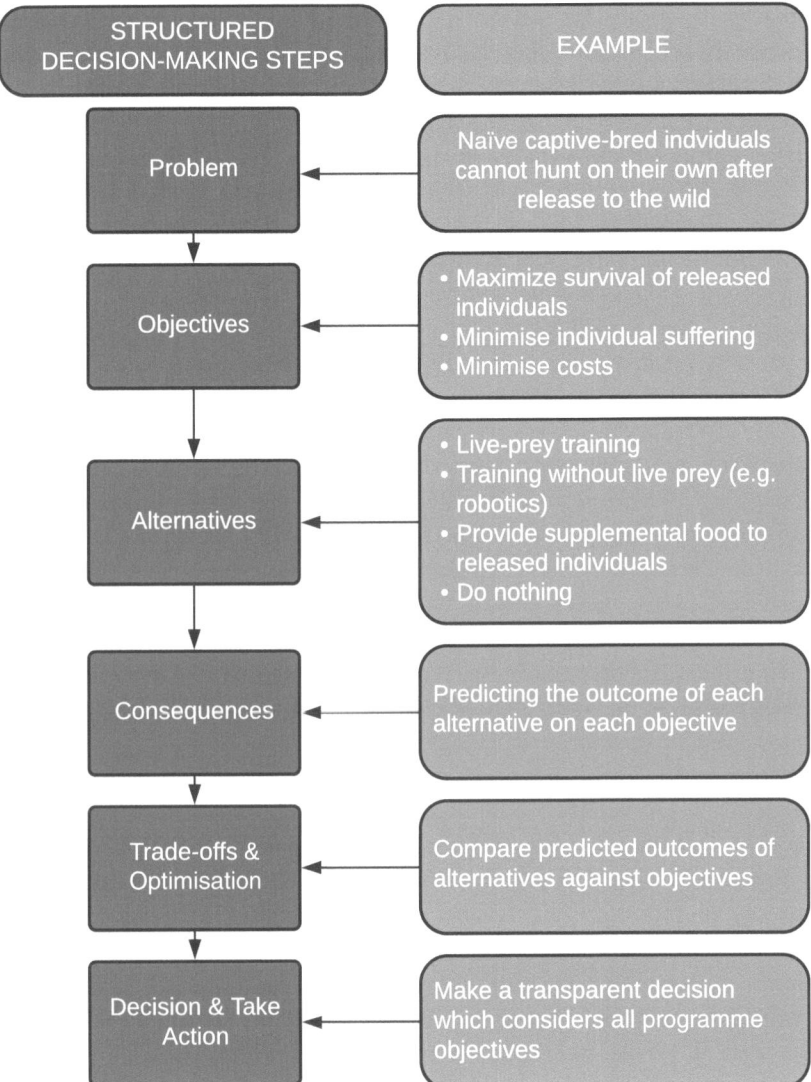

Figure 6.5 Flow chart showing how competing values and objectives can be evaluated, and trade-offs considered in a transparent manner, within a structured decision-making framework. Here we illustrate the contentious example of whether or not to feed captive-bred predators live prey prior to release. However, the process could be adapted to consider any number of trade-offs and ethical dilemmas encountered in animal translocations. Ideally, this would be an iterative process.

example would be selecting appropriate post-release monitoring methods, which could be evaluated to assess the expected value of the information gained balanced against the relative level of animal welfare concern (Runge et al., 2011). This process not only has the potential to help conservationists move forward when faced with dilemmas but would also provide transparency in how these concerns are integrated into and valued in conservation translocations. Open-source software is currently being developed to assist this process (Nagy-Reis et al., 2020). However, where conflict means that one group does not accept the other's values regardless of context then SDM will not help and more specialist conflict resolution techniques may be the only path forward (Glikman et al., this volume).

Crucially, as conservation practitioners keen to see progress, successful execution of conservation actions, and timely reaction to the current and on-going biodiversity crisis (IPBES 2019), we might reasonably ask whether attempting to reconcile these issues will help or hinder conservation. Certainly, the helpfulness of this broader approach can be seen in practical terms: better animal welfare can lead to better conservation outcomes. This can be achieved by using the best methods and ensuring that projects are carried out with the utmost care and consideration of the well-being of the individual animals that are involved (e.g. Beausoleil et al., 2018). Animal welfare review boards play a role here, and where these systems exist, many practitioners believe that individual animals are adequately protected by established animal welfare standards and ethical codes of conduct. Critics disagree; for example, Wallach et al. (2018) suggest that in practice such standards 'afford minimal protection and readily permit strategies that enact varying degrees of violence against wildlife' (see also Vucetich & Nelson, 2007). Certainly there is room for improvement, but one problem is that it is not easy to predict how individual animals will fare post release, and animals often need to be released in a trial, or as part of an experimental approach, to progress. This type of iterative process was highlighted by practitioners in our survey and is a well-established scientific approach, but it may be problematic when it involves live animals. On the other hand, 'compassionate conservationists' call for empathy in all conservation actions (e.g. Wallach et al., 2018), but human empathy can be biased and inconsistent (Griffin et al., 2020). Humans are more likely to feel empathy towards primates, or other mammals, than they are towards snakes, for example. Here, empathy, and an inability to accept risk, can hinder conservation projects, particularly in the face of uncertainty.

Nevertheless, regardless of one's personal values, responsible project management requires that termination of the project remains an option in circumstances where goal achievement becomes unlikely or where the risks outweigh the benefits (Moehrenschlager & Lloyd, 2016) (i.e. in this case, where animal welfare risks outweigh conservation benefits) (Dalrymple & Bellis, this volume). The decision where to draw the line would undoubtedly differ among stakeholders but it must remain an option, be clearly identified, and be agreed upon by all stakeholders. Another problem is that there is little discussion in the scientific literature of some of the more contentious issues (e.g. feeding live prey), which might help inform such judgements (cf. Hampton & Hyndman, 2019). A robust scientific evidence base (regarding the effectiveness or necessity of the procedure) would not resolve the ethical dilemma involved (regarding the life and suffering of the prey species versus that of the predator post release), but it would enable a more informed evaluation of conflicting moral obligations and this could feed into the type of SDM approach referred to above.

To encourage explicit consideration of the issues we urge practitioners to seek to improve methods wherever possible, use good practice guidelines, and share information. They should go beyond what is required by ethical review boards and the law, consider the value of any particular action, and assess the availability of alternatives and the real costs to the individual animals involved (cf. Vucetich & Nelson, 2007).

We suggest that cost-benefit analyses are needed, and that relevant decision theory may be useful in combining uncertain knowledge of consequences (objective evidence) with our values (subjective reactions to this knowledge). Approaches such as SDM provide a powerful platform for clear deliberation, explicitly recognising both objective and subjective elements of decisions, to gain insight and promote balanced choices. Ethical perfection is probably not realistic (Vucetich & Nelson, 2007), but we should strive to become progressively more ethical.

6.5 Key Messages

- Greater consideration of the welfare of individual animals can contribute to population-level conservation success (i.e. healthy individuals = a healthy population)
- Conservation, welfare, and ethical issues do not exist in isolation; conservation translocation practitioners identify them as interlinked.

- There are situations in conservation translocations where individual-level animal welfare concerns and population-level conservation concerns conflict, requiring trade-offs and generating difficult moral dilemmas.
- Conservation translocation practitioners would like more guidance on incorporating welfare considerations into their programmes, highlighting a need for standardised, taxon-specific guidelines, as well as increased communication among practitioners.
- Structured decision-making could be a useful approach for balancing or aligning welfare and conservation considerations and would provide transparency as to how these concerns are addressed.
- Overall, there is demand for increased consideration of individual animal welfare in conservation practice. Open, honest, and critical assessment of the issues is required, together with respectful dialogue, collaboration, and knowledge sharing among stakeholders.

Acknowledgements

Sincere thanks to all the survey participants who offered their time and provided valuable insights, to Paul Johnson for helpful feedback on the ethical issues, and to John Ewen and Martin Gaywood for their critical editing. Thanks also to Jill Hockaday, Fiona La Taro, Tiit Maran, and Andrew 'Harry' Harrington for allowing us to use their photos.

References

Baker, L. (2017) Translocation biology and the clear case for compassionate conservation, *Israel Journal of Ecology and Evolution*. 63, 52–60.

Baker, L., Lawrence, M. S., Toews, M., Kuling, S. & Fraser, D. (2016) Personality differences in a translocated population of endangered kangaroo rats (*Dipodomys stephensi*) and implications for conservation success. *Behaviour*. 153, 1795–1816.

Bateson, P. (2005) Ethics and behavioral biology. *Advances in the Study of Behavior*. 35, 211–233.

Batson, W. G., Gordon, I. J., Fletcher, D. B., Portas, T. J. & Manning, A. D. (2017) The effect of pre-release captivity on the stress physiology of a reintroduced population of wild eastern bettongs. *Journal of Zoology*. 303, 311–319.

Beausoleil, N. J., Mellor, D. J., Baker, L., et al. (2018) "Feelings and fitness" not "feelings or fitness" - the raison d'être of conservation welfare, which aligns conservation and animal welfare objectives. *Frontiers in Veterinary Science*. 5, 296.

Bekoff, M. (2013a) *Ignoring Nature No More: The Case for Compassionate Conservation*. Chicago, IL, University of Chicago Press.

Bekoff, M. (2013b) Compassionate conservation: a green conversation. *Psychology Today*, May 14, 2013.

Berger-Tal, O., Blumstein, D. T. & Swaisgood, R. R. (2020) Conservation translocations: a review of common difficulties and promising directions. *Animal Conservation*. 23, 121–131.

Biggins, E., Godbey, J., Hanebury, L., et al. (1998) The effect of rearing methods on survival of reintroduced black-footed ferrets. *The Journal of Wildlife Management*. 62, 643–653.

Blumstein, D. T., Buckner, J., Shah, S., Patel, S., Alfaro, M. E. & Natterson-Horowitz, B. (2015) The evolution of capture myopathy in hooved mammals: a model for human stress cardiomyopathy? *Evolution, Medicine and Public Health*. 2015, 195–203.

Breed, D., Meyer, L. C. R., Steyl, J. C. A., Goddard, A., Burroughs, R. & Kohn, T. A. (2019) Conserving wildlife in a changing world: Understanding capture myopathy - a malignant outcome of stress during capture and translocation. *Conservation Physiology*. 7, coz027.

Bremner-Harrison, S., Prodohl, P. A. & Elwood, R. W. (2004) Behavioural trait assessment as a release criterion: boldness predicts early death in a reintroduction programme of captive-bred swift fox (*Vulpes velox*). *Animal Conservation*. 7, 313–320.

Bright, P. & Morris, P. (1994) Animal translocation for conservation: performance of dormice in relation to release methods, origin and season. *Journal of Applied Ecology*. 31, 699–708.

Callen, A., Hayward, M. W., Klop-Toker, K., et al. (2020). Envisioning the future with 'compassionate conservation': an ominous projection for native wildlife and biodiversity. *Biological Conservation*. 241, 108365.

Callicott, J. B. (1980). Animal liberation: a triangular affair. *Environmental Ethics*. 2, 311–328.

Callicott, J. B. (1988). Animal liberation and environmental ethics: back together again. *Between the Species*. 4, 163–169.

Callicott, J. B. (1998) 'Back together again' again. *Environmental Ethics*. 7, 461–475.

Callicott, J. B. (2011) Uma palinódia introtório [An introductory palinode]. In Galvão, P. (ed.) *Os Animais têm direitos [Do animals have rights?]: Perspectivas e argumentos*. Lisbon, Dinalivros, pp. 121–131. English translation available from: https://jbcallicott.weebly.com/introductory-palinode.html [Accessed 1 February 2021].

Chauvenet, A., Ewen, J., Armstrong, D., et al. (2012) Does supplemental feeding affect the viability of translocated populations? The example of the hihi. *Animal Conservation*. 15, 337–350.

Coleman, A., Richardson, D., Schechter, R. & Blumstein, D. T. (2008) Does habituation to humans influence predator discrimination in Gunther's dik-diks (Madoqua guentheri)? *Biology Letters*. 4, 250–252.

Converse, S. J., Moore, C. T., Folk, M. J. & Runge, M. C. (2013) A matter of tradeoffs: reintroduction as a multiple objective decision. *The Journal of Wildlife Management*. 77, 1145–1156.

Cottle, L., Tamir, D., Hyseni, M., Bühler, D. & Lindemann-Matthies, P. (2010) Feeding live prey to zoo animals: response of zoo visitors in Switzerland. *Zoo Biology*. 29, 344–350.

Dalrymple, S. E. & Bellis, J. M. (2023) Conservation translocations: planning and the initial appraisal. In Gaywood, M. J., Ewen, J.G., Hollingsworth, P. M. and Moehrenschlager, A. (eds.) *Conservation Translocations*. Cambridge, Cambridge University Press.

Dickens, M. J., Delehanty, D. J. & Romero, L. M. (2010) Stress: an inevitable component of animal translocation. *Biological Conservation*. 143, 1329–1341.

Draper, C. & Bekoff, M. (2013) Animal welfare and the importance of compassionate conservation - A comment on McMahon et al. (2012). *Biological Conservation*. 158, 422–423.

Edwards, H. A., Bidwell, M. T. & Moehrenschlager, A. (2019) A call for structured decision making in conservation programs considering wild egg collection. *Biological Conservation*. 238: 108226.

Ewen, J. G., Walker, L., Canessa, S. & Groombridge, J. J. (2014) Improving supplementary feeding in species conservation. *Conservation Biology*. 29, 341–349.

Ewen, J. G., Canessa, S., Converse, S. J. & Parker, K. A. (2023) Decision-making in animal conservation translocations: biological considerations and beyond. In Gaywood, M. J., Ewen, J.G., Hollingsworth, P. M. and Moehrenschlager, A. (eds.) *Conservation Translocations*. Cambridge, Cambridge University Press.

Fischer, C. & Romero, L. (2019) Chronic captivity stress in wild animals is highly species-specific. *Conservation Physiology*. 7, coz093.

Glikman, J. A., Frank, B., Sandström, C., et al. (2023) The human dimensions and the public engagement spectrum of conservation translocation. In Gaywood, M. J., Ewen, J.G., Hollingsworth, P. M. and Moehrenschlager, A. (eds.) *Conservation Translocations*. Cambridge, Cambridge University Press.

Gray, J. (2018) Challenges of compassionate conservation. *Journal of Applied Animal Welfare Science*. 21(sup1), 34–42.

Gregory, R., Failing, L., Harstone, M., Long, G., McDaniels, T. & Ohlson, D. (2012) *Structured Decision Making: A Practical Guide to Environmental Management Choices*. Oxford, Wiley-Blackwell.

Griffin, A. S., Callen, A., Klop-Toker, K., Scanlon, R. J. & Hayward, M. W. (2020) Compassionate conservation clashes with conservation biology: should empathy, compassion, and deontological moral principles drive conservation practice? *Frontiers in Psychology*. 11, 1139.

Haage, M., Bergvall, U. A., Maran, T., Kiik, K. & Angerbjörn, A. (2013) Situation and context impacts the expression of personality: the influence of breeding season and test context. *Behavioural Processes*. 100, 103–109.

Haage, M., Maran, T., Bergvall, U. A., Elmhagen, B. & Angerbjörn, A. (2017) The influence of spatiotemporal conditions and personality on survival in reintroductions-evolutionary implications. *Oecologia*. 183, 45–56.

Hampton, J. O. & Hyndman, T. H. (2019) Underaddressed animal-welfare issues in conservation. *Conservation Biology*. 33, 803–811.

Hanmer, H. J., Thomas, R. L. & Fellowes, M. D. E. (2017) Provision of supplementary food for wild birds may increase the risk of local nest predation. *IBIS*. 159, 158–167.

Harrington, L. A., Moehrenschlager, A., Gelling, M., Atkinson, R. P. D., Hughes, J. & Macdonald, D. W. (2013) Conflicting and complementary ethics of animal welfare considerations in reintroductions. *Conservation Biology*. 27, 486–500.

Hayward, M. W., Callen, A., Allen, B. L., et al. (2019) Deconstructing compassionate conservation. *Conservation Biology*. 33, 760–768.

Hayek, T., Stanley-Price, M. R., Ewen, J. G., Lloyd, N., Saxena, A. & Moehrenschlager, A. (2016) *An Exploration of Conservation Breeding and Translocation Tools to Improve the Conservation Status of Boreal Caribou Populations in Western Canada*. Calgary, Alberta, Canada, Centre for Conservation Research, Calgary Zoological Society.

IPBES (2019) *Summary for Policymakers of the Global Assessment Report on Biodiversity and Ecosystem Services of the Intergovernmental Science-Policy Platform on Biodiversity and Ecosystem Services*. IPBES.

IUCN (2013) *Guidelines for Reintroductions and Other Conservation Translocations. Version 1.0*. Gland, Switzerland, IUCN Species Survival Commission.

Jachowski, D. S., Bremner-Harrison, S., Steen, D. A. & Aarestrup, K. (2016) Accounting for potential physiological, behavioral and community-level responses to reintroduction. In Jachowski, D. S., Millspaugh, J. L., Angermeier, P. L. and Slotow, R. (eds.) *Reintroduction of Fish and Wildlife Populations*. Berkeley, CA, University of California Press.

Johnson, P. J., Adams, V. M., Armstrong, D. P., et al. (2019) Consequences matter: compassion in conservation means caring for individuals, populations and species. *Animals*. 9, 1115.

Jule, K. R., Leaverand, L. A. & Lea, S. E. G. (2008) The effects of captive experience on reintroduction survival in carnivores: a review and analysis. *Biological Conservation*. 141, 355–363.

Kleiman, D. (1989) Reintroduction of captive mammals for conservation. *BioScience*. 39, 152–161.

Linklater, W. L. & Gedir, J. V. (2011) Distress unites animal conservation and welfare towards synthesis and collaboration. *Animal Conservation*. 14, 25–27.

McMahon, C., Harcourt, R., Bateson, P. & Hindell, M. (2012) Animal welfare and decision making in wildlife research. *Biological Conservation*. 153, 254–256.

McMahon, C., Harcourt, R., Bateson, P. & Hindell, M. (2013) Animal welfare and conservation, the debate we must have: a response to Draper and Bekoff (2012). *Biological Conservation*. 158, 424.

Minteer, B. A. & Collins, J. P. (2005) Ecological ethics: building a new tool kit for ecologists and biodiversity managers. *Conservation Biology*. 19, 1803–1812.

Moehrenschlager, A. & Lloyd, N. (2016) Release considerations and techniques to improve conservation translocation success. In Jachowski, D. S., Millspaugh, J. L., Angermeier, P. L. and Slotow, R. (eds.) *Reintroduction of Fish and Wildlife Populations*. Berkeley, CA, University of California Press.

Moorhouse, T. P. & Macdonald, D. W. (2005) Indirect negative impacts of radio-collaring: sex ratio variation in water voles. *Journal of Applied Ecology*. 42, 91–98.

Moseby, K. E., Hill, B. M. & Lavery, T. H. (2014) Tailoring release protocols to individual species and sites: one size does not fit all. *PLoS ONE*. 9, e99753.

Nagy-Reis, M., Dickie, M., Sólymos, P., et al. (2020) 'WildLift': an open-source tool to guide decisions for wildlife conservation. *Frontiers in Ecology and Evolution*. 8.

Parker, K. A., Dickens, M. J., Clarke, R. H. & Lovegrove, T. G. (2012) The theory and practice of catching, holding, moving and releasing animals. In: Ewen, J. G., Armstrong, D. P., Parker, K. A. and Seddon P. J. (eds.) *Reintroduction Biology*. Oxford, Blackwell Publishing, pp. 105–137.

Reading, R., Miller, B. & Shepherdson, D. (2013) The value of enrichment to reintroduction success. *Zoo Biology*. 32.

Regan, T. (2003). Animal rights: what's in a name? In: Light, A. and Rolston, H. (eds.) *Environmental Ethics*. Oxford, Blackwell Publishing, pp. 65–73.

Richardson, K., Castro, I. C., Brunton, D. H. & Armstrong, D. P. (2013) Not so soft? Delayed release reduces long-term survival in a passerine reintroduction. *Oryx*. 49, 1–7.

Richardson, K. M., Parlato, E. H., Walker, L. K., Parker, K. A., Ewen, J. G. & Armstrong, D. P. (2019) Links between personality, early natal nutrition and survival of a threatened bird. *Philosophical Transactions of the Royal Society B*. 374, 20190373.

Rolston, H. III. (1988) *Environmental Ethics. Duties to and Values in the Natural World*. Philadelphia, PA, Temple University Press.

Ruffell, J. & Parsons, S. (2009) Assessment of the short-term success of a translocation of lesser short-tailed bats *Mystacina tuberculata*. *Endangered Species Research*. 8, 33–39.

Runge, M., Converse, S. J. & Lyons, J. (2011) Which uncertainty? Using expert elicitation and expected value of information to design an adaptive program. *Biological Conservation*. 144, 1214–1223.

Schaub, M., Pradel, R. & Lebreton, J. (2004) Is the reintroduced white stork (*Ciconia ciconia*) population in Switzerland self-sustainable? *Biological Conservation*. 119, 105–114.

Seddon, P. J. & Armstrong, D. P. (2016) Reintroduction and other conservation translocations: history and future developments. In Jackowski, D. S., Millspaugh, J. J., Angermeier, P. L., and Stotow, R. (eds). *Reintroduction of Fish and Wildlife Populations*. Berkeley, CA, University of California Press, pp. 7–27.

Sekar, N. & Shiller, D. (2020) Engage with animal welfare in conservation. *Science*. 369, 629.

Shier, D. M. & Owens, D. H. (2006) Effects of predator training on behavior and post-release survival of captive prairie dogs (*Cynomys ludovicianus*). *Biological Conservation*. 132, 126–135.

Singer, P. (2003) Not for humans only: the place of nonhumans in environmental issues. In Light, A. and Rolston, H. (eds.) *Environmental Ethics*. Oxford, Blackwell Publishing, pp. 55–64.

Smith, B. R. & Blumstein, D. T. (2008) Fitness consequences of personality: a meta-analysis. *Behavioral Ecology*. 19, 448–455.

Soorae, P. S. 2011. *Global Re-Introduction Perspectives, 2011: More Case Studies from Around the Globe*. Gland, Switzerland, IUCN SSC Reintroduction Specialist Group and Abu Dhabi, AE, Environment Agency-Abu Dhabi.

Soorae, P. S. 2013. *Global Re-Introduction Perspectives, 2013: Further Case Studies from Around the Globe.* Gland, Switzerland, IUCN SSC Reintroduction Specialist Group and Abu Dhabi, AE, Environment Agency-Abu Dhabi.

Soorae, P. S. 2016. *Global Re-Introduction Perspectives, 2016: Case Studies from Around the Globe.* Gland, Switzerland, IUCN SSC Reintroduction Specialist Group and Abu Dhabi, AE, Environment Agency-Abu Dhabi.

Soorae, P. S. 2018. *Global Reintroduction Perspectives, 2018: Case Studies from Around the Globe.* Gland, Switzerland, IUCN SSC Reintroduction Specialist Group and Abu Dhabi, AE, Environment Agency-Abu Dhabi.

Soulé, M. E. (1985) What is conservation biology?: a new synthetic discipline addresses the dynamics and problems of perturbed species, communities, and ecosystems. *BioScience.* 35, 727–734.

Sutherland, W. J., Clout, M., Depledge, M., et al. (2015) A horizon scan of global conservation issues for 2015. *Trends in Ecology & Evolution.* 30, 17–24.

Tarszisz, E., Dickman, C. R. & Munn, A. J. (2014) Physiology in conservation translocations. *Conservation Physiology.* 2, cou054.

Teixeira, C. P., De Azevedo, C. S., Mendl, M., Cipreste, C. F. & Young, R. J. (2007) Revisiting translocation and reintroduction programmes: the importance of considering stress. *Animal Behaviour.* 73, 1–13.

Trouwborst, A., Blackmore, A., Blyth, S., Fleurke, F., McCormack, P. & Gaywood, M. J. (2023) Conservation translocations and the law. In Gaywood, M. J., Ewen, J.G., Hollingsworth, P. M. and Moehrenschlager, A. (eds.) *Conservation Translocations.* Cambridge, Cambridge University Press.

van Heezik, Y., Seddon, P. J. & Maloney, R. F. (1999) Helping reintroduced houbara bustards avoid predation: effective anti-predator training and the predictive value of pre-release behaviour. *Animal Conservation.* 2, 155–163.

Vargas, A. & Anderson, S. H. (1998) Ontogeny of black-footed ferret predatory behavior towards prairie dogs. *Canadian Journal of Zoology.* 76, 1696–1704.

Vucetich, J. A. & Nelson, M. P. (2007) What are 60 warblers worth? Killing in the name of conservation. *Oikos.* 116, 1267–1278.

Wallach, A. D., Bekoff, M., Batavia, C., Nelson, M. P. & Ramp, D. (2018) Summoning compassion to address the challenges of conservation. *Conservation Biology.* 32, 1255–1265.

Webster, J. (1994) *Animal Welfare: A Cool Eye towards Eden.* Oxford, Blackwell Publishing.

Yeates, J. W. (2010) Death is a welfare issue. *Journal of Agricultural and Environmental Ethics.* 23, 229–241.

7 · Conservation Translocations for Plants

JOYCE MASCHINSKI AND
MATTHEW A. ALBRECHT

7.1 Introduction

Recent estimates indicate that between 20 and 40 per cent of plant species worldwide are considered at risk of becoming extinct in the wild (Pimm & Raven, 2017; Nic Lughadha et al., 2020). Conservation assessments of 241,910 vascular plants and bryophytes reveal that up to 32,542 species are threatened with extinction (Bachman et al., 2018). Because the conservation status of many plant species is unknown, it is possible that these numbers are even higher. Between 1900 and 2018, 571 modern plant extinctions have occurred, at a rate as much as 500 times faster than background rates (Humphreys et al., 2019). Patterns of extinction show disproportionate extinctions in high diversity regions with tropical or Mediterranean climates, islands, and in woody perennial plant species. Given that the emerging impacts of climate change forebode unprecedented risk and rates of endangerment, traditional conservation measures (i.e. habitat protection and management) alone are unlikely to be able to prevent species extinction.

Touted worldwide as a valuable conservation strategy, in the next century plant reintroduction and other types of conservation translocation will certainly be a solution to stabilise and restore declining plant populations that face global change (Kennedy et al., 2012). In the USA, 77 per cent of 254 recovery plans recommended reintroductions for endangered and threatened plants in the continental USA, and almost all listed Hawaiian species have conservation translocation recommended in recovery plans (M. Albrecht, unpublished data). Reintroductions and other conservation translocations are recommended for 36 per cent of Australian threatened plant species, while seed collection and storage are recommended for 74 per cent (Hoeppner & Hughes, 2018). Given that many threatened plant species occur outside of statutory protected habitats (e.g. cactus status reported by Goettsch et al., 2018), conservation translocations can provide an essential option for plant conservation.

The intention of plant conservation translocation is to ensure the conservation of a species in a natural context where it can undergo evolution, usually within its historical indigenous range. It is never the first action to take for a critically endangered species, even when crisis is imminent. For species highly threatened with extinction, actions to control threats and manage habitat are essential first steps and should be coupled with *ex situ* collection (Guerrant et al., 2004; Robichaux et al., 2017). Conserving wild rare plant populations in natural habitats in as many locations as possible should be our highest priority. *Ex situ* collections not only provide a backup safeguard for wild populations, but can provide propagules for future conservation translocations.

Plant conservation translocation combines the art and science of horticulture, ecology, and evolution. Although early efforts to reintroduce plants to the wild often failed (Falk et al., 1996; Godefroid et al., 2011), more recently practitioners have refined the practice and have shared their experiences of success and failure for the sake of improving future practice (Maschinski & Haskins, 2012; Albrecht et al., 2019; CPC, 2019). Over the past 35 years, the number and persistence of plant conservation translocations have increased. Akin to restoration ecology, the science of conservation translocation ecology is emerging as practitioners conduct translocations as experiments (Abeli & Dixon, 2016). Following guidelines and carefully planning translocation events prior to implementation may be responsible for improved practice. Many fundamentals of conservation translocation outlined in this book and by guidelines (IUCN, 2013; National Species Reintroduction Forum, 2014) apply to animal and plant reintroductions, while specific guidelines for rare plants are also available (Commander et al., 2018; CPC, 2019). Here, we address some special characteristics of plant biology that influence conservation translocations and discuss designs that maximise successful population establishment.

7.2 Plant Characteristics and the Associated Considerations for Conservation Translocations

7.2.1 Individual Plants Are Sedentary

When considering a conservation translocation for a plant species, several general characteristics of plants should influence planning. For a plant, place is paramount. An individual plant will usually spend its entire lifetime in one place. The longer-lived the species, the more above and below ground changes potentially will occur. Site-specific factors (duration of freezing temperature days, light conditions, timing and frequency of precipitation, soil water holding capacity, etc.) influence an individual plant's or a species' ability to grow and persist in a habitat.

In contrast to an individual plant, next generation seeds or clonal propagules may move. Dispersal from parent-specific locations may change the extent of a population or the total range of a plant species. In a conservation translocation context, both individual site and options for expansion must be considered.

7.2.1.1 Consequences of Sedentary Habit that Affect Conservation Translocations

Microsite Is Important Appropriate placement of translocated individuals is critical for long-term persistence; conversely, inappropriate placement can cause a conservation translocation to fail. Prior to installation, practitioners may test microsite characteristics of the target species *in situ* to quantify the parameters needed for a successful translocation (Maschinski et al., 2012; Richardson et al., 2013). Alternatively, conducting a translocation as an experiment with microsite as a variable can determine the optimal translocation microsite (Wendelberger & Maschinski, 2009; Roncal et al., 2012; Maschinski et al., 2012; Peterson et al., 2013).

Microsite is also important for non-vascular plants, fungi, and lichens. The presence of high-quality deadwood and rock outcroppings supports saprophytic fungi (Komonen & Muller, 2018), bryophytes, and lichens (Humphrey et al., 2002). Understanding the role of microsite for non-vascular plant diversity is a ripe opportunity for future experiments.

A great challenge is to understand not only the microhabitat conditions required for rare plant species today, but also the extent to which the microhabitat conditions at recipient sites are likely to change in the future.

Localised Adaptations Are Possible The sedentary nature of plants means that they are often adapted to local environmental conditions. Localised adaptations may occur over short distances, particularly when gene flow is restricted, dispersal is short-range, and a strong selection gradient exists (Bradshaw, 1972).

New molecular genomic techniques afford opportunities to quantify and identify genes within a population that may be associated with local adaptation. For example, in a genomic analysis of 10 populations of the chocolate tree *Theobroma cacao*, Nelson et al. (2020) found that 71.5 per cent of genes showed significant associations with changes in abiotic factors, while 6.5 per cent of genes related to disease resistance. Sensitivity to abiotic change is certainly a concern for a species' persistence in a particular location with rapidly changing climate.

Local adaptation can play a major role in determining the dynamics of range shifting under climate change. Short dispersal distance, high velocity of climate change, and high local density dependent effects increase extinction risk (Atkins & Travis, 2010). These are relevant concerns for selecting appropriate seed sources (see Section 7.5.1) and for selecting the recipient conservation translocation site that will be suitable now and in the future (see Dalrymple & Bellis, this volume). Understanding the breeding biology, the population genetic structure, and, if possible, the genomic structure along with the environmental conditions associated with your target species also have important implications for selecting seed sources for conservation translocations (see Neaves et al., this volume, for more genetic details).

7.2.2 Plants Can Withstand Stress in Multiple Ways

Because an individual plant lacks the ability to escape quickly, plants have evolved several ways to cope with variation in the environment. Coping mechanisms include seed banks, seed or whole plant dormancy, and flexibility to allocate resources for growth, survival, or reproduction in response to changed conditions, such as resprouting after fire. These mechanisms allow for survival during harsh times and a rapid response to favorable conditions. For example, the endangered vine *Jacquemontia reclinata*, introduced to full sun coastal dunes on the shoreward side of primary dunes, rapidly allocated resources to several metres of stem growth within a year to seek light when adjacent vegetation shaded it (J. Maschinski, personal observation).

Seed dormancy allows seeds to persist in soil seed banks for many years. When precipitation is high, some annual seeds rapidly respond by germinating from the soil seed bank (Doak et al., 2002). For example, in a favourable wet year, 800,000 endangered annual San Diego thornmint *Acanthomintha ilicifolia* plants appeared in comparison to 5,000 observed in a drought year (Anderson et al., 2019).

Dormancy of whole plants, vegetative or flowering buds, or below ground structures is another means of escaping unfavourable conditions (Nicole et al., 2005) such as extreme temperatures, or changes in water or nutrient availability (Horvath et al., 2003). Specialised structures such as corms or meristems at the base of a plant also allow escape and recovery from catastrophic events such as fire, storms, or flooding (Bond & Midgley, 2001).

7.2.2.1 Consequences of Multiple Plant Stress Strategies to Conservation Translocations

Dormancy As variation increases in a target species' environment, practitioners should expect greater genetic and phenotypic variation in seed dormancy and timing of germination (Mitchell et al., 2017), particularly in short-lived species (Rees, 1996). Dormancy varies among individuals, populations, and across years due to molecular, genetic, and environmental factors (Penfield, 2017). Environmental variation experienced by maternal plants during seed maturation, the timing of development and position of a seed on the plant or within a fruit (Andersson & Milberg, 1998), as well as environmental signals in the soil seed bank (Finch-Savage & Footitt, 2017), influence dormancy and timing of germination. Changes in temperature as low as 1°C can affect whether a seed develops dormancy or not (Springthorpe & Penfield, 2015). Thus, determining how to break seed dormancy will be an important pre-translocation step. This may require many trials as seed dormancy is complex (Baskin & Baskin, 2014).

Vegetative dormancy refers to plants that are alive but are not visible above ground. In some herbaceous perennial plants, it is hypothesised to be similar to seed dormancy as a bet-hedging response to variable environments (Shefferson, 2009). Plants with subterranean storage structures may have vegetative dormancy, including orchids, some Cyperaceae, Asclepidaceae, Fabaceae, Liliaceae, and others.

Dormancy allows the possibility of sowing seeds or planting propagules while they are dormant and 'safe'. Environmental cues may then determine when emergence from dormancy occurs. Transplanting whole plants during dormancy can greatly reduce the stress of transplant shock. However, it may not be obvious whether a seed or propagule is dormant or dead until it emerges from dormancy or is tested (Kildisheva et al., 2020).

For seed translocations, dormancy may require many years of post-installation monitoring to capture emergence. Yet it may be possible to overcome germination delays of dormant seeds by providing a germination cue prior to sowing seeds at a recipient site (Turner et al., 2013; Maschinski et al., 2018); this will be especially effective if the dormancy of the target seed is understood and tested prior to the conservation translocation (Kildisheva et al., 2020). Pre-treating Florida prairie clover (*Dalea carthagenensis* var. *floridana*) seeds with a freezing treatment prior to sowing improved field germination by 6 per cent and aided overall population establishment. However, should conditions suddenly turn unfavourable, a practitioner must be prepared to water new seedlings.

Specialised Structures and Vegetative Buds Provide Multiple Ways to Propagate Some Rare Species   Compared to most animal translocations, plant translocations can begin with large founding populations, because it is relatively easy to generate many plants if the horticulture and germination requirements are understood. Propagating from seed has the advantages of increased genetic diversity and low cost. In contrast, propagating from cuttings can allow the replication of experimental genetic units across treatments and can accelerate maturation and recruitment. Wendelberger and Maschinski (2009) transplanted 141 replicated clones of *Tephrosia angustissima* var. *corallicola* derived from stem cuttings of 57 maternal plants into three microhabitats. The installed plants produced seeds within three months, whereas seedlings would have taken several months to reach maturity. This gave the translocation a jumpstart on recruiting the next generation of seedlings.

7.3 Abiotic Impacts to Consider for Translocated Populations

Changes in abiotic conditions may contribute to orders of magnitude differences in plant reproduction across years and/or populations. Reproduction is often tied to weather; however, sensitivity to weather varies among species, across populations, or across small spatial scales. Reproduction is not predictable on fine scale, although broad generalisations may be possible. For example, in a drought year, seed production is not likely to be high.

7.3.1 Population Expansion May Be Tied to Environmental Variation

The ability to establish in a place and invade new spaces that allow a population to expand will be subject to environmental variation. Plant traits such as germination vary across populations depending upon site-specific conditions (see the crenulate leadplant case study, Box 7.1). Hobbs et al. (2018) caution that in a changing environment an inverse

Box 7.1 *Case study: Time lag, spatial and temporal variation in recruitment of crenulate leadplant*

The crenulate leadplant *Amorpha herbacea* var. *crenulata* (Figure 7.1) is a USA endangered sub-shrub endemic to 31 km^2 in eastern Miami-Dade County, Florida, USA. It occurs in communities that were historically associated with frequent burning and seasonal flooding. By the 1980s, nearly all such habitat had been developed. Remaining fragments are fire-suppressed and hydrologically altered. Only two wild populations remain today; combined, these contain approximately 500 plants. Severe fire suppression and invasion by non-native plant species actively contribute to its decline.

Figure 7.1 Crenulate leadplants at the Deering Estate, Florida, USA. The species was observed responding with vigorous flowering after fire, followed by high seedling recruitment (photo: J.E. Possley). (A black and white version of this figure will appear in some formats. For the colour version, please refer to the plate section.)

In 1995, Fairchild Tropical Botanic Garden introduced 190 juvenile crenulate leadplants to pine rockland in the Deering Estate, one of the largest preserves in the County, which also had a historical record of the taxon (Gann et al., 2002). Staff grouped plants into seven clusters and spaced them throughout a heterogeneous environment. For 24 years, Fairchild has monitored the population semi-annually, recording outplant survival, seedling recruitment, and environmental conditions.

Observers noted seedling recruitment eight years after installation. Recruitment was strongly clustered both spatially and temporally. Recruitment occurred consistently near some individuals, suggesting low dispersal distances and favourable spatial pockets. Soil moisture and bulk density were the most important factors discriminating high and intermediate outplant survival and recruitment. In 2008, after the area's first fire in 15 years, seedling recruitment peaked significantly (Figure 7.2). However, only 8 per cent of seedlings survived to the next year. Transplant survival from 1995 to 2019 was 16.5 per cent. Of 97 individuals alive in 2019, the majority (60 per cent) were young seedlings under 10 cm tall. There has been only one case to date where an F2 seedling reached maturity and began to produce its own seedlings. Long-term population persistence depends upon reintroducing intermittent fire in the habitat.

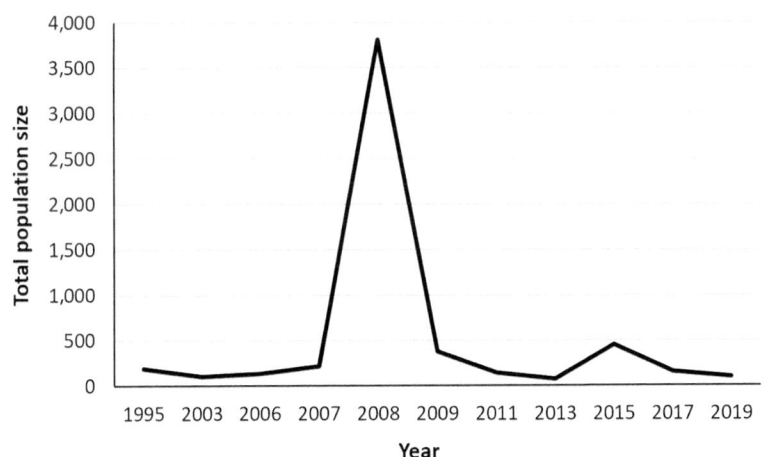

Figure 7.2 Total crenulate leadplants at the Deering Estate, from installation of 190 plants in 1995 to 97 individuals (33 outplants and 64 recruits) in 2019. The 2008 spike represented seedling recruitment following a prescribed fire.

Allee effect may occur, whereby adults may be able to persist but regeneration is reduced or eliminated. Whether a particular population can adapt to changing climate may be related to the heterogeneity of the environment that influences seed dormancy and germination (Finch et al., 2019).

With changing climate, a location that currently supports a rare species population may have mismatched phenology or environmental cues needed for germination in the future (Hobbs et al., 2018). Long-term shifts in precipitation may change the optimal management needed to sustain a rare plant population (Bernardo et al., 2016). As drought frequency increased, less management was required to reduce canopy cover as higher levels of shade improved the persistence of endangered *Astragalus bibullatus* (Bernardo et al., 2016). Practitioners can mitigate high environmental variation by testing plant performance across spatial and temporal variation gradients, by using many founders, and by monitoring long term.

7.3.2 Habitat Maintenance Will Be Necessary to Ensure Translocated Population Health

Threat management post translocation is necessary to ensure the long-term persistence of most translocated species (Bialic-Murphy et al., 2018; Figure 7.3). Identifying which parts of the lifecycle have the greatest impact on population growth rate (vital rates) can provide guidance for land managers. For example, seedling herbivory by molluscs and rodents threatens both wild and reintroduced *Delissea waianaeensis* population persistence. Using transfer function analysis, Bialic-Murphy et al. (2018) examined the comparative benefit versus cost of controlling each herbivore. They found that controlling rodents that ate adult plants increased fertility 5.9-fold at a lower cost than treating mollusc seedling herbivores, thereby resulting in a higher economic efficiency for population sustainability.

For many species, gaps in the plant canopy are essential for survival and recruitment (Richardson et al., 2013). The absence of natural disturbance that periodically opens gaps may prevent growth and recruitment in a translocated population. Because humans often suppress disturbance regimes, disturbance-adapted species may require ongoing management on two fronts: one for promoting high survivorship, and a second for promoting high reproduction and recruitment (Bernardo et al., 2016). (See *Astragalus bibullatus* case study, Albrecht, this volume.)

Conservation Translocations for Plants · 221

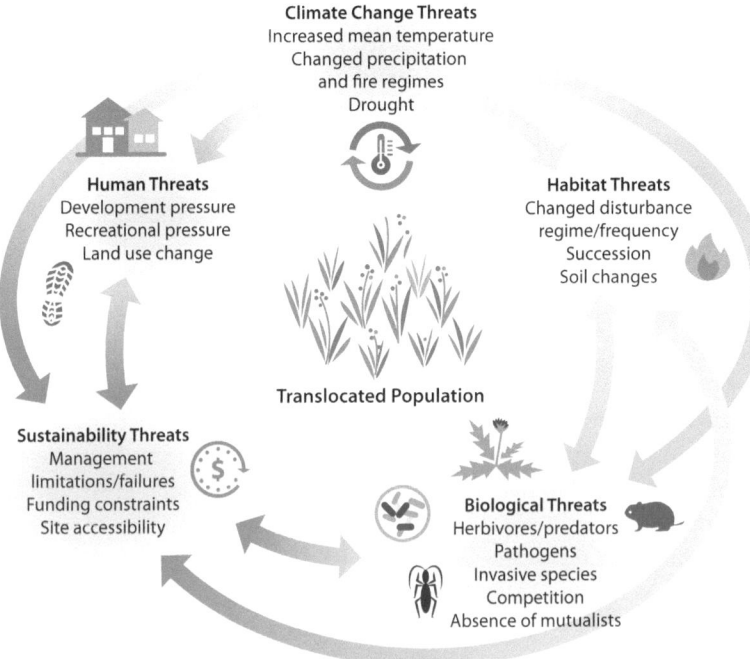

Figure 7.3 Threats a translocated plant population faces include: those imposed directly by humans that cause habitat destruction or degradation; changes in temperature, precipitation, and sea level resulting from climate change; biological threats from insect or mammalian herbivores or pathogens; changes to the condition of the habitat via invasive species incursion, changes in disturbance regime or frequency (i.e. fire, storms, flood events), altered or degraded microclimate conditions due to changes in succession and competition, changes in soil microbial communities; and sustainability threats caused by funding constraints, personnel changes, and changes in land protection, land ownership, or land management (illustration: Joyce Maschinski and Robin Mouat). (A black and white version of this figure will appear in some formats. For the colour version, please refer to the plate section.)

7.4 Mutualistic and Antagonistic Interactions to Consider in Plant Conservation Translocations

7.4.1 Mycorrhizae

Underground, a rich biotic community lives in association with plants. The carbohydrates that plants manufacture feed other organisms, often via physical connections. Fungal mycelia essentially expand the nutrient and water absorbing potential of an individual plant, thus promoting health and root development.

While many plants have mycorrhizal relationships that regulate adult plant growth, orchids also depend on mycorrhizae to facilitate seed germination (Batty et al., 2006). These relationships may be specialist or generalist (Waud et al., 2017). In orchid conservation introductions outside of natural range in China, Downing et al. (2017) found that both narrow- and wide-ranging orchids established relationships with new mycorrhizal fungi when translocated. Species with a wide range tended to have a greater number and more diverse groups of fungi than narrow-range orchid species.

Using mycorrhizae in plant translocations can be beneficial. Propagating plants with native whole soil supplements (collected with landowner permission) from the propagule source site or from the recipient site can provide locally adapted mycorrhizae and improve plant performance (Haskins & Pence, 2012; Maltz & Treseder, 2015; Middleton et al., 2015). However, practitioners should be cognisant that field soil collected from the top 20–30 cm also may contain pathogens (Haskins & Pence, 2012; Mitchell et al., this volume).

The efficacy of mycorrhizal additives in a translocation context varies with soil nutrients, the diversity of the recipient community, and fungal and plant phylogeny (Thompson, 2015; Hoeksema et al., 2018). Plant response to mycorrhizal additives is most positive when plants are phosphorus-limited rather than nitrogen-limited and tends to be more positive when the soil community is complex and diverse (Hoeksema et al., 2010). Successful reintroduction of rare species may require the simultaneous restoration of below-ground microbial communities, particularly if recipient communities are degraded (Koziol & Bever, 2017).

Certainly, the persistence of microbe inoculum in a field setting (Weinbaum et al., 1996), as well as the impacts on the entire plant community over time (Wubs et al., 2016), are vital topics for future research. For species with an obligate reliance on soil microbes, using nurse plants that have been inoculated with soil microbes from remnant sites is a promising method for introducing beneficial soil microbes to translocation sites without having to inoculate every transplant. Studies in grasslands have found that soil microbes can spread up to two metres from nurse plants to non-inoculated neighbouring plants (Middleton & Bever, 2012) and persist in locations over time to benefit next generation seedling recruitment (Middleton & Bever, 2012; Koziol & Bever, 2017).

7.4.2 Pollinators

With pollinators threatened worldwide by broad-spectrum pesticide application, land use change, and habitat loss, it is reasonable to ask whether pollinator specificity will increase extinction risk or whether lack of pollinator presence will limit conservation translocation success. Pollinators are key mutualists that aid plant sexual reproduction. Reviews of plant–pollinator interactions indicate that few plant guilds are served by specialist pollinators, and those that are tend to be tropical species (Johnson et al., 2000). More common, especially in the northern hemisphere, are generalist pollination systems. Even rare plant species share the services of many types of pollinators, which suggests they are linked to a larger community network than sometimes imagined. The Center for Plant Conservation Pollination Database contains records for 2,115 plant taxa and 2,144 pollinators. Associated with 1,217 globally rare plant species are 345 generalist pollinator observations; many generalist pollinators service more than one globally rare plant (CPC, 2021). Using this tool, viewers can see the web of pollinator interactions.

For a rare plant species that has a specialist pollinator, such as *Caladenia hastata*, an endangered sexually deceptive orchid, the presence of its specialist wasp pollinator, *Lestricothynnus hastata*, is required for a successful translocation (Reiter et al., 2017). Out of 233 potential translocation sites, only five harboured the pollinator and only two had vegetation structure and composition similar to that of the wild populations. Consequently, the pollinator was the main limiting factor for *Caladenia hastata* range expansion (Reiter et al., 2017).

7.4.3 Herbivores and Predators

Another consequence of being sessile is that plants are unable to evade herbivores and predators easily. In many terrestrial and aquatic ecosystems, the widespread loss of apex predators has resulted in trophic cascades and increased herbivore abundance (Estes et al., 2011). Due in part to the loss of large predators, overabundant deer herds in many parts of North America and Europe have altered vegetation structure and composition by selectively consuming certain species and functional groups such as tree seedlings and forbs (Côté et al., 2004).

Herbivores and seed predators can negatively impact plant abundance, distribution, and population growth rates (Maron & Crone, 2006). However, species vary widely in their susceptibility to herbivore-mediated declines in fitness depending on site disturbance and the availability of light and resources (Maron & Crone, 2006; Maron et al., 2014). Seed predators are likely to affect the fitness of seed-limited populations (Crawley, 1997). Translocation programmes should consider that herbivore impacts will likely vary among restoration techniques, microhabitats, sites, and years.

Practitioners should always screen *ex situ* grown material for pathogens and pests prior to outplanting to avoid contaminating wild populations (CPC, 2019; Mitchell et al., this volume). Although invertebrate herbivores can negatively impact endangered plants (Joe & Daehler, 2008), excluding vertebrate consumers and predators is more common in plant translocations (Guerrant, 2012). Excluding herbivores increases establishment and survival in most translocations (e.g. *Astragalus bibullatus* case study, Albrecht, this volume). In the jarrah forests of south-western Australia, Daws and Koch (2015) found that protecting reintroduced understory perennials from large mammals with plant guards increased survival, growth, and lateral spread compared to unprotected plantings. However, excluding herbivores or predators across an entire site can increase competing vegetation and thereby reduce the persistence of rare plant populations over time (Wilsey and Martin, 2015). Thus, it is important for practitioners to consider both the direct and indirect effects of herbivore or predator exclusion in plant translocations.

When excluding consumers or seed predators, practitioners should also be aware of the unintended consequences of caging such as modifications to light availability and microclimate. Cage design and size can influence the magnitude of microclimatic alterations: even small, open-topped exclusion cages can increase mean overall temperature relative to uncaged locations (Evans et al. 2018). Changes in microclimate could influence temperature-driven processes such as seed dormancy, bud break, and flowering onset. To minimise alterations to microclimate, small cages with as large a mesh size as possible are recommended (Evans et al., 2018).

Although herbivores and predators can be a source of pressure impacting the outcomes of plant translocations, it is also important to note that successful plant translocations may lead to the beneficial survival of co-threatened dependent species. For example, using microsatellite data, Moir et al. (2016) found that the genetic structures of endangered feather leaf banksia *Banksia brownii* and herbivorous plant louse *Trioza barrettae* showed high congruency. When a herbivore is closely tied to a particular plant species, decline of one could potentially cascade to the other with the dependent herbivore species often being lost to extinction before the host. Conserving a rare host plant may also conserve the biota dependent upon the species.

7.5 Factors Related to Successful Plant Conservation Translocations

The outcome of plant translocations varies greatly among species. Life history traits, abiotic and biotic characteristics of recipient sites, the design of founder populations, planting technique, and genetic issues can interact to influence translocation success. Comprehensive reviews of well-monitored and documented plant translocations shed light on the relative importance of these factors in determining translocation success across a broad array of species and ecological contexts.

7.5.1 Selection of Source Populations

Genetic factors play a significant role in translocation outcomes. Without careful selection of source material and appropriate genetic management, translocated populations may be maladapted, experience inbreeding or outbreeding depression, and lack adaptive potential and resilience to environmental change (Robichaux et al., 1997). While genetic considerations in translocations are often complicated and species-specific, some generalities have emerged from the plant translocation literature.

Plant translocations are more likely to be successful when genetic material comes from locations where the climatic and environmental conditions are similar to recipient sites (Lawrence & Kaye, 2011; Noël et al., 2011). Although conventional thinking is to source only from local populations, there are no simple geographic distance rules for matching source populations with recipient sites. In spatially heterogeneous environments, geographic and genetic distance is often decoupled. Different types of seed sourcing strategies can be applied based on genetic and environmental differences between populations and projected effects of future climate change on the species (Breed et al., 2013).

Knowing the genetic structure prior to translocation can help determine whether to separate or mix population sources. In some reviews of rare plant translocations, short-term success was greater when using material from mixed source populations rather than a single source

(Godefroid et al., 2011; Dalrymple et al., 2012), whereas other studies found no advantage of mixing population sources (Liu et al., 2015). In rare outcrossing species, mixing source material can increase genetic diversity in the founder population and improve the chances of success. For example, to reintroduce the rare outcrossing perennial herb *Centauria corymbosa* successfully, Colas et al. (2008) found that mixing population sources and planting at high densities was required to ensure adequate numbers of compatible mates for seed production.

If genetic information is lacking, there are several ecological and evolutionary factors that should be considered when selecting source populations (Havens et al., 2015). For species with a highly selfing mating system, using a few local populations that are best adapted to the translocation site is recommended to maximise evolutionary potential (Weeks et al., 2011). On the other hand, in species that are self-incompatible or show long-distance gene flow (e.g. wind-dispersed seeds), source populations can often be safely mixed for translocation (Weeks et al., 2011). For extremely rare species with little genetic variation or elevated levels of inbreeding, fitness can often be improved by bringing together genotypes from multiple populations to increase genetic diversity, especially if fragmentation has disrupted historical patterns of gene flow and resulted in population genetic differentiation (Maschinski et al., 2013). Applying the decision tree by Frankham et al. (2011) can help determine whether to mix source populations for translocation.

7.5.2 Founder and Propagule Size

A key feature of a successful translocation involves overcoming the demographic and genetic constraints of small population size. Theory predicts that small populations are at great risk of extinction from demographic and environmental stochasticity (Caughley, 1994). Empirical evidence from both plant and animal reintroductions indicates that the total number of founding individuals is often positively related to the establishment and survival of translocated populations (Godefroid et al., 2011; Silcock et al., 2019; but see Liu et al., 2015), indicating that a sufficient population size is a common factor underlying success across taxa. In a review of 174 plant translocations, Albrecht and Maschinski (2012) found that translocations founded with greater than 50 seedlings or plants had greater survival than those founded with less than 50 individuals.

However, the number of propagules needed to establish viable populations can vary among species, sites, and environmental conditions. Translocations into highly variable environments or with species whose population growth rates fluctuate greatly over time will require large numbers of founder individuals to create a viable population (Knight, 2012) and possibly repeated planting attempts.

Seedling recruitment is often a major bottleneck in population establishment and growth. Translocations can bypass this stage by using greenhouse-grown plants or vegetative cuttings rather than directly sowing seeds. Because germination rates are usually better in nursery than field conditions, using large propagules can further minimise the loss of valuable genetic material for rare species.

In long-lived species, seed translocations often show much lower success rates than seedling or whole plant translocations (Albrecht & Maschinski, 2012) or they will require thousands of seeds to be successful (Knight, 2012). Long-lived species tend to benefit more from large propagule size than short-lived species, as seeds and small propagules often have much lower establishment and survival rates relative to large propagules (Albrecht & Maschinski, 2012).

Large propagules may also reach reproductive maturity sooner than seeds. Seed production and seedling recruitment are often greater in translocations with large founder sizes (Albrecht et al., 2019). Facilitating high growth and reproduction during the early stages of a translocation is important for small populations to overcome vulnerabilities associated with stochastic effects (Iles et al., 2016). Translocations with numerous reproductive adults have increased chances of second generation recruitment (Albrecht et al., 2019).

While the initial number of founders and propagule size are often related to establishment and survival, other factors tend to become more important in determining whether plant translocations are self-sustaining over long time scales. For example, the initial number and size of propagules were less important than life-history traits and habitat type in explaining whether reintroductions produced second generation seedlings (Albrecht et al., 2019). Similarly, life form and species identity were better predictors of second generation seedling recruitment than the number of founder individuals in Australian translocations (Silcock et al., 2019). Founder sizes may be less important in creating self-sustaining populations over the long term if habitat conditions and site management are unable to support high vital rates and positive population growth rate (Knight, 2012).

7.5.3 Site Quality and Management

Habitat quality is considered the most important determinant of success in both plant and animal translocations (Wolf et al., 1996; Godefroid et al., 2011; Dalrymple et al., 2012; Cochran-Biederman, 2015). In plant translocations, a poor understanding of the biology of the focal species and its habitat requirements is a major barrier to long-term success (Godefroid et al., 2011, Dalrymple et al., 2012; Silcock et al., 2019). Knowing the key environmental drivers of population growth prior to translocation can help practitioners choose sites that match a species' habitat requirements and ecological niche (Knight, 2012). If habitat requirements are known, assessments that compare potential recipient sites quantitatively based on their physical, biological, and logistical characteristics can be used to guide site selection (Maschinski et al., 2012). However, in many cases, determining suitable habitat can be challenging when rare species exist in a few remnant populations in degraded habitats that can no longer support viable populations.

When habitat requirements are poorly understood, experimentation and predictive models can help improve the chances of translocation success. Questad et al. (2014) developed habitat suitability models with LiDAR data to identify fine-scale topographic features that differentiated high- and low-quality habitats of at-risk plant species in dryland ecosystems of Hawaii. High-quality sites maintained a buffered microclimate that reduced stress and favoured translocation success.

Using multiple sites can help identify the drivers of translocation outcomes and determine where future translocations are most likely to be successful. In a large-scale experiment, Caughlin et al. (2019) translocated two rare plant species to eight sites that each contained five different patch types that varied in degree of isolation and edge-to-area ratio. Site identity was a better predictor than patch type of long-term population dynamics, suggesting that broad-scale factors associated with particular sites (e.g. soil conditions, elevation, and site matrix conditions) were more important for long-term translocation success than variation in patch-scale attributes (e.g. herbivory, pollination, and fire regimes).

Translocations that manipulate the pre- or post-planting environment tend to show greater success than those without any manipulations or site management (Godefroid et al., 2011). Management manipulations including fencing, reducing competition, and applying appropriate disturbance regimes (e.g. fire) are correlated with increased survival of translocations over the short term (Godefroid et al., 2011; Fenu et al.,

2016). Factors that promote rapid growth and reproduction during the early stages of translocation often increase the chances of second generation recruitment and long-term population persistence (Albrecht et al., 2019).

7.5.4 Monitoring and Long-Term Success

Long-term monitoring should be part of any conservation translocation plan. Clearly, establishing new, rare plant populations requires years, if not decades (Albrecht et al., 2011, 2019). Practitioners can prepare for this long-term engagement by anticipating the possibility of loss. To track individual transplants over many years, label plants or plots with durable tags, use redundancy (e.g. place two markers and record angles to the target plant), and record GPS points. Finding labels again in a dynamic habitat may require the use of metal detectors or pit tags. Practitioners need to be aware that repeated visits to the translocated plants may compact soil and create unintended consequences. Beware of creating compacted paths or trampling target plants during monitoring.

A significant consequence of dormancy is the practical problem of the practitioner's inability to determine how the translocation is faring. If the founding plant or seed is not visible, it is hard to know whether it is alive, dormant, or dead. Without excavating dormant seeds or plant propagules (a practice we do not recommend), practitioners are restricted to assessing plants that have emerged above ground. The solution is careful documentation, flexible data management and analysis, and setting monitoring rules to account for dormancy. If a plant fails to emerge above ground during a monitoring event, do not pull the tags or labels and assume it is dead. Dormancy may last many years. When plants 'resurrect', it is important to distinguish the previously dormant individual from new ones. Conditions for resurrection from dormancy may be unpredictable. For example, in an experimental introduction of crenulate leadplant *Amorpha herbacea* var. *crenulata* to different microhabitats, Roncal et al. (2012) reported that plants installed in the grassy glade had high mortality. However, monitoring in subsequent years showed that many of these plants revived from vegetative dormancy (Lange et al., 2019).

Second generation seedling recruitment demonstrates that a translocated species can produce viable offspring and a recipient site can potentially support a self-sustaining population (IUCN, 2013). However, a lack of second generation seedling recruitment is often cited as a leading cause of translocation failure (Godefroid et al., 2011; Silcock et al., 2019), although many translocations are not monitored over time scales long enough to observe recruitment (Albrecht et al., 2011; Godefroid et al., 2011). In well-monitored plant translocations, lags in seedling recruitment are common, sometimes occurring a decade or more after a reintroduced species reaches reproductive maturity (Albrecht et al., 2019). Lags tend to be more variable across space and time in longer-lived perennials compared to short-lived species (Albrecht et al. 2019). Short-lived species often reach reproductive maturity quickly and tend to allocate resources towards growth and reproduction, while long-lived species often delay reproductive maturity and allocate resources to survival rather than reproduction. Thus, plant life histories are essential to consider and account for when evaluating translocation outcomes.

Plant growth form can also influence patterns of translocation success. In a review of 222 translocations in China, Liu et al. (2015) found that herbaceous species showed greater flowering and fruiting success than woody plants and were more likely to reach reproductive maturity during the time frames the conservation translocations were monitored. However, some short-lived woody plants from subtropical habitats showed rapid population growth and recruitment lags that were considerably shorter than some herbaceous species, when translocated to sites managed with appropriate disturbance regimes (Albrecht et al., 2019). In particular, translocated long-lived herbs of temperate grasslands often exhibited long time lags and low recruitment rates (Albrecht et al., 2019). Overall, poor recruitment rates in translocations are most often linked to inadequate precipitation and/or disturbance regimes (Bowles et al., 2015; Silcock et al., 2019).

Plant reproductive traits can also shape reintroduction outcomes. Rare species with chronically poor seed production and low seed viability or outcrossing mating systems are more likely to show recruitment limitation and may be slow to produce second generation seedlings (Albrecht et al., 2019; Silcock et al., 2019). Translocated species that allocate resources to clonal growth or vegetative reproduction, which enables some individuals to persist for long time periods, may also be slow to recruit seedlings and complete their lifecycle after translocation (Silcock et al., 2019). In such cases, clonal growth should be monitored and considered when evaluating success criteria.

Monitoring plans should account for spatial spread and metapopulation dynamics when evaluating success (Maschinski & Albrecht, 2017). For species with long-distance seed dispersal mechanisms, new recruits may appear outside of tagged plots, and therefore areas outside

Figure 1.2 Rangers in the Sultanate of Oman protect and monitor the first herd of Arabian oryx released in 1982 (photo: Mark Stanley-Price).

Figure 1.3 Beavers are ecosystem engineers and have been used in conservation translocation projects as a means of restoring ecological processes. At this Scottish site a metre-wide stream was dammed by beavers, resulting in an extensive beaver pond and associated wetland habitat (photo: Martin Gaywood).

Figure 1.4 Woolly willow is a montane species that is vulnerable to grazing pressures. In Scotland it is now restricted and at risk, but an ongoing reinforcement programme has taken place over several years to try to restore this and other subarctic willow scrub species (photo: Lorne Gill/NatureScot).

Figure 1.5 The freshwater pearl mussel has been subject to a range of pressures including pollution, river engineering, and pearl fishing, to the extent that it is now critically endangered. A range of conservation interventions is being applied and tested in Scotland, including the use of conservation translocations (photo: Sue Scott/NatureScot).

Figure 1.6 Assisted colonisation of the fruticose terricolous arctic-alpine lichen *Flavocetraria nivalis* has been tested and monitored within the Cairngorm Mountains, Scotland. Each individual transplant was tagged as shown, to assist future identification (photo: Lorne Gill/NatureScot).

Figure 1.7 Following its extirpation in the early twentieth century, the reintroduction of the white-tailed eagle to the west coast of Scotland began in the 1970s, with a second phase in the 1990s. This was followed by the 'East of Scotland Sea Eagle Project' that ran from 2007 to 2012. There are now estimated to be about 150 breeding birds in the country (photo: Lorne Gill).

Figure 4.1 Immediate release of wild-caught kākāriki in Aotearoa New Zealand that also showcases local community involvement (photo: Darren Markin).

Figure 4.5 Male hihi, *Notiomystis cincta* (photo: John Sibley).

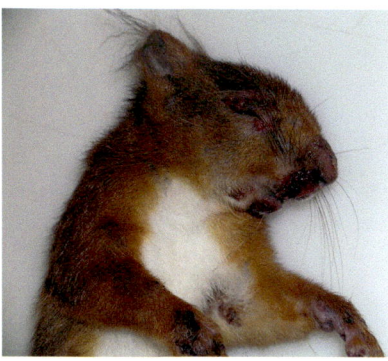

Figure 5.1 A healthy red squirrel *Sciurus vulgaris* in an English woodland (A) (photo: Simon Fraser) and exudative erythematous dermatitis of the eyelids and face in a red squirrel with squirrelpox viral disease (B) (photo: Julian Chantrey).

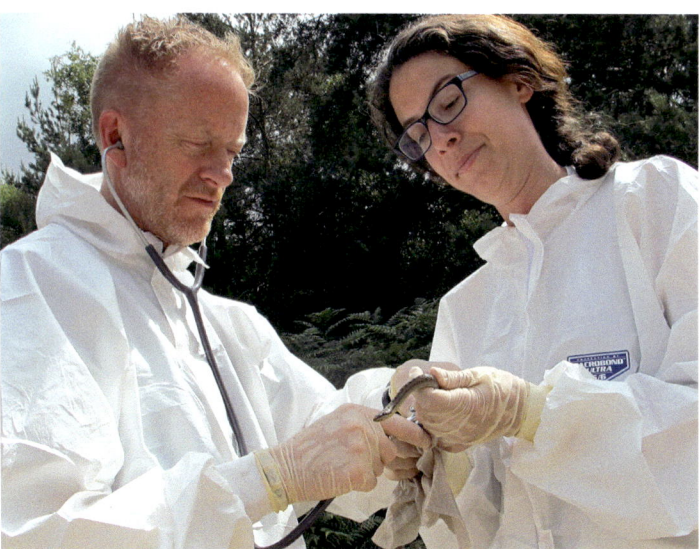

Figure 5.3 A health examination of a smooth snake *Coronella austriaca* as a component of post-release health surveillance. Note the use of personal protective equipment as a component of a biosecurity barrier to reduce the probability of transfer of parasites on to and off the release site (photo: ZSL).

Figure 6.1 Care is taken to provide conditions as close to 'natural' as is possible for northern leopard frogs *Lithobates pipiens* in a captive conservation breeding programme supplying animals for release (photo: Jill Hockaday, Calgary Zoo).

Figure 6.2 (A) Young captive-bred European mink *Mustela lutreola* are placed or are born in (B) enclosures at the release site prior to release to provide them with experience of their natural riparian habitat. Hiiumaa Island, Estonia (photos: (A) Andrew Harrington (B) Tiit Maran).

(B)

Figure 6.2 (cont.)

Figure 6.3 Live trapping and mark-recapture techniques are used to monitor a translocated population of black-tailed prairie dogs *Cynomys ludovicianus*. White plastic covers over the traps provide protection from the sun and therefore aid individual animal welfare during monitoring (photo: Fiona Le Taro, Calgary Zoo).

Figure 6.4 A satellite transmitter carefully attached to a burrowing owl *Athene cunicularia* by trained individuals for monitoring survival on migration routes (photo: Calgary Zoo).

Figure 7.1 Crenulate leadplants at the Deering Estate, Florida, USA. The species was observed responding with vigorous flowering after fire, followed by high seedling recruitment (photo: J.E. Possley).

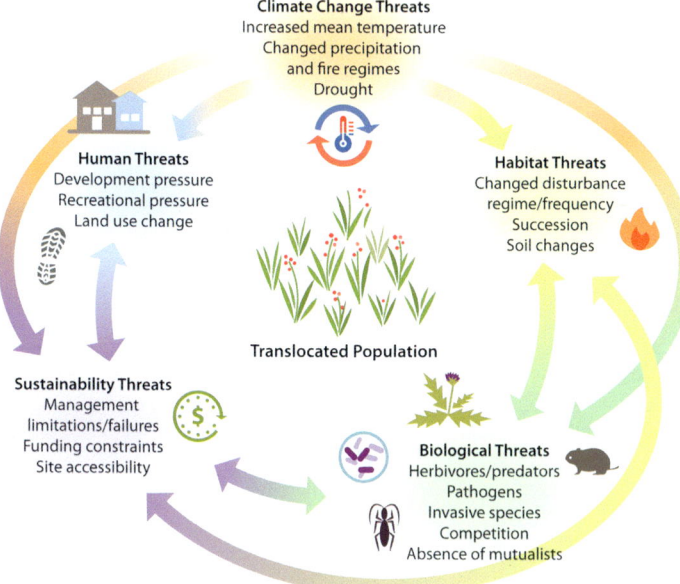

Figure 7.3 Threats a translocated plant population faces include: those imposed directly by humans that cause habitat destruction or degradation; changes in temperature, precipitation, and sea level resulting from climate change; biological threats from insect or mammalian herbivores or pathogens; changes to the condition of the habitat via invasive species incursion, changes in disturbance regime or frequency (i.e. fire, storms, flood events), altered or degraded microclimate conditions due to changes in succession and competition, changes in soil microbial communities; and sustainability threats caused by funding constraints, personnel changes, and changes in land protection, land ownership, or land management (illustration: Joyce Maschinski and Robin Mouat).

Figure 8.1 Juniper *Juniperus communis* infected with *Phytophthora austrocedri* (photo: Sarah Green).

Figure 14.1 Fruits of huagaimu (photo: Ye Chen).

Figure 14.2 Propagation of huagaimu in the nursery of Kunming Botanical Garden, Kunming Institute of Botany (photo: Yuan Zhou).

Figure 15.1 Conservation staff and volunteers carefully transplant a Pyne's ground-plum seedling for a reintroduction experiment. Seedlings were propagated from seed that had been stored in the Missouri Botanical Garden's frozen seed bank (photo: M. Albrecht).

Figure 15.3 A reintroduced Pyne's ground-plum plant (four years old) with fruits that had been caged at the beginning of the reintroduction experiment (photo: M. Albrecht).

Figure 17.1 Species released at Tijuca by REFAUNA: (A) red-rumped agouti; (B) brown howler; (C) yellow-footed tortoise (photos: Marcelo Rheingantz).

Figure 18.1 Conservation translocations of wild pampas deer from neighbouring private lands, where the deer's habitat is being modified by pine plantations, to restored grasslands inside the protected area. This resulted in two thriving populations of this threatened native species (photo: Rafael Abuín Aido – Fundación Rewilding Argentina).

Figure 20.1 Young scouts (A) meeting a golden eagle and (B) climbing trees to get an eagle's eye view of the landscape, to earn their 'Eagle Champions' badges (photo: Phil Wilkinson/SSGEP).

Figure 21.1 Native oysters in (A) the Firth of Clyde and (B) the Loch Sween Marine Protected Area (photos: (A) NatureScot; (B) David Donnan/NatureScot).

Figure 22.1 A two-year-old tortoise, hatched on Round Island, being assessed for microchip insertion and individual recognition (photo: Nik Cole).

Figure 22.2 Surveying the impact of tortoise herbivory in exclosure (pictured) and control plots on Round Island (photo: Nik Cole).

Figure 23.1 A disease-free Tasmanian devil (photo: Carolyn Hogg).

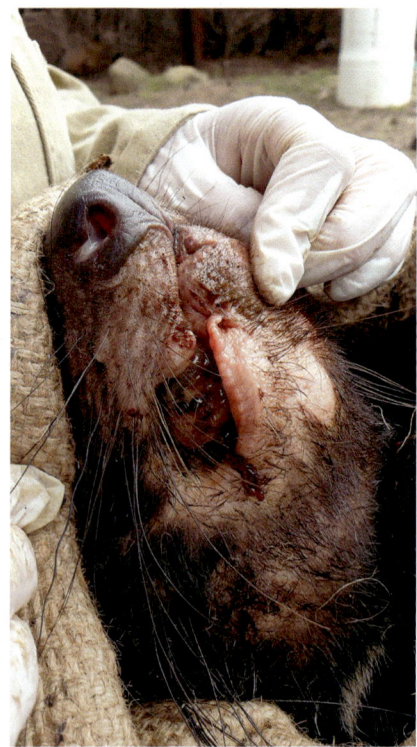

Figure 23.2 Tasmanian devil with devil facial tumour disease (photo: Carolyn Hogg).

such plots should be searched. Photographing seedlings in cultivation can help with field identification. As the population expands it may be necessary to shift the location of monitoring plots/transects over time.

7.6 Conclusion

Recovering rare plant species at the scale needed to prevent extinction will require trained, dedicated personnel, sympathetic and motivated land managers, and public support. Botanical gardens have played a vital role in the understanding of rare species biology, conservation horticulture, reintroduction, and recovery of threatened plant species because of their highly trained staff and intentionally managed collections (Guerrant et al., 2004; Kramer et al., 2010; Havens, 2017; Smith & Pence, 2017). However, obtaining financial support for such efforts is a global challenge (Fenu et al., 2019). In comparison to animal reintroductions, in the USA alone endangered plant species have received less public support and less than 5 per cent of the funding that animal species have received, despite the fact that there are hundreds more plant species listed under protection of the US Endangered Species Act than animals (Negron-Ortiz, 2014). This general phenomenon of a lower societal awareness of plant species presents a challenge (Wandersee & Schussler, 2001) that can be especially detrimental if policymakers, developers, practitioners, the media, and the general public fail to realise the importance of plants to human well-being or that plant translocation is a complex, long-term enterprise.

Involving the public can be critical to expand social and financial support for plant conservation. Citizen science projects, which have a long history of gathering accurate and valuable monitoring data on animal population trends (Tulloch et al., 2013), could be incorporated into plant translocation programmes to generate reliable data on plant population trends, threats, and habitat change. Because there are limited numbers of trained professionals for plants and fungi, citizen scientists may be an effective solution for magnifying our ability to create and monitor persistent translocated populations (Kramer et al., 2010; Irga et al., 2018).

7.7 Key Messages

- Between 20 and 40 per cent of plant species are at risk of extinction in the wild worldwide.
- Conservation translocation is an accepted strategy intended to ensure the conservation of a species in a natural context where it can undergo

evolution – usually, but not exclusively, within its historical indigenous range.
- Because plants are sedentary, practitioners should take care to select an appropriate recipient site, consider the possibility of local adaptation, and choose source material with similar climatic and environmental conditions to the recipient site.
- As variation increases in a target species' environment, practitioners should expect greater variation in seed dormancy and timing of germination, and take account of this in the monitoring plan.
- Prior to translocation, consider the focal species' dependency on below-ground mutualists (e.g. mycorrhizal fungi) and whether recipient sites have appropriate plant pollinator networks.
- When selecting founder populations, use as large a founder size as feasible to increase the chance of establishment and survival, and consider focal species' life history when choosing propagule stage.
- If the biology of the focal species is poorly understood and key environmental drivers of translocation success are unknown, use experiments, predictive models, and multiple recipient sites to test hypotheses and improve translocation outcomes.
- Ensure that threats are known and abated before and after translocation, consider herbivore or predator exclusion, and create a management plan to maintain appropriate habitat conditions or disturbance regimes over the long term.
- Develop a detailed monitoring and data management plan to track translocated individuals and make sure to account for plant life history, vegetative dormancy, and the potential for lags in next generation seedling recruitment.
- Translocation success requires long-term commitment, financial and public support, and collaboration among conservation partners.

References

Abeli, T. & Dixon, K. (2016) Translocation ecology: the role of ecological sciences in plant translocation. *Plant Ecology*. 217, 123–125.

Albrecht, M. A. (2023) Applying adaptive management to reintroductions of Pyne's ground-plum *Astragalus bibullatus*. In Gaywood, M. J., Ewen, J. G., Hollingsworth, P. M. and Moehrenschlager, A. (eds.) *Conservation Translocations*. Cambridge, Cambridge University Press.

Albrecht, M. A. & Maschinski, J. (2012) Influence of founder population size, propagule stages, and life history on the survival of reintroduced plant populations. In Maschinski, J. and Haskins, K. E. (eds.) *Plant Reintroduction in a Changing Climate: Promises and Perils.* Washington, DC, Island Press, pp. 171–188.

Albrecht, M. A., Guerrant Jr., E. O., Kennedy, K., & Maschinski, J. (2011) A long-term view of rare plant reintroduction. *Biological Conservation.* 144, 2557–2558.

Albrecht, M. A., Osazuwa-Peters, O. L., Maschinski, J., et al. (2019) Effects of life history and reproduction on recruitment time lags in reintroductions of rare plants. *Conservation Biology.* 33, 601–611.

Anderson, S., Davitt, J., Weatherson, T., Horn, C., Heineman, K. D. & Maschinski, J. (2019) *Report to San Diego Association of Governments (SANDAG) Seed Collections SANDAG Contract 5004953.* San Diego, CA, San Diego Association of Governments.

Andersson, L. & Milberg, P. (1998) Variation in seed dormancy among mother plants, populations and years of seed collection. *Seed Science Research.* 8, 29–38.

Atkins, K. E. & Travis, J. M. J. (2010) Local adaptation and the evolution of species ranges under climate change. *Journal of Theoretical Biology.* 266, 449–457.

Bachman, S. P., Lughadha, E. M. N. & Rivers, M. C. (2018) Quantifying progress toward a conservation assessment for all plants. *Conservation Biology.* 32, 516–524.

Baskin, C. & Baskin, J. (2014) *Seeds: Ecology, Biogeography, and Evolution of Dormancy and Germination.* 2nd ed. New York, Academic Press.

Batty, A. L., Brundett, M. C., Dixon, K. W. & Sivasithamparam, K. (2006) Symbiotic seed germination and propagation of terrestrial orchid seedlings for establishment at field sites. *Australian Journal of Botany.* 54, 375–381.

Bernardo, H. L., Albrecht, M. A. & Knight, T. M. (2016) Increased drought frequency alters the optimal management strategy of an endangered plant. *Biological Conservation.* 203, 243–251.

Bialic-Murphy, L., Gaoue, O. G. & Knight, T. (2018) Using transfer function analysis to develop biologically and economically efficient restoration strategies. *Scientific Reports.* 8, 2094.

Bond, W. J. & Midgley, J. J. (2001) Ecology of sprouting in woody plants: the persistence niche. *Trends in Ecology and Evolution.* 16, 45–51.

Bowles, M. L., McBride, J. L. & Bell, T. J. (2015) Long-term processes affecting restoration and viability of the federal threatened Mead's milkweed (*Asclepias meadii*). *Ecosphere.* 6, art11.

Bradshaw, A. D. (1972) Some of the evolutionary consequences of being a plant. In Dobzhansky, T., Hecht, M. K. and Steere, W. C. (eds.) *Evolutionary Biology,* vol. 5. New York, Appleton-Century-Crofts, pp. 26–47.

Breed, M., Stead, M., Ottewell, K., Gardner, M. & Lowe, A. (2013) Which provenance and where? Seed sourcing strategies for revegetation in a changing environment. *Conservation Genetics.* 14, 1–10.

Caughley, G. (1994) Directions in conservation biology. *Journal of Animal Ecology.* 63, 215–244.

Caughlin, T. T., Damschen, E. I., Haddad, N. M., Levey, D. J., Warneke, C. & Brudvig, L. A. (2019) Landscape heterogeneity is key to forecasting outcomes of plant reintroduction. *Ecological Applications.* 29, e01850.

Cochran-Biederman, J. L., Wyman, K. E., French, W. E. & Loppnow, G. L. (2015) Identifying correlates of success and failure of native freshwater fish reintroductions. *Conservation Biology.* 29, 175–186.

Colas, B., Kirchner, F., Riba, M., et al. (2008) Restoration demography: a 10-year demographic comparison between introduced and natural populations of endemic *Centaurea corymbosa* (Asteraceae). *Journal of Applied Ecology.* 45, 1468–1476.

Commander, L. E., Coates, D., Broadhurst, L., Offord, C. A. Makinson, R. O. & Matthes, M. (2018) *Guidelines for the Translocation of Threatened Plants in Australia,* 3rd ed. Canberra, Australian Network for Plant Conservation.

Côté, S. D., Rooney, T. P., Tremblay, J. P., Dussault, C. & Waller, D. M. (2004) Ecological impacts of deer overabundance. *Annual Review of Ecology, Evolution, and Systematics.* 35, 11–147.

CPC (Center for Plant Conservation) (2019) *CPC Best Plant Conservation Practices to Support Species Survival in the Wild.* Available from: https://saveplants.org/wp-content/uploads/2020/12/CPC-Best-Practices-5.22.2019.pdf [Accessed 11 May 2021].

CPC (Center for Plant Conservation) (2021) *CPC Pollinators of Rare Plants Database.* Available from: https://saveplants.org/national-collection/pollinator-search/ [Accessed 19 April 2021].

Crawley, M. J. (1997) Plant-herbivore dynamics. In M. J. Crawley (ed.) *Plant Ecology.* Oxford, Blackwell, pp. 401–474.

Dalrymple, S. E. & Bellis, J. M. (2023) Conservation translocations: planning and the initial appraisal. In Gaywood, M. J., Ewen, J. G., Hollingsworth, P. M. and Moehrenschlager, A. (eds.) *Conservation Translocations.* Cambridge, Cambridge University Press.

Dalrymple, S. E., Banks, E., Stewart, G. B. & Pullin, A. S. (2012) A meta-analysis of threatened plant reintroduction from across the globe. In Maschinski, J. and Haskins, K. E. (eds.) *Plant Reintroduction in a Changing Climate: Promises and Perils.* Washington, DC, Island Press, pp. 31–50.

Daws, M. & Koch, J. (2015) Long-term restoration success of re-sprouter understorey species is facilitated by protection from herbivory and a reduction in competition. *Plant Ecology.* 216, 565–576.

Doak, D. F., Thomson, D. & Jules, E. S. (2002) Population viability analysis for plants: understanding the demographic consequences of seed banks for population health. In Beissenger, S. R. and McCullough, D. R. (eds.) *Population Viability Analysis.* Chicago, The University of Chicago Press, pp. 312–337.

Downing, J. L., Liu, H., Shicheng, S., et al. (2017) Contrasting changes in biotic interactions of orchid populations subject to conservation introduction vs. conventional translocation in tropical China. *Biological Conservation.* 212, 29–38.

Estes, J. A., Terborgh, J., Brashares, J. S., et al. (2011) Trophic downgrading of planet Earth. *Science.* 333, 301–306.

Evans, P., Davis, E. L., Gedalof, Z. E. & Brown, C. D. (2018) Small herbivore exclosure cages alter microclimate conditions. *Forest Ecology and Management.* 415–416, 118–128.

Falk, D. A., Millar, C. & Olwell, P. (1996) *Restoring Diversity: Strategies for Reintroduction of Endangered Plants.*, Washington, DC, Island Press.

Fenu, G., Cogoni, D. & Bacchetta, G. (2016) The role of fencing in the success of threatened plant species translocation. *Plant Ecology*. 217, 207–217.

Fenu, G., Bacchetta, G., Charalambos, S. C., et al. (2019) An early evaluation of translocation actions for endangered plant species on Mediterranean islands. *Plant Diversity*. 2, 94–104.

Finch. J., Walck, J. L., Hidayati, S. N., Kramer, A. T., Lason, V. & Havens, K. (2019) Germination niche breadth varies inconsistently among three *Asclepias* congeners along a latitudinal gradient. *Plant Biology*. 21, 425–438.

Finch-Savage, W. E. & Footitt, S. (2017) Seed dormancy cycling and the regulation of dormancy mechanisms to time germination in variable field environments. *Journal of Experimental Botany*. 68, 843–856.

Frankham, R., Ballou, J. D., Eldridge, M. D. B, et al. (2011) Predicting the probability of outbreeding depression. *Conservation Biology*. 25, 465–475.

Gann, G., Bradley, K. A. & Woodmansee, S. W. (2002) *Rare Plants of South Florida: Their History, Conservation and Restoration*. Miami, The Institute for Regional Conservation, 1063 pp.

Godefroid, S., Piazza, C., Rossi, G., et al. (2011) How successful are plant species reintroductions? *Biological Conservation*. 144, 672–682.

Goettsch, B., Duran, A. P. & Gaston, K. J. (2018) Global gap analysis of cactus species and priority sites for their conservation. *Conservation Biology*. 33, 369–376.

Guerrant, E. O., Jr. 2012. Characterizing two decades of rare plant reintroductions. In Maschinski, J. and Haskins, K. E. (eds.) *Plant Reintroduction in a Changing Climate: Promises and Perils*. Washington, DC, Island Press, pp. 9–29.

Guerrant, E. O. Jr., Havens, K. & Maunder, M. (eds.) (2004) *Ex Situ Plant Conservation: Supporting Species Survival in the Wild*. Washington, DC, Island Press.

Haskins, K. E. & Pence, V. (2012) Transitioning plants to new environments: beneficial applications of soil microbes. In Maschinski, J. and Haskins, K. E. (eds.) *Plant Reintroduction in a Changing Climate: Promises and Perils*. Washington, DC, Island Press, pp. 89–107.

Havens, K. (2017) The role of botanic gardens and arboreta in restoring plants from populations to ecosystems. In Blackmore, S. and Oldfield, S. (eds.) *Plant Conservation Science and Practice: The Role of Botanic Gardens*. Cambridge, Cambridge University Press.

Havens, K., Vitt, P., Still, S., Kramer, A. T., Fant, J. B. & Schatz, K. (2015) Seed sourcing for restoration in an era of climate change. *Natural Areas Journal*. 35, 122–133.

Hobbs, R. J., Valentine, L. E., Standish, R. J. & Jackson, S. T. (2018) Movers and stayers: novel assemblages in changing environments. *Trends in Ecology & Evolution*. 33, 116–128.

Hoeksema, J. D., Bever, J. D., Chakraborty, S., et al. (2018) Evolutionary history of plant hosts and fungal symbionts predicts the strength of mycorrhizal mutualism. *Communications Biology*. 1, 116.

Hoeksema, J. D., Chaudhary, V. B., Gehring, C. A., et al. (2010) A meta-analysis of context-dependency in plant response to inoculation with mycorrhizal fungi. *Ecology Letters*. 13, 394–407.

Hoeppner, J. M. & Hughes, L. (2018) Climate readiness of recovery plans for threatened Australian species. *Conservation Biology*. 33, 534–452.

Horvath, D. P., Anderson, J. V., Chao, W. S. & Foley, M. E. (2003) Knowing when to grow: signals regulating bud dormancy. *Trends in Plant Science*. 11, 534–540.

Humphrey, J. W., Davey, S., Peace, A. J., Ferris, R., & Harding, K. (2002) Lichens and bryophyte communities of planted and semi-natural forests in Britain: the influence of site type, stand structure and deadwood. *Biological Conservation*. 107, 165–180.

Humphreys, A. M., Govaerts, R. Ficinski, S. Z., Lughadha, E. N. & Vorontsova, M. S. (2019) Global dataset shows geography and life form predict modern plant extinction and rediscovery. *Nature, Ecology & Evolution*. 3, 1043–1047.

Iles, D. T., Salguero-Gómez, R., Adler, P. B. & Koons, D. N. (2016) Linking transient dynamics and life history to biological invasion success. *Journal of Ecology*. 104, 399–408.

Irga, P. J., Barker, K., & Torpy, F. R. (2018) Conservation mycology in Australia and the potential role of citizen science. *Conservation Biology*. 32, 1031–1037.

IUCN (2013) *Guidelines for Reintroductions and Other Conservation Translocations. Version 1.0*. Gland, Switzerland, IUCN Species Survival Commission.

Joe, S. M. & Daehler, C. C. (2008) Invasive slugs as under-appreciated obstacles to rare plant restoration: evidence from the Hawaiian Islands. *Biological Invasions*. 10, 245–255.

Johnson, S. D., Steiner, K. E., Johnson, S. D. & Steiner, K. E. (2000) Generalization versus specialization in plant pollination systems. *Trends in Ecology & Evolution*. 15, 140–143.

Kennedy, K., Albrecht, M. A., Guerrant Jr., E. O., Dalrymple, S. E., Maschinski, J. & Haskins, K. E. (2012) Synthesis and future directions. In Maschinski, J. and Haskins, K. E. (eds.) *Plant Reintroduction in a Changing Climate: Promises and Perils*. Washington, DC, Island Press, pp. 265–275.

Kildisheva, O. A., Dixon, K. W., Silveira, F. A. O., et al. (2020) Dormancy and germination: making every seed count in restoration. *Restoration Ecology*. 28, S256–S265.

Knight, T. M. (2012) Using population viability analysis to plan reintroductions. In Maschinski, J. and Haskins, K. E. (eds.) *Plant Reintroduction in a Changing Climate: Promises and Perils*. Washington, DC, Island Press, pp. 155–169.

Komonen, A. & Muller, J. (2018) Dispersal ecology of deadwood organisms and connectivity conservation. *Conservation Biology*. 32, 535–545.

Koziol, L. & Bever, J. D. (2017) The missing link in grassland restoration: arbuscular mycorrhizal fungi inoculation increases plant diversity and accelerates succession. *Journal of Applied Ecology*. 54, 1301–1309.

Kramer, A. T., Zorn-Arnold, B. & Havens, K. (2010) Assessing botanical capacity to address grand challenges in the United States. Botanical Capacity Assessment Project. Available from: https://www.bgci.org/wp/wp-content/uploads/2019/06/US-Botanical-Capacity-Report.pdf [Accessed 11 May 2021].

Lange, J., Possley, J., Cuni, L., Wintergerst, S. & Harding, B. (2019) *Conservation of South Florida endangered and threatened flora: 2018-2019 Program at Fairchild Tropical Botanic Garden.* Final report under contract #025243, Florida Department of Agriculture and Consumer Services, Division of Plant Industry, Gainesville, FL.

Lawrence, B. A. & Kaye, T. N. (2011) Reintroduction of *Castilleja levisecta*: effects of ecological similarity, source population genetics and habitat quality. *Restoration Ecology*. 19, 166–176.

Liu, H., Ren, H., Liu, Q., Wen, X., Maunder, M. & Gao, J. (2015) Translocation of threatened plants as a conservation measure in China. *Conservation Biology*. 29, 1537–1551.

Maltz, M. R. & Treseder, K. K. (2015) Sources of inocula influence mycorrhizal colonization of plants in restoration projects: a meta-analysis. *Restoration Ecology*. 23, 625–634.

Maron, J. L. & Crone, E. (2006) Herbivory: effects on plant abundance, distribution and population growth. *Proceedings of the Royal Society B: Biological Sciences*. 273, 2575–2584.

Maron, J. L., Baer, K. C. & Angert, A. L. (2014) Disentangling the drivers of context-dependent plant–animal interactions. *Journal of Ecology*. 102, 1485–1496.

Maschinski, J. & Albrecht, M. A. (2017) Center for Plant Conservation's Best Practice Guidelines for the reintroduction of rare plants. *Plant Diversity*. 39, 390–395.

Maschinski, J. & Haskins, K. E. (eds.) (2012) *Plant Reintroduction in a Changing Climate: Promises and Perils.* Washington, DC, Island Press.

Maschinski, J., Falk, D. A., Wright, S.J., Possley, J., Roncal, J. & Wendelberger, K. S. (2012) Optimal locations for plant reintroductions in a changing world. In Maschinski, J. and Haskins, K. E. (eds.) *Plant Reintroduction in a Changing Climate: Promises and Perils.* Washington, DC, Island Press, pp. 109–129.

Maschinski, J., Wright, S. J., Koptur, S. & Pinto-Torres, E. (2013) When is local the best paradigm? Breeding history influences conservation reintroduction survival and trajectories in times of extreme climate events. *Biological Conservation*. 159, 277–284.

Maschinski, J., Possley, J., Walters, C., Hill, L., Krueger, L. & Hazelton, D. (2018) Improving success of rare plant seed reintroductions: a case study of *Dalea carthagenesis* var. *floridana*, a rare legume with dormant seeds. *Restoration Ecology*. 26, 636–641.

Middleton, E. L. & Bever, J. D. (2012) Inoculation with a native soil community advances succession in a grassland restoration. *Restoration Ecology*. 20, 218–226.

Middleton, E. L., Richardson, S., Koziol, L., et al. (2015) Locally adapted arbuscular mycorrhizal fungi improve vigor and resistance to herbivory of native prairie plant species. *Ecosphere*. 6, art276.

Mitchell, J., Johnston, I. G. & Bassel, G. W. (2017) Variability in seeds: biological, ecological, and agricultural implications. *Journal of Experimental Biology*. 68, 809–817.

Mitchell, R., Green, S. & Hollingsworth, P. M. (2023) Plant health, biosecurity and conservation translocations. In Gaywood, M. J., Ewen, J. G., Hollingsworth,

P. M. and Moehrenschlager, A. (eds.) *Conservation Translocations.* Cambridge, Cambridge University Press.

Moir, M. L., Coates, D. J., Kensington, W. J., et al. (2016) Concordance in evolutionary history of threatened plant and insect populations warrant unified conservation management approaches. *Biological Conservation.* 198, 135–144.

National Species Reintroduction Forum (2014) *The Scottish Code for Conservation Translocations.*, Inverness, Scottish Natural Heritage.

Neaves, L. E., Ogden, R. & Hollingsworth, P. M. (2023) Genomics and conservation translocations. In Gaywood, M. J., Ewen, J. G., Hollingsworth, P. M. and Moehrenschlager, A. (eds.) *Conservation Translocations.* Cambridge, Cambridge University Press.

Negron-Ortiz, V. (2014) Pattern of expenditures for plant conservation under the Endangered Species Act. *Biological Conservation.* 171, 36–43.

Nelson, J. T., Motamayor, J. C. & Cornejo, O. E. (2020) Environment and pathogens shape local and regional adaptations to climate change in the chocolate tree, *Theobroma cacao* L. *Molecular Ecology.* 30, 656–669.

Nic Lughadha, E., Bachman, S. P., Leão, T. C. C., et al. (2020) Extinction risk and threats to plants and fungi. *Plants, People, Planet.* 2, 389–408.

Nicole, F., Brzosko, E. & Till-Bottraud, I. (2005) Population viability analysis of *Cypripedium calceolus* in a protected area: longevity, stability and persistence. *Journal of Ecology.* 93, 716–726.

Noël, F., Prati, D., van Kleunen, M., Gygax, A., Moser, D. & Fischer, M. (2011) Establishment success of 25 rare wetland species introduced into restored habitats is best predicted by ecological distance to source habitats. *Biological Conservation.* 144, 602–609.

Penfield, S. (2017) Seed dormancy and germination. *Current Biology.* 27, R853–R909.

Peterson, C. L., Kaufmann, G. S., Vandello, C. & Richardson, M. L. (2013) Parent genotype and environmental factors influence introduction success of the critically endangered Savannas mint (*Dicerandra immaculata* var. *savannarum*). *PLoS ONE.* 8, e61429.

Pimm, S. & Raven, P. (2017) The fate of the world's plants. *Trends Ecology Evolution.* 32, 317–320.

Questad, E. J., Kellner, J. R., Kinney, K., et al. (2014) Mapping habitat suitability for at-risk plant species and its implications for restoration and reintroduction. *Ecological Applications.* 24, 385–395.

Rees, M. (1996) Evolutionary ecology of seed dormancy and seed size. *Philosophical Transactions of the Royal Society B: Biological Sciences.* 351, 1299–1308.

Reiter, N., Vicek, K., O'Brien, N., et al. (2017) Pollinator rarity limits reintroduction sites in an endangered sexually deceptive orchid (*Caladenia hastata*): implications for plants with specialized pollination systems. *Botanical Journal of the Linnean Society.* 184, 122–136.

Richardson, M. L., Watson, M. L. J. & Peterson, C. L. (2013) Influence of community structure on the spatial distribution of critically endangered *Dicerandra immaculata* var. *immaculata* (Lamiaceae) at wild, introduced, and extirpated locations in Florida scrub. *Plant Ecology.* 214, 443–453.

Robichaux, R. H., Friar, E. A. & Mount, D. W. (1997) Molecular genetic consequences of a population bottleneck associated with reintroduction of the Mauna Kea silversword (*Argyroxiphium sandwicense* ssp. *sandwicense* [Asteraceae]). *Conservation Biology*. 11, 1140–1146.

Robichaux, R. H., Moriyasu, P. Y., Enoka, J. H., et al. (2017) Silversword and lobeliad reintroduction linked to landscape restoration on Mauna Loa and Kīlauea, and its implications for plant adaptive radiation in Hawai'i. *Biological Conservation*. 213, 59–69.

Roncal, J., Maschinski, J., Schaffer, B., Gutierrez, S. M. & Walters, D. (2012) Testing appropriate habitat outside of historic range: the case of *Amorpha herbacea* var. *crenulata* (Fabaceae). *Journal for Nature Conservation*. 20, 109–116.

Shefferson, R. P. (2009) The evolutionary ecology of vegetative dormancy in mature herbaceous perennial plants. *Journal of Ecology*. 97, 1000–1009.

Silcock, J. L., Simmons, C. L., Monks, L., et al. (2019) Threatened plant translocation in Australia: a review. *Biological Conservation*. 236, 211–222.

Smith, P. & Pence, V. (2017) The role of botanic gardens in ex situ conservation. In Blackmore, S. and Oldfield, S. (eds.) *Plant Conservation Science and Practice: The Role of Botanic Gardens*. Cambridge, Cambridge University Press,

Springthorpe, V. & Penfield, S. (2015) Flowering time and seed dormancy control use external coincidence to generate life history strategy. *Plant Biology*. 4, e05557.

Thompson, E. R. (2015) What role do plant-fungal mutualisms play in restoration ecology? Assessing the impacts of coastal dune modification on mycorrhizae, and whether reconnecting mycorrhizal networks can facilitate restoration of dune vegetation. BEnviSci Honors Thesis, School of Earth & Environmental Science, University of Wollongong, Australia. Available from: https://ro.uow.edu.au/thsci/142 [Accessed 19 August, 2019].

Tulloch, A. I. T., Possingham, H. P., Joseph, L. N., Szabo, J. & Martin, T. G. (2013) Realising the full potential of citizen science monitoring programs. *Biological Conservation*. 165, 128–138.

Turner, S. R., Steadman, K. J., Vlahos, S., Koch, J. M. & Dixon, K. W. (2013) Storage for restoration-ready seeds: the feasibility of prestorage dormancy alleviation for mine site revegetation. *Restoration Ecology*. 21, 186–192.

Wandersee, J. & Schussler, E. E. (2001) Toward a theory of plant blindness. *Plant Science Bulletin*. 47, 1.

Waud, M., Brys, R., van Landuyt, W., Lievens, B. & Jacquemyn, H. (2017) Mycorrhizal specificity does not limit the distribution of an endangered orchid species. *Molecular Ecology*. 26, 1687–1701.

Weeks, A. R., Sgro, C. M., Young, A. G., et al. (2011) Assessing the benefits and risks of translocations in changing environments: a genetic perspective. *Evolutionary Applications*. 4, 709–725.

Weinbaum, B. S., Allen, M. F. & Allen, E. B. (1996) Survival of arbuscular mycorrhizal fungi following reciprocal transplanting across the Great Basin, USA. *Ecological Applications*. 6, 1365–1372.

Wendelberger, K. S. & Maschinski, J. (2009) Linking GIS, observational and experimental studies to determine optimal seedling microsites of an endangered plant in a subtropical urban fire-adapted ecosystem. *Restoration Ecology*. 17, 845–853.

Wilsey, B. J. & Martin, L. M. (2015) Top-down control of rare species abundances by native ungulates in a grassland restoration. *Restoration Ecology*. 23, 465–472.

Wolf, C. M., Griffith, B., Reed, C. & Temple, S. A. (1996) Avian and mammalian translocations: update and reanalysis of 1987 survey data. *Conservation Biology*. 10, 1142–1154.

Wubs, E. R. J., van der Putten, W. H., Bosch, M. & Bezemer, T. M. (2016) Soil inoculation steers restoration of terrestrial ecosystems. *Nature Plants*. 2, 16107.

8 · *Plant Health, Biosecurity, and Conservation Translocations*

RUTH J. MITCHELL, SARAH GREEN, AND
PETER M. HOLLINGSWORTH

8.1 Introduction

Conservation translocations, the deliberate movements of organisms from one site to another, are intended to yield a measurable conservation benefit at the level of a population, species, or ecosystem (IUCN, 2013). This usually involves improving the conservation status of the focal species and/or restoring natural habitat, ecosystem functions, or processes. Within this definition there is a broad spectrum of plant conservation translocations. At one extreme is the movement of rare species, usually a few individuals, to reinforce a population or establish a new one (Bell, 2021). At the other end of this spectrum are large scale conservation translocations of plants for habitat restoration such as rewilding or restoration of forest or peatland (Lamb, 2018).

In addition to the conservation benefits that arise from conservation translocations, there is the potential for unintended negative consequences to arise, including the risks of pest and pathogen outbreaks (Simler et al., 2019; Frankel et al., 2020). Disease risk analysis and disease surveillance in animal conservation translocations is relatively well established, especially for vertebrates (Sainsbury & Carraro, this volume). In contrast, for plants, explicit consideration of the risks associated with pests and pathogens in conservation translocation is much rarer.

Plant pests and pathogens may impact on conservation translocations in two main ways. First, when plants are translocated, it is difficult to ensure that only the intended species is translocated. In reality a 'biological package' (*sensu* Davidson & Nettles, 1992) is moved that contains not only the plant but also microorganisms or other organisms that may include species regarded as pests and pathogens. This may impact the translocated species and/or the wider environment, and in some instances introduce novel pests and pathogens to new areas (Simler et al., 2019). Second, plants may be successfully translocated but then become infected by pests and pathogens already present at the destination site, which may impact on their establishment and survival (Silcock et al., 2019).

In this chapter we outline key aspects of conservation translocations where plant health warrants consideration and highlight the importance of increased attention to this topic in routine practice. We identify some key issues to consider when assessing plant health risks during conservation translocations and some practical steps that can be taken to improve biosecurity. Unlike animal health, which includes animal welfare (Harrington et al., this volume), and non-infectious diseases (Sainsbury & Carraro, this volume), our focus is on the plant health consequences of biotic agents (i.e. pests and pathogens). Abiotic drivers of plant fitness are covered by Maschinski and Albrecht (this volume), and the relationships between genetic diversity and fitness are covered by Neaves et al. (this volume). A brief glossary of key terminology is included in Table 8.1, noting that some terminology used with respect to plant health differs from how it is used in relation to animal health (Sainsbury & Carraro, this volume). Our chapter focuses on vascular plants but we include reference to bryophytes as these are increasingly included in conservation translocations (e.g. Wittram et al., 2015; Caporn et al., 2017).

8.2 Major Plant Health Threats

Inadvertent introductions of pests and pathogens into new geographic regions, resulting in infection or infestation of naïve (susceptible) hosts, have led to significant environmental and socio-economic impacts. Examples of catastrophic impacts of plant pathogens include the potato blight, caused by the importation of *Phytophthora infestans* from Central America into Europe, resulting in the devastating Irish potato famine in the 1840s (Goss et al., 2014). The chestnut blight pathogen *Cryphonectria parasitica* killed billions of American chestnut trees following its introduction into North America in the early twentieth century from a probable origin in Asia, and Dutch elm disease, *Ophiostoma novo-ulmi*, which was brought into Britain on Canadian elm logs, killed around 28 million elm trees between 1970 and 1990, altering British landscapes forever (Brasier, 2008). The invasive pathogen *Phytophthora cinnamomi* continues to cause enormous damage to native woody ecosystems on several continents including Australia, South Africa, and Europe (Brasier, 2008). Epidemics caused by clonal lineages of *Phytophthora ramorum*, recently

Table 8.1. *Definition of key terms.*

Term	Meaning
Biosecurity	Actions taken to minimise the risks of pest and pathogen outbreaks
Co-introduction	Accidental introduction of a pest and/or pathogen at a destination site by plant conservation translocation
Destination site	The site into which the plant is translocated
Disease	Any transmissible infection of a pathogen causing damage and/or reducing plant fitness
Donor site	The site from which plant material is taken
Ecosystem processes	The physical, chemical, and biological processes that link organisms and their environment. These may include biogeochemical/nutrient cycling, energy flow, and food web dynamics
Foundation species	A species forming a key component of an ecosystem, such that many species depend on its presence for survival
Naïve host	A host species that has previously not encountered a pest or pathogen, reducing opportunities for the evolution of defence/tolerance mechanisms
Outplanting	Planting translocated plants in destination sites
Pathogen	A virus, bacteria, fungus, or fungus-like organism that causes disease
Pest	An invertebrate such as an arthropod or nematode that causes damage to a plant, either by direct action or by acting as a disease vector
Risk assessment	The process of assessing the likelihood and impact of an event

shown to have an origin in south-east Asia, have devastated native oak forests along the west coast of the USA as well as removing larch as a viable timber species in some areas of the UK (Jung et al., 2021). Most recently, the ash dieback epidemic, caused by the invasive fungal pathogen *Hymenoscyphus fraxineus*, is estimated to cost the UK around £15 billion due to the associated loss of numerous ecosystem services (Hill et al., 2019). Introduced insect pests such as the gypsy moth *Lymantria dispar* subsp. *dispar*, the Asian longhorned beetle *Anoplophora glabripennis*, and the emerald ash borer *Agrilus planipennis* have also had major environmental and economic impacts in different parts of the globe.

Threats from new pest and disease outbreaks continue to increase due to the globalisation of trade and other human-mediated transport of biological material, coupled with climate change and other changes to natural systems (Chapman et al., 2017; Burdon & Zhan, 2020). One of the most concerning current threats comes from the bacterial plant pathogen *Xylella fastidiosa*, which has a broad host range and causes multiple disease outbreaks across a wide range of plant species (Rapicavoli et al., 2018). *Xylella* is currently found in France, Spain, Italy, the Americas, and Taiwan. In Italy, tens of thousands of commercial olive trees have been destroyed in an attempt to stop *Xylella* spreading.

Since many of the pests and pathogens listed above have been detected on plants or plant material being traded internationally, it is the commercial movement of material that is most likely to have been the original source of outbreaks in new regions (Straw et al., 2016; Hill et al., 2019; Zahiri et al., 2019; Spence et al., 2020). As the volume of plants moved for conservation translocation is considerably lower than in forestry, horticulture, or agriculture, the risks and likelihood of transferring plant pests and pathogens may also be perceived to be lower. However, plant conservation translocations are largely unregulated with no consistent risk assessment process, they often involve movement into high conservation value sites, and there are examples of the movement of pests and pathogens (e.g. *Phytophthora* species) during conservation translocation (Rooney-Latham et al., 2015; Frankel et al., 2020).

8.3 Plant Health Risks from Different Types of Plant Translocations

The reintroduction of endangered plant species within their natural range, and/or the reinforcement of extant populations, typically involves the movement of a relatively small number of individuals, with careful attention given to their placement and care at the destination site. In many cases, the habitat at the destination site may be of high conservation value, such as a national park or other type of protected area. Despite the precision of rare plant translocations, and the often high value of the translocated material and the destination site, there is a notable absence of reference to biosecurity in the rare plant conservation translocation literature.

Threatened plant species are also increasingly being considered for assisted colonisation, where populations are translocated outside their indigenous range. This is often to move species to new regions projected to be more climatically favourable as part of a climate adaptation strategy (Simler et al., 2019). Assisted colonisation translocations may also be undertaken to establish plant populations in new locations when there is a disease threat within their existing natural range (e.g. Monks et al., 2019; Summerell & Liew, 2020). Depending on the scale of the assisted colonisation, there is potential for co-introduction of pests and pathogens into new areas, with the potential for major impacts due to a lack of co-evolution between plants at the destination site and any introduced pests and pathogens (Simler et al., 2019). Most of the known examples of major plant health problems arising from movements out of indigenous range are from forestry or horticulture, rather than conservation. However, examples from these sectors (Section 8.2) show the potential severity of pest and pathogen spread. Simler et al. (2019) reviewed assisted colonisation practices in North America and found that although pest and pathogen risks are occasionally mentioned, there is a general lack of regulation or best practice guidance to reduce or mitigate biosecurity risks.

Beyond the translocation of rare species, plants are also frequently translocated for conservation purposes when the populations themselves are not threatened. This type of conservation translocation is for restoring natural habitat or ecosystem functions or processes such as restoration of peatland, grassland, moorland, or woodland and can involve the establishment of considerable numbers of individuals. The scale of these translocations presents a risk of pest and pathogen establishment over large areas, potentially spreading to other host plants (Rooney-Latham et al., 2019). The nature of this risk has only recently been realised. Sims and Garbelotto (2021), for example, published the first controlled survey linking the presence of entire *Phytophthora* species assemblages to failing restoration projects and to the plant production facilities that provided plant stock for restoration, while showing that *Phytophthora* species were absent in neighbouring undisturbed sites.

8.4 Plant Health Risks from Different Plant Translocation Methods

Translocation of plants can use material from different stages of the plant's lifecycle and different planting methods (Maschinski & Albrecht, this volume). These different approaches carry different pest and pathogen risks.

8.4.1 Soil

A key difference between plant and animal conservation translocations is that plant translocations often involve the movement of associated soil and the organisms contained within it. While some of these soil organisms are beneficial to the plant, such as mycorrhizal fungi (Maschinski & Albrecht, this volume), it is very difficult to know what is in the soil, particularly with respect to pests and pathogens (Migliorini et al., 2015). This risk has recently been highlighted in North American nurseries growing native plants for conservation translocation. *Phytophthora tentaculata* was detected for the first time in the USA in native plant nurseries in four California counties and in restoration sites on orange sticky monkey flower *Diplacus aurantiacus*, coffeeberry *Frangula californica*, and sage *Salvia* spp. (Rooney-Latham et al., 2015). Following this discovery, a wider survey found that *Phytophthora* species were common on nursery stock grown for restoration and revegetation purposes in California (Rooney-Latham et al., 2019) and that 25 new *Phytophthora* species, including *P. tentaculata, P. cactorum*, and other new or new hybrid *Phytophthora* species, had been unintentionally but extensively introduced into restoration areas in the greater San Francisco Bay Area (Eshleman et al., 1998; Garbelotto et al., 2018). This led to an extensive response to coordinate efforts to reduce their spread (Frankel et al., 2018).

8.4.2 Mature Plants

Mature plants are typically moved with a considerable amount of attached soil, and in the case of trees, the sheer size of the individuals being moved increases the potential for the co-movement of pests and pathogens. As mature plants are typically held and grown in nursery settings, this offers potential for further pest and pathogen spread. For instance, Osterbauer et al. (2013) surveyed nurseries in Oregon for hazardous conditions that might contribute to the spread of *Phytophthora ramorum*. Hotspots for infection of nursery stock include the use of untreated water for irrigation, plants being left in standing water, and the re-use of unsterilised plant pots (Osterbauer et al., 2013).

8.4.3 Seeds, Bulbs, and Tubers

The use of seed in conservation translocations is considered intrinsically lower risk than translocations involving living vegetative tissue, as many (but not all) plant pathogens are not transmitted by seed (Anderson et al., 2004). Nevertheless, some pests and pathogens can be moved via seed or seed batches. *Hymenoscyphus fraxineus*, which causes ash dieback, is found on seed (McCartan et al., 2015). The weevil *Bradybatus kellneri* was found in the UK on a consignment of Lobel's maple *Acer lobelii* seeds from northern Italy in 2017 and has since been found in the wider environment on planted field maple *Acer campestre* trees (Lane et al., 2020). In addition, pathogenic bacteria colonising seed can result in pathogen transmission to non-host plants (Darrasse et al., 2010). Bulbs and tubers are susceptible to infection by pests and diseases, and this is well documented in the agricultural and horticultural literature (Read, 1989; Strange & Scott, 2005).

8.4.4 Turf

Turf translocation is primarily used when whole plant communities are translocated (Trueman et al., 2007; Pywell et al., 2011; e.g. Le Stradic et al., 2016) and can act as a seed source for multiple species. However, turf translocation also brings with it multiple disease risks: the pests and pathogens associated with all the different plants within the community, plus any within the soil. Fortunately, due to the logistics of movement, turf transplants usually occur within the immediate vicinity of the donor site; where this is the case, the risk of widespread transmission of pests and pathogens not already present at the site is limited, although transmission could still occur through dirty equipment (see Section 8.4.6).

8.4.5 Novel Methods

For the widespread conservation translocation of plants, particularly where there is a wish not to damage the donor population, novel cultivation and planting methods have been developed. For example, to achieve large scale habitat restoration of bare peat and the re-establishment of functioning bogs, new methods were developed for the establishment of *Sphagnum* mosses. In order to overcome issues surrounding a lack of source material, Moors For the Future Partnership (MFFP) worked with Micro-propagation Services to establish Beadamoss® (Wittram et al., 2015). The approach involves micro-propagated *Sphagnum* being chopped into fragments and placed in a special gel bead. Different beads can contain different species of *Sphagnum*, allowing multiple species to be sown simultaneously over a large area (Caporn et al., 2017). An alternative, also developed by MFFP, is SoluMoss®, which uses longer strands of *Sphagnum* suspended in a liquid solution to disperse the plants. Part of the rationale for using Beadamoss® and SoluMoss® is that they are relatively safe in terms of biosecurity risk with the propagules grown within a sterile laboratory environment (Wittram et al., 2015). Alternative methods, such as translocation of clumps of *Sphagnum*, entail a higher biosecurity risk with potential movement of pests and pathogens from the donor to the restoration site; these include bulgy-eye disease *Cryptosporidium baileyi* that affects birds, and heather beetle *Lochmaea suturalis* that mainly affects common heather *Calluna vulgaris* (Wittram et al., 2015).

8.4.6 Contaminated Equipment, Machines, and Tools

In addition to the risks of pest and pathogen transmission from plants, plant propagules, and rooting media, there is an additional biosecurity risk to plant conservation translocations from equipment involved in the translocation (Ranawaka et al., 2020; Wondafrash et al., 2021). At the fine scale, this may include pest and pathogen transport on clothing, footwear, and hand tools used in plantings. For conservation translocations involving larger scale activities and the use of machines at the translocation sites, vehicle tyres and other machine surfaces can be contaminated with non-trivial volumes of soil and other organic matter and represent a potential vector for pest and pathogen spread.

8.5 Developing a Balanced Approach to Plant Health for Conservation Translocations

There are multiple types of plant conservation translocations, and it is clear that some conservation translocations will be riskier than others. Where small amounts of plant material are moved over small distances, the risks may often be low. However, other translocations involve greater risk, including:

- Species known to be susceptible to major pests or pathogens that lead to high rates of mortality and/or damage.
- Species known to be susceptible to pests and pathogens that also impact on a broad range of host species.
- Translocations that involve moving large volumes of plant material and/or soil.
- Translocations that move plant material long distances and/or cross natural ecological barriers.
- Translocations that involve propagation at facilities with poor biosecurity.

There is thus a balance to be achieved, involving greater general awareness of the plant health risks associated with conservation translocations, and embedding appropriate practical and efficient biosecurity practices where required, while acting proportionately and not overburdening low risk conservation translocations such that unnecessary biosecurity measures become a barrier to positive conservation actions.

A general limiting step in the practical implementation of the above philosophy is a lack of guidance targeted at practitioners on how to assess plant health risks during plant conservation translocations and how to make informed decisions about the risks. The IUCN guidance on translocations (IUCN, 2013), and associated national guides such as the Scottish Code for Conservation Translocations (National Species Reintroduction Forum, 2014) identify the risks of introducing new pests or pathogens during conservation translocations. However, the guidance is typically high level and does not provide practical advice on how to (i) assess plant health risks, and (ii) implement appropriate biosecurity measures.

Even where guidance exists, a lack of awareness can lead to limited impact. For example, guidance documents were drawn up by policymakers and research scientists for the conservation planting of juniper *Juniperus communis* in the UK to reduce the risk of further introductions of the invasive pathogen *Phytophthora austrocedri*. This guidance recommended not planting juniper in or near any existing viable juniper populations. Despite this, the number of juniper plantings on such sites increased significantly in the decade following publication of the guidance (Donald et al., 2021). A subsequent survey of the practitioners involved revealed that very few had read or followed the guidance (Donald F., unpublished data). It was recommended that future guidance should be developed in consultation with a selection of key practitioners, who would then disseminate the guidance so that the target audiences were reached more effectively (Donald F., personal communication 2021). Thus, a clear challenge is to consult widely during the development of guidance and to increase awareness of the guidance that does exist and make it more accessible.

8.6 Assessing Biosecurity Risks for the Conservation Translocation of Plant Species

Although there is no formal guidance for plant translocations on how to conduct a disease risk analysis (DRA) similar to that for animal health

> Box 8.1 *Key factors to consider when assessing the biosecurity risks for plant conservation translocations*
>
> - Assess whether a conservation translocation is needed.
> - Assign responsibility for plant health risk assessments and biosecurity.
> - Assess whether there are known pests and pathogens of the focal species.
> - Assess the risks of spreading pests and pathogens at the source site.
> - Assess the risks of pest and pathogen infection during transportation and holding phases.
> - Assess the risk of pest and pathogen escape during the propagation stage.
> - Assess whether the quarantine plan and pest and pathogen monitoring steps are adequate.
> - Assess whether a narrow genetic base of the translocated material may lead to an increased susceptibility to pests and pathogens.
> - Assess whether there is a risk of increased susceptibility due to plants not being adapted to their destination sites.
> - Assess whether the translocated plants may spread disease at the destination site.

(Sainsbury & Carraro, this volume; Ewen et al., this volume), there are some key factors that should be considered in all plant conservation translocations with respect to assessing biosecurity risks, and these contain many of the elements of a DRA (Box 8.1).

An important first step is to assess whether alternatives to conservation translocation would be more appropriate (IUCN, 2013). This involves assessing whether the desired conservation outcomes could be achieved with better management of existing sites to avoid the pest and pathogen risks associated with the movement of plants and soil. Where translocation is the preferred option, it is then important to have clarity at the outset about who is responsible for oversight of biosecurity. Thus, one simple step is to identify a 'biosecurity risk manager' with oversight of the plant health aspects of the translocation (i.e. from initial planning through to post-translocation monitoring). This mitigates the possibility of everyone assuming that someone else is considering the biosecurity risks.

Another important step is to assess whether there are any known pests and pathogens of concern for the focal species and how this fits with the relevant legal framework for the countries concerned. Legislative aspects are of particular importance for the movement of plants across international borders and/or where the plant species concerned is impacted by notifiable pests or pathogens (e.g. those pests and pathogens governed by statutory legislation). Assessing whether there are pests or pathogens of concern to the focal species is typically done by consulting plant health experts (particularly with respect to the legal framework) and/or conducting a literature search covering both the source and destination regions. Where any pests and pathogens are identified as a problem, then the mechanisms and pathways of spread of these key pests and pathogens should be determined. The outcome of this assessment may be used to judge whether it is too risky to translocate a particular species. If the translocation is to proceed, it will also inform the type of material that is translocated (e.g. seed or cuttings versus bareroot or containerised stock) and how/where the plant stock is grown (i.e. the location and type of any propagation facilities).

The sourcing of plant material should be based on good biosecurity procedures. These include taking care not to introduce pests and pathogens to the source sites by minimising the unintended movement of soil and plant debris (i.e. on boots, tools, and equipment). Transportation and holding of material – either en route from the source site to the propagation facilities or from the propagation facilities to the destination site – also presents an opportunity for infections/infestations to occur. Plant material should be protected within a contained environment during transport, where possible, and held on a free-draining surface well away from other plant stock or possible sources of plant pests and pathogens.

Facilities for propagation can include botanic gardens, small scale propagation units managed by conservation organisations or individual enthusiasts (with a range of experience and applications of best practice), through to commercial plant nurseries of varying size and scales. Evaluating the risks associated with the propagation facilities is important as some of the greatest risks associated with plant health occur during propagation. Anecdotal evidence suggests that *Phytophthora austrocedri* has been introduced from infected nursery stock into native juniper populations through reinforcement planting programmes in the UK (Riddell et al., 2020; Box 8.2). Batches of young sessile oak *Quercus petraea* grown by commercial nurseries from seed collected from an ecologically sensitive site in Scotland and destined for translocation back to that site were

Box 8.2 *Case study: pathogen spread associated with juniper conservation plantings in the UK (taken from Riddell et al., 2020)*

The invasive oomycete pathogen *Phytophthora austrocedri*, which has an unknown geographic origin, has recently emerged in wild juniper *Juniperus communis* populations in the UK, causing extensive mortality (Figure 8.1; Green et al., 2015). Prior to the emergence of this pathogen there were concerns over a general decline of juniper populations in Britain over the last seventy years due to overgrazing, burning, and lack of regeneration. These concerns prompted an acceleration of conservation plantings across Britain from the late 1990s onwards, aimed at bolstering locally declining juniper woodlands. These programmes used recommended propagation methods whereby seeds collected from local populations were raised, in some cases by commercial plant nurseries, before being planted back out onto the site. Numerous juniper woodlands in northern England and

Figure 8.1 Juniper *Juniperus communis* infected with *Phytophthora austrocedri* (photo: Sarah Green). (A black and white version of this figure will appear in some formats. For the colour version, please refer to the plate section.)

Scotland were subject to reinforcement planting of this sort prior to the appearance of disease symptoms (Green et al., 2015). Given that *P. austrocedri* has been regularly detected in UK nurseries and intercepted on imported *Juniperus* and Cupressaceae hosts, the planting of nursery-raised junipers infected with *P. austrocedri* is clearly a potential pathway of disease spread. Riddell et al. (2020) proposed that a single genetic lineage of *P. austrocedri* circulating in traded junipers and present in commercial nurseries may indeed have been inadvertently introduced into wild juniper populations in Britain through these restoration plantings. Once a pathogen has established at a site to a sufficient level to raise concern and elicit investigation, eradication is usually impossible.

found to harbour the root pathogen *Phytophthora quercina* (Green et al., 2020a). In the latter case, consideration was subsequently given to future use of small, locally sited, highly biosecure nurseries for growing up stock for conservation purposes.

A corollary to the risk of the translocated material acquiring pests or pathogens in propagation facilities is the risk of the translocated material introducing pests or pathogens to the nursery from the source population. This is particularly important to assess if material is being introduced into propagation facilities that house important conservation collections (e.g. botanic gardens) or hold stock that is otherwise of high value. Many plant pathogens can exist on their hosts without the appearance of visible symptoms, for example as resting spores or latent infections. Insect infestations in the woody parts of tree hosts may also easily escape detection. Therefore, careful inspection of both the source material prior to propagation and the donor material prior to outplanting is needed to check for the presence of foliage, shoot, or stem lesions, other damage symptoms such as stem bleeds, signs of insect feeding on tissues, or the presence of 'exit holes' in bark (Roques et al., 2017). Some highly damaging *Phytophthora* species produce symptoms on the host that might be dismissed as signs of desiccation due to insufficient watering (Green et al., 2020a). Inspection of any suspect material by a plant health expert is a good practical step for reducing risks, and DNA-based diagnostics offer an increasingly accessible screening technology (e.g. Green et al., 2020b; Hayden, 2020).

New plant populations established with a narrow genetic base may be inherently more vulnerable to impacts from pests and pathogens than plantings utilising a broader genetic base (Jump et al., 2009; see also Neaves et al., this volume). An example of this might be new conservation plantings derived from cuttings or seed collected from just a few remaining plants in the source population, or where the only available source populations are genetically depauperate, as is the case for the endangered Wollemi pine *Wollemia nobilis* (Greenfield et al., 2016). The long-term viability of such conservation translocations may be particularly dependent upon strict biosecurity measures when sourcing, growing, and planting donor material (see Section 8.7). Another important consideration is careful selection of destination sites with respect to distance from related species that might host potentially damaging pests and pathogens.

In cases where the planned translocation of plant species involves large geographical or ecological distances, consideration should be given to the likely phenological adaptation of the plants to the climate (current and future) of the destination region as this may affect vulnerability to local pests and pathogens. The planting of non-UK provenances of silver and downy birches *Betula pendula* and *B. pubescens* as part of new native woodland schemes in Scotland resulted in widespread dieback as planted material became severely infected with two native birch pathogens (*Anisogramma virgultorum* and *Marssonina betulae*). These pathogens are rarely known to cause damage to native indigenous birch stands in Scotland (Green, 2005; DeSilva et al., 2008). This example illustrates how 'local' pests and pathogens present at the destination site may negatively impact the outcome of a translocation.

Finally, steps should be taken to assess and minimise the risks of introducing pests and pathogens at the destination site. Many of the relevant issues here have already been noted in relation to donor sites and transportation, but particular attention to biosecurity should be given where the destination site involves plant movements for long distances and/or crossing natural ecological barriers. Where this occurs, and especially where plantings are being made into valuable ecosystems, any known susceptible material should be subject to a quarantine period and monitored regularly for plant health problems. A quarantine period of 6–12 months is suggested, that will be of sufficient duration to allow unobserved pests to emerge and pathogens to cause an obvious impact on plant fitness. It is also important to consider the local flora at the destination site as part of the risk assessment process, including evaluating risks to other proximal conspecific populations and related species (see Box 8.2).

8.7 Practical Steps to Minimise Plant Health Risks during Plant Conservation Translocations

Here we provide details of some practical 'good practice' steps to minimise plant health risks during conservation translocations (summarised in Box 8.3). The aim is not to provide comprehensive coverage but rather to highlight a set of actions that illustrate the steps that can be taken. We particularly focus on biosecurity within nurseries as this can be a major source of pests and pathogen spread during conservation translocations (Frankel et al., 2020). Hayden (2020) provides an example of developing best practice at the Royal Botanic Garden in Edinburgh.

8.7.1 Sourcing Donor Material

It is important to consider where to source donor material from. Material grown within the destination region for the entirety of its life minimises the risk of introducing pests and pathogens not yet present in the region,

---

Box 8.3 *Steps to minimise plant health risks during conservation translocations*

- Choose the source of donor material according to the biosecurity risk: seeds and cuttings are of lowest risk, and bareroot plants present less risk than containerised plants.
- Ensure that donor material is grown at premises operating to a high standard of biosecurity and with a documented plant health management plan, particularly with regard to use of water and growing media.
- Risk assess the transport process and minimise physical damage and opportunity for biotic infestations en route.
- Avoid transport of unnecessary soil, compost, and plant debris when moving plant material.
- Where there are plant health concerns with donor material destined for vulnerable sites, hold the material at an appropriate quarantine location for 6–12 months with regular inspections.
- Outplant using clean, sterile tools, equipment, and footwear, and clean these again before leaving the site.
- Regularly monitor post translocation for symptoms of ill health, paying attention to biosecurity during visits.

albeit with the caveat that donor material raised alongside other plants imported from overseas is still at risk of acquiring imported pests and pathogens.

Seeds or cuttings carry the lowest biosecurity risk, but effective surface sterilisation is recommended as well as the regular assessment of young plants to check for symptoms of ill health. When translocating whole plants, it is important to note that some pathogens can survive for long periods unseen in soil or contaminated growing media. Therefore, bareroot plants have less risk of carrying soil-borne diseases than containerised plants.

8.7.2 Growing Donor Material

When considering the choice of propagation facilities, the following practicalities are relevant for assessment:

- Water source: water is an effective carrier of many pathogens. Mains or borehole water supplies carry least risk. Sourcing water from open reservoirs, ponds, or rainfall butts or extracting from rivers carries a higher risk unless the water is treated using a method proven to kill damaging microorganisms.
- Drainage: puddles and excess run-off can spread water-borne pathogens. Plants should be grown on a free-draining surface, preferably raised above the ground. Persistent puddles at nurseries – especially on roadways – are a high risk.
- Growing media: assess whether the growing medium is sterile. This may be particularly important when using peat-free mixes that can contain local green waste, wood fibre, bark, and coir and hence the potential for pest and pathogen spread.
- Quarantine holding area: imported plants carry a risk of introducing diseases new to the destination region. It is important to establish if there is a quarantine area where imported material is isolated and observed regularly over several months to check for pests and diseases.
- Plant disposal: establish how unhealthy or unwanted plants are dealt with. Dumping such plants close to the nursery premises carries a high risk of pathogen proliferation. Ideally, plants are disposed of through a contained composting system well away from stock or natural ecosystems.
- Surrounding environment: signs of pests and diseases in adjacent areas. In addition to onsite biosecurity, the immediate surrounds of a nursery

are important for the management of pests and diseases, and the surrounding shelterbelt and landscape trees/shrubs growing in and around the nursery premises should be healthy.
- General nursery hygiene: check that the area is free of weeds, spilt soil/potting mix, and piles of soiled pots. General nursery hygiene is a basic step in minimising risks of pests and diseases, and should also include facilities for disinfestation of tools, pots, boots, and vehicles.

One way of establishing the overall biosecurity credentials of a propagation facility is national certification schemes. An example is the UK's new Plant Healthy Certification Scheme (https://planthealthy.org.uk/) based on an analysis of biosecurity risk according to a Plant Health Management Standard. This is in the early stages of deployment but is a good example of a practical biosecurity accreditation scheme.

8.7.3 Transport of Plant Material

Donor material should be transported in a way that minimises damage to foliage and stems (which might allow ingress of pathogens), prevents desiccation, and protects against exposure to pests and pathogens en route. Packaging material can itself be a pathway for pests and should be subject to appropriate phytosanitary checks. Likewise, where possible, the transport of soil, potting compost, and other debris including weeds should be avoided.

8.7.4 Outplanting

Good biosecurity practices should be employed when outplanting donor material, for example by using clean, sterilised tools and footwear. Larger equipment should be free of soil and plant debris before being taken onto the site, and equipment cleaned before leaving the site. Planting the donor material at the most suitable time of year (depending on species and location) is important to give the best chance of good, early root growth and establishment to facilitate production of resilient plants.

8.7.5 Monitoring

Post-outplanting monitoring should be conducted regularly to assess establishment success and for early detection and identification of plant health problems. Any visits to the site should be done under appropriate standards of biosecurity, ensuring that footwear, tools, and equipment are clean.

8.8 The Consequence of Ignoring Plant Health during Conservation Translocations

The consequences of ignoring plant health during conservation translocations may vary. On the one hand, small scale local movement of many plant species is intrinsically low risk, and in many cases unlikely to lead to any notable or detectable problems. However, there are some situations where the impacts could be far more profound.

8.8.1 Establishment Failure

The most obvious consequence of ignoring plant health is establishment failure. This may be due to either the presence of pests and pathogens at the destination site or the plant already being infected when it was translocated. The impacts of any pests and pathogens on the translocated plants may be exacerbated as plants undergoing translocation may experience some degree of transplantation shock and reductions in associated fitness. This may further increase their susceptibility to disease and pest damage. The diversity of species involved in conservation translocations and the limited knowledge of the distribution of pests and pathogens in the natural environment make predictions difficult and under-reporting likely. However, in a review of the success of threatened plant translocation in Australia, disease was listed as the reason for failure in 40 out of 724 translocations (Silcock et al., 2019).

8.8.2 Co-introduction of Pests and Pathogens Impacting Other Individuals or Host Species

If co-introduction of pests and pathogens occurs it may lead to spread to other plants within the wider environment. This may involve damage to individuals of the same species, as occurred across Europe with the spread of ash dieback (Woodward & Boa, 2013; Baral et al., 2014), or to naïve host plant species, as when *Phytophthora tentaculata* spread from native plant nurseries to new hosts (Rooney-Latham et al., 2015).

8.8.3 Secondary Impacts on Species, Ecosystem Processes, and Services

Where pest and pathogen outbreaks associated with translocations impact on foundation species, this can trigger declines in associated species and changes in ecosystem processes and services (Ellison et al., 2005) (Table 8.2). Thus, a decline in one plant species may impact many hundreds of associated species. Mitchell et al. (2014) list 44 species known only to occur on common ash *Fraxinus excelsior* in the UK, and 62 species that are highly associated (rarely found on any other tree species). They predict that these 106 species are most at risk from a decline in the population of ash due to ash dieback. In North America, 43 native arthropod species are known to be associated only with ash *Fraxinus* spp., and a further 30 arthropod species are associated with only one or two host plants in addition to ash, and were therefore identified as being at risk from declines due to the emerald ash borer *Agrilus planipennis* (Cleavitt et al., 2008; Gandhi & Herms, 2010).

The loss or decline of foundation plant species due to pests and pathogens can also impact on wider ecosystem processes and services (Boyd et al., 2013; Freer-Smith & Webber, 2017). Table 8.2 summarises some of these impacts; the examples are all from trees, and not all relate to conservation translocations. However, they were selected to illustrate the consequential impacts of plant pests and pathogens on the wider environment.

8.9 Conclusion

In this chapter we have aimed to draw attention to the risks associated with plant pests and pathogens and the movement of plant material. In doing this, we have signposted some cases where plant pests and pathogens have resulted in major problems. We stress that our aim is not to be alarmist or to create barriers to positive conservation action. We reiterate that many conservation translocations will be low risk. However, we also note the inadequate attention given to plant health and biosecurity in the plant conservation translocation literature, and the need for greater awareness and biosecurity measures. In particular, there is a lack of guidance aimed at practitioners to aid risk assessments and practical biosecurity actions. Development of disease risk assessment guidance similar to that used for animal conservation translocations (Sainsbury & Carraro, this volume) could be a step towards resolving this, along with integration of best practice for risk assessments and biosecurity from other

Table 8.2. (cont.)

Impact	Example pest/pathogen and host	Example impact on wider environment	Country	Reference
Decreased water quality due to changes in nutrient cycling	Ash dieback on ash trees	Total economic cost of £15 billion	UK	Hill et al. (2019)
	Hemlock woolly adelgid in hemlock forests	A nearly 20-fold increase in nitrate levels	Southern Pennsylvania	Cessna and Nielsen (2012)
	Bark beetles in evergreen forests	Increase total organic carbon and total trihalomethane (TTHM) production in stream waters, with levels of TTHM concentration exceeding regulatory maximum contaminant levels at some sites	North America	Brouillard et al. (2016)
Regulating services				
Flood control – increased groundwater flow due to loss of transpiration and canopy cover	Rocky Mountain pine beetle *Dendroctonus ponderosae* in pine forests	Late-summer groundwater levels in catchments with high tree mortality were 30% higher than in neighbouring catchments	North America	Bearup et al. (2014)
	Rocky Mountain pine beetle in pine forests	Decline in snow lie; increased snow melt and downstream water flow	North America	Embrey et al. (2012)
Climate regulation – decline in carbon storage	Review of insect pests on carbon budgets in North American forests	All studies showed a decline in carbon storage, some as large as 70%	North America	Hicke et al. (2012)

Emerald ash borer on ash trees	May release the equivalent to 1.7–2.9 times the quantity of carbon annually sequestered by forest vegetation	USA	Flower et al. (2013)
Ash dieback in ash trees	Disturbance to British forest carbon stocks of around 5 million tonnes of carbon, equivalent to >4% of the total vegetation carbon stock in the UK	UK	Reay (2013)
Mountain pine beetle *Dendroctonus ponderosae* in pine forests	Release of 270 mega-tonnes of carbon between 2000 and 2020, losses which will convert the forest from a small net sink to a large net carbon source	British Columbia, Canada	Kurz et al. (2008)
Emerald ash borer on ash	An additional 6,113 human deaths due to respiratory diseases and 15,080 cardiovascular related deaths, likely due to increased temperature and air pollution	USA	Donovan et al. (2013)
Increased air temperature and pollution linked to increased human deaths			

sectors regularly dealing with plant pests and pathogens, such as agriculture, forestry, and horticulture.

8.10 Key Messages

- Increased global movement of biological materials, coupled with climate change and other environmental pressures, is leading to increasing threats to plants from pests and pathogens.
- These pests and pathogens are relevant to plant conservation translocations as a source of translocation failure, and because the translocation itself can lead to pest and pathogen transmission.
- Many plant conservation translocations are relatively low risk, especially those involving the small scale local movement of plant material between proximal sites.
- In contrast, plant translocations that involve movement of large amounts of material and/or large geographical distances or crossing natural ecological barriers, are intrinsically of higher risk.
- Additional high-risk factors include the potential for pest and pathogen transmission to occur at nursery/propagation facilities, especially if the translocated material is held in close proximity to other plants infected with pests and pathogens and/or material sourced from distant localities.
- Despite the importance of these issues, plant health risks are often not explicitly considered in plant conservation translocations.
- To support greater awareness and the effective uptake of appropriate biosecurity steps in plant conservation translocations, there is a pressing need to develop generally applicable best practice guidelines targeted at translocation practitioners.

Acknowledgements

R. J. M. was partially funded by the 2016–2022 Strategic Research Programme of the Scottish Government. R. J. M., S. G., and P. M. H. acknowledge funding support from the Scottish Plant Health Centre. The Royal Botanic Garden Edinburgh acknowledges funding support from the Scottish Government Rural and Environment Sciences and Analytical Services Division (RESAS). This chapter was considerably improved by comments from John Ewen, Martin Gaywood, Ian Toth, Fiona Burnett, and Chris Quine.

References

Anderson, P. K., Cunningham, A. A., Patel, N. G., et al. (2004) Emerging infectious diseases of plants: pathogen pollution, climate change and agrotechnology drivers. *Trends in Ecology & Evolution*. 19, 535–544.

Aukema, J. E., Leung, B., Kovacs, K., et al. (2011) Economic impacts of non-native forest insects in the continental United States. *PLoS ONE*. 6, e24587.

Baral, H. O., Queloz, V. & Hosoya, T. (2014) *Hymenoscyphus fraxineus*, the correct scientific name for the fungus causing ash dieback in Europe. *Ima Fungus*. 5, 79–80.

Bearup, L. A., Maxwell, R. M., Clow, D. & Mccray, J. E. (2014) Hydrological effects of forest transpiration loss in bark beetle-impacted watersheds. *Nature Climate Change*. 4, 481–486.

Bell, S. A. J. (2021) Successful recruitment following translocation of a threatened terrestrial orchid (*Diuris tricolor*) into mining rehabilitation in the Hunter Valley of NSW. *Ecological Management & Restoration*. 22, 204–207.

Boyd, I. L., Freer-Smith, P. H., Gilligan, C. A. & Godfray, H. C. J. (2013) The consequence of tree pests and diseases for ecosystem services. *Science*. 342, 1235773.

Brasier, C. M. (2008) The biosecurity threat to the UK and global environment from international trade in plants. *Plant Pathology*. 57, 792–808.

Brouillard, B. M., Dickenson, E. R. V., Mikkelson, K. M. & Sharp, J. O. (2016) Water quality following extensive beetle-induced tree mortality: interplay of aromatic carbon loading, disinfection byproducts, and hydrologic drivers. *Science of the Total Environment*. 572, 649–659.

Burdon, J. J. & Zhan, J. (2020) Climate change and disease in plant communities. *PLoS Biology*. 18, e3000949.

Cahill, D. M., Rookes, J. E., Wilson, B. A., Gibson, L. & Mcdougall, K. L. (2008) *Phytophthora cinnamomi* and Australia's biodiversity: impacts, predictions and progress towards control. *Australian Journal of Botany*. 56, 279–310.

Caporn, S. J. M., Rosenburgh, A. E., Keightley, A. T., et al. (2017). *Sphagnum* restoration on degraded blanket and raised bogs in the UK using micropropagated source material: a review of progress. *Mires and Peat*. 20, 17.

Cessna, J. F. & Nielsen, C. (2012) Influences of hemlock woolly adelgid-induced stand-level mortality on nitrogen cycling and stream water nitrogen concentrations in Southern Pennsylvania. *Castanea*. 77, 127–135.

Chapman, D., Purse, B. V., Roy, H. E. & Bullock, J. M. (2017) Global trade networks determine the distribution of invasive non-native species. *Global Ecology and Biogeography*. 26, 907–917.

Cleavitt, N. L., Eschtruth, A. K., Battles, J. J. & Fahey, T. J. (2008) Bryophyte response to eastern hemlock decline caused by hemlock woolly adelgid infestation. *Journal of the Torrey Botanical Society*. 135, 12–25.

Darrasse, A., Darsonval, A., Boureau, T., et al. (2010) Transmission of plant-pathogenic bacteria by nonhost seeds without induction of an associated defense reaction at emergence. *Applied and Environmental Microbiology*. 76, 6787–6796.

Davidson, W. R. & Nettles, V. F. (1992) Relocation of wildlife: identifying and evaluating disease risks. *Transactions of the North American Wildlife and Natural Resources Conference.* 57, 466–473.

De Silva, H., Green, S. & Woodward, S. (2008) Incidence and severity of dieback in birch plantings associated with *Anisogramma virgultorum* and *Marssonina betulae* in Scotland. *Plant Pathology.* 57, 272–279.

Donald, F., Purse, B. V. & Green, S. (2021) Investigating the role of restoration plantings in introducing disease—a case study using *Phytophthora. Forests.* 12, 764.

Donovan, G. H., Butry, D. T., Michael, Y. L., et al. (2013). The relationship between trees and human health evidence from the spread of the Emerald Ash Borer. *American Journal of Preventive Medicine.* 44, 139–145.

Ellison, A. M., Bank, M. S., Clinton, B. D., et al. (2005) Loss of foundation species: consequences for the structure and dynamics of forested ecosystems. *Frontiers in Ecology and the Environment.* 3, 479–486.

Embrey, S., Remais, J. V. & Hess, J. (2012) Climate change and ecosystem disruption: the health impacts of the North American rocky mountain pine beetle infestation. *American Journal of Public Health.* 102, 818–827.

Eshleman, K. N., Morgan, R. P., Webb, J. R., Deviney, F. A. & Galloway, J. N. (1998) Temporal patterns of nitrogen leakage from mid-Appalachian forested watersheds: role of insect defoliation. *Water Resources Research.* 34, 2005–2016.

Ewen, J. G., Canessa, S., Converse, S. J. & Parker, K. A. (2023) Decision-making in animal conservation translocations: biological considerations and beyond. In Gaywood, M. J., Ewen, J. G., Hollingsworth, P. M. and Moehrenschlager, A. (eds.) *Conservation Translocations.* Cambridge, Cambridge University Press.

Flower, C. E., Knight, K. S. & Gonzalez-Meler, M. A. (2013) Impacts of the emerald ash borer (*Agrilus planipennis* Fairmaire) induced ash (*Fraxinus* spp.) mortality on forest carbon cycling and successional dynamics in the eastern United States. *Biological Invasions.* 15, 931–944.

Frankel, S. J., Alexander, J. M., Benner, D. & Shor, A. (2018) Coordinated response to inadvertent introduction of pathogens to California restoration areas. *California Agriculture.* 72, 205–207.

Frankel, S. J., Alexander, J., Benner, D., Hillman, J. & Shor, A. (2020) *Phytophthora* pathogens threaten rare habitats and conservation plantings. *Sibbaldia: The International Journal of Botanic Garden Horticulture.* 18, 53–65.

Freer-Smith, P. H. & Webber, J. F. (2017) Tree pests and diseases: the threat to biodiversity and the delivery of ecosystem services. *Biodiversity and Conservation.* 26, 3167–3181.

Gandhi, K. J. K. & Herms, D. A. (2010) North American arthropods at risk due to widespread *Fraxinus* mortality caused by the alien emerald ash borer. *Biological Invasions.* 12, 1839–1846.

Garbelotto, M., Frankel, S. J. & Scanu, B. (2018) Soil- and waterborne *Phytophthora* species linked to recent outbreaks in Northern California restoration sites. *California Agriculture.* 72, 208–216.

Goss, E. M., Tabima, J. F., Cooke, D. E. L., et al. (2014) The Irish potato famine pathogen *Phytophthora infestans* originated in central Mexico rather than the Andes. *Proceedings of the National Academy of Sciences of the United States of America.* 111, 8791–8796.

Green, S. (2005) *Birch dieback in Scotland*. Forestry Commission Information Note 72, Edinburgh. Available at: www.forestresearch.gov.uk/research/dieback-of-birch/ [Accessed 10 June 2021].

Green, S., Elliot, M., Armstrong, A. & Hendry, S. J. (2015) *Phytophthora austrocedrae* emerges as a serious threat to juniper (*Juniperus communis*) in Britain. *Plant Pathology*. 64, 456–466.

Green, S., Cooke, D. E. L., Dunn, M., et al. (2020a) Global threats from *Phytophthora* spp.; understanding drivers of emergence and opportunities for mitigation through nursery best practice. Final report to THAPBI. Available from: www.forestresearch.gov.uk/research/global-threats-from-phytophthora-spp-phyto-threats/ [Accessed 6 June 2022].

Green, S., Riddell, C. E., Frederickson-Matika, D., et al. (2020b) Diversity of woody-host infecting Phytophthora species in public parks and botanic gardens as revealed by metabarcoding, and opportunities for mitigation through best practice. *Sibbaldia: The International Journal of Botanic Garden Horticulture*. 18, 67–88.

Greenfield, A., Mcpherson, H., Auld, T., et al. (2016). Whole-chloroplast analysis as an approach for fine-tuning the preservation of a highly charismatic but critically endangered species, *Wollemia nobilis* (Araucariaceae). *Australian Journal of Botany*. 64, 654–658.

Harrington, L. A., Lloyd, N. & Moehrenschlager, A. (2023) Animal welfare, animal rights, and conservation translocations: moving forward in the face of ethical dilemmas. In Gaywood, M. J., Ewen, J. G., Hollingsworth, P. M. and Moehrenschlager, A. (eds.) *Conservation Translocations*. Cambridge, Cambridge University Press.

Hayden, K. (2020) Botanic gardens and plant pathogens: a risk based approach at the Royal Botanic Garden Edinburgh. *Sibbaldia: The International Journal of Botanic Garden Horticulture*. 18, 127–139.

Hicke, J. A., Allen, C. D., Desai, A. R., et al. (2012) Effects of biotic disturbances on forest carbon cycling in the United States and Canada. *Global Change Biology*. 18, 7–34.

Hill, L., Jones, G., Atkinson, N., et al. (2019) The £15 billion cost of ash dieback in Britain. *Current Biology*. 29, R315–R316.

I-M-Arnold, A., Gruning, M., Simon, J., et al. (2016) Forest defoliator pests alter carbon and nitrogen cycles. *Royal Society Open Science*. 3, 7.

IUCN (2013) *Guidelines for Reintroductions and Other Conservation Translocations. Version 1.0*. Gland, Switzerland, IUCN Species Survival Commission.

Jump, A. S., Marchant, R. & Peñuelas, J. (2009) Environmental change and the option value of genetic diversity. *Trends in Plant Science*. 14, 51–58.

Jung, T., Horta Jung, M., Webber, J. F., et al. (2021) The destructive tree pathogen *Phytophthora ramorum* originates from the laurosilva forests of East Asia. *Journal of Fungi*. 7, 226.

Kurz, W. A., Dymond, C. C., Stinson, G., et al. (2008) Mountain pine beetle and forest carbon feedback to climate change. *Nature*. 452, 987–990.

Lamb, D. (2018) Undertaking large-scale forest restoration to generate ecosystem services. *Restoration Ecology*. 26, 657–666.

Lane, S. A., Jukes, A., Harrow, M. & Moore, J. (2020) *Bradybatus kellneri* Bach (Curculionidae) established in Britain. *The Coleopterist*. 29, 92.

Le Stradic, S., Seleck, M., Lebrun, J., et al. (2016) Comparison of translocation methods to conserve metallophyte communities in the Southeastern DR Congo. *Environmental Science and Pollution Research*. 23, 13681–13692.

Lõhmus, A. & Runnel, K. (2014) Ash dieback can rapidly eradicate isolated epiphyte populations in production forests: a case study. *Biological Conservation*. 169, 185–188.

Lovett, G. M., Arthur, M. A., Weathers, K. C. & Griffin, J. M. (2010) Long-term changes in forest carbon and nitrogen cycling caused by an introduced pest/pathogen complex. *Ecosystems*. 13, 1188–1200.

Maschinski, J. & Albrecht, M. (2023) Conservation translocations for plants. In Gaywood, M. J., Ewen, J. G., Hollingsworth, P. M. and Moehrenschlager, A. (eds.) *Conservation Translocations*. Cambridge, Cambridge University Press.

McCartan, S. A., Webber, J. F. & Jinks, R. L. (2015) Hot-water treatment as a possible method for eradicating *Chalara fraxinea* (*Hymenoscyphus pseudoalbidus*) infection from ash fruits (*Fraxinus excelsior* L). *Quarterly Journal of Forestry*. 109, 18–23.

Migliorini, D., Ghelardini, L., Tondini, E., Luchi, N. & Santini, A. (2015) The potential of symptomless potted plants for carrying invasive soilborne plant pathogens. *Diversity and Distributions*. 21, 1218–1229.

Mitchell, R. J., Beaton, J. K., Bellamy, P. E., et al. (2014) Ash dieback in the UK: a review of the ecological and conservation implications and potential management options. *Biological Conservation*. 175, 95–109.

Monks, L., Barrett, S., Beecham, B., et al. (2019) Recovery of threatened plant species and their habitats in the biodiversity hotspot of the Southwest Australian Floristic Region. *Plant Diversity*. 41, 59–74.

National Species Reintroduction Forum (2014) *Best Practice Guidelines for Conservation Translocations in Scotland*. Inverness, Scottish Natural Heritage.

Neaves, L. E., Ogden, R. & Hollingsworth, P. M. (2023) Genomics and conservation translocations. In Gaywood, M. J., Ewen, J. G., Hollingsworth, P. M. and Moehrenschlager, A. (eds.) *Conservation Translocations*. Cambridge, Cambridge University Press.

Osterbauer, N. K., Lewis, S., Hedberg, J. & Mcaninch, G. (2013) Assessing potential hazards for *Phytophthora ramorum* establishment in Oregon nurseries. *Journal of Environmental Horticulture*. 31, 133–137.

Pywell, R. F., Meek, W. R., Webb, N. R., Putwain, P. D. & Bullock, J. M. (2011) Long-term heathland restoration on former grassland: the results of a 17-year experiment. *Biological Conservation*. 144, 1602–1609.

Rabenold, K. N., Fauth, P. T., Goodner, B. W., Sadowski, J. A. & Parker, P. G. (1998) Response of avian communities to disturbance by an exotic insect in spruce-fir forests of the southern Appalachians. *Conservation Biology*. 12, 177–189.

Ranawaka, B., Hayashi, S., Waterhouse, P. M. & De Felippes, F. F. (2020) *Homo sapiens*: The superspreader of plant viral diseases. *Viruses*. 12, 1462.

Rapicavoli, J., Ingel, B., Blanco-Ulate, B., Cantu, D. & Roper, C. (2018) *Xylella fastidiosa*: an examination of a re-emerging plant pathogen. *Molecular Plant Pathology*. 19, 786–800.

Read, M. (1989) The bulb trade – a threat to wild plant populations. *Oryx*. 23, 127–134.

Reay, D. S. (2013) New directions: ash dieback and British carbon stocks. *Atmospheric Environment*. 74, 110–111.

Riddell, C. E., Dun, H. F., Elliot, M., et al. (2020) Detection and spread of *Phytophthora austrocedri* within infected *Juniperus communis* woodland and diversity of co-associated Phytophthoras as revealed by metabarcoding. *Forest Pathology*. 50, e12602.

Rooney-Latham, S., Blomquist, C., Swiecki, T., Bernhard, T. E. & Frankel, S. (2015) First detection in the US: new plant pathogen, *Phytophthora tentaculata*, in native plant nurseries and restoration sites in California. *Native Plant Journal*. 16, 23–26.

Rooney-Latham, S., Blomquist, C. L., Kosta, K. L., Gou, Y. Y. & Woods, P. W. (2019) *Phytophthora* species are common on nursery stock grown for restoration and revegetation purposes in California. *Plant Disease*. 103, 448–455.

Roques, A., Cleary, M., Matsiakh, I. & Eschen, R. (eds.) (2017) *Field Guide for the Identification of Damage on Woody Sentinel Plants*. Wallingford, UK, CABI.

Sainsbury, A. W. & Carraro, C. (2023) Animal disease and conservation translocations. In Gaywood, M. J., Ewen, J. G., Hollingsworth, P. M. and Moehrenschlager, A. (eds.) *Conservation Translocations*. Cambridge, Cambridge University Press.

Silcock, J. L., Simmons, C. L., Monks, L., et al. (2019) Threatened plant translocation in Australia: a review. *Biological Conservation*. 236, 211–222.

Simler, A. B., Williamson, M. A., Schwartz, M. W. & Rizzo, D. M. (2019) Amplifying plant disease risk through assisted migration. *Conservation Letters*. 12, e12605.

Sims, L. L. & Garbelotto, M. (2021) *Phytophthora* species repeatedly introduced in Northern California through restoration projects can spread into adjacent sites. *Biological Invasions*. 23, 2173–2190.

Snyder, C. D., Young, J. A., Lemarie, D. P. & Smith, D. R. (2002) Influence of eastern hemlock (*Tsuga canadensis*) forests on aquatic invertebrate assemblages in headwater streams. *Canadian Journal of Fisheries and Aquatic Sciences*. 59, 262–275.

Spence, N., Hill, L. & Morris, J. (2020) How the global threat of pests and diseases impacts plants, people, and the planet. *Plants, People, Planet*. 2, 5–13.

Strange, R. & Scott, P. (2005) Plant disease: a threat to global food security. *Annual Review of Phytopathology*. 43, 83–116.

Straw, N. A., Fielding, N. J., Tilbury, C., Williams, D. T. & Cull, T. (2016) History and development of an isolated outbreak of Asian longhorn beetle *Anoplophora glabripennis* (Coleoptera: Cerambycidae) in southern England. *Agricultural and Forest Entomology*. 18, 280–293.

Summerell, B. & Liew, E. (2020) *Phytophthora* root rot: its impact in botanic gardens and on threatened species conservation. *Sibbaldia: The International Journal of Botanic Garden Horticulture*. 18, 89–104.

Tingley, M. W., Orwig, D. A., Field, R. & Motzkin, G. (2002) Avian response to removal of a forest dominant: consequences of hemlock woolly adelgid infestations. *Journal of Biogeography*. 29, 1505–1516.

Trueman, I., Mitchell, D. & Besenyei, L. (2007) The effects of turf translocation and other environmental variables on the vegetation of a large species-rich mesotrophic grassland. *Ecological Engineering*. 31, 79–91.

Vendettuoli, J. F., Orvvig, D. A., Krumins, J. A., Waterhouse, M. D. & Preisser, E. L. (2015) Hemlock woolly adelgid alters fine root bacterial abundance and mycorrhizal associations in eastern hemlock. *Forest Ecology and Management*. 339, 112–116.

Wittram, B. W., Roberts, G., Buckler, M., King, L. & Walker, J. S. (2015) *A Practitioners Guide to Sphagnum Reintroduction*. Edale, UK, Moors for the Future Partnership.

Wondafrash, M., Wingfield, M. J., Wilson, J. R. U., et al. (2021) Botanical gardens as key resources and hazards for biosecurity. *Biodiversity and Conservation*. 30, 1929–1946.

Woodward, S. & Boa, E. (2013) Ash dieback in the UK: a wake-up call. *Molecular Plant Pathology*. 14, 856–860.

Zahiri, R., Christian Schmidt, B., Schintlmeister, A., Yakovlev, R. V. & Rindoš, M. (2019) Global phylogeography reveals the origin and the evolutionary history of the gypsy moth (Lepidoptera, Erebidae). *Molecular Phylogenetics and Evolution*. 137, 1–13.

9 · *Genomics and Conservation Translocations*

LINDA E. NEAVES, ROB OGDEN, AND
PETER M. HOLLINGSWORTH

9.1 Introduction

Conservation translocation – the deliberate movement of organisms between sites to yield a conservation benefit – fundamentally involves a manipulation of the genetic composition of the translocated species. This is important, as the genetic make-up of the translocated population can influence its fitness and be a contributing factor in the overall outcome of the translocation. Beyond this intrinsic issue, genetic and genomic technologies can themselves provide powerful insights into species biology and ecosystem function, which can further guide decision-making to support conservation translocations and habitat management.

There are numerous syntheses of the key genetic principles guiding conservation translocations (e.g. Frankham et al., 2011; Weeks et al., 2011; Jamieson & Lacy, 2012; Bell et al., 2019; Van Rossum & Hardy, 2022), and genetic principles are embedded in the IUCN translocation guidelines (IUCN, 2013). At the level of genetic management of species during translocation, this fundamentally involves taking steps to minimise loss of genetic variation and the risk of inbreeding depression, promoting the presence of appropriate adaptive variation and evolutionary potential, and minimising the risks of outbreeding depression due to mixing of divergent lineages. In this chapter we briefly summarise the main conceptual areas where genetic information is relevant for conservation translocations (Figure 9.1), and then focus on outlining the emerging perspectives and insights from the rapidly developing field of genomic science and the associated exponential increase in the availability of DNA sequence data. We finish with a brief horizon scan of future prospects and opportunities. Table 9.1 provides a glossary of key terminology relevant to genomics and conservation translocations.

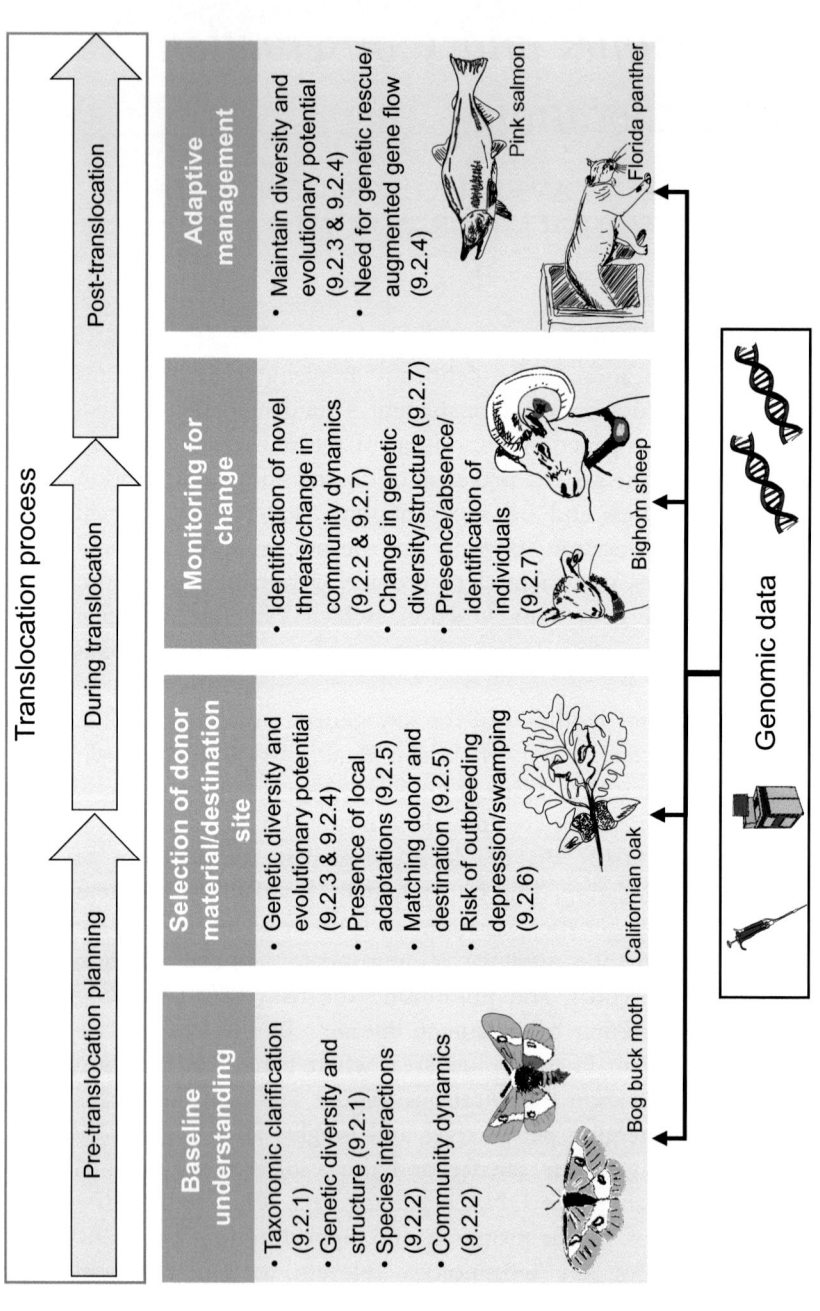

Figure 9.1 The different stages of conservation translocations and areas where genomic data can inform decision-making. The relevant sections in the text of this chapter are marked in parentheses.

Table 9.1. *Glossary of genomic terminology used in this chapter.*

Term	Meaning
Adaptive variation	Genetic variation conferring a fitness advantage with respect to interactions between an organism and its environment
Chromosome level genome sequence	Relatively complete sequence of an organism's genome, such that the genome assembly includes coverage of all or most of the DNA in each chromosome
CRISPR-Cas9	An approach for finding and editing a specific region of DNA within the genome
ddRAD (also RADseq)	A method of shearing and sequencing DNA samples to retrieve comparable data from large numbers (many thousands) of genomic regions
DNA barcoding	Use of short, standardised regions of DNA for telling species apart
eDNA (environmental DNA)	DNA shed by organisms into their environment, allowing for non-invasive genetic assays by sequencing environmental samples such as water, soil, and air
Evolutionary (adaptive) potential	The ability of a population to evolve to cope with environmental changes
Evolutionary rescue	Restoration or enhancement of genetic diversity that allows for adaptive evolutionary change to increase the likelihood of persistence
Evolutionary significant units (ESU)	Divergent intra-specific lineages showing no evidence of recent gene flow
Facilitated adaptation	Targeted reinforcement/augmentation of populations, inter-breeding, or other genetic interventions designed to introduce favourable genetic variants to facilitate a population's ability to adapt to the prevailing (or future) environmental conditions
Fitness	Survival and/or reproductive success of an individual
Founder representation	The degree to which the genetic diversity of a given donor population is maintained post release
Genetic drift	Random changes in the frequency of genetic variants from generation to generation due to chance events; genetic drift typically leads to the loss of genetic diversity in small populations
Genetic rescue	The reinforcement of a population to introduce novel genetic variants to overcome reduced fitness associated with inbreeding depression

(cont.)

Table 9.1. (cont.)

Term	Meaning
Genetic swamping	The dilution or replacement of the genetic variants in one population, by gene flow from another genetically distinct population or taxon, which may lead to outbreeding depression
Genome	The totality of the genetic information in an organism
Heterozygosity	The state of having different genetic variants at a given locus. High *genome wide heterozygosity* implies that a given individual has lots of different genetic variation at multiple sites scattered throughout its genome
Homozygosity	The state of having identical genetic variants at a given locus. High homozygosity occurs when the parents are closely related and is indicative of inbreeding
High throughput sequencing (HTS)	HTS is sometimes referred to as massively parallel sequencing (MPS), or second/third/next generation sequencing. It broadly describes the capacity to simultaneously sequence very large numbers of individuals and/or genetic regions
Hybridisation	Genetic exchange between divergent populations or species
Inbreeding depression	Decrease in fitness stemming from mating among closely related individuals, often associated with low levels of genetic diversity
Introgression	Transfer of genetic material from one species or population to another by hybridisation
Locus/loci	A region/regions of the genome
Major histocompatibility complex (MHC) genes	A closely linked set of genes found in higher vertebrates that code for proteins that confer resistance against pathogens
Meta-barcoding	DNA barcoding of mixtures of samples to identify the different species present in a mixture (e.g. species identification from faecal samples or pollen loads)
Meta-genomics	Non-targeted DNA sequencing of composite mixtures of organisms to detect the species and genes present in the sample
Nuclear genome	The majority of an organism's DNA, packaged into chromosomes
Outbreeding depression	Decrease in fitness stemming from mixing of genetically divergent lineages, due to intrinsic genetic incompatibilities or differential adaptations

Table 9.1. (*cont.*)

Term	Meaning
Predictive provenancing	Introduction of genotypes into populations that are predicted to be pre-adapted to current or future conditions
RADseq	See ddRAD
Runs of homozygosity (ROH)	Continuous sections of the genome, where the genetic types (DNA sequences) inherited from either parent are identical; long ROH are indicative of recent inbreeding
Single nucleotide polymorphism (SNP)	Minimal description of genetic variation (a difference at a single nucleotide site); SNP assays often screen hundreds or thousands of genetic variants simultaneously
Whole genome sequencing	Sequencing the entire genome of an organism (in practice this means either the entire genome, or more usually a significant portion of it)

9.1.1 The Conceptual Relevance of Genetics for Conservation Translocations

9.1.1.1 Using Genetic Data to Understand the Study System

Understanding taxonomy and population structure: In many cases, morphological characters are adequate for species discrimination, and the selection of donor populations for translocation is not confounded by taxonomic ambiguity. However, this is not always the case, and the use of genetic data to provide taxonomic clarification can support the selection of donor populations. Likewise, where there is intra-specific structure such as subspecies or distinct geographic-genetic lineages, genetic studies can be used to guide the selection of appropriate donor populations and destination sites to support successful translocations.

Understanding ecosystem dynamics: Understanding multi-trophic interactions and key interdependencies among species can be challenging in complex systems. DNA-based methods can provide fundamental insights into species interactions such as diet analysis, pollination networks, key mutualisms, and pathogen infection; these insights can guide translocation strategies and habitat management.

9.1.1.2 Genetic Management of Focal Species

Maintaining genetic variation and evolutionary potential: There are many stages in the translocation process that can affect the genetic diversity of a translocated population, including the selection of donor sites and individuals, the dynamics of any *ex situ* holding phase or population, and the potential for differential survival, establishment, and reproduction at the destination site. A primary concern is an inadvertent genetic bottleneck, where genetic diversity is lost, increasing the likelihood of immediate reductions in fitness due to mating among close relatives (inbreeding depression), and a longer-term constraint on the ability of the translocated population to adapt to changing conditions (evolutionary potential).

Genetic rescue: An explicit outcome of some conservation translocations is population reinforcement to increase genetic diversity in small and/or genetically impoverished populations. The introduction of translocated individuals (or gametes) from elsewhere can result in 'genetic rescue' of the recipient population: an increase in fitness by introducing genetic variation to overcome intrinsic genetic problems due to inbreeding depression. An extension of genetic rescue is 'evolutionary rescue': the reinforcement of a population to provide additional genetic variation to facilitate adaptation to new or changing environmental conditions.

Matching donor populations and destination sites: Many species show marked intra-specific adaptive differences between populations, with populations often differentially adapted to local conditions. Understanding genetically determined adaptive differences among populations can be important in selecting appropriate donor material for a given destination site to maximise survival and establishment prospects. This includes deciding whether to use single or mixed sources, and whether to factor in projected environmental change when selecting donor populations.

Minimising risks of outbreeding depression: Where populations have been isolated for long periods of time, they can develop genetic incompatibilities and/or show divergent adaptations to their local environmental conditions. Mixing divergent populations during a translocation process (e.g. when reinforcing an extant population or using multiple sources) can in some circumstances lead to a reduction in fitness (outbreeding depression), and an important consideration when mixing divergent populations is to evaluate whether the genetic benefits from mixing override concerns about outbreeding depression.

Genetic monitoring: Following a translocation, it is important to monitor the affected populations over time to assess the immediate impacts and long-term outcomes of the intervention. Genetic monitoring can provide data on the demographic results of the translocation, such as estimates of geographic dispersal and census size. It can also provide insights into the population genetic consequences of translocation, such as levels of diversity, introgression, genetic bottlenecks, or long-term issues such as adaptation of subsequent generations to the destination habitat.

9.2 Applications of Genomics to Conservation Translocations

The use of genetic information in conservation translocations is being revolutionised due to developments in high throughput sequencing (HTS) platforms and bioinformatic pipelines which now provide cost effective access to vast quantities of DNA sequence data from wild species. This step change to larger scale *genomic* approaches is enabling transformative insights into the biology and dynamics of species and ecosystems, and can guide the design, execution, and monitoring of conservation translocations. In this chapter we adopt a broad definition of genomics, which includes HTS approaches for species identification and ecosystem characterisation that are based on sequencing a small part of the genome from large sample sizes of individuals, as well as the stricter definition of genomics involving recovery of very large amounts of sequence data from the genomes of individual samples.

9.2.1 Understanding Taxonomy and Population Structure

High throughput sequencing enables high-resolution insights into taxonomic and population genetic structure, which can guide the design of translocation projects. Key areas of application include screening for the presence of cryptic species and/or clarifying species boundaries, elucidating the genetic make-up of previously extirpated populations, and the detection of major intra-specific lineages. In all of these cases, the overriding principle is obtaining data to inform the translocation process, including appropriateness of whether to translocate, and matching appropriate donor material to destination sites.

DNA barcoding of museum specimens of the regionally extinct Esper's marbled white butterfly *Melanargia russiae* revealed an east–west split in mitochondrial DNA lineages and, coupled with climate matching, was used to recommend the sourcing of donor material from Ukraine for translocation to re-establish the species in Hungary (Dincă et al., 2018). Likewise, Dupuis et al. (2020) used ddRAD sequencing to obtain high-resolution insights into the taxonomy and regional population structure of threatened bog buck moths *Hemileuca* spp. to guide conservation and reintroduction strategies. The data clarified previous taxonomic uncertainty and showed that the declining populations in New York State were genetically distinct from, yet conspecific with, populations from Ontario, but a different species to ecologically similar populations in the western Great Lakes Region, informing potential sources for future translocations.

The reintroduction programme of the Eurasian beaver *Castor fiber* to west Scotland was supported by a genomic analysis of population structure and diversity across Eurasian sample localities (Senn et al., 2014). Campbell-Palmer et al. (2020) subsequently undertook a retrospective analysis of unofficial releases in the east of Scotland to elucidate their origins, confirming the taxonomy of the individuals as Eurasian beavers (as opposed to non-native North American beavers *Castor canadensis*, which look almost identical), and assigning the majority to source populations from Germany. Confirming the taxonomic status and genetic make-up of this unofficial release population has helped guide decision-making around its conservation status in the wild and suitability as donor material for subsequent translocations.

A common challenge for studies focusing on genomic differentiation of taxa and populations is translating the output into clear guidance for translocation management. Where research is driven by a specific management question, then data interpretation can be conceptually straightforward (e.g. whether a given population belongs to the focal species for the translocation or a different species; Mikheyev et al., 2017). In contrast, where there is no a priori question identified, and/or where the biological system itself shows a complicated continuum of variation, it can often be difficult to move from data to decision-making. One example of this challenge is a study of taxon differentiation and hybridisation between wildcats *Felis silvestris* and domestic cats *Felis catus* in Scotland to inform species restoration plans (Senn et al., 2019). A combination of direct SNP assays and ddRAD detected a complete continuum between the Scottish wildcat and domestic cats due to extensive introgressive hybridisation. The absence of clear discontinuities in the data necessitates subjective value decisions from the conservation community to define what constitutes a wildcat in Scotland, and hence the unit that is the focus for species conservation and restoration (Senn et al., 2019).

A different type of emergent challenge with genomic scale data relates to the sheer power of the approach. With access to vast amounts of DNA sequence data, detecting ever finer scale differences between populations is the norm. However, deciding whether or not statistically significant differentiation among populations is also biologically meaningful is difficult, and this will often require additional contextual data. Coates et al. (2018) provide a thoughtful overview of the practical implications of the new resolving power of genomic data, outlining the criteria for selecting appropriate designation of divergent genetic lineages as new species, versus evolutionary significant units (ESUs), versus management units, and make the case for new guidelines to support a more consistent approach to interpretation and translation into conservation recommendations.

9.2.2 Ecosystem Assessments: Understanding Community Dynamics, Species Interactions, and Dependencies

High throughput sequencing platforms enable species detection and identification from composite samples (i.e. mixtures of samples) and from environmental samples (i.e. soil, water, faeces), which were not feasible with 'traditional' sequencing approaches. This allows assessment of the organisms in the community and the identification of key interdependencies and trophic interactions, which may influence translocation success. Approaches include meta-barcoding (the recovery of selected gene regions to tell species apart – and meta-genomics which sequences all of the recoverable DNA of all the organisms in a sample).

9.2.2.1 Food Webs and Trophic Interactions

Meta-barcoding studies are transforming understanding of diet (Pompanon et al., 2012; Kartzinel et al., 2015), and understanding food webs can be important for translocations, both from the perspective of ensuring that the dietary requirements of the translocated species are met, and in terms of understanding their role in trophic interactions and predicting potential downstream impacts of their predation on the ecosystem. Ducotterd et al. (2020) used meta-barcoding of faecal samples of the European pond turtle *Emys orbicularis* to assess whether reintroduced populations predated on co-occurring threatened amphibian species. Diet analysis of faecal samples from four populations (including two reintroduction sites) revealed a highly diverse opportunistic and omnivorous diet including 270 different species of vertebrates, invertebrates, and plants. The overriding conclusion was that plants formed the major component of the diet, and predation threats to rare amphibians were marginal (Ducotterd et al., 2020).

DNA-based diet analysis can support the design of translocation programmes for species that are reclusive and difficult to observe, such as the giant wall gecko *Tarentola gigas* whose diet was analysed as part of a feasibility study for its planned reintroduction to Santa Luzia Island in Macronesia (Pinho et al., 2018). Diet analysis can also be used to understand predation risk to the translocated species. For example, Emery et al. (2021) used DNA barcoding of gut contents to show that the translocation failure of the extinct-in-the-wild blue-tailed skink *Cryptoblepharus egeriae* to Christmas Island was likely due to predation from the invasive giant centipede *Scolependra subspinipes*.

Beyond diet analysis, another area where genomics is shedding light on multi-trophic interactions is in the understanding of plant–pollinator relationships (Vamosi et al., 2017). There is great potential here to inform the design and monitoring of conservation translocations, particularly where plants have specialist pollinator requirements (Reiter et al., 2017), and/or where the plant–pollinator relationships are difficult to observe and remain cryptic. This includes DNA barcoding of pollen loads on pollinators to establish which plant species a known pollinator pollinates, and potentially DNA barcoding of flowers to detect traces of DNA left behind by pollinators to establish which species they are visited by (Vamosi et al., 2017; Thomsen & Sigsgaard, 2019).

9.2.2.2 Microbial Symbioses
Meta-barcoding and meta-genomic studies enhance understanding of symbiotic relationships that impact on the survival and fitness of many species (West et al., 2019; Francioli et al., 2021). Microbes living on and inside organisms form intimate partnerships that influence processes such as acquisition of nutrients (e.g. plant–mycorrhizal partnerships), digestion of food (e.g. plant fibre digestion), and resistance to disease (Mueller et al., 2020). The process of holding organisms *ex situ* prior to a translocation, and then releasing them into a novel environment, has the potential to uncouple symbiotic relationships and may thus impact on the outcome of the translocation.

In animals, the gut microbiome influences individual health and digestion (Ley et al., 2008). In the koala *Phascolarctos cinereus*, meta-barcoding showed that the microbiome composition affects which *Eucalyptus* species can be digested (Blyton et al., 2019), which has implications for the selection of donors for translocations. The microbiome of captive animals is frequently different and less diverse compared to wild animals. While the microbiome can equilibrate post release (Chong et al., 2019), the time required and health consequences during equilibration are not well understood. In the Fijian crested iguana *Brachylophus vitiensis*, differences were still evident two months post release (Eliades et al., 2021). Meta-barcoding and meta-genomic insights can thus identify microbiome dynamics during different stages of a translocation, and for instance identify the degree to which delayed-release strategies enable microbiome adjustment, as was recently shown for Przewalski's horse *Equus ferus przewalskii* (Tang et al., 2020).

The importance of the relationship between plants and mycorrhizal partners is well established (Koziol et al., 2018), and DNA barcoding studies have been used to evaluate whether the absence of appropriate mycorrhizal partners limits the establishment of rare species (Egidi et al., 2018). Less well characterised, and still in its infancy, is the importance of the wider plant microbiome for influencing plant translocations. Zahn and Amend (2017) used DNA barcoding to monitor the dynamics of transplanted leaf endophytes from con-generic relatives of the extinct-in-the-wild *Phyllostegia kaalaensis* to improve disease resistance in the remaining *ex situ* populations of this Hawaiian endemic. Significant reduction in disease prevalence was achieved in treated plants, and a small population subsequently translocated into the wild remained disease free in the year following outplanting (Zahn & Amend, 2017).

9.2.3 Maintaining Genetic Variation and Evolutionary Potential

A key concern when undertaking translocations is that the established population is viable in the long term and includes sufficient diversity to avoid inbreeding depression and allow adaptation to environmental conditions. Genomic scale approaches can provide increasingly high-resolution insights assessing variation and differentiation to inform the selection of donor sources, including functional and adaptive diversity.

The selection of donors (and their contribution to the translocated population) is critical for maximising diversity in translocated populations. Numerous translocations of Alpine ibex *Capra ibex* have enabled a dramatic demographic recovery. Yet, genome wide analyses have revealed that repeated and serial reintroductions using captive-bred animals from a single source have resulted in genetic differentiation, recent inbreeding, and low diversity genome wide and at specific immune system related genes (Brambilla et al., 2018; Grossen et al., 2018; Kessler et al., 2021). This places the long-term persistence of these translocated populations at risk, as low levels of diversity and inbreeding are associated with reduced population growth rates (Bozzuto et al., 2019), inbreeding depression in males (Brambilla et al., 2015), and increased susceptibility to infectious diseases, notably keratoconjunctivitis, which is a key disease in ibex populations (Brambilla et al., 2018). In contrast, populations of bighorn sheep *Ovis canadensis* appear to have retained much of their genomic diversity despite translocations being largely sourced from a single site, albeit not involving serial translocations (Flesch et al., 2020) (see also Section 9.2.7).

Genomic diversity may also be maximised through the selection of donors from multiple sources. White et al. (2018) used large-scale SNP datasets to show that despite relatively limited total founders (17–122 individuals) from small, low diversity sources, a multi-source approach resulted in lower levels of inbreeding and similar or greater diversity than the original source populations up to 18 years after the reintroduction in four Australian mammals. For example, 14 years after the reintroduction of 30 boodie or burrowing bettong *Bettongia lesueur* from two sources, there was significantly less inbreeding and higher diversity compared with either source. The high resolution of the genomic data, however, showed that while mixed ancestry was dominant, the representation of the different sources was asymmetrical. One source (Bernier Island) was more poorly represented than the other (Heirisson Prong), despite contributing twice as many founders. One likely cause was that Heirisson Prong animals were established in the reserve a year prior to the Bernier Island founders, providing the opportunity to acclimatise and establish territories, leading to asymmetrical success among the sources. This illustrates the degree to which genetic diversity may be lost at multiple stages during translocations and how each stage requires consideration and monitoring.

Ex situ/captive populations represent a key source of individuals for translocations. However, genetic changes associated with limited founders, genetic drift, and adaptation to captive conditions can compromise their usefulness. Genomic data can inform captive breeding and selection of individuals for reintroduction to ensure that all the available diversity is represented. To inform the reintroduction of scimitar-horned oryx *Oryx dammah* to Tunisia and Chad, analysis of genome wide SNPs (Ogden et al., 2020) combined with chromosomal level resequencing (Humble et al., 2020) was conducted across global captive sources and reintroduced individuals. The data showed significant variation in candidate source diversity, and managers used this information to balance the genetic contribution of founders from multiple captive sources including Europe, the USA, and the United Arab Emirates, resulting in reintroduced populations displaying genomic variation representative of global diversity.

Genetic drift or inadvertent selection pressures on captive animals can lead to maladaptation and reduce the success of translocations. Genomic comparison between isolated (without gene flow) and integrated (with wild gene flow) hatchery lineages of Chinook salmon *Oncorhynchus tshawytscha* found rapid genetic divergence between the two lines, suggesting potential adaptation to captive conditions in the isolated lineage (Waters et al., 2018). Understanding genetic changes in captive populations and their divergence from wild populations is important for the selection of individuals for reintroductions, as hatchery-bred fish frequently show reduced survival and productivity in the wild (e.g. O'Sullivan et al., 2020).

The ability of genomic approaches to identify changes at functional genes provides the potential to select individuals or to direct captive breeding in order to focus solely on the maintenance of beneficial functional genetic variants, such as at MHC genes that are involved in responses to disease. In the Tasmanian devil *Sarcophilus harrisii*, wild populations have been decimated by the devil facial tumour disease (DFTD) (Hogg & Wise, this volume). In this situation, the intentional selection of genetic variants associated with resistance to the disease could make these captive animals better suited when released to the wild (Hamede et al., 2021). However, this needs to be balanced against the risk that this focus might lead to a loss of genetic diversity at other genes that might create a different set of fitness problems. This means that continued monitoring of genome wide diversity, and not just disease resistance genes, remains essential to the management of captive devil colonies acting as a source for translocations (Wright et al., 2020).

9.2.4 Genetic and Evolutionary Rescue: Reinforcement

Genetic rescue is a widely applied translocation strategy for alleviating inbreeding depression, and its benefits have been demonstrated in plants, vertebrates, and invertebrates (Whiteley et al., 2015; Frankham, 2016; Bell et al., 2019). The Florida panther *Puma concolor couguar* is a classic example of genetic rescue, with the addition of eight female Texas pumas to an extant population of about 26 adults in 1995 resulting in an seven-fold increase in population size by 2017 and significant reduction in the frequency of abnormalities, including tail kinks and undescended testes (Johnson et al., 2010). Analysis of puma genomes demonstrated extremely low levels of genome wide diversity and long runs of homozygosity (ROH), indicative of inbreeding, prior to the rescue, while the following generation possessed a three-fold increase in genome wide heterozygosity. These data also showed that inbreeding can (re)occur rapidly if populations remain small and highlights the importance of ongoing genetic restoration to maintain genetic diversity in small populations (Hedrick, 2005).

Evolutionary rescue is similar to genetic rescue but focuses on the introduction of genetic diversity, including potentially adaptive variants, to facilitate future evolutionary change (Carlson et al., 2014; Bell, 2017). The introduction of novel diversity to provide this benefit, without disrupting or displacing any unique or useful locally adaptive variants, is a balancing act, which can be informed by the high resolution of genomic approaches (Box 9.1). Genomic approaches can also inform the selection of source material to enable the introduction of specific adaptive variants suitable to the prevailing, or future, conditions (see also Section 9.2.5 and Box 9.2).

Box 9.1 *Connectedness, genetic rescue, and local adaptations in Trinidadian guppies* Poecilia reticulata *(Fitzpatrick et al., 2020)*

Reinforcement can increase diversity and introduce potentially adaptive genes, but genetic swamping of local adaptive variation is also possible (i.e. the potential to cause harm, rather than benefits). Multiple approaches were used to assess the impacts of a reinforcement of two Trinidadian guppy *Poecilia reticulata* populations from a source under differing conditions (i.e. from high to low predation pressure; Fitzpatrick et al. (2020)). Prior to the translocation the destination sites were distinct from each other and the source site, and exhibited low levels of genetic diversity. Following the translocations, levels of genetic diversity increased ten-fold. Hybrid and immigrant genotypes exhibited higher fitness (for longevity and reproductive success) and the population size increased. High-resolution RADseq data showed that overall the genomes of source and destination sites became more similar after the translocation, suggesting the potential for genetic swamping in the destination site. However, these data also had the power to show that genetic variants unique to the destination site and likely representing locally adapted variants (to the different predation pressure) were not impacted in the same way. In contrast to the pattern overall, these genes retained the same levels of divergence between source and destination sites that were present prior to the translocation (Fitzpatrick et al., 2020). Thus, in these guppies, the recipient population benefited from genetic rescue effects of the introduction of new individuals, increasing overall genetic diversity, without suffering loss or swamping of local adaptation as a result. These studies suggest that with careful management, genetic restoration and increases in levels of diversity can be achieved without the loss of important locally adapted variation.

Box 9.2 *Adaptation to local conditions and genome-assisted selection of sources*

Even where local adaptations are present, local sources may not represent the individuals most suited to the prevailing conditions, particularly under changing climatic conditions. Browne et al. (2019) investigated local adaptation in valley oaks *Quercus lobata* in California, using a large dataset comprising over 12,000 SNPs and data on growth traits from common garden experiments. Their findings indicated that the oaks were already maladapted to current climatic conditions, experiencing slower growth rates than that possible at cooler temperatures. Combining information on growth with genomic data they were able to assess genotype–phenotype–environment associations and identify genotypes likely to promote fast growth under warmer temperatures. This could enable genome-assisted selection of seed sources for translocations (i.e. predictive provenancing or facilitated adaptation) to match individuals to future climates/conditions. Further, their data indicated that this genome-assisted approach provided better adjusted growth rates compared with source selection based on climate matching alone.

9.2.5 Matching Donor Populations and Destination Sites

Adaptations to local environmental conditions are widespread in many different species and well documented using a range of approaches, including common garden experiments and reciprocal transplants (Leimu & Fischer, 2008; Fraser et al., 2011; Halbritter et al., 2018). The high resolution of genomic approaches and the ability to target functional genes are providing additional details on when, how, and where populations are adapted to local conditions. For example, Ingvarsson and Bernhardsson (2020), using a large dataset of over 4 million SNPs, found that climate-associated SNPs showed 30 times greater differentiation compared with the genome wide average in European aspen *Populus tremula*.

Genomic data have also demonstrated that some populations are not well adapted to the current climatic conditions due to an adaptational lag, often associated with long-lived species (see Box 9.2). This means that local genetic diversity may not represent the ideal source material for current or future climates. This, combined with a greater understanding of the consequences of fragmentation and small population size, has led to a shift away from the traditional assumptions that local provenances are always best. Instead, mixed and targeted provenancing are also options to broaden the genetic composition of populations and include genotypes adapted to likely future conditions (Diallo et al., 2021).

9.2.6 Minimising Risks of Problems from Population Mixing/Hybridisation

Where a translocation involves the mixing of individuals from different locations (e.g. reinforcement or multi-source translocation) there is some risk of outbreeding depression due to the breakup of co-adapted gene complexes, disruption of local adaptations, or swamping of native genotypes. Genomic-level data provide greater power to detect the extent of genetic mixing, not only through increased sensitivity for detecting introgression, but by providing insights as to where introgression occurs in an individual's genome. De Cahsan et al. (2021) used mitochondrial genomes and the sequences from expressed genes to assess evidence for swamping of local genetic diversity resulting from the translocations of endangered European fire-bellied toads *Bombina bombina* from the genetically divergent southern populations in Austria to northern Germany, at least ten years earlier. They detected widespread introgression (four of five sites) but there was variation in the extent of introgression across the genome. Some introgressed variants were found in all four sites, consistent with genetic swamping. However, there was also evidence that local variants were maintained at similar frequencies, suggesting local adaptations were retained. That said, large scale introductions of individuals with unsuitable/incompatible genomic diversity can have detrimental consequences, as is seen in the mixing of odd- and even-year lineage of pink salmon *Oncorhynchus gorbuscha* (Gharrett et al., 1999).

Genomic data can also provide insights into the genetic incompatibilities that lead to outbreeding depression. For example, the copepod *Tigriopus californicus* is a well-known example of outbreeding depression, where mixing of individuals from different isolated populations resulted in a reduction in fitness (Edmands, 2007). Genome sequencing demonstrates extensive divergence between populations (Barreto et al., 2018), well beyond that recommended for mixing by Frankham et al. (2011). Further, the additional information from genomic data helped pinpoint the nature of the genetic incompatibilities, predominantly the breakup of co-adapted gene complexes, leading to outbreeding depression (Lima et al., 2019; Healy & Burton, 2020), which may inform translocations to minimise their impacts.

Overall, there is an emerging perspective that the risks associated with outbreeding depression have been overstated, and that outbreeding depression is largely predictable through consideration of genetic and ecological differences (Box 9.3; Frankham et al., 2011, 2017). Genomic

---

Box 9.3 *Assessing benefits and risks of mixing/hybridisation*

When multiple populations are mixed there is a risk of detrimental consequences such as outbreeding depression, swamping of adaptive diversity, or loss of genetic integrity. Critically, these risks must be weighed against the risks associated with low diversity, inbreeding, and limited adaptive potential, and the possible benefits that mixing may provide. Current information suggests that the detrimental consequences of outbreeding depression are largely predictable (Figure 9.2; Frankham et al., 2017). Mixing populations is most likely to be recommended over separate management units where the risks of outbreeding are low; this is particularly true for populations where isolation is anthropogenic in nature (Figure 9.2). The risks increase and become more complicated as levels of divergence increase, and mixing is usually avoided between highly divergent lineages and/or different taxa (and especially so between different species). In these circumstances, the likelihood of genetic incompatibilities is increased, and value judgements may also be relevant regarding the merging/erosion of genetically distinct taxa. The framework developed and refined by Frankham et al. (2017) provides conceptual guidance for decision-making. Genomic data can support the implementation of this approach by quantifying levels of divergence and identifying situations where the translocation would lead to mixing of highly divergent lineages that may pose a risk of outbreeding depression, swamping of adaptive variation, and/or loss of taxonomic integrity, versus situations when hybridisation is likely to be beneficial (Ottenburghs, 2021)

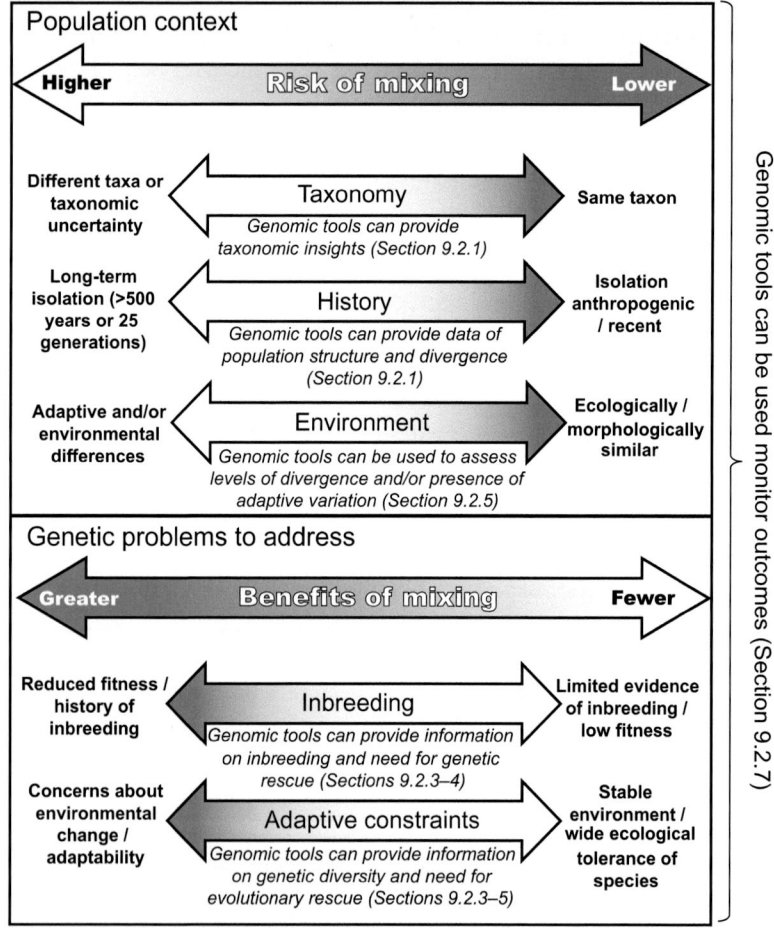

Figure 9.2 Summary of the situations where there are (a) risks of outbreeding depression from mixing populations during translocation, and (b) benefits from mixing. Darker grey shading indicates situations with lower risks and/or greater benefits. The risks of outbreeding depression are based on the framework of Frankham et al. (2011, 2017).

analysis of wild ass *Equus hemionus* revealed high levels of admixture and no outbreeding depression when a mix of two subspecies (*E. h. onager* and *E. h. kulan*) were used in reintroductions (Zecherle et al., 2021). In some instances, such as the brook trout *Salvelinus fontinalis*, even populations that had been separated for tens of thousands of years/generations showed little evidence of outbreeding depression when mixed (Wells et al., 2019). Thus, it is important, particularly for small, isolated populations, that the relative risks of outbreeding depression are weighed against the risk of the loss of diversity and inbreeding. Indeed, when the relative risks of inbreeding and outbreeding depression are considered, mixing is more likely to be recommended than separate management (Liddell et al., 2021).

9.2.7 Genomic Monitoring of Translocations

The role of genomics in monitoring translocations can be divided into approaches that utilise DNA analysis as an identification tool and those that reveal changes to the genetic make-up of the resulting populations.

In terms of taxonomic identification, the use of genetic data to monitor the outcome of species translocations is fairly well established, through post-release analysis of non-invasive samples, such as scats, or the use of eDNA, particularly in freshwater habitats, to detect the presence and distribution of reintroduced species (Hempel et al., 2020). Typically, these forms of genetic monitoring are conducted using traditional low throughput DNA-based approaches. However, the advent of genomics has increased the range of questions and applications that may be addressed during monitoring work.

Meta-genomic analysis of faecal samples can be used to monitor diet in translocated individuals, which informs understanding of post-release behaviour and ecology and may potentially help to explain the success or failure of a translocation. A study of the brush-tailed possum *Trichosurus vulpecula* in South Australia (Bannister et al., 2020) used DNA barcoding to monitor changes in diet following reintroduction. Immediately after release, a wide range of plant species were consumed, which over time decreased to a much smaller number of preferred plant genera. This decrease in dietary diversity was attributed to post-release acclimatisation as the possums discovered the location of their preferred food plants at the destination site. Similar genomic approaches to characterise the microbiome of individuals post release have the potential to further inform post-release monitoring studies (see Section 9.2.2.2).

A further application of meta-genomic identification in translocations is to monitor outcomes at a community level. Monitoring of fungal diversity over a 10-year restoration project of a grassy woodland community in South Australia (Yan et al., 2018) used community meta-barcoding to reveal a significant shift in fungal community composition towards the desired restored condition. As availability of genomic tools increases, these more complex taxonomic identification approaches are expected to broaden the array of ecological information available to include in monitoring project outcomes.

In terms of understanding changes to population genetic diversity, a broad suite of genomic applications can be deployed to monitor population structure, gene flow, admixture, and inbreeding. Following a successful captive breeding programme, the translocation of Tasmanian devils to Maria Island, Tasmania, included deliberate mixing of genetically distinct wild and captive-bred populations. Genomic monitoring, using a genome wide SNP panel, was applied to monitor diversity at release in 2012 and over the next six years post release (McLennan et al., 2020). These data were used to reconstruct the post-release pedigree and assess genetic representation of the founders to assess reproductive success, admixture, and the changes in genetic diversity over time. The results demonstrated that genetic admixture had led to an increase in genetic diversity of the new populations over any of the source populations, with no measurable impact of outbreeding depression on reproductive success.

Arguably, the two largest scale post-translocation genomic monitoring studies conducted to date involve the Rocky Mountain bighorn sheep and the Alpine ibex, and this monitoring has provided significant information on approaches for source selection (as outlined in Section 9.2.3). In Nevada, USA, the genetic results of multiple statewide translocations of bighorn sheep implemented across five decades have been investigated using over 17,000 DNA markers across the genome (Jahner et al., 2019). By combining knowledge of source population history with contemporary population genomic data, it was demonstrated that very little gene flow exists among the current fragmented populations and that, instead, local population genetic diversity is driven primarily by the number and distinctiveness of the source populations that contributed to the founders in each area. Where the combined founder diversity was highest, translocated populations have been able to maintain levels of heterozygosity observed in remnant populations elsewhere. In Switzerland, a similar study applied over 100,000 genomic markers to monitor multiple populations of reintroduced ibex, indicating clear reductions in diversity compared to the source population, with diversity indices (genome wide SNP density and heterozygosity) showing up to 25 per cent loss of variation (Grossen et al., 2018). Monitoring of inbreeding (via ROH) has led to recommendations for the future management of alpine ibex reintroductions and translocations to reverse this trend.

Beyond direct monitoring of translocation outcomes, genomic data can also be used to inform models that describe the long-term outcomes of past, current, or future translocations. The development of forecasting models informed by empirical genomic data generated pre- and post-release has been shown to potentially improve translocation success (Seaborn et al., 2021). By enabling predictions of change over time at both neutral and adaptive genetic markers, iterative models can be used to assess likely outcomes relating to population diversity and structure and to guide management interventions.

9.3 Future Prospects

As advances in genomic science continue to develop it is worth considering the future deployment of genomic approaches to conservation translocations. Here we take a brief look at some of the themes set to develop over the coming decade.

9.3.1 A Genome-Enabled Biota

The availability of DNA data to both identify and characterise global biodiversity is set to increase exponentially. One major initiative is the International Barcode of Life project, aiming to produce a comprehensive DNA barcode reference library of all multicellular plants, animals, and fungi (Hobern, 2021; https://ibol.org/). Likewise, just two decades since the human genome was first sequenced, the Earth BioGenome Project is now underway, aiming to establish reference genome sequences for all plants, animals, and fungi (Lewin et al., 2018; www.earthbiogenome.org/). These initiatives will generate major baseline-enabling resources for the genetic management and genetic monitoring of translocation programmes.

9.3.2 Predicting Adaptation from Sequence Data

Understanding the relationships between DNA sequences and how organisms interact with their environment remains a challenge. Although major advances have been made in this area, the complicated nature of the issue, typically involving multiple interacting genes, makes simple predictions difficult. Nevertheless, the growing ability to understand the key elements of genetic variation that confer local adaptation has considerable potential to inform the selection of source populations and to enhance the management of adaptive variation in conservation translocations in a changing environment (Flanagan et al., 2018).

9.3.3 Gene Editing and Designer Translocations: Manipulating DNA to Improve Outcomes

The ability to edit genes and alter their function through technologies such as CRISPR-Cas9 enables geneticists to perform precisely targeted genetic modifications in living organisms, altering the DNA and affecting the expression or function of specific genes. Its application to natural populations is certainly likely to be contentious, but the possibility of engineering endangered populations to be, for example, resistant to invasive pathogens, is a concept many may find attractive (Novak et al., 2018). On the Hawaiian Islands, where invasive avian malaria has decimated many bird species, a strategy known as *facilitated adaptation* has been considered, in which gene-edited malaria-resistant honeycreepers *Drepanis coccinea* would be released to develop population-wide immunity (Samuel et al., 2020); similar pathogen-resistant gene editing is being evaluated in plants (Barrett et al., 2019). Although there are numerous social and ethical considerations regarding the acceptability of gene editing in wild populations, the development of 'designer translocations', in which pre-adapted individuals are used as founders to increase the likelihood of translocation success, is now within the realms of technical possibility. The factors limiting the deployment of this technology are likely to be social acceptability and/or the need to develop appropriate risk assessment procedures and an operational legislative framework.

9.3.4 Gene Drives: Releasing Organisms to Reduce Threats

Gene drives are mechanisms that skew the inheritance ratio of an allele, or copy of DNA, such that it is preferentially inherited. CRISPR-based artificial gene drives combine gene editing with a gene drive mechanism to manipulate the composition and fitness of a population. Thus, gene drives can spread 'self-destruct' variants through a target population, decreasing its fitness and/or controlling reproductive success. There is great interest in the potential use of gene drives to control invasive species and diseases (Moro et al., 2018; Rode et al., 2019). Although this does not fall within the conventional definition of a conservation translocation, there is a natural conceptual extension, in the sense of gene drives as translocations involving the release of modified organisms into the wild to achieve a positive conservation outcome. This may involve, for instance, releasing pathogens or individuals of an invasive species modified with gene drive mechanisms to reduce or eliminate the source of a pressure threatening an endangered species. Compared with gene editing (Section 9.3.3), this is undertaking a translocation of a genetically modified version of the 'problem species' to decrease its fitness rather than a translocation of a genetically modified version of the 'threatened species' to increase its fitness. With increasing scientific support for the potential viability of such an approach (Faber et al., 2021), attention is also now turning to risk assessment of cross-species contamination (Courtier-Orgogozo et al., 2020) and promoting public debate about the ethics of gene editing wildlife and releases into the natural environment (MacDonald et al., 2021).

9.3.5 Cloning and De-extinction: Translocation of Recreated Species

Where species reach extremely small numbers of individuals and can no longer be recovered by conventional interventions, the transfer of the genetic material of one individual into the egg cell of another offers the potential for replication of individuals to bolster numbers and 'buy time', to avoid extinction, and to provide material for population augmentation (Gross, 2021). Technologies for cloning are already established for some species in a laboratory setting, and the potential exists for their deployment to conservation programmes, albeit with considerable further research and development to achieve practical and affordable implementation along with associated development of appropriate ethical and legal frameworks (Sandler et al., 2021). Likewise, for species that have already gone extinct, various biotechnological strategies are being explored for de-extinction (Shapiro, 2017). These including cloning (transfer of DNA from preserved tissues into the egg cell of a related species), and cloning with gene editing to correct DNA errors where the genetic material of the extinct species has degraded (see also Seddon, this volume). An alternative conceptual approach is to use reference genome sequences from extinct species as a template to guide gene editing of a related species to 'copy-correct' enough nucleotides to modify individuals of an extant species into an approximation of an extinct taxon (e.g. by changing enough genes to alter the phenotype to resemble the extinct species). The technological and social challenges to make de-extinction a practical reality are considerable by any of these approaches (IUCN, 2016). However, from a genomic perspective, the rapid rate in production of genome sequence data is a useful enabler, as it provides more reference data on what constitutes a functioning genome that de-extinction is aiming to approximate. For both de-extinction, and cloning of surviving individuals of an extant species, an important consideration will be avoidance of creating populations that suffer problems associated with limited genetic diversity, and the principles for maintaining genetic diversity in such heavily manipulated interventions are the same as for managing more conventional translocations (Steeves et al., 2017). Another important issue is the cost-benefit analysis of such intensive interventions, set against what else a finite amount of conservation funding might achieve for existing, viable populations (Moehrenschlager et al., this volume; IUCN, 2016).

9.4 Considerations for Applying Genomics to Translocations

Conservation genomics straddles the divide between conservation practice and academic research. One of the key barriers to wider uptake of genomics is mutual understanding between the conservation science and management communities. Programmes that educate wildlife managers and policymakers in conservation genetics are few and far between, with geneticists perhaps having even less exposure to the process of conservation translocation planning and the associated technical, practical, and ethical complexities, as well as the influence of the wider public. The need to bridge these gaps is well recognised (Shafer et al., 2015) but the solutions require continued effort.

Despite the reducing costs and increasing availability of genomics, conservation budgets are finite and deployment of genomic approaches to support translocations requires objective cost-benefit analysis. Providing conservation managers with impartial advice on when genomic information is and is not likely to be informative for planning translocations is crucial. Formal cost-benefit analyses for the application of genetics to natural resource management have been developed, for example for fisheries (Martinsohn et al., 2019), but in practice, for most conservation scenarios, the net value of genomic information must be assessed on a case-by-case basis.

The way in which genomic data are generated for a project will determine both its cost and its utility. Costs can be reduced through partnerships with research institutions that may be able to offset staff time and equipment access, and utilise students to generate data for both applied and academic scientific outputs. Such a model can be very effective and is often necessary where state-of-the-art technologies are not yet commercially available, but it is important to recognise and manage for an inevitable bifurcation in project interests, with conservation managers seeking answers to practical questions while academics are rewarded for scientific novelty and publications. Alternative conservation genomic service models range from applied research contracts with scientific institutions through to commercial contracts with private service providers. The availability of commercial services for techniques such as eDNA analysis is growing, but for many other conservation genomics applications dedicated collaborations remain the default mechanism for bringing together the necessary different skill sets to link genomics to conservation translocation management.

9.5 Key Messages

- Advances in genomic science are providing high-resolution insights into the diversity of species and populations, and increased understanding of how they function and interact.
- The application of genomic data to conservation translocations is now widespread and there are many examples of genomic data being used to guide the implementation of translocations, ranging from selecting donor individuals/populations, to understanding the dynamics of interspecific interactions, and to the design and monitoring of population reinforcements to achieve genetic and/or evolutionary rescue.
- The rapidly accelerating generation of genomic data from the world's species will lead to further major advances in understanding biodiversity at the genomic level, with associated benefits for translocation management and monitoring.
- However, genomic data and genomic technologies are not a panacea; despite the power of these approaches, uncertainties can remain in data interpretation and translation into practical management actions.
- As the science at the interface of genomics and conservation translocations continues to develop, there is a pressing need to focus continually

on translating data to support practical decision-making, and, at least in the short term, to develop further guidance and thinking that allows extrapolation from well-resourced studies with extensive genomic data to guide actions and decisions in translocations where generating genomic data is not yet feasible.
- As genetic/genomic technologies enable greater technological interventions for conservation translocations, the need to extend multi-stakeholder dialogue will continue and grow; this ranges from promoting informed dialogue between geneticists and conservationists to ensure effective deployment of approaches and resources, and wider societal engagement in setting the agenda for if, when, and how approaches involving genetic modification should be deployed.

Acknowledgements

The Royal Botanic Garden Edinburgh acknowledges funding support from the Scottish Government Rural and Environment Sciences and Analytical Services Division (RESAS). P. M. H. and L. E. N. acknowledge funding support from the Leverhulme Trust (Grant RPG-2015-273 to P. M. H.). This chapter was considerably improved by comments from Martin Gaywood and Axel Moehrenschlager.

References

Bannister, H., Croxford, A., Brandle, R., Paton, D. C. & Moseby, K. (2020) Time to adjust: changes in the diet of a reintroduced marsupial after release. *Oryx*, 55, 1–10.

Barreto, F. S., Watson, E. T., Lima, T. G., et al. (2018) Genomic signatures of mitonuclear coevolution across populations of *Tigriopus californicus*. *Nature Ecology & Evolution*. 2, 1250–1257.

Barrett, L. G., Legros, M., Kumaran, N., et al. (2019) Gene drives in plants: opportunities and challenges for weed control and engineered resilience. *Proceedings of Royal Society of London B, Biological Sciences*. 286, 20191515.

Bell, D. A., Robinson, Z. L., Funk, W. C., et al. (2019) The exciting potential and remaining uncertainties of genetic rescue. *Trends in Ecology & Evolution*. 34, 1070–1079.

Bell, G. (2017) Evolutionary rescue. *Annual Review of Ecology, Evolution, and Systematics*. 48, 605–627.

Blyton, M. D. J., Soo, R. M., Whisson, D., et al. (2019) Faecal inoculations alter the gastrointestinal microbiome and allow dietary expansion in a wild specialist herbivore, the koala. *Animal Microbiome*. 1, 6.

Bozzuto, C., Biebach, I., Muff, S., Ives, A. R. & Keller, L. F. (2019) Inbreeding reduces long-term growth of Alpine ibex populations. *Nature Ecology & Evolution*. 3, 1359–1364.

Brambilla, A., Biebach, I., Bassano, B., Bogliani, G. & Von Hardenberg, A. (2015) Direct and indirect causal effects of heterozygosity on fitness-related traits in Alpine ibex. *Proceedings of Royal Society of London B, Biological Sciences*. 282, 20141873.

Brambilla, A., Keller, L., Bassano, B. & Grossen, C. (2018) Heterozygosity–fitness correlation at the major histocompatibility complex despite low variation in Alpine ibex (*Capra ibex*). *Evolutionary Applications*. 11, 631–644.

Browne, L., Wright, J. W., Fitz-Gibbon, S., Gugger, P. F. & Sork, V. L. (2019) Adaptational lag to temperature in valley oak *Quercus lobata* can be mitigated by genome-informed assisted gene flow. *Proceedings of the National Academy of Sciences of the United States of America*. 116, 25179–25185.

Campbell-Palmer, R., Senn, H., Girling, S., et al. (2020) Beaver genetic surveillance in Britain. *Global Ecology and Conservation*. 24, e01275.

Carlson, S. M., Cunningham, C. J. & Westley, P. a. H. (2014) Evolutionary rescue in a changing world. *Trends in Ecology & Evolution*. 29, 521–530.

Chong, R., Grueber, C. E., Fox, S., et al. (2019) Looking like the locals - gut microbiome changes post-release in an endangered species. *Animal Microbiome*. 1, 8.

Coates, D. J., Byrne, M. & Moritz, C. (2018) Genetic diversity and conservation units: dealing with the species-population continuum in the age of genomics. *Frontiers in Ecology and Evolution*. 6, https://doi.org/10.3389/fevo.2018.00165.

Courtier-Orgogozo, V., Danchin, A., Gouyon, P.-H. & Boëte, C. (2020) Evaluating the probability of CRISPR-based gene drive contaminating another species. *Evolutionary Applications*. 13, 1888–1905.

De Cahsan, B., Westbury, M. V., Paraskevopoulou, S., et al. (2021) Genomic consequences of human-mediated translocations in margin populations of an endangered amphibian. *Evolutionary Applications*. 14, 1623–1634.

Diallo, M., Ollier, S., Mayeur, A., et al. (2021) Plant translocations in Europe and the Mediterranean: geographical and climatic directions and distances from source to host sites. *Journal of Ecology*. 109, 2296–2308.

Dincă, V., Bálint, Z., Vodă, R., et al. (2018) Use of genetic, climatic, and microbiological data to inform reintroduction of a regionally extinct butterfly. *Conservation Biology*. 32, 828–837.

Ducotterd, C., Crovadore, J., Lefort, F., et al. (2020) The feeding behaviour of the European pond turtle (*Emys orbicularis*, L. 1758) is not a threat for other endangered species. *Global Ecology and Conservation*. 23, e01133.

Dupuis, J. R., Geib, S. M., Schmidt, C. & Rubinoff, D. (2020) Genomic-wide sequencing reveals remarkable connection between widely disjunct populations of the internationally threatened bog buck moth. *Insect Conservation and Diversity*. 13, 495–500.

Edmands, S. (2007) Between a rock and a hard place: evaluating the relative risks of inbreeding and outbreeding for conservation and management. *Molecular Ecology*. 16, 463–475.

Egidi, E., May, T. W. & Franks, A. E. (2018) Seeking the needle in the haystack: undetectability of mycorrhizal fungi outside of the plant rhizosphere associated with an endangered Australian orchid. *Fungal Ecology*. 33, 13–23.

Eliades, S. J., Brown, J. C., Colston, T. J., et al. (2021) Gut microbial ecology of the Critically Endangered Fijian crested iguana (*Brachylophus vitiensis*): effects of captivity status and host reintroduction on endogenous microbiomes. *Ecology & Evolution*. 11, 4731–4743.

Emery, J.-P., Valentine, L. E., Hitchen, Y. & Mitchell, N. (2021) Survival of an extinct in the wild skink from Christmas Island is reduced by an invasive centipede: implications for future reintroductions. *Biological Invasions*. 23, 581–592.

Faber, N. R., Mcfarlane, G. R., Gaynor, R. C., et al. (2021) Novel combination of CRISPR-based gene drives eliminates resistance and localises spread. *Scientific Reports*. 11, 3719.

Fitzpatrick, S. W., Bradburd, G. S., Kremer, C. T., et al. (2020) Genomic and fitness consequences of genetic rescue in wild populations. *Current Biology*. 30, 517–522.e5.

Flanagan, S. P., Forester, B. R., Latch, E. K., Aitken, S. N. & Hoban, S. (2018) Guidelines for planning genomic assessment and monitoring of locally adaptive variation to inform species conservation. *Evolutionary Applications*. 11, 1035–1052.

Flesch, E. P., Graves, T. A., Thomson, J. M., et al. (2020) Evaluating wildlife translocations using genomics: A bighorn sheep case study. *Ecology and Evolution*. 10, 13687–13704.

Francioli, D., Lentendu, G., Lewin, S. & Kolb, S. (2021) DNA metabarcoding for the characterization of terrestrial microbiota—pitfalls and solutions. *Microorganisms*. 9, 361.

Frankham, R. (2016) Genetic rescue benefits persist to at least the F3 generation, based on a meta-analysis. *Biological Conservation*. 195, 33–36.

Frankham, R., Ballou, J., Eldridge, M., et al. (2011) Predicting the probability of outbreeding depression. *Conservation Biology*. 25, 465–475.

Frankham, R., Ballou, J., Ralls, K., et al. (2017) *Genetic Management of Fragmented Animal and Plant Populations*. Oxford, Oxford University Press.

Fraser, D. J., Weir, L. K., Bernatchez, L., Hansen, M. M. & Taylor, E. B. (2011) Extent and scale of local adaptation in salmonid fishes: review and meta-analysis. *Heredity*. 106, 404–420.

Gharrett, A. J., Smoker, W. W., Reisenbichler, R. R. & Taylor, S. G. (1999) Outbreeding depression in hybrids between odd- and even-broodyear pink salmon. *Aquaculture*. 173, 117–129.

Gross, M. (2021) In vitro conservation. *Current Biology*. 31, R1065–R1068.

Grossen, C., Biebach, I., Angelone-Alasaad, S., Keller, L. F. & Croll, D. (2018) Population genomics analyses of European ibex species show lower diversity and higher inbreeding in reintroduced populations. *Evolutionary Applications*. 11, 123–139.

Halbritter, A. H., Fior, S., Keller, I., et al. (2018) Trait differentiation and adaptation of plants along elevation gradients. *Journal of Evolutionary Biology*. 31, 784–800.

Hamede, R., Madsen, T., McCallum, H., et al. (2021) Darwin, the devil, and the management of transmissible cancers. *Conservation Biology.* 35, 748–751.

Healy, T. M. & Burton, R. S. (2020) Strong selective effects of mitochondrial DNA on the nuclear genome. *Proceedings of the National Academy of Sciences of the United States of America.* 117, 6616–6621.

Hedrick, P. (2005) Genetic restoration: a more comprehensive perspective than 'genetic rescue'. *Trends in Ecology & Evolution.* 20, 109–109.

Hempel, C. A., Peinert, B., Beermann, A. J., et al. (2020) Using environmental DNA to monitor the reintroduction success of the Rhine sculpin (*Cottus rhenanus*) in a restored stream. *Frontiers in Ecology and Evolution.* 8, 81 https://doi.org/10.3389/fevo.2020.00081.

Hobern, D. (2021) BIOSCAN: DNA barcoding to accelerate taxonomy and biogeography for conservation and sustainability. *Genome Biology*, 64, 161–164.

Hogg, C. & Wise, P. (2023) Assisted colonisation as a conservation tool: Tasmanian devils and Maria Island. In Gaywood, M. J., Ewen, J. G., Hollingsworth, P. M. and Moehrenschlager, A. (eds.) *Conservation Translocations.* Cambridge, Cambridge University Press.

Humble, E., Dobrynin, P., Senn, H., et al. (2020) Chromosomal-level genome assembly of the scimitar-horned oryx: insights into diversity and demography of a species extinct in the wild. *Molecular Ecology Resources.* 20, 1668–1681.

Ingvarsson, P. K. & Bernhardsson, C. (2020) Genome-wide signatures of environmental adaptation in European aspen (*Populus tremula*) under current and future climate conditions. *Evolutionary Applications.* 13, 132–142.

IUCN (2013) *Guidelines for Reintroductions and Other Conservation Translocations. Version 1.0.* Gland, Switzerland, IUCN Species Survival Commission.

IUCN (2016) *Guiding Principles on Creating Proxies of Extinct Species for Conservation Benefit.* Gland, Switzerland, IUCN Species Survival Commission.

Jahner, J. P., Matocq, M. D., Malaney, J. L., et al. (2019) The genetic legacy of 50 years of desert bighorn sheep translocations. *Evolutionary Applications.* 12, 198–213.

Jamieson, I. G. & Lacy, R. C. (2012) Managing genetic issues in reintroduction biology. In Ewen, J. G., Armstrong, D. P., Parker, K. A. and Seddon, P. (eds.) *Reintroduction Biology.* Oxford, Wiley Blackwell.

Johnson, W. E., Onorato, D. P., Roelke, M. E., et al. (2010) Genetic restoration of the Florida panther. *Science.* 329, 1641–1645.

Kartzinel, T. R., Chen, P. A., Coverdale, T. C., et al. (2015) DNA metabarcoding illuminates dietary niche partitioning by African large herbivores. *Proceedings of the National Academy of Sciences of the United States of America.* 112, 8019–8024.

Kessler, C., Brambilla, A., Waldvogel, D., et al. (2021) A robust sequencing assay of a thousand amplicons for the high-throughput population monitoring of Alpine ibex immunogenetics. *Molecular Ecology Resources.* 22, 66–85.

Koziol, L., Schultz, P. A., House, G. L., et al. (2018) The plant microbiome and native plant restoration: the example of native mycorrhizal fungi. *BioScience.* 68, 996–1006.

Leimu, R. & Fischer, M. (2008) A meta-analysis of local adaptation in plants. *PLoS ONE.* 3, e4010.

Lewin, H. A., Robinson, G. E., Kress, W. J., et al. (2018) Earth BioGenome Project: sequencing life for the future of life. *Proceedings of the National Academy of Sciences of the United States of America.* 115, 4325–4333.

Ley, R. E., Hamady, M., Lozupone, C., et al. (2008) Evolution of mammals and their gut microbes. *Science.* 320, 1647–1651.

Liddell, E., Sunnucks, P. & Cook, C. N. (2021) To mix or not to mix gene pools for threatened species management? Few studies use genetic data to examine the risks of both actions, but failing to do so leads disproportionately to recommendations for separate management. *Biological Conservation.* 256, 109072.

Lima, T. G., Burton, R. S. & Willett, C. S. (2019) Genomic scans reveal multiple mito-nuclear incompatibilities in population crosses of the copepod *Tigriopus californicus*. *Evolution.* 73, 609–620.

MacDonald, E. A., Edwards, E. D., Balanovic, J. & Medvecky, F. (2021) Scientifically framed gene drive communication perceived as credible but riskier. *People and Nature.* 3, 457–468.

Martinsohn, J. T., Raymond, P., Knott, T., et al. (2019) DNA-analysis to monitor fisheries and aquaculture: too costly? *Fish and Fisheries.* 20, 391–401.

McLennan, E. A., Grueber, C. E., Wise, P., Belov, K. & Hogg, C. J. (2020) Mixing genetically differentiated populations successfully boosts diversity of an endangered carnivore. *Animal Conservation.* 23, 700–712.

Mikheyev, A. S., Zwick, A., Magrath, M. J. L., et al. (2017) Museum genomics confirms that the Lord Howe Island stick insect survived extinction. *Current Biology.* 27, 3157–3161.e4.

Moehrenschlager, A., Soorae, P. & Steeves, T. E. (2023) From genes to ecosystems and beyond: addressing eleven contentious issues to advance the future of conservation translocations. In Gaywood, M. J., Ewen, J. G., Hollingsworth, P. M. and Moehrenschlager, A. (eds.) *Conservation Translocations.* Cambridge, Cambridge University Press.

Moro, D., Byrne, M., Kennedy, M., Campbell, S. & Tizard, M. (2018) Identifying knowledge gaps for gene drive research to control invasive animal species: the next CRISPR step. *Global Ecology and Conservation.* 13, e00363.

Mueller, E. A., Wisnoski, N. I., Peralta, A. L. & Lennon, J. T. (2020) Microbial rescue effects: how microbiomes can save hosts from extinction. *Evolution and Ecology of Microbiomes.* 34, 2055–2064.

Novak, B. J., Maloney, T. & Phelan, R. (2018) Advancing a new toolkit for conservation: from science to policy. *The CRISPR Journal.* 1, 11–15.

Ogden, R., Chuven, J., Gilbert, T., et al. (2020) Benefits and pitfalls of captive conservation genetic management: evaluating diversity in scimitar-horned oryx to support reintroduction planning. *Biological Conservation.* 241, 108244.

O'Sullivan, R. J., Aykanat, T., Johnston, S. E., et al. (2020) Captive-bred Atlantic salmon released into the wild have fewer offspring than wild-bred fish and decrease population productivity. *Proceedings of Royal Society of London B, Biological Sciences.* 287, 20201671.

Ottenburghs, J. (2021) The genic view of hybridization in the Anthropocene. *Evolutionary Applications.* 14, 2342–2360.

Pinho, C. J., Santos, B., Mata, V. A., et al. (2018) What is the giant wall gecko having for dinner? Conservation genetics for guiding reserve management in Cabo Verde. *Genes*. 9, 599.

Pompanon, F., Deagle, B. E., Symondson, W. O. C., et al. (2012) Who is eating what: diet assessment using next generation sequencing. *Molecular Ecology*. 21, 1931–1950.

Reiter, N., Vlcek, K., O'Brien, N., et al. (2017) Pollinator rarity limits reintroduction sites in an endangered sexually deceptive orchid (*Caladenia hastata*): implications for plants with specialized pollination systems. *Botanical Journal of the Linnean Society*. 184, 122–136.

Rode, N. O., Estoup, A., Bourguet, D., Courtier-Orgogozo, V. & Débarre, F. (2019) Population management using gene drive: molecular design, models of spread dynamics and assessment of ecological risks. *Conservation Genetics*. 20, 671–690.

Samuel, M. D., Liao, W., Atkinson, C. T. & Lapointe, D. A. (2020) Facilitated adaptation for conservation – can gene editing save Hawaii's endangered birds from climate driven avian malaria? *Biological Conservation*. 241, 108390.

Sandler, R. L., Moses, L. & Wisely, S. M. (2021) An ethical analysis of cloning for genetic rescue: case study of the black-footed ferret. *Biological Conservation*, 257, 109118.

Seaborn, T., Andrews, K. R., Applestein, C. V., et al. (2021) Integrating genomics in population models to forecast translocation success. *Restoration Ecology*, 29, e13395.

Seddon, P. J. (2023) The role of conservation translocations in rewilding and de-extinction. In Gaywood, M. J., Ewen, J. G., Hollingsworth, P. M. and Moehrenschlager, A. (eds.) *Conservation Translocations*. Cambridge, Cambridge University Press.

Senn, H., Ogden, R., Frosch, C., et al. (2014) Nuclear and mitochondrial genetic structure in the Eurasian beaver (*Castor fiber*) – implications for future reintroductions. *Evolutionary Applications*, 7, 645–662.

Senn, H. V., Ghazali, M., Kaden, J., et al. (2019) Distinguishing the victim from the threat: SNP-based methods reveal the extent of introgressive hybridization between wildcats and domestic cats in Scotland and inform future in situ and ex situ management options for species restoration. *Evolutionary Applications*. 12, 399–414.

Shafer, A. B. A., Wolf, J. B. W., Alves, P. C., et al. (2015) Genomics and the challenging translation into conservation practice. *Trends in Ecology & Evolution*. 30, 78–87.

Shapiro, B. (2017) Pathways to de-extinction: how close can we get to resurrection of an extinct species? *Functional Ecology*. 31, 996–1002.

Steeves, T. E., Johnson, J. A. & Hale, M. L. (2017) Maximising evolutionary potential in functional proxies for extinct species: a conservation genetic perspective on de-extinction. *Functional Ecology*. 31, 1032–1040.

Tang, L., Li, Y., Srivathsan, A., et al. (2020) Gut microbiomes of endangered Przewalski's horse populations in short- and long-term captivity: implication for species reintroduction based on the soft-release strategy. *Frontiers in Microbiology*. 11, 363 https://doi.org/10.3389/fmicb.2020.00363.

Thomsen, P. F. & Sigsgaard, E. E. (2019) Environmental DNA metabarcoding of wild flowers reveals diverse communities of terrestrial arthropods. *Ecology and Evolution.* 9, 1665–1679.

Vamosi, J. C., Gong, Y.-B., Adamowicz, S. J. & Packer, L. (2017) Forecasting pollination declines through DNA barcoding: the potential contributions of macroecological and macroevolutionary scales of inquiry. *New Phytologist.* 214, 11–18.

Van Rossum, F. & Hardy, O. J. (2022) Guidelines for genetic monitoring of translocated plant populations. *Conservation Biology.* 36, e13670.

Waters, C. D., Hard, J. J., Brieuc, M. S. O., et al. (2018) Genomewide association analyses of fitness traits in captive-reared Chinook salmon: applications in evaluating conservation strategies. *Evolutionary Applications.* 11, 853–868.

Weeks, A. R., Sgro, C. M., Young, A. G., et al. (2011) Assessing the benefits and risks of translocations in changing environments: a genetic perspective. *Evolutionary Applications.* 4, 709–725.

Wells, Z. R. R., Bernos, T. A., Yates, M. C. & Fraser, D. J. (2019) Genetic rescue insights from population- and family-level hybridization effects in brook trout. *Conservation Genetics.* 20, 851–863.

West, A. G., Waite, D. W., Deines, P., et al. (2019) The microbiome in threatened species conservation. *Biological Conservation.* 229, 85–98.

White, L. C., Moseby, K. E., Thomson, V. A., Donnellan, S. C. & Austin, J. J. (2018) Long-term genetic consequences of mammal reintroductions into an Australian conservation reserve. *Biological Conservation.* 219, 1–11.

Whiteley, A. R., Fitzpatrick, S. W., Funk, W. C. & Tallmon, D. A. (2015) Genetic rescue to the rescue. *Trends in Ecology & Evolution.* 30, 42–49.

Wright, B. R., Farquharson, K. A., Mclennan, E. A., et al. (2020) A demonstration of conservation genomics for threatened species management. *Molecular Ecology Resources.* 20, 1526–1541.

Yan, D., Mills, J. G., Gellie, N. J. C., et al. (2018) High-throughput eDNA monitoring of fungi to track functional recovery in ecological restoration. *Biological Conservation.* 217, 113–120.

Zahn, G. & Amend, A. S. (2017) Foliar microbiome transplants confer disease resistance in a critically-endangered plant. *PeerJ.* 5, e4020.

Zecherle, L. J., Nichols, H. J., Bar-David, S., et al. (2021) Subspecies hybridization as a potential conservation tool in species reintroductions. *Evolutionary Applications.* 14, 1216–1224.

10 · The Human Dimensions and the Public Engagement Spectrum of Conservation Translocation

JENNY A. GLIKMAN, BEATRICE FRANK, CAMILLA SANDSTRÖM, SAMANTHA MEYSOHN, MICHELLE BOGARDUS, FRANCINE MADDEN, AND ALEXANDRA ZIMMERMANN

10.1 Introduction

Conservation translocations involve the intentional movement of living organisms from one area to another, in order to provide a conservation benefit (IUCN, 2013). They span a range of activities, including reintroducing species where they have been lost, reinforcing populations with additional individuals, assisting colonisation of populations in new places where threats are lower, or replacing extirpated species to fulfil an ecological role (IUCN, 2013; Gaywood & Stanley-Price, this volume). The last two conservation translocations are strategies that involve moving species outside their indigenous range to proactively take actions to reduce the combined pressures of environmental change, land use, and extinctions. The concept of rewilding can include one or more of these types of conservation translocations (Seddon, this volume).

Even if the biological component of a translocation appears justifiable, its long-term resiliency may require support and durable partnership with and among stakeholders. While the translocation of iconic, non-threatening, and/or non-controversial species is often supported by local communities, as showcased by the scarlet macaw *Ara macao* reintroduction on the Nicoya Peninsula of Costa Rica (Williams & Haines, 2021), this is often not the case for those species whose potential for impact on human livelihoods and well-being is high. Opposition to conservation translocations can indeed arise when the species translocated may have 'real or perceived negative biological, economic, social, or political interactions' with humans (Frank, 2016, p. 738). Such human–wildlife interactions are defined as conflict and include wildlife impacting humans (e.g. livestock predation, humans injured by wildlife), humans impacting wildlife (e.g. poaching/killing wildlife), and conflicts between humans over wildlife (e.g. disagreement between stakeholders over a species' management, or the species becoming a symbol of government) (Frank, 2016; Frank & Glikman, 2019).

Conservation translocation has the potential to arouse conflict, especially when the species concerned raise fear and/or other concerns (Coz & Young, 2020; Gaywood & Stanley-Price, this volume). For example, conflicts can be magnified and intensified when the conservation translocation of carnivores is proposed. This is especially evident in areas where such species have been extirpated or reduced to extremely small populations through human persecution over centuries. For example, the reintroduction of a large-bodied predator such as the wolf *Canis lupus* in Yellowstone National Park in the USA has exacerbated existing conflicts and generated new clashes between members of different interest groups (Clark et al., 2005; Vaske et al., 2013). These dynamics are not unique to the USA. Similarly, in Norway and Sweden, the wolf has returned after being declared functionally extinct in the 1960s, and this has become a highly politicised issue in the public debate as well as in local and national political arenas (Eriksson, 2016). Although the first natural colonisation of wolves in Scandinavia in the 1980s took place without human involvement, since then authorities have actively translocated wolves to maintain biological and genetic diversity and ecosystem function (Swedish Environmental Protection Agency (SEPA), 2015; Elofsson & Häggmark, 2021). The return of the wolf has sparked debates over the role of these carnivores in Scandinavian landscapes, as well as introduced new dynamics in the discussion concerning the socio-economic cost and the impacts of conservation measures (SEPA, 2015).

Conservation translocations that are met with high conflict, lack of public support, and other factors, can undermine the public's perceptions of the value or success of conservation as a whole, and thereby affect trust in conservation actors and organisations. As stressed by Wilson (2018a; see also Riley & Sandström, 2016; Butler et al., 2019), successful carnivore reintroduction relies on a series of well-designed and supported strategies, including: understanding the biology of a species; developing meaningful management plans; implementing well-planned communication strategies; having institutional, political, and financial support; gaining buy-in of local communities and the wider public; and fostering inclusive governance. Some of these aspects have been largely addressed in conservation translocation research, especially the biology of the species and its capacity to form viable populations at potential release sites as well as the technical aspects tailored to address impacts related to negative human–wildlife interactions (e.g. Mukesh et al., 2015; Brichieri-Colombi & Moehrenschlager, 2016; Díaz et al., 2018). However, this is not the case for the human dimensions' components of conservation translocations.

While the IUCN Guidelines for Reintroductions and other Conservation Translocations (IUCN, 2013) cover some of the socio-economic considerations of conservation translocations, they are more focused on biological ones. As such, some have argued for the development of new and more overarching guidelines covering environmental and social considerations with the same weight (Butler et al., 2019). In line with this need for more social science in conservation translocation in general, and especially for highly controversial species, in this chapter we pinpoint the need to analyse the capacities of governance and institutional arrangements to reduce conflicts and enhance legitimacy (Sjölander-Lindqvist et al., 2015) at the collective level, starting with the planning process (e.g. see Riley & Sandström, 2016). We also highlight the importance of analysing and engaging individuals' perspectives on translocations to understand public as well as stakeholder responses to these types of management efforts. At the same time, we need to understand the social conflict in the system where translocation may take place, and even among the actors involved in translocation. Independently from the species translocated, and whether such an event is supported or opposed by local communities, the social aspects need to be addressed through a comprehensive engagement process because every conservation translocation is driven by human decisions and human interventions. The long-term outcomes of the translocation project depend upon the support and durability of partnerships with and among people who live with, are affected by, and are willing to coexist with the species translocated (e.g. Barlow, this volume). For example, the lack of prior engagement with stakeholders for the reintroduction of Eurasian beaver *Castor fiber* in Tayside, Scotland, led to negative impacts (Campbell-Palmer et al., this volume). This is why we argue that the engagement process should be developed as a long-term commitment and continuous process rather than being a stand-alone event. Doing so should help improve conservation translocation success over the long term.

Understanding how human dimensions affect the acceptance of a species and what triggers people to be willing to live – and even better coexist – with a translocated population are the first building blocks toward the long-term success of a conservation translocation project. However, to be fully meaningful, these building blocks need to be nested in a social engagement process, something that is often overlooked in translocation planning. To help address this gap, we discuss participatory processes in general and in conservation translocation in particular. Furthermore, we focus on how participatory processes can support our understanding and ability to prevent and address different levels of conflicts among people and with wildlife, especially with the more controversial species such as large herbivores and large carnivores. Engagement, when done well and planned early enough in the process, can reduce the risk of destructive conflict arising. We explore how participatory processes are not all created equal and, depending on their quality, can make the difference between coexistence and destructive conflicts. We conclude by offering recommendations on conflict dialogue processes designed to re-build trust and develop collaborative relationships among stakeholders involved in the translocation.

10.2 Overview of Conservation Translocation of Different Taxa

Conservation translocations have become increasingly used to boost species conservation (Seddon et al., 2014). By far the majority of conservation translocations focus on species that are usually not contentious, for example many plants and invertebrates, and some birds and mammals. A minority focus on charismatic and high conflict species such as wolves and lynxes (Breitenmoser & Haller, 1993; Linnell et al., 2009; Ripple & Beschta, 2012; Wilson, 2018b; Kubala et al., 2019). Whatever species is translocated, the human dimensions or social aspects of such events are often overlooked or inadequately addressed, yet every conservation translocation is driven by human decisions and human interventions.

While the human dimension context of translocations is important across species, social studies and projects featuring smaller herbivores, birds, and marine species are barely mentioned in conservation translocation literature reviews (Seddon et al., 2007; McMurdo Hamilton et al., 2021). An interesting case study that describes social aspects of conservation translocation is the South of Scotland Golden Eagle Project, which is a collaborative project between land managers and conservationists working to increase the breeding population and range of golden eagles *Aquila chrysaetos* in southern Scotland through education and engagement (Barlow, this volume). The social component of this project is in its infancy, yet the vision for success entails developing strong partnerships where equal input from the land management sectors and conservationists is provided. Other examples are the reintroduction of white-tailed eagles *Haliaeetus albicilla* in Scotland (Arts et al., 2012) and of Murray crayfish *Euastacus armatus* in south-eastern Australia. The latter was successful because it led to the long-term establishment of crayfish thanks to the engagement of recreational fishers in supporting reintroduction (Whiterod et al., 2021).

A lot of attention, however, has been paid to large mammal reintroductions. For example, the potential reintroduction of grizzly bears *Ursus arctos horribilis* into the Selway-Bitteroot area of the Rocky Mountains (Dak, 2015) as well as the much-debated proposal to reintroduce the lynx in Britain (for an overview see Neilson, 2021, and Johnson & Greenwood, 2020) have been discussed at length in the literature. In the latter case, the first licence application to reintroduce lynx to England was eventually refused by the UK Secretary of State for Environment, Food and Rural Affairs (DEFRA, 2018), because the project concerned did not meet the necessary standards set out in the IUCN Guidelines for Reintroduction and other Conservation Translocations (IUCN, 2013), such as an ecological feasibility study in England (Johnson & Greenwood, 2020). In addition, the statutory licensing authority found that the plan was missing organisational resilience, with insufficient evidence of a sufficient budget, and it was unclear to what extent the plan would meet the stated aims (DEFRA, 2018). However, given the strong interest from environmental organisations, in combination with the government's commitment to international conventions, the discussion over the reintroduction of lynx will probably continue and evolve in the future.

The conservation of threatened and endangered mega-herbivores, such as elephants (Goldenberg et al., 2019) and rhinos (Sheil & Kirkby, 2018), also receives great attention among academics and the media. Mega-herbivores play an important role in their ecosystems, such as through herbivory and seed dispersal, and have the potential to benefit a number of other species and facilitate habitat restoration if translocated (Campos-Arceiz & Blake, 2011). Additionally, these animals often enjoy international public support, which may be an important factor in their translocation (Riley & Sandström, 2016). Despite this support and the potential benefits of these translocation projects, for example through economic benefits from tourism, such species are challenging to translocate, especially elephants. Conflict surrounding elephant conservation is a significant problem in both Africa and Asia (Wilson et al., 2015). In Sri Lanka, the translocation of elephants *Elephas maximus maximus* identified as 'problem-elephants' into national parks caused the intensification of conflicts and increased elephant mortality at the release sites (Goldenberg et al., 2019), or the elephants returned to the sites from which they were captured (Fernando et al., 2012). The latter authors therefore suggested shifting the management focus from moving elephants to trying to prevent animals becoming 'problem-elephants'. These translocations were motivated by management challenges, and were not conservation translocations, but they illustrate the complex issues surrounding the movement of such species.

No matter what species are involved, or the motivation for translocation (conservation or management), such actions need to be combined with meaningful, high-quality participatory engagement among stakeholders in order to address the likely social conflict within the context and identify other issues.

10.3 Overview of the Societal Side of Conservation Translocation Conflicts

Where efforts have been made to understand and address the human dimensions of conservation translocation projects, researchers have mainly focused on the governance and associated decision-making aspects. As highlighted by Butler et al. (2019) through four case studies, some developed countries and regions have considered and, to a certain extent, integrated human capacity in conservation translocation process and management. From white-tailed eagles in Ireland, to bison *Bison bison* in Canada, sika deer *Cervus nippon taiouanus* in Taiwan, and tigers *Panthera tigris* in Cambodia, the success of rewilding involves engaging stakeholders in decision-making processes. This allows the development of social acceptance of unexpected benefits, costs, and conflicts generated by the translocated species, including the positive or negative interactions between stakeholders with different interests in the species. This study discusses the broader societal context of conservation translocation (i.e. policy, institutions) as well as focusing on the qualitative and quantitative studies of values, attitudes, beliefs, and feelings held by local people and the wider public towards the species concerned.

In southern Portugal, researchers studied local key actors' perceptions of the Iberian lynx *Lynx pardinus* in relation to translocation and towards carnivores in general. This study revealed a diversity of values among rural key actors and showed how environmental discourses were shaped by the local culture dominated by utilitarian values. Nevertheless, the mainly utilitarian views people held towards the environment and carnivores did not necessarily influence support or resistance towards the translocation of the Iberian lynx. The study showed that individual value orientations are not always coherent and predictable, but rather complex. To avoid presenting stereotypes among individuals in an urban or rural context, there is a need to incorporate local perspectives into management decisions because these may have implications for successful translocation projects (Lopes-Fernandes & Frazão-Moreira, 2017). Furthermore, it is essential to combine different ways of knowing, such as Western science and Indigenous knowledge, practices, and processes through the co-production of, or Indigenous-led, approaches that may enable more nuanced conservation translocation decisions (IPBES, 2018; McMurdo Hamilton et al., 2021; Ewen et al., this volume).

Conservation translocations are assumed to be more socially sustainable if they consider the human dimensions, from both a planning and implementation standpoint. Riley and Sandström (2016) suggested using the concept of 'wildlife acceptance capacity', which is extended from the biological carrying capacity theory to the social dimension of human–wildlife relationships. Such concepts allow for the understanding of factors affecting stakeholders' perceptions about species' impacts on natural and human-modified environments, other species, and/or people's well-being. The authors argue that studies of biological carrying capacities should take into account wildlife acceptance capacity when considering translocation efforts.

Importantly, acceptance of a species often is not related to the species itself or its biology, but to the perception that the translocated population may result in limitations in land use, access, and sovereignty. The conservation translocation of protected species may also imply government or authority overreach. For example, when a protected species is translocated to protected land surrounded by private properties, the surrounding landowners may become concerned that the government body managing the species will demand access and rights over their land or that additional legal restrictions will be put in place for the species' protection. In the case of the white-tailed eagle in Ireland, fears around lamb predation, possible restriction in land uses, and income loss have resulted in the illegal persecution of these reintroduced predators (Slocum, 2012; Viney, 2018). In a study of public attitudes towards the Swedish wolf policy, Eriksson (2016) found that politically alienated individuals, often situated in rural areas, favoured a more restrictive wolf policy. The wolf policy in Sweden has therefore become a symbolic issue around which rural citizens rally their fight against urban interests for political autonomy. Additionally, in rural/poor communities, conservation translocations can be seen as expensive excesses and symbols of inequality where money is spent to recover a wild species rather than supporting the livelihoods of local communities (Hernández et al., 2021). The reinforcement of the wild population of the Orinoco crocodile *Crocodylus intermedius* through translocation in Venezuela is an example. Political and financial instability, extreme poverty, and the expropriation of local ranchers near three important captive-breeding facilities led to unlawful hunting of released crocodiles and nest poaching. Implementation of the species translocation without the government providing any socio-economic alternative for local inhabitants was thought to be a key constraint for the recovery of this species. A similar case study is presented by Hernández and Ferrer (2021) for the Orinoco turtle *Podocnemis expansa* in Venezuela. In such cases, conservation translocation can exacerbate deep-seated divisions within communities resulting from distrust towards the government and outsiders, the lack of similar values and priorities between interested parties, and the perception that an endangered species has more rights and value than the local community. These governance issues and psychological dispositions, among many others, are often the basis of deep-rooted conflicts with wildlife and between interest groups, frequently resulting in sections of the public, communities, or specific groups dominating the political and media debate, as well as jeopardising conservation project outcomes (Young et al., 2016).

10.4 Human Dimensions of Conservation Translocation

Overall, the success of any conservation translocation strategy depends on understanding and effectively engaging with the perceptions, attitudes, and beliefs of people in relation to wildlife and each other. The human dimensions of wildlife conservation draw on several of the social sciences disciplines (e.g. social psychology, anthropology, and sociology) to understand how humans value wildlife, how they affect and are affected by wildlife, and how decisions are made to manage wildlife (Decker et al., 2012).

One theoretical framework that describes human psychological predispositions and behaviours toward wildlife is the cognitive hierarchy (Fulton et al., 1996), which postulates how different cognitions (e.g. attitudes, beliefs, and norms) drive human behaviours (Vaske & Donnelly, 1999; Jacobs & Vaske, 2019). In order to properly understand, and potentially predict, how an individual behaves and makes decisions towards wildlife, it is necessary to understand all the cognitions by themselves, and how they interrelate. Like an iceberg (Figure 10.1), behaviours are what we can observe; however, how an individual behaves depends on a series of latent, non-observable cognitions that are built upon each other. Immediately below the surface lies behavioural intention (i.e. the commitment to engage in a specific behaviour), which is the most direct predictor of behaviour (Vaske & Donnelly, 1999).

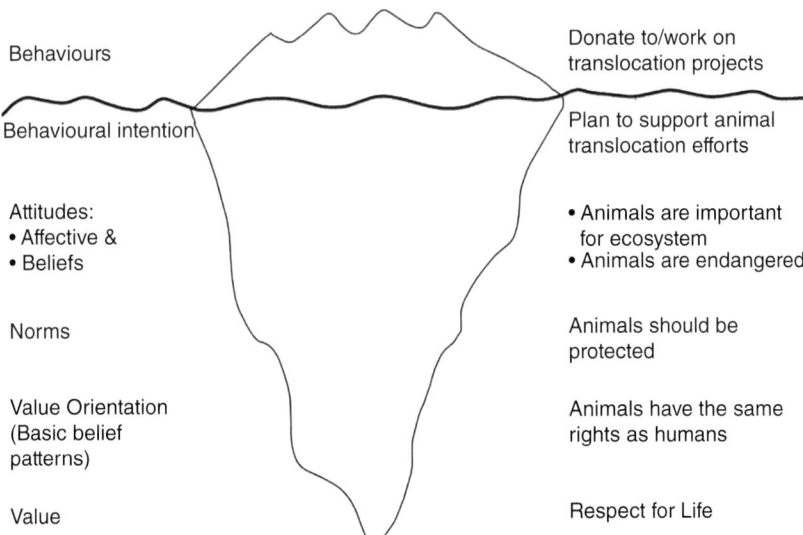

Figure 10.1 An adaptation of the cognitive hierarchy framework (Fulton et al., 1996; Vaske & Donnelly, 1999). The left side of the iceberg shows how the elements are related to each other; on the right side of the iceberg is an example of each cognition.

Attitudes are positive or negative evaluations of objects or actions and have cognitive and affective components (Ajzen & Fishbein, 1980; Eagly & Chaiken, 1993; Verplanken et al., 1998). The affective component of attitudes includes feelings, moods, and emotions about an object or behaviour (Eagly & Chaiken, 1993). The cognitive component, beliefs, are associations or linkages that people establish based on attributes related to the object (Eagly & Chaiken, 1993; Glikman et al., 2010). Norms are individual or shared standards that guide actions. Attitudes and norms are influenced by patterns of basic beliefs, which form value orientations and give meaning to fundamental values that are the core of an individual belief system (Vaske & Donnelly, 1999) (Figure 10.1).

Despite the increased recognition that is necessary to integrate human dimensions in wildlife conservation strategies, the application of social science has been limited to carrying out surveys on quantifying attitudes and perceptions of damages, or by consulting the public sporadically and superficially. This limitation is clearly demonstrated by Tosi et al. (2015, p. 12): 'Within the study area of the reintroduction project, the attitude of the resident population has been surveyed three times. In the preliminary phase of the reintroduction project (1997), the survey involved only the residents of the study area of the project... The subsequent interviews, conducted in 2003 and 2011... involved the whole Province of Trento.'

The human dimensions approach applied in this case study was mainly to understand communities' attitudes towards the translocation of the species. No effort was made to engage communities in the decision-making process, which is a key step in fostering support for any conservation project. Questionnaire surveys do not allow for an in-depth and transparent engagement and can even cause an estrangement of communities from conservation projects (Gray et al., 2020). While human dimensions research is necessary to understand the extent to which a species is accepted, this is only the first building block and a vital step towards the long-term success of a conservation translocation project. However, to be fully meaningful, this building block needs to be nested in an engagement process that is tailored to the species and the social context in which the reintroduction, assisted colonisation, reinforcement, or ecological replacement happens. Such processes can range from providing the necessary information to those who are involved (participants) to ensure they are engaged in a meaningful way, to communicating to the participants how their input will be used, to sharing or delegating power between decision makers and participants.

10.5 A Continuum of Public Engagement

Public engagement processes have become crucial for conservation success because of their power to shed light on deep-rooted problems and to foster dialogues between right-holders and individuals, specialists, project partners, citizen groups, stakeholders, and communities among others (hereafter referred as the public). Together, these processes can deal with complex issues and generate shared solutions for multifaceted issues. When the public is meaningfully engaged, knowledge sharing and learning are facilitated, trust is built between parties, generating common ground, and a sense of 'ownership' is gained by those likely to benefit from, be affected by, or be interested in the conservation project outcomes. Creating a shared vision between interested parties is pivotal to avoiding confrontation and delays in decision-making. A well-developed engagement process has the power to increase the transparency, credibility, and legitimacy of a decision-making process, thus fostering support for a conservation policy, plan, or project – including a conservation translocation process.

However, often engagement processes are poorly designed or only partially implemented, generating more harm than good. This is especially true when the participatory activities planned are demanding, the public expectations are not kept realistic, there is a lack of continuity in the engagement process, and/or the process itself becomes endless. Failure can also be fostered by not considering limitations in participation capacity, power imbalances, and the complexity of the political and public environment in which the engagement process is nested. Another precaution is to consider what is culturally appropriate in different communities. It is always important to avoid being too prescriptive in an engagement process because the context in which a project takes place, and the associated social interests, will influence how the engagement process finally looks.

A wide array of established literature exists on frameworks and toolboxes that help users design, plan, and carry out public engagement processes beyond conservation-related topics (Arnstein, 1969; Creighton, 2005; Dovers et al., 2015; Riley & Sandström, 2016). While these guidelines may differ in their outlook, they all emphasise the importance of determining four key components before starting an engagement process (Dovers et al., 2015):

1. What is the purpose of the engagement?
2. Who should be engaged?

3. When should participants be engaged (e.g. how often and for how long)?
4. How should participants be engaged?

The flow diagram in Figure 10.2 illustrates the four key components of the public engagement process and how these questions lead to choosing a meaningful engagement process for a specific project. A main prerequisite for starting a meaningful process of public engagement is that government, institutions, organisations, managers, and whoever is the decision maker, can exercise their capacity but are able to delegate and share some of their power at the same time. As defined by Arnstein (1969), power is delegated when the decision maker shares some degree of control, management, decision-making authority, or funding with the public. Such power sharing helps develop accountability for the project among all entities involved. Such an approach is key to finding middle ground and over time building trust and respect between all parties involved in the project. There are three other requirements that must also be considered to ensure successful public engagement (Fleming, 1997): time and money, fairness, and inclusion. If all these requirements are met, such a process can lead to a successful conservation outcome and a balance of power and natural resource management.

Based on these components it is possible to understand the degree to which the public is engaged and influences decisions relating to the goals, time scale, resources, and levels of concern in a decision-making process. Participants can be engaged by providing information, offering feedback on a project (information giving and receiving), sharing information to make sure all the concerns and aspirations for a project are taken into account, working collaboratively to find common ground for action and solution between participants, and by delegating decisions to them.

Nevertheless, planning and running a tailored engagement process are not the final steps. Reporting back and evaluating the process is essential to ensure the success of any public involvement activity. Such an approach keeps the public engaged over time, increases the transparency of the process, and legitimises the decision-making process. Establishing methods for decision makers and right-holders to effectively report back ensures that those involved see their input was received, understood, and valued. If some input provided by participants is not taken forward, the rationale behind such a decision needs to be offered to avoid distrust arising from the public towards the decision-making body (e.g. a decision might be beyond an agency's jurisdiction, or there may be no current

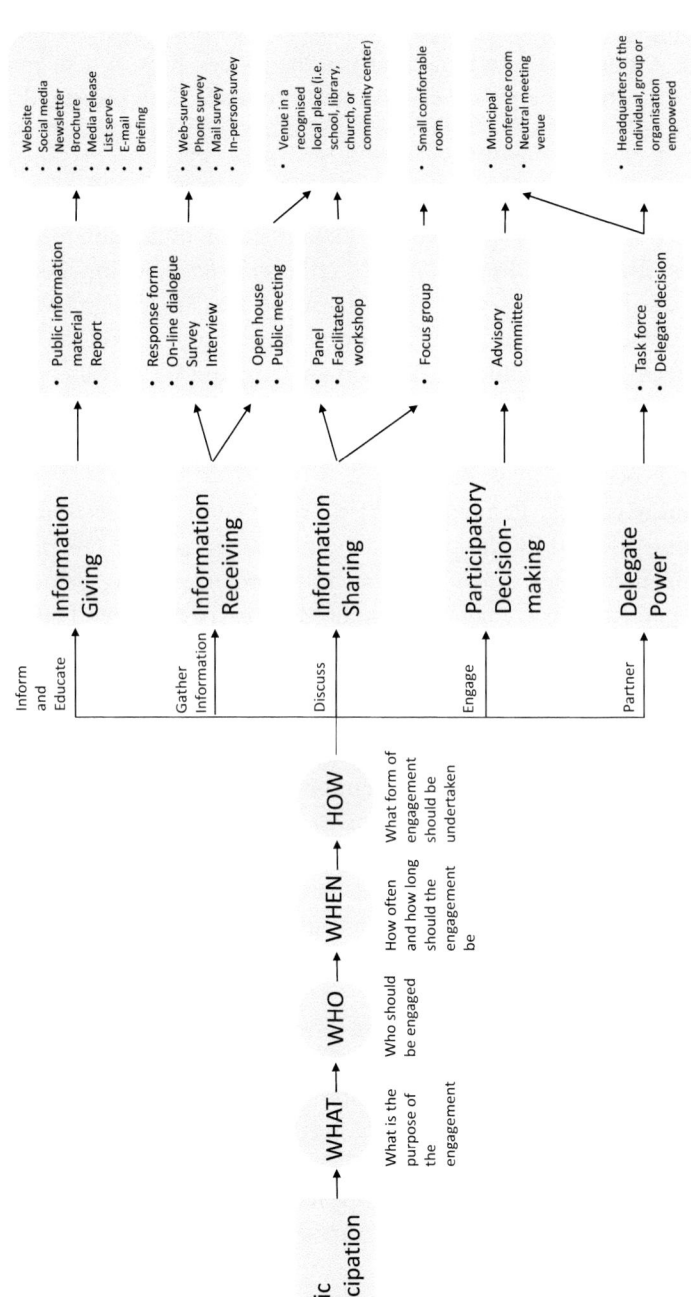

Figure 10.2 Components of an engagement process and the degree to which the public is engaged and influences decisions.

policy relating to a particular suggestion). Since reporting back can take many forms, it should be determined early in the process how, when, and to whom results from the public engagement process will be presented.

10.5.1 Criteria to Design, Plan, and Evaluate the Engagement Process

The final step of a good engagement process is an evaluation of its effectiveness (Rowe & Frewer, 2000; Abelson & Gauvin, 2006; Warburton et al., 2012). Several criteria can be used to design, plan, and evaluate whether an engagement process or activity has produced its intended effects (i.e. to influence decision-making and raise awareness) (Rowe & Fewer, 2000; Warburton et al., 2012). All or a combination of the criteria below can be used to design, plan, and evaluate a public participation process or activity.

Representativeness: Participants should represent a cross-section of the population affected by the project or issue. When participants provide multiple and diverse views about an issue/project, an engagement approach is perceived as credible. A stakeholder matrix compiled before the engagement exercise can help the convenor(s) evaluate if all the identified individuals, groups, and organisations are present and in what capacity. Ethnicity, age, sex, and other socio-demographic characteristics can be collected through response forms, and surveys or meeting evaluation forms can further establish if an event or process was representative.

Independence: The engagement process should be conducted in an independent, unbiased way. This can be done by involving a mutually agreed-upon management team, consultant, or neutral third-party organisation in the design, implementation, and evaluation of an engagement process. Media monitoring can also help to assess if the process is perceived as independent.

Early involvement: The key players should be involved as early as possible in the engagement process. The design of an engagement plan should include monitoring during and after the engagement process to help assess whether the public was involved early, on time, or later than expected.

Influence: The insights and feedback provided by participants should inform decision-making. It is important to assess if and how information has been given to the public on why their input has been considered or not in the decision-making process. Explaining why some suggestions have not been brought forward allows the influence of an engagement process to be evaluated.

Transparency: The process should enable participants to understand what is going on and how decisions are made. Transparency can be evaluated by assessing how information was disseminated prior, during, and after the project and/or the engagement process. Media monitoring can be used to assess if the process was perceived as transparent.

Accessibility: Participants should have access to the resources that enable them to participate in the engagement process or activity. Accessibility can be evaluated by assessing if information made available to the participants was easy to obtain, if the venue of the engagement activity or process was well served by public transport, if funding was made available to participants to attend the event, and if the times and dates selected resulted in a representative sample of participants. Other criteria can be used to identify and evaluate the accessibility of the engagement process.

Cost-effectiveness: The cost of the engagement process or activity should be proportionate to the importance and extent of engagement needed for the project. Cost-effectiveness can be measured as the costs of running the engagement process versus the non-monetary and monetary costs of not running such a process. If the public is not engaged, costs can range from lack of trust and implications for future projects and initiatives, to not identifying potentially serious or perceived management issues, to missing out on socio-economic opportunities that the public may wish to access. Other criteria to consider when evaluating the effectiveness of an engagement process or activity are the extent to which the objective of the project or event has been met and the lessons learned by conducting the engagement process or activity (i.e. what went well, what went wrong, what can be done better the next time?).

10.6 Conflicts over Conservation Translocations

Too often a 'check box' approach is applied when engaging with the public. Decision makers, project proposers, or convenors, after a superficial engagement that satisfies their obligations around involving the public, develop plans and then submit them for approval instead of co-producing those plans with stakeholders. However, these cursory or 'one size fits all' approaches typically are unable to reflect the specific needs of the unique social dynamics within the given situation and fall short of building or repairing relationships and addressing the underlying histories or mistrust that continue to shape the conflict. For instance, what may seem to be a conflict about a species causing damage to livestock may instead reflect a deeper issue between landowner and authorities over power, status, autonomy, recognition, or identity (Madden & McQuinn, 2014). As a result, conservation decisions may face resistance and even retaliation. The co-production of plans, instead, recognises participants as experts of their own circumstances and involves them in generating research knowledge alongside the researcher, which helps to bridge gaps in understanding between different parties and to tailor the plan to the specific context of the conservation translocation event.

The challenge, however, is that each conflicts is unique (Zimmermann et al., 2021) and many are weighed down by very deep-rooted hidden issues that may take time to uncover, while others are comparatively less complicated and more solvable. A professional trained and working in one of the conflict fields (Moffitt & Bordone, 2012) may be able to see relatively quickly how deep-rooted a conflict might be. Yet, among the great diversity of conflict practitioners, not all have the capacity to design and guide processes that reconcile these deeper-rooted conflicts. Project leaders, decision-makers, and others not specifically trained in these matters could unintentionally cause harm if a process is not designed to match the needs of the unique context (Madden & McQuinn, 2014).

The field of conflict and peace studies offers many conceptual models for understanding conflicts between groups of people, including tools that analyse the sources, cycles, patterns, and types of conflicts (Ramsbotham et al., 2011). As these are usually designed for social or violent conflicts, recent efforts have been made to adapt some of them for use in wildlife conservation. Here we present a useful model for orienting the reader to the potential depth of conflict that may exist in any proposed conservation translocation.

The idea that there are different depths or severities of conflict is common among various conceptual models for understanding conflict, including the concept of three levels. One such model is the 'Levels of Conflict' framework, which orientates conservation practitioners and stakeholders to the types and depths of conflict in a given situation (Canadian Institute for Conflict Resolution (CICR), 2000; Madden & McQuinn, 2014, 2017; Zimmerman et al., 2020). Levels of Conflict is a useful starting place to begin analysing the complexity, depth, and scope of a conflict. The model classifies three conflict levels: dispute, underlying conflict, and identity-based or deep-rooted conflict. The first level, disputes, are the surface, immediate, often tangible, manifestations of a conflict that are seemingly at the 'centre' of the conflict (Figure 10.3). The middle level – underlying conflict – occurs when there is a history of

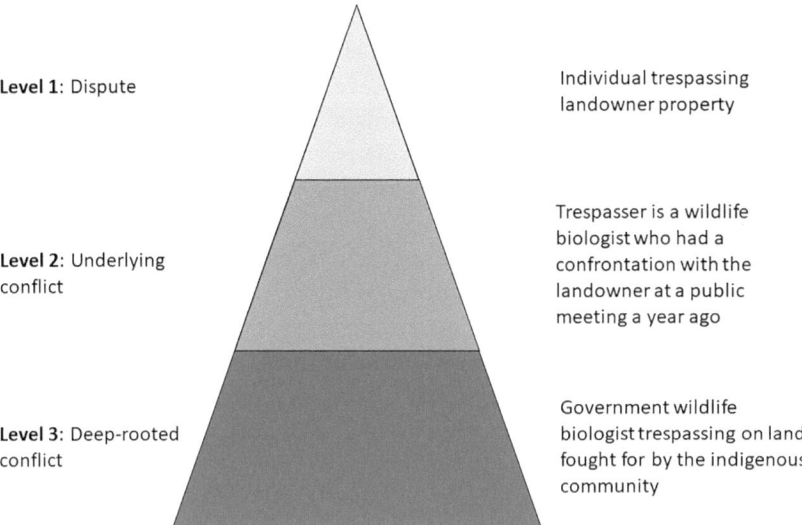

Figure 10.3 Adaptation of Levels of Conflict (Madden & McQuinn, 2014; Zimmermann et al., 2020). On the left side of the triangle are the different levels of conflict and on the right side of the triangle is an example of each level of conflict in a hypothetical conservation conflict scenario.

unresolved disputes. Past interactions or decisions made by the same parties can intensify or aggravate the current situation. Any new or recent dispute carries added significance that is not necessarily evident from the bare facts of the current dispute alone. The third level, identity or deep-rooted conflict, occurs when one experiences or perceives prejudicial assumptions or threats towards or by another based on their group affiliation. Identity conflict is typically thought of as an 'us versus them' conflict but can also be evident as an 'us versus us' conflict, or 'same side' conflict.

Although Levels of Conflict may seem like a simple model, the interplay between levels makes an analysis of conflict much more complex. For example, imagine a private landowner finds a trespasser walking across their property in an effort to reach a wetland. While the landowner may be annoyed, this dispute is unlikely to affect the landowner's attitude towards a conservation translocation (Level 1; Figure 10.3). Now, instead of an unknown entity, imagine the trespasser is a wildlife biologist who had a confrontation with the landowner at a public meeting a year ago over a protected bird species on that wetland. The wildlife biologist is accessing the wetland to survey for the presence of that endangered bird, and this is not the first time the landowner has witnessed someone from the agency that this biologist represents accessing their property. It is plausible to believe that this will have 'underlying' implications for how the landowner might react and form opinions about the endangered bird and potential conservation translocation in the area based on their past experience with this wildlife biologist (Level 2; Figure 10.3). Lastly, now imagine that the landowner is a member of an Indigenous community that has fought for the last century to regain the rights to their land and autonomy in managing it. This time the wildlife biologist is new to the government agency and has never met the landowner, but their uniform or vehicle bears the symbol of that agency. Under these circumstances, the landowner may make prejudicial assumptions about the government wildlife biologist and translocation project based on their negative relationship and perceptions of the government agency. Suddenly, this conservation translocation is no longer about the actual bird or the logistics of the translocation effort, but instead about the perception of the bird as a physical embodiment of a centuries old struggle against racism or colonialism and for a family, a community, and/or a people to manage their own lands and determine their own fates (Level 3; Figure 10.3). Moreover, the long-term outcomes of the translocation project now hinge on the reconciliation of this deeper-rooted conflict before the agency can hope for genuine buy-in from the landowner.

To assist the decision maker or conservation scientist in identifying the depth of the situation, Zimmermann et al. (2020) offered a list of potential questions and typical symptoms that may provide clues as to what may be going on in a given situation and with respect to the three levels of conflict. This makes it possible for the user to get some initial indication of the level and begin to think about whether and what type of external conflict specialist to bring into the case. For example, asking a landowner about a given species could elicit anything ranging from a mild or indifferent reaction to hostile and polarised language – all of which provide clues to the extent of deeper grievances.

Conservation translocations may accompany, provoke, or exacerbate existing social and conservation conflicts. Most conservation projects are preceded by a history of conservation action or research within or near human communities, making it highly likely that deeper conflicts may exist in the socio-political context (Cusack et al., 2021; Dietsch et al., 2021). In many conservation cases, conflicts also involve deep-seated values, needs, and beliefs in which one group's identity may be defined in opposition to another's due to perceived threats to their identity or way of life (Madden & McQuinn, 2014; Jacobsen & Linnell, 2016; Dietsch et al., 2021). Conservation conflicts often have a contentious history that contributes meaning and emotion to every new dispute, deepening both sides' entrenchment in their positions and their negative perceptions of each other; there is often a history of inequity where one or some low-power groups have been systematically disempowered (Coleman, 2006; Madden & McQuinn, 2014). To be sure, the group or groups who are being or have been marginalised or disempowered vary considerably depending on the particular conflict context and even depending on the point in time. In one conflict context a group may have high power, while that same group may have low power elsewhere. Putting in place a process that ignores or avoids engaging with existing deep-rooted or identity conflict may unintentionally escalate or aggravate a conflict. If conservationists' efforts do not account for what is more deeply at stake for the participants, they may too narrowly diagnose or misdiagnose the causes of the conflict and inadvertently perpetuate the conditions and structures that gave rise to the current manifestation of the conflict (Innes & Booher, 1999; Madden & McQuinn, 2017). Those in power may be reluctant to change a decision-making process in a way that necessitates giving up some control or power. However, a more inclusive decision-making process that addresses the deep-rooted conflict dynamics and balances power may expand receptivity and buy-in, as well as the range of acceptable solutions (Madden & McQuinn, 2017). A good example of inclusive co-management and multi-stakeholder translocation planning has been implemented in New Zealand (McMurdo Hamilton et al., 2021), and further details on a similar decision-making process are covered in this volume (e.g. Ewen et al., this volume). It is important to remember that for any method of decision-making it is fundamental that actors understand and are willing to embrace multiple values and perspectives; when this does not occur, destructive conflict arises or deepens and additional resources, skills, and processes are required. Addressing deep-rooted conflict in a transformative way does not guarantee that no conflict will ever occur, but rather enables all parties to engage more constructively with conflict when it does occur.

Indeed, one fundamental objective of conservation translocations is grounded in a biological perspective, where conservationists and biologists identify a need to improve the conservation status of a species or restore ecosystem processes. Even when the reasoning behind a translocation decision seems to be grounded in biology, often the reality is that conservationists are aiming to restore and protect species from the impacts human behaviours have on their survival. Thus, the impediment to success is in not effectively addressing the human–human conflict that underlies the conservation concern.

Most conservation professionals are trained from early in their careers in biology and ecological processes. Some are trained in social sciences, mostly from a research perspective. However, many are not formally taught the methods for changing a destructive conflict into an opportunity for positive change that benefits both the people and the resource, reconcile deeply entrenched conflicts, nurture common ground among people who may be deeply opposed to conservation goals, and change processes to lead to more sustainable solutions and genuine coexistence (Madden & McQuinn, 2017).

Box 10.1 provides details of one approach to dealing with conflict issues called 'Conservation Conflict Transformation' (CCT).

10.7 Conclusion

Conservation translocations are most likely to succeed if they are considered as multidisciplinary endeavours that deepen understanding and forge valuable relationships among resource managers, scientists, government, non-governmental organisations, and communities. Increasing and improving the quality of community participation, awareness, and potential benefits should be viewed as a primary translocation output rather than just a component of the planning process. The general public is critical of the success of conservation and restoration and yet often detached and isolated from the natural world. However, translocations provide a unique opportunity to reconnect. Translocations and restoration can be better planned, funded, implemented, and maintained by increasing their broad support base. To increase the likelihood of success, before a translocation is considered, relevant parties should analyse the social dynamics and ensure the process for engagement is an appropriate match with the social conditions surrounding and underlying the overall project. This should include accounting for any potential conflict, not just over a potential translocation, but also the conflicts existing in the social environment surrounding the translocation.

> **Box 10.1** *Conservation Conflict Transformation (CCT): An approach*
>
> Conservation conflicts are microcosms of larger societal conflicts, and Conservation Conflict Transformation (CCT) uses disputes about conservation issues as entry points to address these dynamics in society. CCT is an approach that adapts the theories, principles, processes, and skills of conflict transformation used in peacebuilding, and adapts them to the conservation context (Madden & McQuinn, 2014; www.cpeace.ngo). Building on conflict transformation principles (Lederach, 2003; Austin et al., 2011), this body of work is informed by neurology, psychology, anthropology, sociology, behavioural science, behavioural economics, complexity theory, systems thinking, and other disciplines.
>
> Conflict is conceptualised as a natural, potentially constructive, and even creative element of human relationships and processes. The goal of CCT is not to stop conflicts, but to harness the energy of the ebbs and flows of conflict to sustain problem-solving within a given context (Deutsch, 1973; Lederach, 2003; Austin et al., 2011; Madden & McQuinn, 2014, 2017). The practice of CCT provides a way of thinking about, understanding, and actively addressing such conflicts. Practitioners of CCT consider disputes as opportunities to engage constructively with the underlying relationships, decision-making processes and structures, and social systems (Lederach et al., 2007; Madden & McQuinn, 2014, 2017) to create an enabling social environment for effective, lasting, broadly supported conservation efforts. In fact, transforming these conservation conflicts can positively address social conflicts in the larger society as well. For example, a CCT intervention led by the Center for Conservation Peacebuilding (CPeace, www.cpeace.ngo) around wolf conflict in the Pacific Northwest of the USA led to positive advancements in the coexistence of landowners with wolves and overall social conflict around wolves and wolf management, as well as advancements in wildfire prevention, preventing and reducing domestic animal abuse, and improving workplace safety and equity for women, among others. This positive ripple effect within society is a hallmark of CCT. Another example from CPeace's work involved a CCT intervention in the Galápagos that created unprecedented and broadly supported advances in ecological restoration, while also addressing needs for climate change adaptation, education, and sustainable development (CPeace, unpublished data).

10.8 Key Messages

- Understanding the 'human dimensions of wildlife' – how humans value wildlife, how they affect and are affected by wildlife, and what triggers people to be willing to live with a translocated species – is the first building block and vital step towards the long-term success of a conservation translocation project.
- However, to be fully meaningful, understanding of the human dimensions needs to be nested in a social engagement process, which has too often been overlooked or poorly designed by conservation practitioners. A well-developed engagement process has the power to increase the transparency, credibility, and legitimacy of a decision-making process, thus fostering support for a conservation policy, plan, or project – including a conservation translocation.
- Nevertheless, planning and running a tailored engagement process are not the final steps. Reporting back and evaluating the process is key to ensuring the success of any public involvement. Such an approach keeps the public engaged over time, increases transparency, and legitimises the decision-making process.
- 'One size fits all' stakeholder engagement approaches typically fail to reflect the specific needs of the unique social dynamics within the system and fall short of reconciliation of the relationships and disentanglement of the deeper roots of conflict. The 'Levels of Conflict' model is one tool used to orientate conservation practitioners and stakeholders to the types and depths of conflict in a given situation.
- Conservation conflicts are microcosms of larger societal conflicts, and Conservation Conflict Transformation (CCT) provides a way of thinking about, understanding, and actively addressing such conflicts. Practitioners of CCT consider disputes as opportunities to engage constructively with the underlying relationships, decision-making processes, and social systems to create an enabling social environment for effective, lasting, broadly supported conservation efforts.

References

Abelson, J. & Gauvin, F. P. (2006) Assessing the impacts of public participation: Concepts, evidence and policy implications. Available from: www.cprn.org/documents/42669_fr.pdf [Accessed 15 October 2020].

Ajzen, I. & Fishbein, M. (1980) *Understanding Attitudes and Predicting Social Behavior*. Englewood Cliffs, NJ, Prentice-Hall.

Arnstein, S. R. (1969) A ladder of citizen participation. *Journal of the American Institute of Planners*. 45, 216–224.

Arts, K., Fischer, A. & Van der Wal, R. (2012) Common stories of reintroduction: a discourse analysis of documents supporting animal reintroductions to Scotland. *Land Use Policy*. 29, 911–920.

Austin, B., Fischer, M. & Giessmann, H. J. (2011) *Advancing Conflict Transformation*. Barbara Budrich Publishers.

Barlow, C. (2023) The role of community engagement in conservation translocations: the South of Scotland Golden Eagle Project (SSGEP). In Gaywood, M. J., Ewen, J. G., Hollingsworth, P. M. and Moehrenschlager, A. (eds.) *Conservation Translocations*. Cambridge, Cambridge University Press.

Breitenmoser, U. & Haller, H. (1993) Patterns of predation by reintroduced European lynx in the Swiss Alps. *The Journal of Wildlife Management*. 57, 135–144.

Brichieri-Colombi, T. A. & Moehrenschlager, A. (2016) Alignment of threat, effort, and perceived success in North American conservation translocations. *Conservation Biology*. 30, 1159–1172.

Butler, J. R. A, Young, J. C. & Marzano, M. (2019) Adaptive co-management and conflict resolution for rewilding across development contexts. In Pettorelli, N., Durant, S. M. and du Toit, J. T. (eds). *Rewilding*. Cambridge, Cambridge University Press, pp. 386–412.

Campbell-Palmer, R., Bauer, A., Jones, S., Ross, B. & Gaywood, M. J. (2023) The return of the Eurasian beaver to Britain: the implications of unplanned releases and the human dimension. In Gaywood, M. J., Ewen, J. G., Hollingsworth, P. M. and Moehrenschlager, A. (eds.) *Conservation Translocations*. Cambridge, Cambridge University Press.

Campos-Arceiz, A. & Blake, S. (2011) Megagardeners of the forest – the role of elephants in seed dispersal. *Acta Oecologica*. 37, 542–553. Canadian Institute for Conflict Resolution (CICR) (2000) *Becoming a Third-Party Neutral: Resource Guide*. Ottawa, Canada, Ridgewood Foundation for Community Based Conflict Resolution (Int'l).

Clark, T. W., Rutherford, M. B. & Casey, D. (2005) *Coexisting with Large Carnivores. Lessons from Greater Yellowstone*. Washington, DC, Island Press.

Coleman, P. T. (2006) Intractable conflict. In Deutsch, M., Coleman, P. T. and Marcus, E. C. (eds.) *Handbook of Conflict Resolution*. San Francisco, CA, Jossey-Bass, pp. 533–559.

Coz, D. M. & Young, J. C. (2020) Conflicts over wildlife conservation: learning from the reintroduction of beavers in Scotland. *People & Nature*. 2, 406–419. Creighton, J. L. (2005) *The Public Participation Handbook: Making Better Decisions Through Citizen Involvement*. San Francisco, CA, Jossey-Bass.

Cusack, J. J., Bradfer-Lawrence, T., Baynham-Herd, Z., et al. (2021) Measuring the intensity of conflicts in conservation. *Conservation Letters*, e12783.

Dak, M. J. (2015) *West: A Failed Attempt to Reintroduce Grizzly Bears in Mountain West*. Lincoln, NE, University of Nebraska Press.

Decker, D. J., Riley, S. J. & Siemer, W. F. (eds.) (2012) *Human Dimensions of Wildlife Management*, 2nd ed. Baltimore, MD, Johns Hopkins University Press.

DEFRA (2018) Letter from Secretary of State for Environment, Food and Rural Affairs to Lynx UK Trust about lynx reintroduction in Kielder Forest, Northumberland. Available from: https://assets.publishing.service.gov.uk/government/uploads/system/uploads/attachment_data/file/761267/letter-from-sos-to-lynx-uktrust-181203.pdf [Accessed 3 January 2020].

Deutsch, M. (1973) *The Resolution of Conflict*. New Haven, CT, Yale University Press.

Díaz, M., Anadón, J. D., Tella, J. L., Giménez, A. & Pérez, I. (2018) Independent contributions of threat and popularity to conservation translocations. *Biodiversity Conservation*. 27, 1419–1429.

Dietsch, A. M., Wald, D. M., Stern, M. J. & Tully, B. (2021) An understanding of trust, identity, and power can enhance equitable and resilient conservation partnerships and processes. *Conservation Science and Practice*, e421.

Dovers, S., Feary, S., Martin, A., McMillan, L., Morgan, D. & Tollefson, M. (2015) Engagement and participation in protected area management: who, why, how and when? In Worboys, G. L., Lockwood, M., Kothari, A., Feary S. and Pulsford I. (eds.) *Protected Area Governance and Management*. Canberra, ANU Press, pp. 413–440.

Eagly, A. H. & Chaiken, S. (1993) *The Psychology of Attitudes*. Fort Worth, TX, Harcourt.

Elofsson, K. & Häggmark, T. (2021) The impact of lynx and wolf on roe deer hunting benefits in Sweden. *Environmental Economics and Policy Studies*. 23, 683–719.

Eriksson, M. (2016) Changing attitudes to Swedish wolf policy: wolf return, rural areas, and political alienation. PhD dissertation, Umeå universitet, Umeå. Available from: http://urn.kb.se/resolve?urn=urn:nbn:se:umu:diva-128861 [Accessed 2 February 2020].

Ewen, J. G., Canessa, S., Converse, S. J. & Parker, K. A. (2023) Decision-making in animal conservation translocations: biological considerations and beyond. In Gaywood, M. J., Ewen, J. G., Hollingsworth, P. M. and Moehrenschlager, A. (eds.) *Conservation Translocations*. Cambridge, Cambridge University Press.

Fernando, P., Leimgruber, P., Prasad, T. & Pastorini, J. (2012) Problem-elephant translocation: translocating the problem and the elephant? *PLoS ONE*. 7, e50917.

Fleming, T. (1997) *The Environment and Canadian Society*. Windsor, Canada, University of Windsor, International Thomson Publishing.

Frank, B. (2016) Human–wildlife conflicts, the need to include tolerance and coexistence: an introductory comment. *Society & Natural Resources*, 29, 738–743.

Frank, B. & Glikman, J. A. (2019) Human–wildlife conflicts and the need to include coexistence. In Frank, B., Glikman, J. A. and Marchini, S. (eds.) *Human-Wildlife Interactions: Turning Conflict into Coexistence*. Cambridge, Cambridge University Press, pp. 439–452.

Fulton, D. C., Manfredo, M. J. & Lipscomb, J. (1996) Wildlife value orientations: a conceptual and measurement approach. *Human Dimension of Wildlife*. 1, 24–47.

Gaywood, M. J. & Stanley-Price, M. (2023) Moving species: reintroductions and other conservation translocations. In Gaywood, M. J., Ewen, J. G.,

Hollingsworth, P. M. and Moehrenschlager, A. (eds.) *Conservation Translocations*. Cambridge, Cambridge University Press.

Glikman, J. A., Bath, A. J. & Vaske, J. J. (2010) Segmenting normative beliefs regarding wolf management in central Italy. *Human Dimensions of Wildlife*. 15, 347–358.

Goldenberg, S. Z., Owen, M. A., Brown, J. L., Wittemyer, G., Min Oo, Z. & Leimgruber, P. (2019) Increasing conservation translocation success by building social functionality in released populations, *Global Ecology and Conservation*. 18, e00604.

Gray, S. M., Booher, C. R., Elliott, K. C., et al. (2020) Research-implementation gap limits the actionability of human-carnivore conflict studies in East Africa. *Animal Conservation*. 23, 7–17.

Hernández, O. & Ferrer, A. (2021) Population reinforcement for the recovery of the Orinoco turtle in Venezuela. In Soorae, P. S. (ed.) *Global Conservation Translocation Perspectives: 2021. Case Studies from around the Globe*. Gland, Switzerland, IUCN SSC Conservation Translocation Specialist Group, Environment Agency - Abu Dhabi and Calgary Zoo, Canada, pp. 93–97.

Hernández, O., Velasco, A., Núñez, R. & Babarro, R. (2021) Population reinforcement for the recovery of the Orinoco crocodile in Venezuela. In: Soorae, P. S. (ed.) *Global Conservation Translocation Perspectives: 2021. Case Studies from around the Globe*. Gland, Switzerland, IUCN SSC Conservation Translocation Specialist Group, Environment Agency - Abu Dhabi and Calgary Zoo, Canada, pp. 87–92.

Innes, J. E. & Booher, D. E. (1999) Consensus building and complex adaptive systems. *Journal of the American Planning Association*. 65, 412–423.

IPBES (2018) The IPBES regional assessment report on biodiversity and ecosystem services for Europe and Central Asia. Rounsevell, M., Fischer, M., Torre-Marin Rando, A. and Mader, A. (eds.) *Secretariat of the Intergovernmental Science-Policy Platform on Biodiversity and Ecosystem Services*, Bonn, Germany, 892 pp. Available from: https://doi.org/10.5281/zenodo.3237428 [Accessed 2 February 2020].

IUCN (2013) *Guidelines for Reintroductions and Other Conservation Translocations. Version 1.0.* Gland, Switzerland, IUCN Species Survival Commission.

Jacobs, M. & Vaske, J. J. (2019) Understanding emotions as opportunities for and barriers to coexistence with wildlife. In Frank, B. F., Glikman, J. A. and Marchini, S. (eds.) *Human-Wildlife Interactions: Turning Conflict into Coexistence*. Cambridge, Cambridge University Press, pp. 65–84.

Jacobsen, K. S. & Linnell, J. D. (2016) Perceptions of environmental justice and the conflict surrounding large carnivore management in Norway—implications for conflict management. *Biological Conservation*. 203, 197–206.

Johnson, R. & Greenwood, S. (2020) Assessing the ecological feasibility of reintroducing the Eurasian lynx (Lynx lynx) to southern Scotland, England and Wales. *Biodiversity and Conservation*. 29, 771–797.

Kubala, J., Guimarães, N. F., Brndiar, J., et al. (2019) Monitoring of Eurasian Lynx (*Lynx lynx*) in the Vepor Mountains and its importance for the national and European management and species conservation. Technical report. Available from: www.lifelynx.eu/wp-content/uploads/2018/06/Monitoring-of-Eurasian-Lynx-Lynx-lynx-in-the-Vepor-Mountains.pdf [Accessed 2 February 2020].

Lederach, J. P. (2003) *Little Book of Conflict Transformation*. New York, NY, Good Books.

Lederach, J. P., Neufeldt, R. & Culbertson, H. (2007) *Reflective Peace Building: A Planning, Monitoring and Learning Toolkit*. Mindanao, Philippines, The Joan B. Kroc Institute for International Peace Studies, University of Notre Dame and Catholic Relief Services.

Linnell, J., Breitenmoser, U., Breitenmoser-Würsten, C., Odden J. & von Arx, M. (2009) Recovery of Eurasian lynx in Europe: what part has reintroduction played? In Hayward M. W. and Somers, M. (eds.) *Reintroduction of Top-Order Predators*. Hoboken, NJ, John Wiley & Sons, pp. 72–91.

Lopes-Fernandes, M. & Frazão-Moreira A. (2017) Relating to the wild: key actors' values and concerns about lynx reintroduction. *Land Use Policy*. 66, 278–287.

Madden, F. & McQuinn, B. (2014) Conservation's blind spot: the case for conflict transformation in wildlife conservation. *Biological Conservation*. 178, 97–106.

Madden, F. & McQuinn, B. (2017) Conservation conflict transformation: the missing link in conservation. In Hill, C. M., Webber, A. D. & Priston, N. E. C. (eds.) *Understanding Conflicts About Wildlife: A Biosocial Approach*. Vol. 9. New York, Berghahn Books, pp. 148–169.

McMurdo Hamilton, T., Canessa, S., Clarke, K., et al. (2021) Applying a values-based decision process to facilitate co-management of threatened species in Aotearoa New Zealand. *Conservation Biology*. 35, 1162–1173.

Moffitt, M. L. & Bordone, R. C. (2012) *The Handbook of Dispute Resolution*. John Wiley & Sons.

Mukesh, S. L. K., Charoo, S. A. & Sathyakumar, S. (2015) Conflict bear translocation: investigating population genetics and fate of bear translocation in Dachigam National Park, Jammu and Kashmir, India. *PLoS ONE*. 10, e0132005.

Neilson, A. (2021) Disenchanted natures: a critical analysis of the contested plan to reintroduce the Eurasian lynx into the Lake District National Park. *Capitalism Nature Socialism*. 32, 107–125.

Ramsbotham, O., Miall, H. & Woodhouse, T. (2011) *Contemporary Conflict Resolution*. Cambridge, Polity.

Riley, S. J. & Sandström, C. (2016) Human dimensions insights for reintroductions of fish and wildlife populations. In Jachowski, D. S., Millspaugh, J. J., Angermeier, P. and Slotow R. (eds.) *Reintroduction of Fish and Wildlife Populations*. Oakland, CA, University of California Press, pp. 55–77.

Ripple, W. J. & Beschta, R. L. (2012) Trophic cascades in Yellowstone: the first 15 years after wolf reintroduction. *Biological Conservation*. 145, 205–213.

Rowe, G. & Frewer, L. J. (2000) Public participation methods: a framework for evaluation. *Science, Technology, & Human Values*. 35, 3–29.

Seddon, P. J. (2023) The role of conservation translocations in rewilding and de-extinction. In Gaywood, M. J., Ewen, J. G., Hollingsworth, P. M. and Moehrenschlager, A. (eds.) *Conservation Translocations*. Cambridge, Cambridge University Press.

Seddon, P. J., Armstrong, D. P. & Maloney, R. F. (2007) Developing the science of reintroduction biology. *Conservation Biology*. 21, 303–312.

Seddon, P. J., Griffiths, C. J., Soorae, P. S. & Armstrong, D. P. (2014) Reversing defaunation: restoring species in a changing world. *Science*. 345, 406–412.

Sheil, D. & Kirkby, A. E. (2018) Observations on southern white rhinoceros *Ceratotherium simum simum* translocated to Uganda. *Tropical Conservation Science*. 11, 1–7.

Sjölander-Lindqvist, A., Johansson, M. & Sandström, C. (2015) Individual and collective responses to large carnivore management: the roles of trust, representation, knowledge spheres, communication and leadership. *Wildlife Biology*. 21, 175–185.

Slocum, N. (2012) Sea eagles poisoned and shot in Ireland. Available from: https://focusingonwildlife.com/news/sea-eagles-poisoned-and-shot-in-ireland/ [Accessed 2 February 2020].

Swedish Environmental Protection Agency (SEPA) (2015) Skrivelse regeringsuppdrag att utreda gynnsam Analys och redovisning av hur socioekonomin påverkas av en vargpopulation som har gynnsam bevarandestatus i Sverigebevarandestatus för varg (M2015/1573/Nm), NV-02945-15. Available from: Analys och redovisning av hur socioekonomin påverkas av en vargpopulation som har gynnsam bevarandestatus i Sverige (naturvardsverket.se) [Accessed 2 February 2020].

Tosi, G., Chirichell, R., Zibordi, F., et al. (2015) Brown bear reintroduction in the Southern Alps: to what extent are expectations being met? *Journal for Nature Conservation*. 26, 9–19.

Vaske, J. J. & Donnelly, M. P. (1999) A value-attitude-behavior model predicting wildland preservation voting intentions. *Society & Natural Resources*. 12, 523–537.

Vaske, J. J., Roemer, J. M. & Taylor, J. G. (2013) Situational and emotional influences on the acceptability of wolf management actions in the Greater Yellowstone Ecosystem. *Wildlife Society Bulletin*. 37, 122–128.

Verplanken, B., Hofstee, G. & Janssen, H. J. W. (1998) Accessibility of affective versus cognitive components of attitudes. *European Journal of Social Psychology*. 28, 23–35.

Viney, M. (2018) Poisoned and preyless: obstacles to rewilding Ireland with eagles. Available from: www.irishtimes.com/news/environment/poisoned-and-preyless-obstacles-to-rewilding-ireland-with-eagles-1.3570338 [Accessed 2 February 2020].

Warburton, D., Wilson, R. & Rainbow, E. (2012) Making a difference: A guide to evaluating public participation in central government. Available from: www.involve.org.uk/wp-content/uploads/2011/03/Making-a-Difference-.pdf [Accessed 5 December 2020].

Whiterod, N. S., Asmus, M., Zukowski, S., Gilligan, D. & Daly, T. (2021) Reintroduction to re-establish locally extirpated populations of the Murray crayfish – second largest freshwater crayfish in the world – in S.E. Australia. In Soorae, P. S. (ed.) *Global Conservation Translocation Perspectives: 2021. Case Studies from around the Globe*. Gland, Switzerland, IUCN SSC Conservation Translocation Specialist Group, Environment Agency - Abu Dhabi and Calgary Zoo, Canada, pp. 6–10.

Williams, S. & Haines, J. (2021) Scarlet macaw reintroduction on the Nicoya Peninsula of Costa Rica. In Soorae, P. S. (ed.) *Global Conservation Translocation Perspectives: 2021. Case Studies from around the Globe*. Gland, Switzerland, IUCN SSC Conservation Translocation Specialist Group, Environment Agency - Abu Dhabi and Calgary Zoo, Canada, pp. 133–136.

Wilson, N. (2018a) Reintroduction or recolonization? How the Fisher came back. *BioScience*. 68, 232.

Wilson, S. (2018b) Lessons learned from past reintroduction and translocation efforts with an emphasis on carnivores. Report compiled within the Action A.4: Elaboration of plans for Guidelines for Lynx Reinforcement (LIFE16 NAT/SI/000634). 43 pg. Available from: www.lifelynx.eu/wp-content/uploads/2018/10/Lessons-Carnivore-Reintroduction-Efforts-Final-Version-4.0-2018.pdf [Accessed 7 April 2020].

Wilson, S., Davies, T. E., Hazarika, N. & Zimmermann A. (2015) Understanding spatial and temporal patterns of human-elephant conflict in Assam, India. *Oryx*. 49, 140–149.

Young, J. C., Thompson, D., Moore, P., MacGugan, A., Watt, A. D. & Redpath, S. M. (2016) A conflict management tool for conservation agencies. *Journal of Applied Ecology*. 53, 705–711.

Zimmermann, A., McQuinn, B. & Macdonald, D. W. (2020) Levels of conflict over wildlife: understanding and addressing the right problem. *Conservation Science and Practice*. 2, e259.

Zimmermann, A., Johnson, P., de Barros, A.E., et al. (2021) Every case is different: cautionary insights about generalisations in human-wildlife conflict from a range-wide study of people and jaguars. *Biological Conservation*. 260, 109185.

11 · *Assisted Colonisation and Ecological Replacement*

MARIA HÄLLFORS AND
SARAH E. DALRYMPLE

11.1 Introduction

The natural environment is changing profoundly as a result of anthropogenic drivers. For most species, this is manifested as increasing levels of habitat degradation and fragmentation. As a consequence of climate change, conservation efforts in the original habitats of a species may lose efficacy as the climatic suitability of the habitat is reduced (Bellis et al., 2020). Simultaneously, new climatically suitable habitat may form elsewhere. Local extinctions of species may also lead to loss of ecosystem functions, with no other species naturally fulfilling the lost ecological role.

To combat the impending extinctions of species and their ecological functions, and to exploit the potential of newly available habitat, conservationists have suggested a suite of interventions that have come to be referred to as *conservation introductions* – introductions of individuals to habitat in areas not previously occupied by the species. These introductions might be conducted in order, for example, to help a species colonise climatically suitable regions made tolerable by climate change, to escape disease in the indigenous range, or to restore a degraded habitat by introducing a functionally important species to ensure continued functionality of that ecosystem.

These interventions have in common the translocation of organisms to regions beyond the recognised boundaries of their indigenous range, and as such are commonly perceived to be a relatively high-risk form of conservation translocation. This chapter describes the two different types of conservation introductions recognised by the IUCN (2013), the underpinnings driving their perceived need, the risks and concerns involved, and methods for identifying suitable species for such introductions.

11.1.1 The Goals of Assisted Colonisation and Ecological Replacement

The IUCN identifies two kinds of conservation introductions: *assisted colonisation* and *ecological replacement* (IUCN, 2013). In both methods, a species is introduced, for conservation purposes, outside its indigenous range. The main goal for both methods is to maintain biological diversity, either regionally or globally, and they often form part of a wider management tool kit for conservation. The two methods differ in the intended conservation outcomes, that is, whether the focus of the intervention is to improve the conservation status of a specific *species* or to fulfil an *ecological function*.

In assisted colonisation, individuals of a *focal species* threatened by, for example, climate change or an emerging disease, are translocated to suitable habitats outside of the species' indigenous range. The recipient sites are those that are currently suitable or, in cases where releases are planned for the future, will become suitable and provide better conditions than the original habitats within the indigenous range. Assisted colonisation is also commonly referred to by various synonyms including assisted migration, managed relocation, and benign introduction, amongst many others.

Ecological replacement, on the other hand, is directed towards *a focal ecosystem* and aims to replace a functional role that was lost with the extirpation of a native species, or alternatively, to deliver the same function as an extant species that cannot be translocated into its former range. Synonyms for ecological replacement are taxon substitution, ecological surrogates, and analogue species.

11.2 A History of Debate and Inconsistent Terminology

During the rather short history of scientific discussions on conservation introductions, ecological replacement and, especially, assisted colonisation, have been intensely debated with several different terms and conceptualisations suggested. The development of the assisted colonisation concept has been covered elsewhere (Hewitt et al., 2011; Hällfors et al., 2014, 2018; Torreya Guardians, 2020). Here, we offer a brief synopsis of the debate and an overview of the sometimes confusing terminology, thereby providing sufficient background for the reader to seek out additional information on the topic.

11.2.1 New Concepts Are Born

The idea of moving species into areas that are, or will soon become, suitable due to anthropogenic climate change was first suggested in the 1980s (Peters & Darling, 1985). As far as we are aware, it took almost ten years for the idea to reappear in the literature (Taylor & Hamilton, 1994). It is nonetheless likely that during that time and after, several ecologists and conservationists would have pondered and individually arrived at the same idea of translocation as a means to conserve species during climate change. Indeed, assisted colonisation involving the movement of species for reasons other than climate change–induced declines was undertaken under the broader term of 'translocation' (e.g. kakapo translocations, which attempted to create populations free from introduced predators; Lloyd & Powlesland, 1994). In the 2000s more attempts at assisted colonisation were carried out and the idea started to gain wide recognition among the general scientific audience, as evidenced by a proliferation of descriptions in the scientific literature (Hewitt et al., 2011; Hällfors et al., 2014).

Despite the relatively short history, the discussion around conservation introductions has been populated with a variety of terms and definitions. In a review study on assisted colonisation conducted in 2013 (Hällfors et al., 2014), 40 terms and 75 definitions were identified for the concept. Here, we follow the terminology suggested by the IUCN (2013), where assisted colonisation is defined as 'the intentional movement and release of an organism outside its indigenous range to avoid extinction of populations of the focal species'. Another, less radical, version of assisted colonisation is assisted gene flow (Aitken & Whitlock, 2013), also called intra-specific assisted colonisation, where, for example, warm-adapted genotypes are translocated to historically cooler parts of the species' range to enhance evolutionary rescue of the species under climate change.

Ecological replacement has fewer synonyms than assisted colonisation, but there are still a number of equivalent terms including taxon substitution and subspecific substitution. The focal species have variously been referred to as ecological substitutes, ecological proxies, ecological surrogates, or analogue species (IUCN, 2013). However, there is potential for confusion in using the term ecological replacement in different contexts. Whilst ecological replacement has a fairly straightforward definition within the conservation translocation literature, the same term is also used to describe autonomous or natural ecological replacement, such as when one species replaces another along ecological gradients or in an invasive context, for example grey squirrels replacing red squirrels (Tompkins et al., 2003). Biological control – translocations of organisms to control an invasive species – might be erroneously included in the IUCN definition above but does not meet the qualifying statement of ecological replacement as an attempt to 're-establish an ecological function lost through extinction' (IUCN, 2013).

Some authors have attempted to conceptually reconcile assisted colonisation and ecological replacement. Schwartz et al. (2012) suggested that the commonly used synonym for assisted colonisation, *managed relocation*, could include both practices to describe 'the movement of species, populations, or genotypes to places outside the areas of their historical distributions to maintain biological diversity or ecosystem functioning with changing climate'. Other authors have described ecological replacement as a particular case of assisted colonisation: *pull* assisted colonisation where the recipient ecosystem is in need of a certain species, as opposed to *push* assisted colonisation where the species needs to arrive in a new ecosystem (Lunt et al., 2013).

11.2.2 Finding Their Place in the Conservation Toolbox

In the 2000s, the planting of commercially available *Torreya taxifolia* saplings on private land in the USA (Barlow & Martin, 2004) and the experimental translocation of individuals of two butterfly species in the UK (Willis et al., 2009) became the most high-profile examples of conservation introductions in practice. These conservation introductions attracted interest across the wider media (e.g. Zimmer, 2007; Marris, 2008) and conceivably prompted the intensive debate on the pros and cons of moving species outside their indigenous range (Hewitt et al., 2011). The main points of discussion concerned the risk of creating invasive species and the ethics of translocating species across their range boundaries. Paradoxically, however, translocations that could arguably be classed as conservation introductions have long been routine in the practice of restoration ecology (see discussion in Butt et al., 2021).

The disciplines of restoration ecology and reintroduction biology (*sensu* Seddon et al., 2007) arose broadly in parallel, albeit each with a different focus. While restoration ecology employs the recovery and repair of vegetation to produce functioning ecosystems, reintroduction biology attends to the recovery of threatened species. This difference is evident in the divergent modes of practice, distinct practitioner communities, and adherence to different statutory tools. However, there are currently signs of convergence of the two disciplines. Flagship restoration projects have increasingly revolved around the reintroduction of keystone species such as grey wolf *Canis lupus* in Yellowstone National Park, USA (Linnell & Jackson, 2019), jaguar *Panthera onca* in Iberá Natural Reserve, Argentina (Zamboni et al., 2017; Donadio et al., this volume), and European bison *Bison bonasus* to various European countries (Seddon & Armstrong, 2019). However, the restoration of key ecological processes sometimes needs to look beyond reinstating past species assemblages and instead employ ecological replacement to fulfil ecosystem function when extirpated species cannot be reintroduced.

Most recently, conservation translocations have been recognised as an important component of rewilding, having been incorporated in its broad definition as "'the repair or refurbishment of an ecosystem's functionality through the (re-)introduction of selected species' (Pettorelli et al., 2019). Where rewilding projects (Seddon, this volume) aim to bring in species that were not previously extant in that region, this constitutes ecological replacement. Much of the literature on rewilding has concentrated on the controversial reinstatement of apex predators (Seddon & Armstrong, 2019) whilst the more left-field proposals have suggested Pleistocene rewilding using completely unrelated species as proxies for long-extinct species (du Toit, 2019; Seddon, this volume). However, rewilding often employs less controversial ecological replacements, bringing in horse and cattle breeds to fill the role left vacant by the loss of wild ancestors (Delibes-Mateos et al., 2019). More on the role of conservation translocations in rewilding and de-extinction is described in Seddon (this volume) and Gaywood and Stanley-Price (this volume).

Assisted colonisation and ecological replacement have consequently been adopted, at least conceptually, in the toolbox of many conservationists and ecologists. Conservation introductions are included in the IUCN guidelines (IUCN, 2013) and are often mentioned in scientific papers as a possible way of safeguarding particular species (Fordham et al., 2013; Ferrarini et al., 2016). Although practical cases to date remain few, the shifting conceptions in conservation biology and the increasing pressures from anthropogenic threats are likely to inspire new projects (Brodie et al., 2021; Butt et al., 2021). In Figure 11.1 we present some published examples of assisted colonisation and ecological replacements conducted to date.

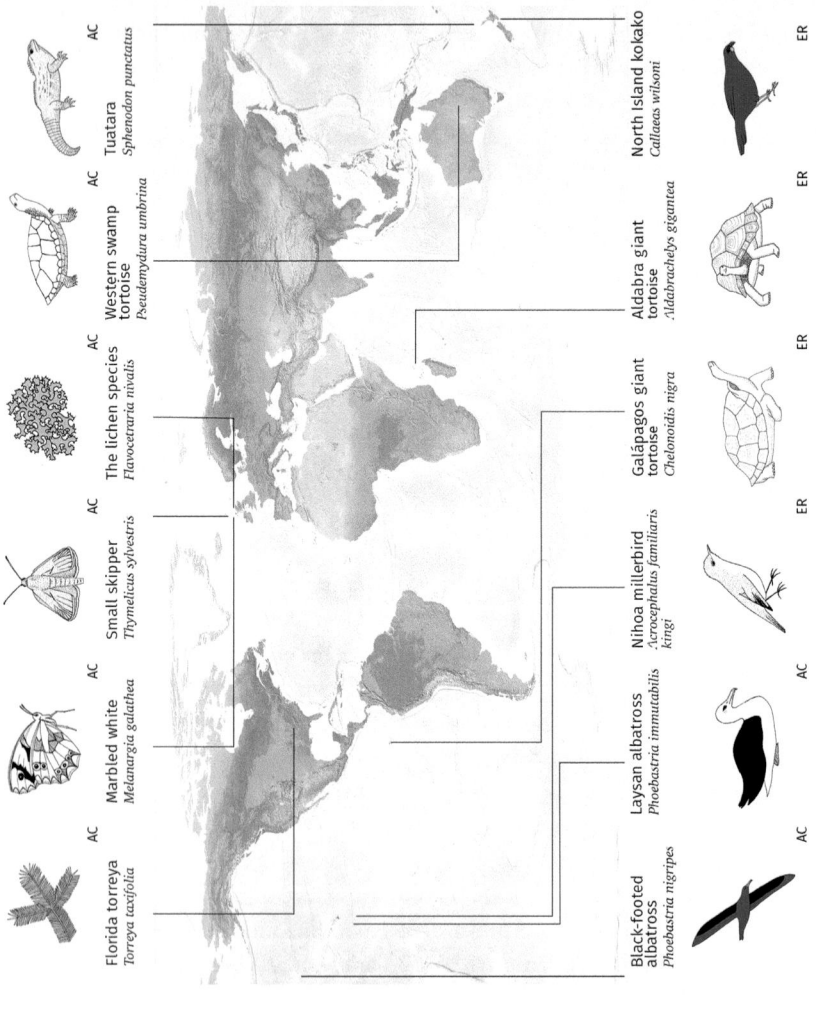

Figure 11.1 A selection of published cases of conservation introductions. Arrows indicate locations where the translocation project occurred. AC, assisted colonisation; ER, ecological replacement. Examples of assisted colonisation: Laysan and black-footed albatrosses in Hawaii (VanderWerf et al., 2019); conifer *Torreya taxifolia* in Florida, USA (Barlow & Martin, 2004); marbled white *Melanargia galathea* and

11.3 Identifying the Need for Conservation Introductions

11.3.1 The Justification for Assisted Colonisation

Assisted colonisation is driven by the need to deal with extensive threats across a species' range by exploiting available habitat that has not previously been occupied by the focal species. The intervention is most frequently cited as a response to climate change, the direct or indirect impacts of which serve to render the species' range partly or wholly unsuitable. As a threat, climate change is unique in having the potential to simultaneously render current occupied habitats unsuitable for certain species while also creating new suitable areas outside their indigenous range (Hällfors et al., 2017). Therefore, in assisted colonisation, the conservationist may, for example, use translocation to mimic the gradual natural dispersal of a species that would hypothetically have occurred during a less rapid change in climatic conditions. During past changes in the climate, the most common successful strategy for adjusting to novel conditions would have been dispersal to new areas to follow the gradual spatial shift in suitable conditions (Brook & Barnosky, 2012), while evolutionary change likely played a minor role. The rapidity of current climatic change thus hinders many species from using this essential strategy to reach emerging favourable areas.

However, the threat posed by climate change is not the sole reason for assisting a species in its colonisation of unoccupied habitat outside its current and past range. Other threats, such as disease, may warrant introducing a species outside of its indigenous range. One example of this is the assisted colonisation of Tasmanian devils to Maria Island to create a disease-free population (Thalmann et al., 2016; Hogg & Wise, this volume). Increasingly, we may see assisted colonisation undertaken in response to the synergistic impacts of climate and disease in the future as altered climatic conditions can be a major factor in altering host–pathogen interactions (Gallana et al., 2013). As an example, the Monteverde golden toad

---

Figure 11.1 (cont.) small skipper *Thymelicus sylvestris* butterflies in the UK (Willis et al., 2009); lichen *Flavocetraria nivalis* in the UK (Brooker et al., 2017); western swamp tortoise *Pseudemydura umbrina* in Australia (Lewis, 2016); and tuatara *Sphenodon punctatus* in New Zealand (Miller et al., 2012). Examples of ecological replacements: kokako *Callaeas wilsoni* in New Zealand (Seddon & Armstrong, 2019); Aldabra giant tortoise *Aldabrachelys gigantea* in the Seychelles (Moorhouse-Gann, 2017); Galápagos giant tortoise in the Galápagos Islands (Hunter et al., 2013); and millerbird *Acrocephalus familiaris kingi* in Hawaii (Freifeld et al., 2016). (Drawings: Sarah Dalrymple. Layout: Manuel Frias).

Bufo periglenes went extinct in Costa Rica in 1989 (Pounds et al., 2006) – the first known extinction that can be linked to climate change. Altered climatic conditions induced outbreaks of the pathogen chytrid fungus *Batrachochytrium dendrobatidis* and caused a complete eradication of the toad population. Assisted colonisations to disease-free areas or areas outside the climatic optimum of the disease could potentially have saved this species.

11.3.2 The Justification for Ecological Replacement

For ecological replacement, the motivation for translocation is not closely tied to particular threats. Any (anthropogenic) environmental change that leads to the loss of an ecological function would warrant the consideration of ecological replacement when it is impossible or unfeasible to restore the species that originally performed the function.

In principle, as long as the desired function is being delivered, the taxon selected is not important; introduced organisms that are entirely unrelated to the extirpated species might deliver the ecological roles. However, in practice, taxa that are closely related are more likely to be ecologically analogous and have the necessary adaptations and traits to sustain the lost functions in the ecosystem. As with assisted colonisation, the rationale behind ecological replacement recognises the inability of species to disperse or evolve rapidly enough to fill the functional roles that have been left vacant due to extinctions. Consequently, species that are closely related and fulfil the same functions are unlikely to co-occur in space and time (Hardin, 1960). Thus, any related analogue species may be geographically distant and unable to disperse to fill the vacant ecological roles. Instead, and as is the case for assisted colonisation, the species needs to be moved from somewhere else, implying a conservation introduction to a site outside its indigenous range.

A good example is the trial ecological replacement of the extinct giant tortoise *Chelonoidis abingdonii* from Pinta Island, Galápagos, with closely related species from other islands in the Galápagos archipelago. In 2010, 39 sterilised individual tortoises representing the two main morphotypes – saddleback and domed shell – were introduced to Pinta Island (Hunter et al., 2013). The two morphotypes are associated with divergent feeding behaviour and consume different vegetation. Unsurprisingly, the saddleback individuals occupied similar feeding niches as the extinct species of the same morphotype, and now contribute to seed dispersal of a variety of species on Pinta Island (Hunter et al., 2013). Also see Jones et al. (this volume) for an example of ecological replacement using giant tortoises in Mauritius.

11.3.3 Identifying Species for Assisted Colonisation

The identification of species that could benefit from assisted colonisation has tended to happen on a case-by-case basis when practitioners identify climate change or other threats that render the indigenous area uninhabitable for the species while recognising suitable habitat outside it. A more top-down approach could involve screening multiple species and identifying which of them might need assisted colonisation in the near future, thereby allowing more lead time for planning, resource acquisition, and sourcing, holding, and storing of organisms or their propagules (in the case of plants).

Species for which assisted colonisation could be a suitable conservation approach would typically meet all of these three criteria:

1. Climate change or another threat rendering the indigenous area uninhabitable but other areas favourable, has been determined to have a detrimental effect on the species and cannot be adequately addressed *in situ*.
2. The species is unable to adjust *in situ* (evolutionarily or plastically) fast or comprehensively enough.
3. The species is unable to disperse fast enough to reach favourable areas before current habitat becomes unsuitable.

Evaluations using the above points are often part of decision-making frameworks for conservation translocations. Simultaneously, the assessment of whether these criteria are met is likely to be unreliable for most species because of our limited knowledge on biodiversity as a whole and species' autecologies in particular. There are also major taxonomic biases in our knowledge (Godet & Devictor, 2018) and the conservationist often needs to rely on approximations and general rules of thumb. Examples include traits that are common to poor dispersers, such as small wing size in moths, as opposed to scientifically measured dispersal distances. Another option is to use information on closely related or functionally similar species (Forbes et al., 2020).

Species distribution models and other predictive methods for estimating niche space can be used to identify species with an increased probability of losing existing suitable habitat whilst also gaining areas of newly available climate space (Hällfors et al., 2016a; Casazza et al., 2021). Similar methods can be employed to estimate the impact of assisted colonisation on the recipient ecosystems (Peterson and Bode, 2021). Box 11.1 describes one particular issue, intra-specific local adaptation, which should be taken into account when using predictive spatial modelling to assess future habitat space.

> Box 11.1 *Considering local adaptation in predictive modelling*
>
> Individuals of a species are not the same across their range but are often adapted to local conditions. This is often not adequately addressed in species distribution models (Atkins & Travis, 2010). The choice of taxonomic entities modelled (i.e. a whole species, subspecies, ecotype, or populations separately) produces different results (Valladares et al., 2014; Hällfors et al., 2016b) with large consequences for decisions such as from where to source and where to move individuals for conservation introductions (Frankham et al., 2012; Hällfors et al., 2016b). In some cases, taxonomic levels, such as subspecies or populations, may provide more informative estimates of the future location of suitable habitat. The challenge is knowing when this is important, as detecting local adaptation can be difficult (Kawecki & Ebert, 2004).
>
> A better understanding of intra-specific local adaptation can be obtained through various experimental means such as reciprocal translocation experiments, or laboratory or greenhouse trials. In the short term, however, generalisations from common traits can be used to approximate whether a species is likely to have locally adapted populations. For example, it is likely that a species has locally adapted populations if: it shows reproductive isolation (Frankham et al., 2012); it occurs on steep environmental gradients; there is taxonomic confusion (Kawecki & Ebert, 2004); it has been divided into subspecies (Oney et al., 2013); it has a discontinuous distribution across its range. Another indication that a species may be locally adapted is that the populations inhabit separate climatic environments across the species range. In cases where local adaptation is suspected to affect the intra-specific niche, modelling populations separately, in addition to modelling the species as a whole, is recommended. The results of both approaches should be compared and decisions accounting for both scenarios on sourcing and recipient site choice need to be made (Hällfors et al., 2016b).

In addition to identifying the spatial prerequisites for a translocation, species distribution models and allied spatial methods may also be valuable in suggesting the time scales over which sites will stay suitable. Conservation introductions may in practice come to include step-wise translocations at several points in time, as climate changes gradually. This adds an additional layer of uncertainty as it is not only difficult to predict where the optimal habitats lie but also requires estimations of how long they will remain suitable. Some tools have been developed that specifically address this (McDonald-Madden et al., 2011; Ferrarini et al., 2016).

11.3.4 Identifying Species for Ecological Replacement

Before identifying candidates for ecological replacement, an initial task is to confirm that none of the remaining species can deliver the ecological function of a species that has been lost from an ecosystem. For keystone species, the ecological function can be identified by its absence, but for most species it is more difficult to pinpoint. Rare species may often be functionally distinctive (Violle et al., 2017; Chapman et al., 2018) and as a result fulfil ecological roles that are more valuable than simple measures of abundance or biomass might suggest. The concept of functional rarity has obvious relevance to ecological replacement because it validates the view that the function of a taxonomic group is more important (or sometimes just more attainable) than the restoration of the species that once fulfilled that role. However, the approach has value in all types of conservation translocation because it promotes the identification of community-level function. In doing so, it is more likely to result in conservation benefit to both focal species and recipient ecosystems.

Quantitative methods for identifying distinctive functional roles rely on the comprehensive description of a set of traits for the range of species in a community (Box 11.2). Fortunately, data on species' traits are becoming increasingly available through online databases such as the TRY Trait Database (Kattge et al., 2011), and in some cases the functional role of a species can be reliably inferred from observed proxies. However, the analytical packages that have been developed to produce quantitative metrics of functional distinctiveness are probably perceived to be inaccessible to many practitioners as they are reliant on the ability to use programming languages such as R. Given the complexities of analysing multivariate data (many species versus many traits), the lack of capacity to undertake these analyses presents the main obstacle to widespread application of functional description. The research community, where the expertise in such approaches currently resides, must take some responsibility for making these analyses more accessible in order to exploit and test the claims made for functional rarity as a rationale for conservation translocations.

> Box 11.2 *Identifying lost ecological function*
>
> When a keystone species is lost from an ecosystem, it is relatively easy to identify the function that the species fulfilled by recognising how the system changed in parallel with its decline and eventual extirpation. In the case of ecological replacement, experimental translocations can also be used to confirm that a substitute species will actually perform the required ecological role (Hunter et al., 2013). However, the impact of most other species extinctions on ecological processes is often more difficult to identify and describe.
>
> When identifying ecological function in the context of species recovery, the starting point is to list the interactions between the focal species and its ecosystem (Akçakaya et al., 2020). These can be quantified using estimates of population size multiplied by the typical individual impact of the focal species. The individual impact can be quantified as, for example, the consumption of seeds or prey items, the effects on abiotic processes such as soil erosion or pedogenesis, or measurable changes in the behaviour of other species. This assessment also needs to be applied to proxy species that might be brought in as the replacement. These comparisons then form the basis for further description of the contribution of a focal species to ecosystem processes and will indicate whether the analogue species will perform the same type and extent of functions compared to the extirpated species.
>
> While such assessments might typically be undertaken by species experts, there are growing numbers of quantitative approaches that can be applied to translocations to identify functionally similar species. Examples include the 'funrar' R statistical package designed to quantify functional rarity (Grenié et al., 2017) by calculating species distinctiveness based on frequency of occurrence within a sample and traits. The trait data can be selected by the user and thus include field-derived measurements, data stored in online databases, or both.

11.4 Risks, Uncertainties, and Challenges

All kinds of species translocations, including those undertaken for conservation, are associated with risks of undesired outcomes and challenges in achieving project goals and targets. The risks involved with proceeding with a project need to be identified, evaluated, and compared to the risks of *not* pursuing the project (Ahteensuu & Lehvävirta, 2014; Phelan et al., 2021). In addition, they should be continuously re-evaluated as data from monitoring and research accumulate. This calls for supportive decision-making frameworks that incorporate stakeholder inclusion and monitoring (see Ewen et al., this volume). The risks involved in conservation translocations in general are covered by Dalrymple & Bellis (this volume) and the IUCN (2013). An example of practical approaches to assessing risk is set out in the Scottish Code for Conservation Translocations (National Species Reintroduction Forum, 2014).

In this section, we discuss how risks and uncertainties can be accounted for in the planning and management phase of conservation introductions (Section 11.4.1). We also cover the three concerns that are perhaps most prominently discussed: the potential of creating invasive species, disruption of natural processes, and ethical and societal views (Sections 11.4.2–11.4.4). Other relevant biological risks that are covered in other chapters of this book include disease transmission (Sainsbury & Carraro, this volume; Mitchell et al., this volume) and the maintenance of genetic diversity (Neaves et al., this volume).

11.4.1 Setting the Stage

There will often be more risks connected to conservation introductions than other conservation translocations; these can be biological or socioeconomic risks (Dalrymple & Bellis, this volume) and can be contextualised as both a risk of negative effects resulting from a successful translocation, and a risk that the translocation fails to meet the project's objectives. Specific risks need to be identified and carefully evaluated, and appropriate mitigation practice put into place. It might be useful to look at analogous situations to better anticipate the intended and unintended outcomes of a planned project (Forbes et al., 2020). However, whilst this can be feasible for reintroductions, an extremely common intervention, finding ecological or geographical analogies for conservation introductions may be impossible because published cases are rare (Butt et al., 2021).

If the risks are deemed to be high and cannot be mitigated, or cannot be satisfactorily evaluated, the conservationist needs to consider whether and with which precautionary measures the translocation can proceed. The assessment of risk and acceptance of what is tolerable will always be relative to the range of available options and their anticipated outcomes. For example, where an assisted colonisation attempt is being compared to a feasible reintroduction, the risk of the assisted colonisation will probably be deemed to be unacceptable given that it incurs higher uncertainty than the reintroduction. However, if assisted colonisation is deemed to be the only feasible action, with inaction resulting in probable extinction, the risks might be worth taking.

Several decision-making frameworks for conservation introductions have been presented and can be used as an aid in defining risks and ensuring transparency in decision-making (Hoegh-Guldberg et al., 2008; Richardson et al., 2009; Vitt et al., 2010; IUCN, 2013; McDonald-Madden et al., 2011; Pérez et al., 2012; Chauvenet et al., 2013; Rout et al., 2013; Schwartz and Martin, 2013; Shoo et al., 2013). The use of decision-making frameworks in translocation project planning is covered by Ewen et al. (this volume).

11.4.2 Risk of Causing Invasive Species

Perhaps the mostly widely voiced concern associated with conservation introductions is the risk of producing invasive species that detrimentally affect other (native) species in the receptor area (e.g. Fazey & Fischer, 2009; Ricciardi & Simberloff, 2009a, 2009b). The potential for invasiveness in a candidate species for a conservation introduction consequently needs to be thoroughly evaluated using a variety of approaches, and a realistic exit plan should always be established (Dalrymple and Bellis, this volume).

One avenue for assessing the likelihood of a species to become invasive is to compare its traits to those of known invasive species (Mathakutha et al., 2019). Invasive plant species tend to have consistent and high seed production, effective long-distance seed dispersal, and long flowering and fruiting periods; they also reproduce vegetatively if they are perennials. They may also be drought tolerant or retain their leaves for a longer time than other species. Such traits are less frequent in rare species that are most likely to be candidates for assisted colonisation, but not necessarily the species proposed for ecological replacement (Chapman et al, 2018). It is worth noting, however, that a species that is threatened in its indigenous range will not automatically be risk free when it comes to introduction, and there are several examples of restricted species that are invasive outside their range (Rocha & Bergallo, 2012).

A study on the invasive potential of species originating from the same continent versus those from other continents demonstrated that there are no known cases of intracontinental invasions of plant species (Mueller & Hellmann, 2008). Therefore, one could make the careful conclusion that short-distance and intracontinental translocations of threatened plants would pose the least risk of introducing invasive non-native species or disrupting the functions of the ecosystem in the new site. For other taxa, intracontinental invasions were rare, but, where they did occur, were equally likely to have severe effects as intercontinental invasions.

Post-release monitoring should plan for a lag between the initial introduction and first indications that a species might become invasive (Aikio et al., 2010). The length of this phase and the likelihood of problematic population growth rates and spread depend on factors such as the number of released individuals, the number of separate introduction events, and the spatial distribution of the individuals.

In addition to species having varying likelihoods of becoming invasive, habitats can also be differentially prone to invasiveness (Aikio et al., 2012). Much research is taking place to investigate what makes communities resilient to novel species invasiveness, how the resilience of the ecosystem can either be strengthened or weakened by the introduction of new species (Chaffin et al., 2016), and how the impact of non-native species on recipient ecosystems can be estimated (Blackburn et al., 2014).

11.4.3 Risks of Disrupting Natural Processes

Ecosystems and species communities are dynamic entities that change constantly as a consequence of eco-evolutionary and abiotic processes. In addition, anthropogenic change is causing substantial shifts in species distributions and abundances, and even affects the course of evolution (Franks et al., 2007). Some species benefit from the changes, or avoid detrimental effects, by adjusting adaptively to the new conditions. The species that are not able to adjust adequately become the focus of conservation action.

When humans target these species and functions through various conservation methods, including assisted colonisation and ecological replacement, we are simultaneously affecting the natural dynamics of these changing communities. This can alter both the antagonistic and facilitative species interactions as we manage for the benefit of one or a few species. Even if the focal species does not become invasive, individual organisms are not passive occupants of ecosystems, and introduced plants and animals are bound to have some effect on the community dynamics. Such impacts on the recipient ecosystems might include the failure of the introduced species to fulfil ecological roles such as providing seed dispersal or nectar provision. Changes in competitive dynamics may also occur, for example through priority effects, where the presence and abundance of the first arriving species affects the colonisation potential of later arriving species (Weslien et al., 2011). When individuals are translocated outside the indigenous range of the species, there is also a risk that they reproduce with closely related species, which can threaten genetic integrity. Other genetic effects that need to be considered when planning, conducting, and monitoring conservation translocations are described by Neaves et al. (this volume).

11.4.4 Ethical and Societal Concerns

Human interests and the societal view of our relationship with nature will affect all conservation introductions incorporating social, ethical, economic, and political factors, as well as indigenous values (McMurdo Hamilton et al., 2021). Our values affect how we decide to govern our society, and these values are therefore mirrored in our laws and regulations. Legal control, aiming to prevent the spread of invasive alien species, will be appropriately strict in many countries. Thus, proposals for conservation introductions may be prohibited or need to go through extensive risk assessment and evaluation before approval.

In conservation translocations in general, and conservation introductions in particular, humans are taking on the role of nature's custodians. Many have asked whether we have the right to intervene in nature, or whether we are obliged to intervene to mitigate for human impacts on the environment. And even if we do find it morally justifiable to move species outside their indigenous range, there is a risk that the integrity of the species is compromised when it becomes detached from its original distribution area (Sandler, 2010). This is because the species' intrinsic value may be perceived as being diminished by being translocated outside its range because its worth is partly defined by its area of occurrence. However, Siipi and Ahteensuu (2016) argue that if the predicted future range is defined as part of the indigenous range were it not for limited time and constrained dispersal, this perceived loss of value is not as extensive.

Another potent aspect to consider is the possibility that the legal status of a species may change after it has been translocated, for example from being defined as a threatened species in its original area to being considered a non-native species in its new area. For more information on planning and monitoring of a translocation project in general, see Dalrymple and Bellis (this volume), on societal and human aspects of conservation translocations in general see Glikman et al. (this volume), and on legal considerations see Trouwborst et al. (this volume).

11.5 Conclusion

Our environment is changing at a rapid pace and the risk of species becoming locally extinct, maladapted, and losing genetic variability in their current locations is growing. The choice of appropriate action to address declines at all levels of biodiversity can range anywhere between inaction, *in situ* management, enhancement of gene flow, and large scale assisted colonisation and ecological replacement. Conservation translocations overall, even those that are considered to be relatively conservative such as reintroductions, are often seen as an option of last resort (Hoegh-Guldberg et al., 2008). Relying on these interventions only once everything else has been attempted could be to the detriment of the success of the translocation, and ultimately the species itself. Learning from past translocation projects provides a crucial tool to inform upcoming conservation translocation projects (Bellis et al., 2020; Dalrymple et al., 2021).

In the near future, assisted colonisation and ecological replacements will probably contribute lifelines for certain species and ecosystems. The approaches will not only apply to highly threatened species as a last resort at the brink of extinction, but also target species that are likely to become critically threatened in the future. Nevertheless, conservation introductions are not to be undertaken lightly and every attempt needs to be made rigorously to minimise negative impacts and learn from the process to ensure that conservation benefit remains the key outcome of translocations.

11.6 Key Messages

- The main goal of conservation introductions is to maintain biological diversity.
- In assisted colonisation, individuals of a *threatened focal species* are translocated to suitable habitats outside the species' indigenous range.
- Ecological replacement, on the other hand, is directed towards *a focal ecosystem* and aims to replace a functional role that has been lost.
- The most widely recognised concerns with conservation introductions are the risk of causing invasive species, the risk of disrupting natural processes, and moral justification.
- The expected beneficial outcome of any conservation translocation needs to be measurable and relate to whole populations or ecosystems.
- In the near future, conservation introductions will probably contribute lifelines for some species and ecosystems.

References

Ahteensuu, M. & Lehvävirta, S. (2014) Assisted migration, risks and scientific uncertainty, and ethics: a comment on Albrecht et al.'s review paper. *Journal of Agricultural and Environmental Ethics*. 27, 471–477.

Aikio, S., Duncan, R. P. & Hulme, P. E. (2010) Lag-phases in alien plant invasions: separating the facts from the artefacts. *Oikos.* 119, 370–378.

Aikio, S., Duncan, R. P. & Hulme, P. E. (2012) The vulnerability of habitats to plant invasion: disentangling the roles of propagule pressure, time and sampling effort. *Global Ecology and Biogeography.* 21, 778–786.

Aitken, S. N. & Whitlock, M. C. (2013) Assisted gene flow to facilitate local adaptation to climate change. *Annual Review of Ecology, Evolution, and Systematics.* 44, 367–388.

Akçakaya, H. R., Rodrigues, A. S. L., Keith, D. A., et al. (2020) Assessing ecological function in the context of species recovery. *Conservation Biology.* 34, 561–571.

Atkins, K. E. & Travis, J. M. J. (2010) Local adaptation and the evolution of species' ranges under climate change. *Journal of Theoretical Biology.* 266, 449–457.

Barlow, C. & Martin, P. S. (2004) Bring Torreya taxifolia north—now. *Wild Earth.* Winter/Spring, 52–56.

Bellis, J., Bourke, D., Maschinski, J., Heineman, K. & Dalrymple, S. (2020) Climate suitability as a predictor of conservation translocation failure. *Conservation Biology.* 34, 1473–1481.

Blackburn, T. M., Essl, F., Evans, T., et al. (2014) A unified classification of alien species based on the magnitude of their environmental impacts. *PLoS Biology.* 12, e1001850.

Brodie, J.F., Liebermann, S., Moehrenschlager, A., et al. (2021). Global policy for assisted colonization of species. *Science.* 372, 456–458.

Brook, B. W. & Barnosky, A. D. (2012) Quaternary extinctions and their link to climate change. In Hannah, L. (ed.) *Saving a Million Species: Extinction Risk from Climate Change.* Washington, DC, Island Press/Center for Resource Economics, pp. 179–198.

Brooker, R. W., Brewer, M., Britton, A. J., et al. (2017) Feasibility study: translocation of species for the establishment or protection of populations in northerly and/or montane environments. Scottish Natural Heritage Commissioned Report number 913.

Butt, N., Chauvenet, A. L. M., Adams, V. M., et al. (2021) Importance of species translocations under rapid climate change. *Conservation Biology.* 35, 775–783.

Casazza, G, Abeli, T, Bacchetta, G, et al. (2021) Combining conservation status and species distribution models for planning assisted colonisation under climate change. *Journal of Ecology.* 109, 2284–2295.

Chaffin, B. C., Garmestani, A. S., Gunderson, L. H., et al. (2016) Transformative environmental governance. *Annual Review of Environment and Resources.* 41, 399–423.

Chapman, A. S. A., Tunnicliffe, V. & Bates, A. E. (2018) Both rare and common species make unique contributions to functional diversity in an ecosystem unaffected by human activities. *Diversity and Distributions.* 24, 568–578.

Chauvenet, A. L. M., Ewen, J. G., Armstrong, D. P., Blackburn, T. M. & Pettorelli, N. (2013) Maximizing the success of assisted colonizations. *Animal Conservation.* 16, 161–169.

Dalrymple, S. E. & Bellis, J. M. (2023) Conservation translocations: planning and the initial appraisal. In Gaywood, M. J., Ewen, J. G., Hollingsworth, P. M. and

Moehrenschlager, A. (eds.) *Conservation Translocations*. Cambridge, Cambridge University Press.

Dalrymple, S. E., Winder, R., & Campbell, E. M. (2021) Exploring the potential for plant translocations to adapt to a warming world. *Journal of Ecology*. 109, 2264–2270.

Delibes-Mateos, M., Barrio, I. C., Barbosa, A. M., Martínez-Solano, Í., Fa, J. E. & Ferreira, C. C. (2019) Rewilding and the risk of creating new, unwanted ecological interactions. In Pettorelli, N., Durant, S. M. and du Toit J. T. (eds.) *Rewilding*. Cambridge, Cambridge University Press, pp. 355–374.

Donadio, E., Zamboni, T. & Di Martino, S. (2023) Bringing jaguars and their prey base back to the Iberá Wetlands, Argentina. In Gaywood, M. J., Ewen, J. G., Hollingsworth, P. M. and Moehrenschlager, A. (eds.) *Conservation Translocations*. Cambridge, Cambridge University Press.

du Toit, J. T. (2019) Pleistocene rewilding: an enlightening thought experiment. In Pettorelli, N., Durant, S. M. and du Toit J. T. (eds.) *Rewilding*. Cambridge, Cambridge University Press, pp. 355–374.

Ewen, J. G., Canessa, S., Converse, S. J. & Parker, K. A. (2023) Decision-making in animal conservation translocations: biological considerations and beyond. In Gaywood, M. J., Ewen, J. G., Hollingsworth, P. M. and Moehrenschlager, A. (eds.) *Conservation Translocations*. Cambridge, Cambridge University Press.

Fazey, I. & Fischer, J. (2009) Assisted colonization is a techno-fix. *Trends in Ecology & Evolution*. 24, 475.

Ferrarini, A., Selvaggi, A., Abeli, T., et al. (2016) Planning for assisted colonization of plants in a warming world. *Scientific Reports*. 6, 28542.

Forbes, E., Alagona, P. S., Adams, A. J., et al. (2020) Analogies for a no-analog world: tackling uncertainties in reintroduction planning. *Trends in Ecology & Evolution*. 35, 551–554.

Fordham, D. A., Akçakaya, H. R., Brook, B. W., et al. (2013) Adapted conservation measures are required to save the Iberian lynx in a changing climate. *Nature Climate Change*. 3, 899–903.

Frankham, R., Ballou, J. D., Dudash, M. R., et al. (2012) Implications of different species concepts for conserving biodiversity. *Biological Conservation*. 153, 25–31.

Franks, S. J., Sim, S. & Weis, A. E. (2007) Rapid evolution of flowering time by an annual plant in response to a climate fluctuation. *Proceedings of the National Academy of Sciences*. 104, 1278–1282.

Freifeld, H. B., Plentovich, S., Farmer, C., et al. (2016) Long-distance translocations to create a second millerbird population and reduce extinction risk. *Biological Conservation*. 199, 146–156.

Gallana, M., Ryser-Degiorgis, M.-P., Wahli, T. & Segner, H. (2013) Climate change and infectious diseases of wildlife: altered interactions between pathogens, vectors and hosts. *Current Zoology*. 59, 427–437.

Gaywood, M. J. & Stanley-Price, M. (2023) Moving species: reintroductions and other conservation translocations. In Gaywood, M. J., Ewen, J. G., Hollingsworth, P. M. and Moehrenschlager, A. (eds.) *Conservation Translocations*. Cambridge, Cambridge University Press.

Glikman, J. A., Frank, B., Sandström, C., et al. (2023) The human dimensions and the public engagement spectrum of conservation translocation. In Gaywood,

M. J., Ewen, J. G., Hollingsworth, P. M. and Moehrenschlager, A. (eds.) *Conservation Translocations*. Cambridge, Cambridge University Press.

Godet, L. & Devictor, V. (2018) What conservation does. *Trends in Ecology & Evolution*. 33, 720–730.

Grenié, M., Denelle, P., Tucker, C. M., Munoz, F. & Violle, C. (2017) funrar: an R package to characterize functional rarity. *Diversity and Distributions*. 23, 1365–1371.

Hällfors, M.H., Vaara, E. M., Hyvärinen, M., et al. (2014) Coming to terms with the concept of moving species threatened by climate change – a systematic review of the terminology and definitions. *PLoS ONE*. 9, e102979.

Hällfors, M. H., Aikio, S., Fronzek, S., Hellmann, J. J., Ryttäri, T. & Heikkinen, R. K. (2016a) Assessing the need and potential of assisted migration using species distribution models. *Biological Conservation*. 196, 60–68.

Hällfors, M. H., Liao, J., Dzurisin, J., et al. (2016b) Addressing potential local adaptation in species distribution models: implications for conservation under climate change. *Ecological Applications*. 26, 1154–1169.

Hällfors, M. H., Aikio, S. & Schulman, L. E. (2017) Quantifying the need and potential of assisted migration. *Biological Conservation*. 205, 34–41.

Hällfors, M. H., Vaara, E. M., Ahteensuu, M. T., Kokko, K. T., Oksanen, M. & Schulman, L. E. (2018) Assisted migration as a conservation approach under climate change. *Encyclopedia of the Anthropocene: Reference Module in Earth Systems and Environmental Sciences, 2018*. https://doi.org/10.1016/B978-0-12-409548-9.09750-5

Hardin, G. (1960) The competitive exclusion principle. *Science*. 131, 1292–1297.

Hewitt, N., Klenk, N., Smith, A. L., et al. (2011) Taking stock of the assisted migration debate. *Biological Conservation*. 144, 2560–2572.

Hoegh-Guldberg, O., Hughes, L., McIntyre, S., et al. (2008) Assisted colonization and rapid climate change. *Science*. 321, 345–346.

Hogg, C. & Wise, P. (2023) Assisted colonisation as a conservation tool: Tasmanian devils and Maria Island. In Gaywood, M. J., Ewen, J. G., Hollingsworth, P. M. and Moehrenschlager, A. (eds.) *Conservation Translocations*. Cambridge, Cambridge University Press.

Hunter, E. A., Gibbs, J. P., Cayot, L. J. & Tapia, W. (2013) Equivalency of Galápagos giant tortoises used as ecological replacement species to restore ecosystem functions. *Conservation Biology: The Journal of the Society for Conservation Biology*. 27, 701–709.

IUCN/SSC (2013) *Guidelines for Reintroductions and Other Conservation Translocations. Version 1.0*. Gland, Switzerland, IUCN Species Survival Commission.

Jones, C. J., Tatayah, V., Moorhouse-Gann, R., Griffiths, C., Zuël, N. & Cole, N. (2023) Slow and steady wins the race: using non-native tortoises to rewild islands off Mauritius. In Gaywood, M. J., Ewen, J. G., Hollingsworth, P. M. and Moehrenschlager, A. (eds.) *Conservation Translocations*. Cambridge, Cambridge University Press.

Kattge, J., Díaz, S., Lavorel, S., et al. (2011) TRY – a global database of plant traits. *Global Change Biology*. 17, 2905–2935.

Kawecki, T. J. & Ebert, D. (2004) Conceptual issues in local adaptation. *Ecology Letters*. 7, 1225–1241.

Lewis, D. (2016) Relocating Australian tortoise sets controversial precedent. *Science*. Available from: www.sciencemag.org/news/2016/08/relocating-australian-tortoise-sets-controversial-precedent [Accessed 27 May 2022].

Linnell, J. D. C. & Jackson, C. R. (2019) Bringing back large carnivores to rewild landscapes. In Pettorelli, N., Durant, S. M. and du Toit J. T. (eds.) *Rewilding*. Cambridge, Cambridge University Press, pp. 248–279.

Lloyd, B. D. & Powlesland, R. G. (1994) The decline of kakapo Strigops habroptilus and attempts at conservation by translocation. *Biological Conservation*. 69, 75–85.

Lunt, I. D., Byrne, M., Hellmann, J. J., et al. (2013) Using assisted colonisation to conserve biodiversity and restore ecosystem function under climate change. *Biological Conservation*. 157, 172–177.

Marris, E. (2008) Moving on assisted migration. *Nature Climate Change*. 1, 112–113.

Mathakutha, R., Steyn, C., Roux, P. C. le, et al. (2019) Invasive species differ in key functional traits from native and non-invasive alien plant species. *Journal of Vegetation Science*. 30, 994–1006.

McDonald-Madden, E., Runge, M. C., Possingham, H. P. & Martin, T. G. (2011) Optimal timing for managed relocation of species faced with climate change. *Nature Climate Change*. 1, 261–265. https://doi.org/10.1038/nclimate1170

McMurdo Hamilton, T., Canessa, S., Clarke, K., et al. (2021) Applying a values-based decision process to facilitate comanagement of threatened species in Aotearoa New Zealand. *Conservation Biology*. 35, 1162–1173.

Miller, K. A., Miller, H. C., Moore, J. A., et al. (2012) Securing the demographic and genetic future of tuatara through assisted colonization. *Conservation Biology*. 26, 790–798.

Mitchell, R., Green, S. & Hollingsworth, P. M. (2023) Plant health, biosecurity and conservation translocations. In Gaywood, M. J., Ewen, J. G., Hollingsworth, P. M. and Moehrenschlager, A. (eds.) *Conservation Translocations*. Cambridge, Cambridge University Press.

Moorhouse-Gann, R. (2017) Ecological replacement as a restoration tool: Disentangling the impacts of Aldabra giant tortoises (*Aldabrachelys gigantea*) using DNA metabarcoding. Phd thesis, Cardiff University. Available from: http://orca.cf.ac.uk/111409/ [Accessed 27 May 2022].

Mueller, J. M. & Hellmann, J. J. (2008) An assessment of invasion risk from assisted migration. *Conservation Biology*. 22, 562–567.

National Species Reintroduction Forum (2014) *The Scottish Code for Conservation Translocations*. Inverness, Scottish Natural Heritage.

Neaves, L. E., Ogden, R. & Hollingsworth, P. M. (2023) Genomics and conservation translocations. In Gaywood, M. J., Ewen, J. G., Hollingsworth, P. M. and Moehrenschlager, A. (eds.) *Conservation Translocations*. Cambridge, Cambridge University Press.

Oney, B., Reineking, B., O'Neill, G. & Kreyling, J. (2013) Intraspecific variation buffers projected climate change impacts on *Pinus contorta*. *Ecology and Evolution*. 3, 437–449.

Pérez, I., Anadón, J. D., Díaz, M., Nicola, G. G., Tella, J. L. & Giménez, A. (2012) What is wrong with current translocations? A review and a decision-making proposal. *Frontiers in Ecology and the Environment*. 10, 494–501.

Peters, R. L. & Darling, J. D. S. (1985) The greenhouse effect and nature reserves. *BioScience*. 35, 707–717.

Peterson, K. & Bode, M. (2021) Using ensemble modeling to predict the impacts of assisted migration on recipient ecosystems. *Conservation Biology*. 35, 678–687.

Pettorelli, N., Durant, S. M. & Toit, J. T. du (eds.). (2019) *Rewilding*, 1st ed. Cambridge, Cambridge University Press.

Phelan, R., Kareiva, P., Marvier, M., Robbins P. & Weber, M. (2021) Why intended consequences? *Conservation Science and Practice*. 3, e408.

Pounds, A. J., Bustamante, M. R., Coloma, L. A., et al. (2006) Widespread amphibian extinctions from epidemic disease driven by global warming. *Nature*. 439, 161–167.

Ricciardi, A. & Simberloff, D. (2009a) Assisted colonization is not a viable conservation strategy. *Trends in Ecology & Evolution*. 24, 248–253.

Ricciardi, A. & Simberloff, D. (2009b) Assisted colonization: good intentions and dubious risk assessment. *Trends in Ecology & Evolution*. 24, 476–477.

Richardson, D. M., Hellmann, J. J., McLachlan, J. S., et al. (2009) Multidimensional evaluation of managed relocation. *Proceedings of the National Academy of Sciences*. 106, 9721–9724.

Rocha, C. & Bergallo, H. de. (2012) When invasive exotic populations are threatened with extinction. *Biodiversity and Conservation*. 21, 3729–3730.

Rout, T. M., McDonald-Madden, E., Martin, T. G., Mitchell, N. J., Possingham, H. P. & Armstrong, D. P. (2013) How to decide whether to move species threatened by climate change. *PLoS ONE*. 8, e75814.

Sainsbury, A. W. & Carraro, C. (2023) Animal disease and conservation translocations. In Gaywood, M. J., Ewen, J. G., Hollingsworth, P. M. and Moehrenschlager, A. (eds.) *Conservation Translocations*. Cambridge, Cambridge University Press.

Sandler, R. (2010) The value of species and the ethical foundations of assisted colonization. *Conservation Biology*. 24, 424–431.

Schwartz, M. W. & Martin, T. G. (2013) Translocation of imperiled species under changing climates. *Annals of the New York Academy of Sciences*. 1286, 15–28.

Schwartz, M. W., Hellmann, J. J., McLachlan, J. M., et al. (2012) Managed relocation: integrating the scientific, regulatory, and ethical challenges. *BioScience*. 62, 732–743.

Seddon, P. J. (2023) The role of conservation translocations in rewilding and de-extinction. In Gaywood, M. J., Ewen, J. G., Hollingsworth, P. M. and Moehrenschlager, A. (eds.) *Conservation Translocations*. Cambridge, Cambridge University Press.

Seddon, P. J. & Armstrong, D. P. (2019) The role of translocation in rewilding. In Pettorelli, N., Durant, S. M. and du Toit J. T. (eds.) *Rewilding*. Cambridge, Cambridge University Press, pp. 303–324.

Seddon, P. J., Armstrong, D. P. & Maloney, R. F. (2007) Developing the science of reintroduction biology. *Conservation Biology*. 21, 303–312.

Shoo, L. P., Hoffmann, A. A., Garnett, S., et al. (2013) Making decisions to conserve species under climate change. *Climatic Change*. 119, 239–246.

Siipi, H. & Ahteensuu, M. (2016) Moral relevance of range and naturalness in assisted migration. *Environmental Values*. 25, 465–483.

Taylor, D. & Hamilton, A. (1994) Impact of climatic change on tropical forests in Africa: Implications for protected area planning and management. In Pernetta, J., Leemans, R., Elder, D. and Humphrey, S. (eds.) *Impacts of Climate Change on Ecosystems and Species: Implications for Protected Areas*. Gland, Switzerland, IUCN.

Thalmann, S., Peck, S., Wise, P., Potts, J. M., Clarke, J. & Richley, J. (2016) Translocation of a top-order carnivore: tracking the initial survival, spatial movement, home-range establishment and habitat use of Tasmanian devils on Maria Island. *Australian Mammalogy*. 38, 68–79.

Tompkins, D. M., White, A. R. & Boots, M. (2003) Ecological replacement of native red squirrels by invasive greys driven by disease. *Ecology Letters*. 6, 189–196.

Torreya Guardians (2020) *Assisted migration (assisted colonization, managed relocation, translocation, helping forests walk) and rewilding of plants and animals in an era of rapid climate change*. Available from: http://torreyaguardians.org/assisted-migration.html [Accessed 3 February 2020].

Trouwborst, A., Blackmore, A., Blyth, S., Fleurke, F., McCormack, P. & Gaywood, M.J. (2023) Conservation translocations and the law. In Gaywood, M. J., Ewen, J. G., Hollingsworth, P. M. and Moehrenschlager, A. (eds.) *Conservation Translocations*. Cambridge, Cambridge University Press.

Valladares, F., Matesanz, S., Guilhaumon, F., et al. (2014) The effects of phenotypic plasticity and local adaptation on forecasts of species range shifts under climate change. *Ecology Letters*. 17, 1351–1364.

VanderWerf, E. A., Young, L. C., Kohley, C. R., et al. (2019) Establishing Laysan and black-footed albatross breeding colonies using translocation and social attraction. *Global Ecology and Conservation*. 19, e00667.

Violle, C., Thuiller, W., Mouquet, N., et al. (2017) Functional rarity: the ecology of outliers. *Trends in Ecology and Evolution*. 32, 356–367.

Vitt, P., Havens, K., Kramer, A. T., Sollenberger, D. & Yates, E. (2010) Assisted migration of plants: changes in latitudes, changes in attitudes. *Biological Conservation*. 143, 18–27.

Weslien, J., Djupström, L. B., Schroeder, M. & Widenfalk, O. (2011) Long-term priority effects among insects and fungi colonizing decaying wood: Species interactions during wood decay. *Journal of Animal Ecology*. 80, 1155–1162.

Willis, S. G., Hill, J. K., Thomas, C. D., et al. (2009) Assisted colonization in a changing climate: a test-study using two U.K. butterflies. *Conservation Letters*. 2, 46–52.

Zamboni, T., Di Martino, S. & Jiménez-Pérez, I. (2017) A review of a multispecies reintroduction to restore a large ecosystem: the Iberá Rewilding Program (Argentina). *Perspectives in Ecology and Conservation*. 15, 248–256.

Zimmer, C. (2007) A Radical Step to Preserve a Species: Assisted Migration. *The New York Times*, January 23. Available from: www.nytimes.com/2007/01/23/science/23migrate.html [Accessed 10 July 2022].

12 · The Role of Conservation Translocations in Rewilding and De-extinction

PHILIP J. SEDDON

12.1 Introduction

> *If native large carnivores have been extirpated from a region, their reintroduction and recovery is central to a conservation strategy.*
>
> Foreman (1999).

The relatively recent roots of the rewilding movement reach back only to 1992 and the coining of the term by Dave Foreman, a co-founder of the Wildlands Project, an ambitious scheme to create a network of large protected areas across North America. In an issue of the *Wild Earth* newsletter, Foreman wrote that, 'It is time to rewild North America; it is past time to reweave the full fabric of life on our continent' (Foreman, 1992). This stirring, if somewhat vague vision was reframed in operational terms by the conservation biology pioneers Michael Soulé and Reed Noss (1998) as large scale conservation based on core protected areas linked by corridors, within which top-order predators have been restored: 'The rewilding argument posits that large predators are often instrumental in maintaining the integrity of ecosystems; in turn, the large predators require extensive space and connectivity.' The tidy shorthand for this approach to '...restoring big wilderness based on the regulatory role of large predators' became 'the three C's: Cores, Corridors, and Carnivores' (Soulé & Noss, 1998).

Twenty years later, reading the literature it might seem that rewilding has become all things to all people (Johns, 2019; Carver et al., 2021) – a term used to unite a wide variety of conservation projects, or to rebrand existing approaches. This diversity, while arguably generating much needed new enthusiasm for large-scale conservation, has been criticised for creating some confusion over exactly what is being or could be done, how, and to what end (Nogues-Bravo et al., 2016). The original vision of Foreman, Soulé, and Noss placed the reintroduction of functionally important species as a central defining feature of rewilding. This chapter explores the role of reintroductions, or more broadly the full spectrum of conservation translocations, in contemporary rewilding projects, and posits the idea that although the original conception of rewilding might be misunderstood, disregarded, and disagreed upon amongst practitioners, or even lost by some rewilding proponents, reintroduction and other forms of conservation translocation remain central to many conservation programmes branded as rewilding (Seddon & Armstrong, 2019). To test this idea I first have to derive a simple framework for classifying rewilding projects. I can then consider the relationship between rewilding and conservation translocations before taking a slight detour to see how the proposals for so-called de-extinction might fit into this schema. All of this should then enable me to tackle my more ambitious goal of defining what might be the useful intersection between conservation translocations, ecological restoration, and rewilding.

Wish me luck.

12.2 An Abundance of Definitions

Rewilding as a concept does not suffer from being undefined, in fact there is no shortage of rewilding definitions and instead we are confounded by a confusing richness of rewilding types, versions, and subsets. Dolly Jorgensen found six different meanings for rewilding in the literature up to 2013, and also expressed concern that the rewilding concept was being used to separate humans from nature (Jorgensen, 2015). A few years later, Hayward and colleagues bemoaned the existence of 14 identifiable distinct definitions of rewilding in its different incarnations (Hayward et al., 2019a) and controversially, and perhaps hopelessly, called for the removal of the term 'rewilding' from the conservation lexicon on the grounds that the more established term 'restoration' was sufficient (Hayward et al., 2019a, 2019b). For better or for worse, depending on your perspective, rewilding is here to stay, but the task remains to figure out what exactly it is and how it relates to other conservation approaches.

Nathalie Pettorelli and colleagues cut through the fog by identifying five main broad definitions of rewilding published over the last five years, and identified three main themes: (1) restoration of biodiversity in degraded areas with minimal human inference or requirement for human benefit; (2) reintroduction of key extirpated species, or their ecological replacements, to restore ecosystem functionality with potential human benefits; and (3) creation of ecosystem self functionality, not necessarily to any past state, to deliver ecosystem services (Pettorelli et al., 2019). Rewilding has been criticised for excluding people from nature, but two of the three themes above highlight human benefit as a core feature; minimising human agency is not the same as isolating humans from the restoration of nature, in which they are inextricably embedded. Sarah Durant and colleagues resolve this concern by neatly characterising rewilding interventions as seeking to: 'move the biotas of defined spaces, including their human inhabitants, along trajectories of increasing wildness towards becoming self-organising and sustainable social-ecological systems' (Durant et al., 2019). By explicitly including the social dimension, rewilding recognises people as part of, not apart from, nature, a concept that is central to the lore of many Indigenous Peoples and one that should strike a chord with those of us in even the most developed nations.

But we are still left with an embarrassing oversupply of rewilding definitions. Pettorelli et al. (2018) attempted a unifying definition of rewilding as:

The reorganisation of biota and ecosystem processes to set an identified social–ecological system on a preferred trajectory, leading to the self-sustaining provision of ecosystem services with minimal ongoing management.

This neatly sidesteps the issue of selecting arbitrary historical target states that is currently exercising restoration ecologists – those concerned with the process and practice of renewing and restoring degraded, damaged, or destroyed ecosystems and habitats in the environment by human action. So, unlike the purposeful intervention that characterises ecological restoration, rewilding, as proposed by the definition above, acknowledges benefits for humans without either excluding them or requiring their close stewardship, and is general enough to encompass a wide range of activities.

Most recently Steve Carver and colleagues (2021), at risk of adding to the embarrassment of definitional riches, defined rewilding as:

The process of rebuilding, following major human disturbance, a natural ecosystem by restoring natural processes and the complete or near complete food web at all trophic levels as a self-sustaining and resilient ecosystem with biota that would have been present had the disturbance not occurred.

They further suggested that:

The ultimate goal of rewilding is the restoration of functioning native ecosystems containing the full range of species at all trophic levels while reducing human control and pressures.

This seems a sensibly concise, if ambitious, vision for rewilding to work with for now – restoration of natural processes and the full range of species, to create a resilient and self-sustaining natural ecosystem. But perhaps another way of understanding the diversity of what is being done or proposed is to consider much broader categories of rewilding projects, a bit like organising a menu of restaurant main courses into four categories: red meat, poultry, fish, and vegetarian options.

Happily for me, commentators have already variously proposed four distinct framings of rewilding; I would like to try to collapse these into just two for simplicity, like lumping vegetarian and non-vegetarian menu options in my previous analogy. Let's see how we go. The four most common types of rewilding proposed are: trophic, Pleistocene, passive, and ecological (Corlett, 2016).

12.2.1 Trophic Rewilding

Trophic rewilding is defined as conservation translocations to restore top-down trophic interactions and associated trophic cascades to promote self-regulating biodiverse ecosystems, and focuses on the role of apex consumers in shaping ecosystems (Mills et al., 1993; Svenning et al., 2016). Apex consumers are most often the relatively large-bodied herbivores and predators (van Wieren, 1995; Sergio et al., 2008) that can have strong effects on vegetation structure and on the density and behaviour of herbivores and mesopredators, respectively (Hobbs, 1996; Hansen & Galetti, 2009; Ripple et al., 2014). Trophic rewilding addresses the prehistoric human-driven defaunation of large-bodied species (Sandom et al., 2014), and recognises that such species have significant effects on ecosystem structure (Dirzo et al., 2014; Ripple et al., 2014), including increasing environmental heterogeneity and consequent species richness (Stein et al., 2014). Candidate species for re-establishment in trophic rewilding can be the missing native species sourced from populations elsewhere, that is, a reintroduction – the movement and release of an organism into part of its indigenous range from which it has been extirpated (IUCN, 2013; see also Table 12.1, and Gaywood & Stanley-Price, this volume, for definitions of terms used throughout the text). Alternatively they could be ecological replacements that might be related to the extinct species, including de-domesticated forms (Seddon & Armstrong, 2019), or they could be unrelated taxa (Bowman, 2012), or extant species genetically modified

Table 12.1. *Definitions of terms used in the text (see also Gaywood & Stanley-Price, this volume).*

Term	Meaning
De-extinction	The creation of functional proxies of extinct species using selective breeding, cloning, or genomic engineering technology (IUCN, 2016)
Ecological replacement	Intentional movement and release of an organism outside its indigenous range to perform a specific ecological function (IUCN, 2013)
Ecosystem engineer	A species that can affect other organisms by creating, modifying, maintaining, or destroying habitats (Byers et al., 2006)
Indigenous range	Known or inferred distribution generated from historical (written or verbal) records or physical evidence of the species' occurrence (IUCN, 2013)
Keystone species	Species that are so important in determining the ecological functioning of a community that they warrant special conservation efforts, and whose loss would precipitate further extinctions; for restoration, keystone species are necessary to help re-establish and sustain ecosystem structure and stability (Mills et al., 1993)
Translocation	Human-mediated movement of organisms from one site for release in another (IUCN, 2013)
Conservation translocation for species conservation	Intentional movement and release where the primary objective is to improve the conservation status of the focal species (this chapter)
Conservation translocation for rewilding	Intentional movement and release of an organism where the primary objective is to restore natural ecosystem functions and processes (this chapter)

with extinct gene sequences intended to result in functional proxies of the extinct species (IUCN, 2016). Trophic rewilding has been called a future-oriented, process-oriented, and non-static restoration strategy, but one that carries a risk of causing harm through exotic species proliferation (Bakker & Svenning, 2018). The process of trophic rewilding is open-ended, with goals framed more around trajectories than stable endpoints. It could be applied at a range of scales to deliver ecological services of benefit to humans, from large remote areas to smaller island systems and even in regions with agricultural intensification or high-density conurbations (Svenning et al., 2019).

12.2.2 Pleistocene Rewilding

Pleistocene rewilding was an idea instigated by Josh Donlan and colleagues (2005, 2006) as a provocative thought experiment addressing the loss of ecosystem functions following the extirpation of megafaunal species in the late Pleistocene (13,000 years BP) due to climate change and/or over-harvest (Nogues-Bravo et al., 2008). The idea of introducing species from Asia and Africa into North American landscapes was greeted with concern and scepticism by a number of groups, not least the invasion biologists. However, this idea did land at a time when both rewilding was gaining greater recognition, albeit as a vaguely defined version of ecological restoration, and reintroduction biologists were starting to consider a conservation role for introductions (i.e. conservation introductions – the movement and release of organisms outside their indigenous range for conservation purposes (IUCN, 2013)). Currently Pleistocene rewilding is considered a useful concept to frame debate, but unfeasible outside small reserves (du Toit, 2019). One such small reserve, and perhaps a poster child for Pleistocene rewilding, is the Pleistocene Park, effectively a fenced reserve of about 20 km^2 in the Republic of Yakutia in northern Siberia. The brainchild of Serge Zimov, this is an experiment in the restoration of the mammoth steppe ecosystem, lost following the extinction of mammoths and the consequent invasion by mosses and shrubs (Zimov et al., 1995). The project uses a range of medium-sized ecological replacements, including musk oxen, moose, horses, and bison, but has no analogue for the woolly mammoth (Zimov et al., 2012); or not yet anyway. The highest profile of the so-called de-extinction projects (see also Section 12.4) is the attempt by George Church to genetically modify Asian elephants using resurrected mammoth gene sequences designed to create a hybrid mammo-phant able to restore grazing and trampling processes in Siberia (Shapiro, 2015), an approach not considered by some to be viable (McCauley et al., 2017). Pleistocene rewilding can be viewed, therefore, as a specialised subset of trophic rewilding, whereby the proposed ecological replacement species are functional proxies for megafauna that went extinct in the late Pleistocene.

12.2.3 Passive Rewilding

Passive rewilding has been most associated with the rewilding of abandoned agricultural landscapes, whereby cessation of farming allows the colonisation of biotic elements without, or with minimal, human management intervention or influence (Carver, 2019). Termed by some the 'purest form of rewilding', passive rewilding allows the spontaneous restoration of (mainly native) vegetation and the return of trophic processes with the natural arrival of a range of animal taxa. Not restricted to abandoned farmland, passive rewilding has been the approach for most core zones in strictly protected areas such as national parks, and the expectation is that biodiversity will increase and ecosystem services might be improved, including reduced soil erosion, reduced flood risks, and enhanced carbon storage (Carver, 2019). Passive rewilding is low cost, but also has no defined endpoints and no target composition outcomes.

12.2.4 Ecological Rewilding

Ecological rewilding is hard to define as something different from passive rewilding and it seems to me the two terms are interchangeable. Petorelli et al. (2019) consider ecological rewilding to be conceptually close to passive rewilding, but suggest only that the former is characterised by limited active management to facilitate natural processes and allow them to regain dominance; although I feel this does not distinguish it from passive rewilding. James Miller and Richard Hobbs (2019) cannot see any difference between the two, lumping them together as ecological rewilding and framing this as a vegetation-centric approach that focuses on succession and follows trajectories not rooted in history. Clearly then, a focus on functional goals rather than the re-creation of some historical species composition opens the way for novel ecosystems, with new combinations of plants and animals.

With the above in mind we can readily combine the four types of rewilding into just two types based on the degree of active management, whereby Pleistocene rewilding becomes a form of trophic rewilding that actively introduces key species to restore lost functions, in which the key megafaunal species to be restored are represented by functional surrogates of those that went extinct 13,000 years ago. Similarly I have chosen to subsume ecological rewilding under passive rewilding since this seems best to capture the essence that there is minimal human activity to facilitate the restoration of natural processes without a target end state. Let us take our two broad categories, *trophic rewilding* and *passive rewilding*, and tackle the next challenge – what role do conservation translocations play in the rewilding enterprise?

12.3 Rewilding as a Driver for Conservation Translocations

Although the original conception of rewilding entailed restoration of key animal species, we have seen that reintroductions are not a ubiquitous feature of everything calling itself rewilding. Rewilding efforts can be broadly placed into two types: trophic and passive. The active management aspects and the critical importance of keystone species that characterise trophic rewilding suggest conservation translocations will remain a key part of at least some rewilding projects. So how important are conservation translocations in modern rewilding endeavours?

In a previous publication, Doug Armstrong and I (Seddon & Armstrong, 2019) assessed the prevalence of rewilding as a driver of reintroductions by reviewing 242 terrestrial faunal translocation projects reported in the IUCN/SSC Conservation Translocation Specialist Group (CTSG) *Reintroduction Perspectives* series (Soorae, 2008, 2010, 2011, 2013, 2016). We sought to identify those projects that highlighted the restoration of ecosystem functions or processes as at least one of the project goals or justifications. Only 15 of 242, 6 per cent, of projects could be termed rewilding under this criterion, but we speculated that this could be slightly misleading because rewilding-focused translocations might not find their way into a series called *Reintroduction Perspectives*. Nevertheless, the dearth of rewilding-oriented translocations was a surprise, especially given the historical emphasis on reintroduction as a key component of rewilding.

To revisit the prevalence of reintroduction as a key component of rewilding I conducted a search using Google Scholar (GS) and the search term (rewilding + reintroduction) to track general trends in published outputs combining these terms. I did separate counts by year from 1987 to 2020, and included only those listings that used 'rewilding' and some variant of 'reintroduction' (i.e. 'reintroducing', 'reintroduce', 're-introduction') in the title, abstract, or core text. I did not include the conservation introduction 'ecological:replacement' term because this only really became established as a form of conservation translocation since the revision of the IUCN Guidelines (IUCN, 2013), but I acknowledge that both reintroduction and conservation introduction can be legitimate approaches to trophic rewilding. I excluded references where one or other of these terms appeared only in cited material, or alone in the text. Generally I could glean this information from the GS summary listing, but sometimes needed to download the source to check the text. I included peer-reviewed papers, books and book chapters, reports, and theses. I excluded sources where the two terms were used in relation to gut microbiota. This was not a systematic review, but I believe it is at least indicative of publication trends over the last 30 years.

The first reference combining the terms rewilding and reintroduction appears to be in 1994, two years after the coining of the term 'rewilding' by Dave Foreman (1992), when John Davis, in discussing the Wildlands Project in North America, stated in reference to the rewilding venture that this 'will entail active restoration which may include reintroducing species' (Davis, 1994). Since that time the number of reintroduction and rewilding outputs has grown exponentially, with four phases evident: 1994–1998, characterised by virtually no outputs; 1999–2006, with a slow trickle of fewer than 10 per year up to the time of publication of the paper on Pleistocene rewilding by Josh Donlan and colleagues (Donlan et al., 2006); 2007–2013, when annual outputs built to about 30, until George Monbiot published *Feral* (Monbiot, 2013) and energised a nascent rewilding movement in Europe (e.g. Rewilding Europe 2017); and, finally, 2014 to the recent time, during which outputs ballooned up to the 144 counted by September of 2020 (Figure 12.1).

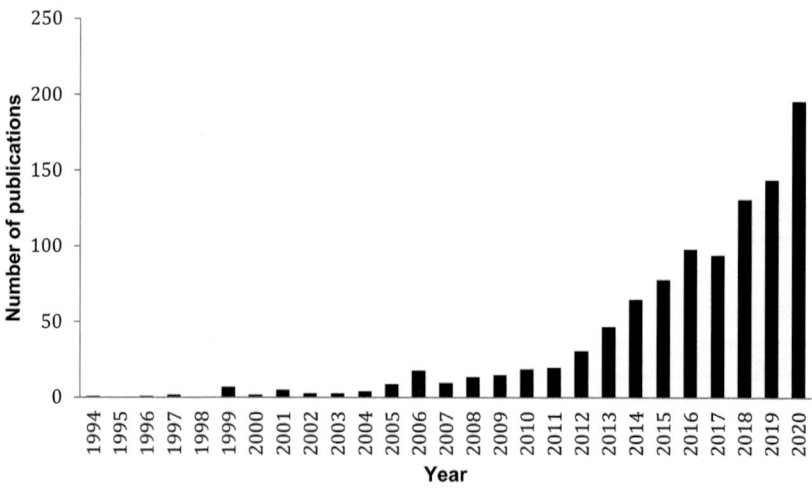

Figure 12.1 Annual growth in the number of published outputs, derived from a Google Scholar search for the terms <*rewilding* + *reintroduction*> appearing in title, abstract, or core text, and including peer-reviewed journal papers, books, book chapters, theses, and reports. Data for 2020 are to the beginning of December only.

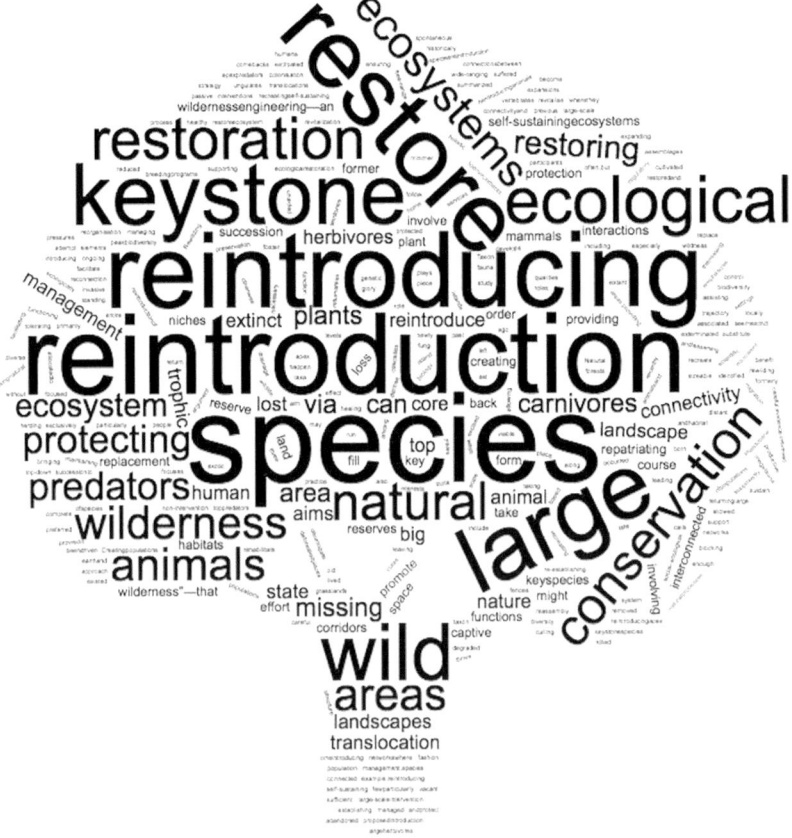

Figure 12.2 Word cloud generated from 1,022 words used in 74 published statements defining rewilding.

The role of reintroductions in rewilding is visibly evident, if slightly less rigorously quantitative, in the word cloud (Figure 12.2) constructed from 1,022 words used in 74 published statements I could find from my search that aimed to define or explain rewilding. Reintroduction and its variants feature prominently among other key words, including *restore*, *keystone*, *wild*, and *natural*. The emphasis on wilderness and large animal species is also apparent. All in all, it is reasonable to conclude that reintroductions are a significant part of many of the diverse projects labelled as rewilding, which happily will, I hope, expose rewilding practitioners to the comprehensive reintroduction planning guidelines produced by the IUCN SSC CTSG (IUCN, 2013). In fact, the role of conservation translocations in rewilding will be even greater than is suggested by this focus on reintroductions alone. Recent reviews (e.g. Swan et al., 2016) suggest that many reintroductions, of invertebrates and plants in particular, are done for ecosystem-related, as opposed to species-related goals. In addition, as indicated earlier, ecological replacements are also a possible means of restoring lost ecosystem functions. Both reintroductions and ecological replacements are key components of two (principles 1 and 3, respectively) of the 10 guiding principles for rewilding put forth by Carver et al. (2021). One extreme form of ecological replacement is that promised by the prospect of de-extinction.

12.4 The Special Case of De-extinction

While rewilding might be a somewhat fraught and loaded term, it is gaining significant traction and consensus. However, the new concept of de-extinction – the recreation of extinct species using genetic and genomic tools – remains mired in controversy. 'De-extinction' is a compelling term that was formally introduced to the public consciousness by the conservation champion Stewart Brand in 2013 (TED, 2013). The rapid development of new genetic and genomic tools seemed to raise the prospect of reversing what we currently consider to be irreversible loss. But the immediate question that was asked was, can we? What exactly are the likely resulting products from each of the three currently known de-extinction pathways of back-breeding, cloning, and genomic engineering (IUCN, 2016; Seddon, 2017a)?

Back-breeding is a bit of a de-extinction cheat that involves the artificial selection of domestic animals to produce a wild-type phenotype, and carries the assumption that such descendant forms have the potential to express the phenotype of an extinct ancestor. This has been the approach in selective breeding of domestic breeds of cattle to produce a beast that has the size and colouration of the aurochs *Bos primigenius* (Stoktstad, 2015), from which cattle descended (Ajmone-Marsan et al., 2010). Clearly, the best that could be produced by this pathway is some phenotypic proxy of an extinct form. Cloning has the potential to come closer to a lost form, but perhaps something a bit different in critical ways because it entails interspecies cloning. For mammals, for example, you need an appropriate surrogate host that is ideally a close relative of the extinct species and able to carry an embryo to term. However, you also need an embryo of the extinct species, and unless you have cryopreserved gametes (eggs and sperm), you will need a host cell in which to place the genetic material from a suitably preserved somatic (body) cell of your extinct form. This restricts cloning as a de-extinction pathway to species that went globally extinct only very recently and from which cells had already been preserved. Even with the right cells, the need to use a surrogate host means that there will be genetic components inherited from the host, there will be epigenetic effects whereby the host environment might turn on or off the activity of some genes, and there will be inevitable postnatal differences from the original extinct species due to learning, the rearing environment, diet, and the resulting microbiome (Shapiro, 2017). Without cryopreserved cells, things get harder but not impossible, thanks to rapid advances in the ability to read and write DNA sequences (Piaggio et al., 2017). The genome of the extinct form must be deciphered from any available tissue, and any gaps filled with the best approximation of the extinct sequences, probably from a nearest living relative, which can be used to create modified cell lines by replacing DNA sequences of the extant species with synthesised DNA in the extinct species sequence. Nuclei from such cells could then be used in cloning. Clearly, however, the result is the creation of a hybrid form with some expression of hybrid traits. There is, therefore, arguably (genetically, epigenetically, ecologically, and evolutionarily) no such thing as true de-extinction (Robert et al., 2017; Shapiro, 2017; Steeves et al., 2017; Wood et al., 2017b). Rather, the current 'de-extinction' pathways could usefully seek to produce functional proxies of extinct forms (IUCN, 2016).

The rationale for these approaches is the recognition of past and ongoing species extinctions and the gaps such extinctions left in ecosystems. It is known that the loss of vital functions, such as predator–prey interactions, herbivory, and seed dispersal, reduces inter-species connectivity and complexity, leading to decreased ecosystem stability and resilience. The argument for de-extinction aligns closely with that of trophic rewilding – that ecosystem repair is possible if these missing functions can be restored – and what better way to restore them than by re-establishing populations of once-extinct species (Seddon, 2017b)? Axel Moehrenschlager, John Ewen, and I drew a parallel between these conservation objectives for de-extinction and the aims of reintroductions and ecological replacements (Seddon et al., 2014b). 'De-extinction' therefore encompasses a range of genetic, genomic, and reproductive technology tools applied to the creation of a functional proxy of an extinct species, with the intent that a suitable founder group of such proxies can be released into appropriate habitat to restore ecosystem functionality. This has been a driving motivation behind a number of recent or ongoing 'de-extinction' projects entailing back-breeding (aurochs, Stoktstad, 2015) or genetic engineering (passenger pigeons (Blockstein, 2017) and mammo-phants (McCauley et al., 2017; Kornfeld, 2018; Herridge 2021)). 'De-extinction' could thus be viewed as just one of the ways of obtaining founders for an ecological replacement form of conservation translocation that could be undertaken under the label of trophic rewilding.

12.5 Harmonising Rewilding, Conservation Translocations, and Ecological Restoration

Clearly there is now a wide range of projects marketing themselves as rewilding, and, perhaps as is to be expected for a relatively new and contested conservation approach, there is variation in the extent to which organisms are being moved and released to achieve rewilding project aims. Some projects have adopted the optimistic ecological restoration principle of 'build it and they will come' by trying to create the conditions for self-assembly of faunal communities, whereas others adhere more closely to the original conception of rewilding and make reintroduction of key missing animals a primary focus. These latter projects seem to be a mélange of ecological restoration and conservation translocations under the label of rewilding. Given that both restoration ecology and reintroduction biology are established applied disciplines, how do or should they interact?

It is surprising perhaps to realise that restoration ecology and reintroduction biology have happily developed along similar but separate pathways (Ewen et al., 2012). Intuitively you might expect there to be a significant degree of overlap and complementarity – after all both seem to be in the business of addressing ecosystem degradation. However, there are two reasons why these two fields have not been fully integrated, and one compelling reason why they are now coming together (Seddon et al., 2007; Seddon, 2010). The first reason for this separate development is that, by definition, ecological restoration is focussed on degraded habitats, whereas historically reintroductions have usually aimed to release organisms into the best, most undamaged and intact areas of habitat. The second reason is that ecological restoration has traditionally concentrated on the recreation of abiotic features and processes, such as water flow regimes; where it has considered biotic elements this has almost exclusively taken the form of replanting of vegetation, but sometimes also with the management of grazing animals. The expectation has been that if the core vegetative and abiotic features are in place then the site-appropriate animals will recolonise. This might well be the case in many situations, but reintroductions explicitly seek to re-establish viable populations of species that have been lost from an area that is otherwise suitable habitat for the focal species and, importantly, where the likelihood of natural recolonisation is negligible due to dispersal barriers or the absence of natural source populations. While plant reintroductions are increasingly common (Dalrymple et al., 2012), there has been a marked historical taxonomic bias towards animal reintroductions, and even with animals a bias towards birds and mammals, and even towards the most charismatic of these (Seddon et al., 2005).

The single compelling reason for the growing ties between ecological restoration and reintroduction (and rewilding, for that matter) is the thorny implications of the 're' part of these words, suggesting a return to some past state. But which state, which point in the past, is desirable or even achievable? The answer to date has depended on where you are in the world; for example, some European restoration time frames often go back a few hundreds of years, but, some New Zealand projects look back to pre-colonisation periods of 500 years or more. Any past target state for restoration will be somewhat arbitrary, and all suffer from the unstated and unwarranted assumption that natural systems have some ideal, static, past condition that can be recreated (Seddon, 2010). Past and ongoing natural and anthropogenic change, invasive species, and global climate change make a nonsense of such assumptions (Seddon et al., 2014a), giving rise to novel ecosystems that need to be acknowledged and managed (Hobbs et al., 2006). The acceptance by both conservation translocation practitioners and restoration ecologists that a full return to some arbitrary former state is generally an unrealistic goal brings both fields into alignment with rewilding, where the focus has always been on functions and processes, and ecosystem redevelopment under changing environmental conditions, rather than historical reconstructions (du Toit & Pettorelli, 2019).

Reintroduction biology's response to this reality has been to expand the spectrum of conservation translocations beyond the release of species within their indigenous range to include two types of conservation introduction entailing releases of organisms into areas they have never before occupied: assisted colonisation and ecological replacement (IUCN, 2013; Table 12.1; Gaywood & Stanley-Price, this volume).

Assisted colonisation is the release to establish a new population of an organism outside its indigenous range to prevent population extinction at any scale, the most compelling rationale for this being shifting distribution ranges due to climate change and barriers to natural dispersal. Other drivers for an assisted colonisation might be to avoid extinction risks following invasion of habitat by exotic predators or disease (Seddon et al., 2015).

The second type of conservation introduction, ecological replacement (Hällfors & Dalrymple, this volume), creates a bridge to ecological restoration in seeking to restore an ecosystem process or function that has been lost through the global extinction of a species by releasing a non-native species that is expected to perform the same ecological role. This is quite new territory for conservation translocations, but has been applied with impressive results through the replacement of one giant tortoise species with another on Indian ocean island systems, enabling the restoration of trampling and grazing functions that create 'tortoise turf' and encouraging native vegetation while discouraging exotic weedy species (Griffiths et al., 2011; Hansen et al., 2010; Jones et al., this volume).

Traditionally, reintroductions have been motivated by the dire conservation status of single species and the need to re-establish populations of such species that have a high probability of persistence. However, the expansion of the translocation spectrum to include ecosystem-related drivers, and the increasing prevalence of multi-species reintroductions, is slowly shifting the emphasis from single-species conservation only, to ecological restoration whereby missing faunal and floral components, or their surrogates, are placed in areas of suitable habitat. Chris Sandom and colleagues frame this as a 'pillar' of rewilding, 'species reintroduction to restore ecosystem functioning', while also acknowledging that in some cases ecological replacement might be required to restore the roles of globally extinct species (Sandom et al., 2013). We can therefore broadly distinguish between translocations for species conservation that primarily seek to restore focal populations and tend to focus on improving the conservation status of a few threatened species, and translocations for rewilding that primarily seek to restore the persistence of natural biodiversity (Figure 12.3). This is not an absolute distinction; there will be cases where, for example, assisted colonisation to benefit a focal species can do double duty as an ecological replacement. Similarly, reinforcement of extant populations could be a component of rewilding to address historic declines in keystone species. Nevertheless, ecological restoration and reintroduction biology are therefore unified by the understanding

Rewilding and De-extinction · 369

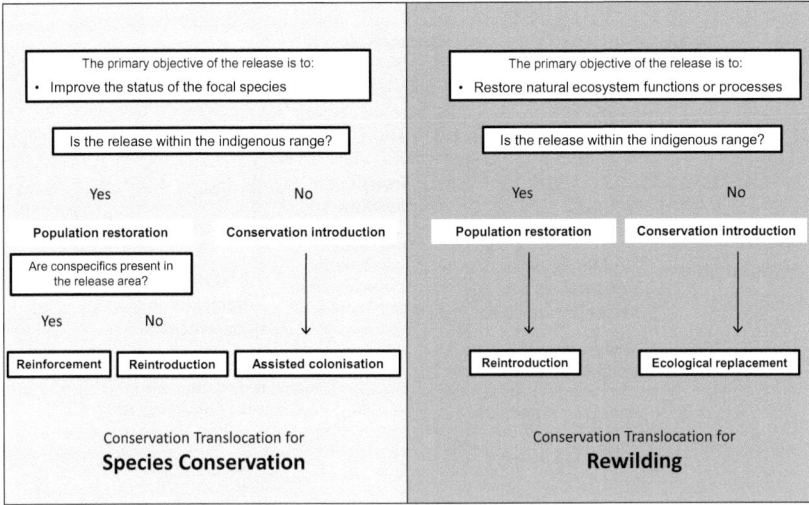

Figure 12.3 Conservation translocations can be categorised as translocations for species conservation (including reinforcements, reintroductions, and assisted colonisations) that primarily aim to improve the conservation of the focal species, and as translocations for rewilding (including reintroductions and ecological replacements) that primarily aim to restore natural ecosystem functions and processes (after Seddon et al. 2014a).

that these endeavours seek to result in stable and resilient natural systems supporting healthy populations of native biodiversity. This can be achieved using past states as a reference rather than a target and in some cases nicely brings the two fields together in something we now know as rewilding.

Figure 12.4 depicts the interaction of the realms of conservation translocations, ecological restoration, and rewilding. Each field has expressions that are separate. Single-species reintroductions, reinforcements, and assisted colonisation tend to be centred on un-degraded habitat and seek primarily to avert extinctions and to improve the conservation status of the focal species. There are many examples of species-focussed reintroductions, including the iconic projects on Arabian oryx *Oryx leucoryx* in Oman, and peregrine falcon *Falco peregrinus* in North America (review in Seddon & Armstrong, 2016), where the intent was to rescue a species from near extinction or to reverse precipitous declines. In contrast, much ecological restoration activity is focussed

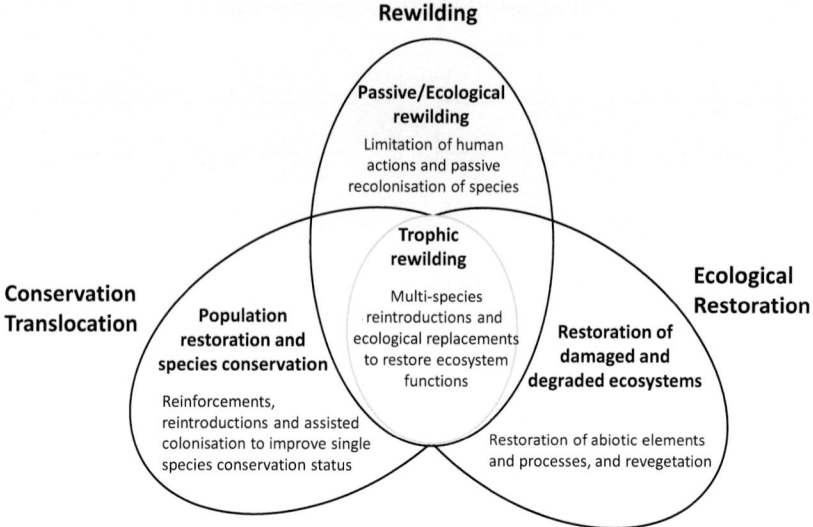

Figure 12.4 The applied fields of conservation translocation, ecological restoration, and rewilding intersect in the endeavour known as trophic rewilding, where the reintroduction or conservation introduction (ecological replacement) of species or their functional proxies is used to restore lost ecosystem functions and processes.

on the most degraded sites and does not actively restore faunal elements, instead seeking to re-vegetate and to recreate abiotic elements. High-profile examples of the restoration of degraded systems include the recovery of landscapes damaged by surface mining activity (Bandyopadhyay & Maiti, 2019). Passive (or ecological) rewilding shares some features of ecological restoration in hoping for recolonisation by animal species without the need for translocations, but differs in aiming to limit human actions. Passive rewilding has been one of the major approaches to the transition from abandoned farmland to semi-natural land in many regions of Europe (Pereira & Navarro, 2015) The three come together as trophic rewilding, where the active reintroduction of one or more key species, or the introduction of their surrogates (ecological replacements using extant forms, e.g. Pleistocene rewilding, or their functional proxies, e.g. de-extinction) is used to restore ecosystem functions and processes, without any explicit historical state as a target endpoint, and with the ambition of minimising or reducing the need for human intervention. Conservation translocations that would fit within this trophic rewilding category include the return of wolves to Yellowstone National Park (a reintroduction), and the introduction of exotic giant tortoises on Indian Ocean islands (ecological replacements) (Jones et al., this volume). The reintroduction of avifauna to predator-free islands (Wood et al., 2017a), and reseeding of seagrass (Reynolds et al., 2016), would similarly serve as diverse examples sitting in the nexus between conservation translocations, ecological restorations, and rewilding.

There is one issue to deal with here, however, since trophic rewilding has been most commonly defined as the active restoration of top-down trophic interactions that focuses on the role of apex consumers in shaping ecosystems (Mills et al., 1993; Svenning et al., 2016). Clearly some conservation translocations seeking to restore ecosystem processes will entail the release of organisms that are not apex consumers, but which might be keystone species, such as ecosystem engineers effecting bottom-up processes. An example would be the reintroduction of prairie dogs, *Cynomys* spp. (Hale and Koprowski, 2018), whose burrow colonies create habitat for a range of species, and additionally aerate and fertilise the soil. I am reluctant to generate yet more terminology to cover all possible variations, and instead propose that we might readily expand the scope of trophic rewilding to encompass the conservation translocation of any key ecosystem element, other than only apex consumers. This is the stance also adopted by Carver et al. (2021) and I believe this is justified given that, although also originally focused on top predators, the keystone species concept has evolved to include species at different trophic levels that maintain ecosystem stability and function, and have disproportionately large impacts on their ecosystems (reviewed in Hale & Koprowski, 2018). A similar expansion in the scope of trophic rewilding to include a range of trophic levels seems sensible.

12.6 Conclusion

In common with disruptive technologies, disruptive conservation concepts can shake things up in a number of ways, both good and bad. There will always be pushback from advocates of the status quo, and their sceptical eyes will be invaluable for shaping these new tools and approaches into something useful. So it is with the seemingly nebulous concept of rewilding – a term that has sparked the interest of a public wearied by environmental warnings and jaded by a litany of conservation bad news stories – and this sparking is a good and timely thing. But there is still work to be done to harness this new rewilding beast, to understand it, to shape it and tame it, and to put it to work towards the best possible outcomes. Part of that work is allowing the rewilding venture to settle into a natural place amongst the older, but still youthful, endeavours of ecological restoration and conservation translocation. In this chapter I have tried to sketch a rough map for how these three close relatives might find each other and most productively interact. I think I have identified a space called trophic rewilding where the careful reintroduction or ecological replacement of key species can be used to restore vital ecological functions to enable the (re)creation of stable, resilient, biodiverse, useful-to-humans, and natural (if sometimes novel) ecosystems.

12.7 Key Messages

- The rewilding concept was born only about 20 years ago as large scale conservation based on core protected areas linked by corridors, within which top-order predators have been restored.
- The original vision of rewilding has become somewhat blurred amongst a plethora of projects branding themselves as rewilding.
- Broadly two types of rewilding can be recognised: trophic rewilding, entailing the active restoration of key species, and passive rewilding, the spontaneous and undirected return of vegetation and ecological processes.
- Conservation translocations, in the form of reintroduction and ecological replacement, are a key component of trophic rewilding, with the prevalence of reintroductions in rewilding projects increasing in recent years.
- Trophic rewilding is therefore the intersection between rewilding, conservation translocations, and ecological restoration, where reintroduction or ecological replacement of key species is used to restore lost ecological functions.

References

Ajmone-Marsan, P., Fernado Garcia, J. & Lenstra, J. A. (2010) On the origin of cattle: how aurochs became cattle and colonized the world. *Evolutionary Anthropology*. 19, 148–157.

Bakker, E. S. & Svenning, J.-C. (2018) Trophic rewilding: impact on ecosystems under global change. *Philosophical Transactions of the Royal Society B*. 373, 20170432.

Bandyopadhyay, S. & Maiti S. K. (2019) Evaluation of ecological restoration success in mining-degraded lands. *Environmental Quality Management*. 29, 89–100.

Blockstein, D. E. (2017) We can't bring back the passenger pigeon: the ethics of deception around de-extinction. *Ethics Policy & Environment*. 20, 33–37.
Bowman, D. (2012) Bring elephants to Australia? *Nature*. 482, 30.
Byers, J. E., Cuddington, K., Jones, C. G., et al. (2006) Using ecosystem engineers to restore ecological systems. *Trends in Ecology and Evolution*. 21, 493–500.
Carver, S. (2019) Rewilding through land abandonment. In Pettorelli, N., Durant, S. M. and Du Toit, J. T. (eds.) *Rewilding*. Cambridge, Cambridge University Press, pp. 99–122.
Carver, S., Convery, I., Hawkins, S., et al. (2021) Guiding principles for rewilding. *Conservation Biology*. 35, 1882–1893.
Corlett, R. T. (2016) Restoration, reintroduction, and rewilding in a changing world. *Trends in Ecology and Evolution*. 31, 453–462.
Dalrymple, S. E., Banks, E., Stewart, G. B. & Pullin, A. S. (2012) A meta-analysis of threatened plant reintroductions from across the globe. In Maschinski, J. & Haskins, K. E. (eds.) *Plant Reintroduction in a Changing Climate. The Science and Practice of Ecological Restoration book series*. Washington, DC, Island Press, pp. 31–50.
Davis, J. (1994) A sidelong glance at The Wildlands Project. In: Burks, D. C. (ed.) *Place of the Wild: a Wildlands Anthology*. Washington, DC, Island Press, pp. 236–245.
Dirzo, R., Young, H. S., Galetti, M., Ceballos, G., Isaac, N. J. B. & Collen, B. (2014) Defaunation in the Anthropocene. *Science*. 345, 401–406.
Donlan, C. J., Berger, J., Bock, C. E., et al. (2005) Re-wilding North America. *Nature*. 436, 913–914.
Donlan C. J., Berger, J., Bock, C. E., et al. (2006) Pleistocene rewilding: an optimistic agenda for twenty-first century conservation. *American Naturalist*. 168, 660–681.
Durant, S. M., Pettorelli, N. & Du Toit, J. T. (2019) The future of rewilding: fostering nature and people in a changing world. In Pettorelli, N., Durant, S. M. and Du Toit, J. T. (eds.) *Rewilding*. Cambridge, Cambridge University Press, pp. 413–425.
Du Toit, J. T. (2019) Pleistocene rewilding: an enlightening thought experiment. In Pettorelli, N., Durant, S. M. and Du Toit, J. T. (eds.) *Rewilding*. Cambridge, Cambridge University Press, pp. 55–72.
Du Toit, J. T. & Pettorelli, N. (2019) The differences between rewilding and restoring an ecologically degraded landscape. *Journal of Applied Ecology*. 56, 2467–2471.
Ewen, J. G., Armstrong, D. P., Parker, K. A. & Seddon, P. J. (eds.) (2012) *Reintroduction Biology: Integrating Science and Management*. Conservation Science and Practice, no. 9. Chichester, Wiley-Blackwell.
Foreman, D. (1992) *Wild Earth* 2, no. 3 (Fall 1992). Republished by the Environment & Society Portal, Multimedia Library. Available from: www.environmentandsociety.org/mml/wild-earth-2-no-3 [Accessed 30 May 2022].
Foreman, D. (1999) The wildlands project and the rewilding of North America. *Denver University Law Review*, 76, 535–554.
Gaywood, M. J. & Stanley-Price, M. (2023) Moving species: reintroductions and other conservation translocations. In Gaywood, M. J., Ewen, J. G.,

Hollingsworth, P. M. and Moehrenschlager, A. (eds.) *Conservation Translocations*. Cambridge, Cambridge University Press.

Griffiths, C. J., Hansen, D. M., Jones, C. G., Zuel, N. & Harris, S. (2011) Resurrecting extinct interactions with extant substitutes. *Current Biology*. 21, 762–765.

Hale, S. L. & Koprowski, J. L. (2018) Ecosystem-level effects of keystone species reintroduction: a literature review. *Restoration Ecology*. 26, 439–445.

Hällfors, M. & Dalrymple, S. E. (2023) Assisted colonisation and ecological replacement. In Gaywood, M. J., Ewen, J. G., Hollingsworth, P. M. and Moehrenschlager, A. (eds.) *Conservation Translocations*. Cambridge, Cambridge University Press.

Hansen, D. M. & Galetti, M. (2009) The forgotten megafauna. *Science*. 324, 42–43.

Hansen, D. M., Donlan, C. J., Griffiths, C. J. & Campbell, K. J. (2010) Ecological history and latent conservation potential: large and giant tortoises as a model for taxon substitutions. *Ecography*. 33, 272–284.

Hayward, M. W., Scanlon, R. J., Callen A., et al. (2019a) Reintroducing rewilding to restoration – rejecting the search for novelty. *Biological Conservation*. 233, 255–259.

Hayward, M. W., Jachowski, D., Bugir, C. K., et al. (2019b) The search for novelty continues for rewilding. *Biological Conservation*. 236, 584–585.

Herridge, V. (2021) Before making a mammoth, ask the public. *Nature*. 598, 387.

Hobbs, N. T. (1996) Modification of ecosystems by ungulates. *Journal of Wildlife Management*. 60, 695–713.

Hobbs, R. J., Arico, S. & Aronson, J., et al. (2006) Novel ecosystems: theoretical and management aspects of the new ecological world order. *Global Ecology and Biogeography*. 15, 1–7.

IUCN (2013) *Guidelines for Reintroductions and Other Conservation Translocations*. Gland, Switzerland, IUCN/SSC Re-introduction Specialist Group. Available from: www.iucnsscrsg.org/ [Accessed 6 June 2022].

IUCN (2016) *IUCN SSC Guiding Principles on Creating Proxies of Extinct Species for Conservation Benefit. Version 1.0*. Gland, Switzerland, IUCN Species Survival Commission. Available from: https://portals.iucn.org/library/sites/library/files/documents/Rep-2016-009.pdf [Accessed 28 May 2022].

Johns, D. (2019) History of rewilding: ideas and practice. In Pettorelli, N., Durant, S. M. and Du Toit, J. T. (eds.) *Rewilding*. Cambridge, Cambridge University Press, pp.12–33.

Jones, C. J., Tatayah, V., Moorhouse-Gann, R., Griffiths, C., Zuël, N. & Cole, N. (2023) Slow and steady wins the race: using non-native tortoises to rewild islands off Mauritius. In Gaywood, M. J., Ewen, J. G., Hollingsworth, P. M. and Moehrenschlager, A. (eds.) *Conservation Translocations*. Cambridge, Cambridge University Press.

Jorgensen, D. (2015) Rethinking rewilding. *Geoforum*. 65, 482–488.

Kornfeldt, T. (2018) *The Re-Origin of Species*. Melbourne, Scribe.

McCauley, D., Hardesty-Moore, M., Halpern, B. S. & Young, H. (2017) A mammoth undertaking: harnessing insight from functional ecology to shape de-extinction priority setting. *Functional Ecology*. 31, 1003–1011.

Miller, J. R. & Hobbs, R. J. (2019) Rewilding and restoration. In Pettorelli, N., Durant, S. M. and Du Toit, J. T. (eds.) *Rewilding*. Cambridge, Cambridge University Press, pp. 123–141.

Mills, L. S., Soulé, M. E. & Doak, D. F. (1993) The keystone species concept in ecology and conservation. *Bioscience*. 43, 219–224.

Monbiot, G. (2013) *Feral: Searching for Enchantment of the Frontiers of Rewilding*. London, Allen Lane, Penguin Press.

Nogues-Bravo, D., Rodriguez, J., Hortal, J., Batra, P. & Araujo, M. B. (2008) Climate change, humans, and the extinction of the woolly mammoth. *PLoS Biology*. 6, e79.

Nogues-Bravo, D., Simberloff, D., Rahbek, C. & Sanders, N. J. (2016) Rewilding is the new Pandora's box in conservation. *Current Biology*. 26, R83–R101.

Pereira, H. M. & Navarro, L. M. (eds.) (2015) *Rewilding European Landscapes*. Springer Open.

Pettorelli, N., Barlow, J., Stephens, P. A., et al. (2018) Making rewilding fit for policy. *Journal of Applied Ecology*. 55, 1114–1125.

Pettorelli, N., Durant, S. M., & Du Toit, J. T. (2019) Rewilding: a captivating, controversial, twenty-first-century concept to address ecological degradation in a changing world. In Pettorelli, N., Durant, S. M. and Du Toit, J. T. (eds.) *Rewilding*. Cambridge, Cambridge University Press, pp. 1–11.

Piaggio, A. J., Segelbacher, G., Seddon, P. J., et al. (2017) Is it time for synthetic biodiversity conservation? *Trends in Ecology and Evolution*. 32, 97–107.

Rewilding Europe 2017. www.rewildingeurope.com [Accessed 10 August 2017].

Reynolds, L. K., Mycott, M., McGlathery, K. J. & Orth, R. J. (2016) Ecosystem services returned through seagrass restoration. *Restoration Ecology*. 24, 583–588.

Ripple, W. J., Estes, J. A., Beschta, R. L., et al. (2014) Status and ecological effects of the World's largest carnivores. *Science*. 343, 1241484.

Robert, A., Thevenin, C., Prince, K., Sarrazin, F. & Clavel, J. (2017) De-extinction and evolution. *Functional Ecology*. 31, 1021–1031.

Sandom, C., Donlan, C. J., Svenning, J.-C. & Hansen, D. (2013) Rewilding. In MacDonald, D. W. and Willis, K. J. (eds.) *Key Topics in Conservation Biology 2*. Chichester, John Wiley & Sons Ltd., pp. 430–451.

Sandom, C., Fuarby, S., Sandel, B. & Svenning, J.-C. (2014) Global late Quaternary megafauna extinctions linked to humans, not climate change. *Proceedings of the Royal Society of London B*. 281, 20133254.

Seddon, P. J. (2010) From re-introduction to assisted colonization: moving along the conservation translocation spectrum. *Restoration Ecology*. 18, 796–802.

Seddon, P. J. (2017a) The ecology of de-extinction. *Functional Ecology*. 31, 992–995.

Seddon, P. J. (2017b) De-extinction and barriers to the application of new conservation tools. *Hastings Center Report*, 47, S5–S8. Available from: https://doi.org/10.1002/hast.745 [Accessed 27 May 2022].

Seddon, P. J. & Armstrong, D. P. (2016) Reintroduction and other conservation translocations: history and future developments. In Jachowski,D. S., Slowtow, R., Angermeier, P. L. & Millspaugh, J. J. (eds.) *Reintroduction of Fish and Wildlife Populations*. Oakland, CA, University of California Press, pp. 7–28.

Seddon, P. J. & Armstrong, D. P. (2019) The role of translocation in rewilding. In Pettorelli, N., Durant, S. M. and Du Toit, J. T. (eds.) *Rewilding*. Cambridge, Cambridge University Press, pp. 303–324.

Seddon, P. J., Soorae, P. S. & Launay, F. (2005) Taxonomic bias in reintroduction projects. *Animal Conservation*. 8, 51–58.

Seddon, P. J., Armstrong, D. P. & Maloney, R. F. (2007) Combining the fields of reintroduction biology and restoration ecology. *Conservation Biology*. 21, 1387–1390.

Seddon, P. J., Griffiths, C. J. Soorae, P. S. & Armstrong, D. P. (2014a) Reversing defaunation: restoring species in a changing world. *Science*. 345: 406–412.

Seddon, P. J., Moehrenschlager, A. & Ewen, J. (2014b) Reintroducing resurrected species: selecting deextinction candidates. *Trends in Ecology and Evolution*. 29, 140–147.

Seddon, P. J., Moro, D., Mitchell, N. J., Chauvenet, A. L. M. & Mawson, P. R. (2015) Proactive conservation or planned invasion? Past, current and future use of assisted colonisation. In Armstrong, D. P., Hayward, M. W., Moro, D., and Seddon, P. J. (eds.) *Advances in Reintroduction Biology of Australian and New Zealand Fauna*. Victoria, Australia, CSIRO Publishing, pp. 105–126.

Sergio, F., Caro, T., Brown, D., et al. (2008) Top predators as conservation tools: ecological rationale, assumptions, and efficacy. *Annual Review of Ecology Evolution and Systematics*. 39, 1–19.

Shapiro, B. (2015) *How to Clone a Mammoth: The Science of De-extinction*. Princeton, NJ, Princeton University Press.

Shapiro, B. (2017) Pathways to de-extinction: how close can we get to resurrections of an extinct species? *Functional Ecology*. 31, 996–1002.

Soorae, P. S. (ed.) (2008) *Global Reintroduction Perspectives: Re-Introduction Case Studies from Around the Globe*. Abu Dhabi, UAE, IUCN/SSC Re-introduction Specialist Group, viii + 284 pp.

Soorae, P. S. (ed.) (2010) *Global Reintroduction Perspectives: Additional Case-Studies from Around the Globe*. Abu Dhabi, UAE, IUCN/ SSC Re-introduction Specialist Group, xii + 352 pp.

Soorae, P. S. (ed.) (2011) *Global Reintroduction Perspectives: More Case-Studies from Around the Globe* Abu Dhabi, UAE, IUCN/ SSC Re-introduction Specialist Group, xiv + 250 pp.

Soorae, P. S. (ed.) (2013) *Global Re-Introduction Perspectives: 2013. Case-Studies from Around the Globe*. Gland, Switzerland, IUCN/SSC Re-introduction Specialist Group, xiv + 282 pp.

Soorae, P. S. (ed.) (2016) *Global Re-Introduction Perspectives: 2016. Further Case Studies from Around the Globe*. Gland, Switzerland, IUCN/SSC Re-introduction Specialist Group, xiv + 276 pp.

Soulé, M. & Noss, R. (1998) Rewilding and biodiversity: complimentary goals for continental conservation. *Wild Earth*. 8, 19–28.

Steeves, T. E., Johnson, J. A. & Hale, M. L. (2017) Maximising evolutionary potential in functional proxies for extinct species: a conservation genetic perspective on de-extinction. *Functional Ecology*. 31, 1032–1040.

Stein, A., Gerstner, K. & Kreft, H. (2014) Environmental heterogeneity as a universal driver of species richness across taxa, biomes, and spatial scales. *Ecology Letters*. 17, 866–880.

Stokstad, E. (2015) Bringing back the aurochs. *Science*. 350, 1144–1147

Svenning, J.-C., Pedersen, P. B. M., Donlan, C. J., et al. (2016) Science for a wilder Anthropocene: synthesis and future directions for trophic rewilding research. *Proceedings of the National Academy of Sciences*. 113, 898–906.

Svenning, J.-C., Munk, M. & Schweiger, A. (2019) Trophic rewilding: ecological restoration of top-down trophic interactions to promote self regulating biodiverse ecosystems. In Pettorelli, N., Durant, S. M. and Du Toit, J. T. (eds.) *Rewilding*. Cambridge, Cambridge University Press, pp. 73–98.

Swan, K. D., McPherson, J. M., Seddon, P. J. & Moehrenschlager, A. (2016) Managing marine biodiversity: the rising diversity and prevalence of marine conservation translocations. *Conservation Letters*. 9, 239-251.

TED (2013) Deextinction. Available from: www.ted.com/topics/deextinction [Accessed 25 September 2019].

van Wieren, S. E. (1995) The potential role of large herbivores in nature conservation and extensive land use in Europe. *Biological Journal of the Linnaean Society*. 56, 11–23.

Wood, J. R., Alcover, J. A., Blackburn, T. M., et al. (2017a) Island extinction: processes, patterns, and potential for ecosystem restoration. *Environmental Conservation*. 44, 348–358.

Wood, J. R., Perry, G. L. W. & Wilmshurst, J. M. (2017b) Using paleoecology to determine baseline ecological requirements and interaction networks for de-extinction candidate species. *Functional Ecology*. 31, 1012–1020.

Zimov, S. A., Chuprynin, V. I., Oreshko, A. P., Chapin III, F. S., Reynolds, J. F. & Chapin, M. C. (1995) Steppe-tundra transition: a herbivore-driven biome shift at the end of the Pleistocene. *American Naturalist*. 146, 765–794.

Zimov, S. A., Zimov, N. S. & Chapin III, F. S. (2012) The past and future of the Mammoth Steppe Ecosystem. In Louys, J. (ed.) *Paleontology in Ecology and Conservation*. Berlin, Springer-Verlag, pp. 193–225.

Part III

Conservation Translocations: Looking to the Future

13 · *From Genes to Ecosystems and Beyond: Addressing Eleven Contentious Issues to Advance the Future of Conservation Translocations*

AXEL MOEHRENSCHLAGER, PRITPAL SOORAE, AND TAMMY E. STEEVES

13.1 Contentious Issues as Opportunities

Conservation translocations have become increasingly widespread to prevent extinction, recover populations, and restore ecological function across terrestrial, freshwater, and marine biomes around the world. Used for thousands of species, this conservation tool is particularly effective in combination with complementary approaches such as protected areas, habitat management, or mitigation of invasive non-native species. The frequency of conservation translocations has increased thirty-fold in thirty years (Armstrong et al., 2019). Future projections suggest a further two- to three-fold escalation over ten years (Swan et al., 2018), and such efforts will be a key component in the 2021–2030 United Nations Decade on Ecosystem Restoration.

Growth in conservation translocations has not only occurred in terms of the number of involved species, ecosystems, and human communities, but also in terms of the degree of innovation, the breadth of ambitions, and the depth of philosophical perspectives. Current expansion in terms of concept and magnitude has tremendous potential to push conservation translocations to an even higher level of relevance and conservation impact. Nevertheless, reintroductions, reinforcements, assisted colonisations, and ecological replacements are not without risk. Both the means and the ends of conservation translocations are often hotly debated, even on issues or approaches that appear to be clearly desirable or inherently good. Would the world want to mitigate human impacts upon species,

genuine engagement with IPLC
synthetic biology compassionate conservation
biodiversity paradox   mitigation translocation   *extinct in the wild*
contentious issues   climate change adaptation
conservation translocations
temporarily displaced species   *future opportunities*
assisted colonisation of humans   IUCN CTSG Guidelines
rewilding   conservation genomics

Figure 13.1 Word cloud of terms derived from this chapter related to the nature and associated future opportunities of contentious topics regarding conservation translocation.

make nature wilder again, be more compassionate in terms of conservation, and be more inclusive of human diversity? Many might think that these ambitions would be obviously desirable for everyone. But would they necessarily? Would it be universally understood what these value statements mean, what strategies would be required, and what trade-offs would be necessary to achieve them? Perhaps not, as even laudable approaches can be contentious depending on details or implications.

As authors, we are sufficiently involved in discussions and initiatives around the world that we frequently encounter contentious issues, which we see as tremendous growth opportunities for conservation translocations. These entail a broad perspective of issues discussed in this chapter, captured in Figure 13.1. We hope that giving what we see as an insider perspective will help to pull some of the often unpublicised issues out from the shadows so that they can be addressed by the global community for the maximum benefit of species, ecosystems, and humanity around the world.

13.2 Genes

While conservation translocations are often associated with the release or planting of individuals and the subsequent growth of populations, some of the greatest potential innovations lie within the organisms themselves on a genetic or genomic level.

13.2.1 Conservation Genomics

Until the early 2000s, conservation geneticists relied almost exclusively on methods that generate a handful of neutral genetic markers (e.g. microsatellites) to inform conservation translocations (Jamieson & Lacy, 2012). Since then, the emergence of high throughput sequencing has enabled access to tens to hundreds of thousands of genomic markers (e.g. single nucleotide polymorphisms, or SNPs; Allendorf et al., 2010). As outlined in Neaves et al. (this volume), genetic markers continue to provide important conservation tools (e.g. Overbeek et al., 2020), but the increased resolution of genome-wide diversity offered by genomic markers presents exciting opportunities to revisit old questions (e.g. detecting inter-specific hybridisation and introgression to reassess extinction risk; Forsdick et al., 2021) and explore new ones (e.g. integrating genomics to forecast translocation success under climate change; Hoffman et al., 2021; Seaborn et al., 2021).

13.2.1.1 What Aspects Are Contentious for Conservation Translocations?
A central tenet of conservation genetics is to minimise loss of genome-wide diversity over time to maintain the ability of threatened species to adapt to changing environments (adaptive potential). Despite abundant empirical and theoretical evidence that genome-wide diversity contributes to population persistence, adaptive potential, and species resilience (Frankham et al., 2017; Ralls et al., 2020; de Woody et al., 2021; García-Dorado & Caballero, 2021; Kardos et al., 2021), recent simulation-based studies have challenged the significance of genome-wide diversity in conservation and advocate for the management of functional genetic variation associated with fitness-related traits (Robinson et al., 2018; Hansson et al., 2021; Kyriazis et al., 2021; Teixeira & Huber, 2021a, 2021b). Beyond their potential to discourage the application of sound conservation genetic management recommendations based on genome-wide diversity – which includes both neutral and non-neutral (functional) genetic variation – these studies may lead to unrealistic expectations regarding the ability to characterise functional genetic variation in most species of conservation concern (Kardos & Shafer, 2018; Kardos et al., 2021). Further, even if fitness-related traits of interest are well known, and the genetic basis of these traits can be readily characterised, managing genetic variation at a small number of fitness-related traits risks losing genetic variation elsewhere in the genome (Kardos & Shafer, 2018).

13.2.1.2 Future Opportunities

Despite – and perhaps, because of – these complexities, there is good reason to explore how holistic approaches that include the characterisation of genome-wide diversity can better support conservation translocation decisions. For example, the conservation genetics community has reignited dialogue around genetic rescue (i.e. the introduction or restoration of new genetic material to small, isolated populations to reduce high genetic load; Whiteley et al., 2015). Genetic rescue was initially limited due to concerns of outbreeding depression (problems arising from the mixing of genetically different lineages). However, accumulating empirical evidence indicates that these risks are low compared to the benefits of addressing low fitness and/or high inbreeding (Ralls et al., 2018, 2020). Despite growing interest in its use as a routine conservation management tool, we still have a limited understanding of how genetic rescue works, especially in animal populations (Undin et al., 2021). Namely, does the introduction of new genetic material primarily increase fitness through the alleviation of high genetic load or the introgression of adaptive genetic variants? Empirical evidence for a small number of well-studied vertebrates suggests that it may be both (e.g. Miller et al., 2012; Fitzpatrick et al., 2020), but it remains to be seen whether these findings are broadly applicable.

13.2.2 Synthetic Biology

Synthetic biology is a fast-moving field where engineering principles are applied to the design of biological parts and systems to result in new and desired traits (Piaggio et al., 2017). Here, we build on recent perspectives that consider the future of synthetic biology, including de-extinction, in conservation (e.g. Johnson et al., 2016; Piaggio et al., 2017; Gaywood & Stanley-Price, this volume). We restrict our considerations to the two approaches most likely to impact conservation translocations in the foreseeable future: gene editing to mitigate extrinsic threats (e.g. pathogens) to population persistence in the wild (Kosch et al., 2019; Powell et al., 2019; Samuel et al., 2020), and cloning to restore lost genetic diversity (Wisely et al., 2015; Tunstall et al., 2018). Whereas gene editing targets functional genes of interest (e.g. genes underlying pathogen resistance) for precision editing (e.g. using CRISPR/Cas9 technology to insert synthetic alleles to enhance pathogen resistance) (Piaggio et al., 2017), cloning targets cryopreserved cell lines sourced from individuals harbouring genetic diversity no longer found in contemporary populations for somatic cell nuclear transfer (e.g. see Figure 1, Wisely et al., 2015).

13.2.2.1 What Aspects Are Contentious for Conservation Translocations?
To date, dialogue regarding the tangled web of scientific, ethical, legal, political, and economic issues raised by synthetic biology has been largely restricted to de-extinction, or the resurrection of functional proxies of extinct species for conservation benefit (e.g. IUCN, 2016; Seddon, 2017; Seddon, this volume). However, early indicators suggest that many concerns regarding de-extinction are broadly applicable to synthetic biology (Taylor et al., 2017; Sandler et al., 2021). Further, like de-extinction, synthetic biology 'fixes' such as gene editing and cloning raise interconnected issues related to the management of extant threatened species including conservation translocations (Seddon et al., 2014a; Seddon, 2017; Seddon, this volume) and conservation genetics (Steeves et al., 2017). There is also concern regarding the diversion of support, including funds, away from more 'traditional' approaches to improve conservation outcomes (e.g. Bennett et al., 2017; Iacona et al., 2017).

13.2.2.2 Future Opportunities
We are cautiously optimistic about the opportunities presented by gene editing, especially for extrinsic threats that cannot be readily mitigated in the wild like chytrid fungus *Batrachochytrium dendrobatidis* in extinct-in-the-wild southern corroboree frogs *Pseudophryne corroboree* (Kosch et al., 2019) and chestnut blight *Cryphonectria parasitica* in American chestnut trees *Castanea dentata* (Powell et al., 2019, but see Westbrook et al., 2020). There is also growing interest in gene editing for climate change mitigation for species such as corals (Cleves et al., 2020) that cannot be managed using 'traditional' approaches like assisted colonisation (Seddon et al., 2014b) and adaptive introgression (e.g. Hamilton & Miller, 2016). However, gene editing will be only effective for well-characterised traits controlled by few genes of large effect, and most traits of conservation interest are likely to be polygenic (i.e. controlled by many genes of small effect).

For a select group of genetically depauperate species with cryopreserved cell lines and appropriate surrogates, cloning is becoming a viable option to resurrect single individuals (e.g. black-footed ferret *Mustela nigripes*, Phelan & Ribbons, 2021; Przewalski's horse *Equus ferus przewalskii*, Aridi, 2020; and northern white rhinoceros *Ceratotherium simum cottoni*, Tunstall et al., 2018). However, it remains to be seen whether the introduction of lost genetic diversity from a small number of individuals will enhance the long-term recovery of these species.

Given the growing momentum for bold innovation in conservation (e.g. 'intended consequences'; Phelan et al., 2021a, 2021b), the answers to some of these questions are to be revealed in due course, beginning with charismatic species at risk. However, from a conservation genetics perspective, in a rapidly changing world, the most effective conservation translocation strategies for the vast majority of threatened species will be those that focus on preventative measures to reduce the loss of remaining genome-wide diversity over time.

13.3 Species

While conservation translocations aim to create benefits at a population or species level, important decisions are made at the level of individual organisms that may not only yield profound benefits but also risks for conservation.

13.3.1 Compassionate Conservation

Compassionate conservation deals with the interface of animal welfare and animal rights through an emphasis on guiding principles to do no harm, that individuals matter, to value all wildlife, and to pursue peaceful coexistence (Bekoff, 2013; Harrington et al., this volume). Compassionate conservation argues that emphasis on the well-being of collectives, such as populations, is untenable if the needs of individuals to achieve such aims are not taken into sufficient consideration (Wallach et al., 2018).

13.3.1.1 What Aspects Are Contentious for Conservation Translocations?
The welfare of individual animals and the status of populations for species recovery are naturally interconnected. The well-being of wild individuals is likely to improve survival and reproduction, which in turn can benefit the growth, expanse, or viability of populations. Indeed, animal welfare and conservation can complement each other perfectly when objectives and techniques to benefit individuals align fully with objectives and techniques that benefit populations.

Nevertheless, problems can arise if values, objectives, or strategies surrounding animal welfare conflict with conservation goals on a population or species level. Such conflict can be rooted in a difference of ethical perspectives, as animal ethics will emphasise the individual while environmental ethics emphasise the persistence and function of species in ecosystems (Harrington et al., 2013). These challenges come to a head in the context of animal rights, which differ from animal welfare in the sense that animals are attributed fundamental rights that are beyond compromise regardless of conservation context (Harrington et al., this volume).

Individual animals can experience increased stress or injury at differing stages of the conservation translocation process. These can relate, for example, to capture of source animals from wild populations, housing and husbandry conditions in captive-breeding programmes, transport to release sites, or suboptimal conditions at release sites. Affected species are not only the release candidates, but also animals used to train carnivores in hunting techniques before release, or predators that are sometimes deterred or killed at release sites to improve the survival of released animals (Harrington et al., 2013; Harrington et al., this volume).

The evolution of compassionate conservation has grappled with differing perspectives regarding fundamental values surrounding animal welfare, animal rights, and conservation. Some will frequently prioritise the needs of individuals above those of conservation outcomes (Wallach et al., 2018), while others simply advocate for conservation to be done more compassionately with increasing consideration of individual welfare (Baker, 2017). More extreme compassionate conservation views, aligned fundamentally with animal rights advocacy, are increasingly perceived as obstructing conservation tools, including conservation translocations, to prevent biodiversity loss, recover species, and restore ecological function (Hayward et al., 2019; Johnson et al., 2019).

13.3.1.2 Future Opportunities
Perceptions, perspectives, and values regarding the treatment of animals change over time and differ among societies with disparate cultural perspectives. Animal welfare will likely receive ever greater attention on a global level, and be of increasing interest for conservation in general and for conservation translocations in particular. Practitioners should embrace increased care and consideration for individual animals at every opportunity, and continuous refinements regarding animal welfare should be an ongoing priority through all stages of conservation translocations.

Conservation translocation practitioners should be clear about the fundamental objectives of their programmes in terms of desired benefits for populations, species, or ecological function. If increasing emphasis on animal rights comes to hinder the pursuit of those objectives, practitioners should be clear and transparent about the ramifications that an individual-only focus would have upon the fate of wild animal populations and their likelihood of persistence. Given the current biodiversity crisis and urgent calls for courageous action to avert further losses in species diversity and ecosystem processes, solutions must be developed that embrace and prioritise improvements for animal welfare instead of misrepresenting animal rights to block important conservation tools such as conservation translocations.

Aside from improving welfare and, potentially, conservation outcomes, focus upon individual animals could have an additional benefit in terms of attracting support for conservation translocation programmes. While lion numbers decreased by tens of thousands of individuals with relatively little attention in a 43 per cent decline across two decades (Lindsey et al., 2018), the illegal killing of one trophy-hunted lion called 'Cecil' caused worldwide outrage that dramatically promoted lion conservation in general. Although it may seem counter-intuitive or even 'unscientific', conservation translocation practitioners should increasingly engage the public with inspiring stories of individual animals instead of only trying to harness enthusiasm around population-level metrics such as growth rates or extinction probabilities.

13.3.2 Mitigation Translocation

Mitigation translocation 'involves the removal of organisms from habitat due to be lost through anthropogenic land use change and release at an alternate site' (IUCN, 2013). If conservation benefit is planned for, mitigation translocation can constitute one of the four conservation translocation types depending on the location and purpose (IUCN, 2013). Germano et al. (2015) characterise mitigation-driven translocations as occurring 'in response to legislation or governmental regulation, with the intent of reducing a development project's effects on animals or plants'. That said, mitigation translocations can also be conducted simply out of concern for organisms without mandated legislation.

13.3.2.1 What Aspects Are Contentious for Conservation Translocations?
Sometimes the extent or frequency of mitigation translocations is stunning. For example, over 70,000 IUCN Red List 'vulnerable' gopher

tortoises *Gopherus polyphemus* have been moved in the USA alone (Germano et al., 2015), and nearly 24,000 reptiles comprised of four species were moved for one port development in the UK. Thirty per cent of plant translocations for 376 taxa in Australia were mitigation translocations (Silcock et al., 2019), and 26 per cent of 154 plant species translocated for conservation purposes in China were due to the Three-Gorge Dam and a related hydropower project (Liu et al., 2015). The numbers of individuals moved in mitigation translocations vastly outnumber those in other planned conservation translocation programmes (Germano et al., 2015). These programmes are controversial in many respects. Some believe they should not be needed for imperiled species, because developments that could harm imperiled species should simply be prevented. Theoretically, the apparent 'mitigation' can be seen as a perfect trade-off where impacts at one site are compensated for through transferred benefits to another site. However, this is rarely the case, and an alternative view is that these activities paint a prettier picture of apparent conservation than the evidence corroborates. Even investments of up to 43 million British pounds annually for a single species (the great crested newt *Triturus cristatus*) can yield little evidence of benefiting populations (Lewis et al., 2017). Finally, mitigation translocations can be controversial because they are not necessarily desirable or defensible, even in light of certain development impacts on imperiled species. In Canada, collaborations with government for the development of amphibian mitigation guidelines were stalled by over a year because it became clear that habitat, genetic, or disease concerns could be so great that extraction into *ex situ* facilities or euthanasia would be preferable alternatives to translocation (Randall et al., 2018).

13.3.2.2 Future Opportunities

Because mitigation translocations are normally associated with economically advantageous industrial developments, and because governments often require associated environmental mitigations or offsets, allocated funds can far surpass those of typical conservation translocations. Germano et al. (2015) pointed out that expenditures for mitigation translocations of green and golden bell frogs *Ranoidea aurea*, categorised as 'vulnerable' on the IUCN Red List, were four times greater over 15 years than those for all Australian amphibian conservation translocation programmes combined. Post-release data for mitigation translocations often do not exist, or if they do, are proprietary for industry-hired environmental consultants or poorly tracked by government agencies (Germano et al., 2015). Collaborations should be established between governments, consultants, and academics to improve the conduct and transparency of these activities. Indeed, the resources and sample sizes associated with many mitigation translocations could yield innovations for translocation approaches and procedures. Finally, integration of mitigation translocations on a species- instead of a site-specific level could potentially yield opportunities where translocated individuals are integrated into broader recovery programmes. For example, situations may exist where the use of some captured individuals as donors would benefit the genetics or demography of *ex situ* programmes that are run for conservation translocations into the wild.

13.3.3 Temporarily Displaced Species

Species can be displaced from natural habitats as a result of anthropogenic activities such as illegal trade, human–wildlife conflict, injury, or merely a perceived need for rescue. However, displacement can also result from non-anthropogenic activities such as environmental disasters like floods, hurricanes, volcanic eruptions, or tsunamis. In some cases, individuals that are brought into human care for a variety of reasons can be released quickly without prolonged confinement, while in other situations rehabilitation or euthanasia needs to be explored.

13.3.3.1 What Aspects Are Contentious for Conservation Translocations?
The scale at which species are taken into human care can be staggering. In New York, USA, between 2012 and 2014, a total of 59,370 wildlife rehabilitation cases were recorded (Hanson et al., 2019). In Italy, over a period spanning six years between 2015 and 2020, a total of 5,881 wild animals belonging to 162 species were transferred to a rehabilitation centre, with 8 per cent of the species listed under CITES and/or the IUCN Red List (Dessalvi et al., 2021). Indeed for 2010 and 2014 combined, a total of 64,143 individual animals from 359 species were confiscated by 54 countries that are party to CITES. These numbers are vast underestimates as reporting by countries is not strictly enforced and many lucrative species may enter the illegal trade again (D'Cruze & Macdonald, 2016). The IUCN guidelines clearly state that surplus animals are not a reason to conduct conservation translocations and should not be the driver (IUCN, 2013, p. 7). Individuals that are placed

into rescue and/or rehabilitation facilities may eventually be unreleasable because of disease, genetic, or behaviour concerns for the release candidates themselves or for potential recipient ecosystems. For some individuals, there could be the possibility of quick release within hours or days at proximate, clearly suitable release sites. For other individuals, including those from species at risk, it can remain perpetually unclear whether release is possible, desirable, responsible, or permitted. These complexities amplify when organisms are moved across international boundaries. Although ghost orchids *Dendrophylax lindenii* collected from the site of a low-cost housing development scheme in the Grand Cayman Islands were destined for Kew Gardens in the UK, insufficient paperwork resulted in confiscation by Dutch authorities and subsequent housing at a botanical garden in Holland (Connolly, 2009); as such, the national jurisdiction inadvertently changed for potential conservation translocations back into the wild. Frequently, individuals remain under care for extensive periods of time, often in suboptimal conditions, until they eventually succumb to poor conditions, illness, or injury. In the case of plants like orchids, which can have a lifespan of more than 100 years, it is important to ensure that they are housed in facilities with secure funding and long-term viability.

13.3.3.2 Future Opportunities
The opportunities are two-fold. First, rehabilitation and confiscation centres should be integrated more closely with zoological or botanical institutions, government agencies, and academics to determine when individuals of imperiled species could become part of responsible conservation release programmes. Given the massive number of individuals from endangered species that come into human hands every year, the inclusion of even a small proportion could yield large gains for the recovery of many wild populations. Second, there should be clear guidance that would facilitate quick releases of endangered species, particularly when these are recently displaced and when they are sourced near suitable habitat. With sufficient monitoring, even opportunistic trial releases could be used to test release methodologies that could prevent the waste of precious resources and/or individual organisms.

13.4 Ecosystems

Conservation translocations are escalating in frequency, and so are the associated benefits on an ecosystem scale. Whether, where, and how interventions occur is a reflection of values and prioritisation made by involved stakeholders, but key sectors of society or geographic regions are sometimes marginalised on local or global scales. Conservation translocations should integrate more diverse perspectives, involve well-planned and monitored actions on an ecosystem level, and focus on global regions where such actions are needed but currently under-represented.

13.4.1 Genuine Engagement with Indigenous Peoples and Local Communities

Recent conservation research, policy, and practice reflect the need for biocultural approaches that link biological, cultural, and linguistic systems to restore and enhance biodiversity in a rapidly changing world (Díaz et al., 2019; Verschuuren et al., 2021). There is also growing recognition by western-trained researchers and practitioners that weaving diverse ways of knowing and seeing the world can lead to better conservation outcomes (Tengö et al., 2017; Herse et al., 2020), especially for culturally significant species (Hayek et al., 2016; Hamilton et al., 2020; Herse et al., 2021; Rayne et al., 2020).

13.4.1.1 What Aspects are Contentious for Conservation Translocations?
Hayek et al. (2016) and Hamilton et al. (2020) are exemplars of genuine engagement with Indigenous Peoples to integrate Indigenous values in structured decision-making (SDM) to inform conservation translocations of culturally significant species. Indeed, step changes like these reflect a growing responsiveness to the needs and aspirations of Indigenous Peoples.

However, even values-based approaches like SDM embed western science and are centred on western perspectives. For example, these approaches are generally initiated and led by western-trained researchers and practitioners, while decision makers are often government agencies that subsequently lead conservation planning and implementation.

We suspect that the general concept of greater Indigenous involvement in conservation translocations will be widely applauded. However, we challenge the conservation translocation community to look beyond the integration of Indigenous values into existing western frameworks and towards the co-adaptation of existing approaches or the co-development of new approaches that weave western science and Indigenous knowledge systems, including customary practices, processes, and language (Rayne et al., 2020). We anticipate that such a transformative change – from stakeholder engagement to partnership – may be contentious for some western-trained scientists, in part, because Indigenous knowledge systems will be weighed equally with western science and practice, and for some government managers, because genuine co-management will entail power-sharing in decision-making (Pelai et al., 2021).

13.4.1.2 Future Opportunities
Rayne et al. (2020) present a novel framework for co-developing conservation translocations that centre Indigenous Peoples and knowledge systems – including customary practices, processes, and language – through the Mi'kmaq principle of Etuaptmumk, or 'two-eyed seeing', described by Mi'kmaq Elder Dr. Albert Marshall as 'learning to see from one eye with the strengths of Indigenous knowledge and ways of knowing, and from the other eye with the strengths of Western knowledge and ways of knowing ... and learning to use both these eyes together, for the benefit of all' (Reid et al., 2021).

The framework presented by Rayne et al. (2020) extends the Conservation Translocation Specialist Group (CTSG) guidelines to include co-designed objectives and success indicators, co-designed translocation strategies, collective implementation, and sustained post-release monitoring and iterative co-management. Genuine partnership lies at the core of the framework, where relationships built on mutual trust and respect are nurtured, iterative collective decision-making is embedded, and equitable benefit-sharing is prioritised. Indeed, we anticipate that such approaches may be readily extended to conservation translocations co-developed with non-Indigenous knowledge holders with intergenerational and/or custodial connections to place (Hill et al., 2020; Pelai et al., 2021).

13.4.2 Biodiversity Paradox

Biodiversity needs to be conserved on a global level for nature and for human benefits, such as ecosystem services that help economies, and healthy ecosystems that help reduce the incidence of disease such as the current COVID-19 pandemic. Habitat degradation or loss can accelerate climate change (Turner et al., 2007), while stable ecosystems in turn can be important for climate change mitigation (Shaw, 2018). According to Mittermeier et al. (2011), 35 global regions now meet the biodiversity hotspot criteria of holding at least 1500 endemic plant species while having lost over 70 per cent of their original habitat extent. A paradox lies in the fact that biodiversity is generally lower, while financial and institutional capacity is generally higher in western countries than in tropical countries (Rodriguez, 2017; Rodriguez et al., 2022).

13.4.2.1 What Aspects Are Contentious for Conservation Translocations?

We analysed flora and fauna case studies from the seven issues of the *Global Conservation Translocation Perspectives* (Soorae, 2008, 2010, 2011, 2013, 2016, 2018, 2021); covering more than 400 distinct species, these data are likely to be a strong indicator of conservation translocation distribution in general. Overall analyses show that the recorded conservation translocations are widely distributed on a global level for animals and plants. However, the majority of plant projects are reported from Australia, followed by the USA (Figure 13.2A). Other areas with high numbers of projects are Western Europe, Brazil, and parts of Central Asia into China, while there is a remarkable absence of projects from parts of Africa and Central Asia. Animal translocations follow a similar trend to those of plants, with a particular scarcity of initiatives in Africa (Figure 13.2B). Some would argue that protected area strategies should be pre-eminent in biodiverse regions instead of more species-focused tools like conservation translocations. However, trends of some 3,000 wild populations show a consistent decline in average species abundance of about 40 per cent between 1970 and 2000, including in protected areas (UNEP, 2006). Many of the biodiversity hotspots are not covered by the projects reported in our analysis (Mittermeier et al., 2011) and more efforts are needed to bridge this gap between area-focused and species-focused conservation. Tropical developing countries have fewer resources and funding for biodiversity related research and recovery work. Partly as a result, we suspect that there may also be a bias in terms of the types of animals that are translocated in tropical countries. While more affluent regions have a high proportion of invertebrate conservation translocations for example (Swan et al., 2016; Nason et al., 2021), we suspect that tropical countries are likely to focus on larger and higher profile species that are comparatively well studied. As is the case for many other conservation strategies, the global distribution of conservation translocations tends to reflect resource availability instead of conservation need.

From Genes to Ecosystems and Beyond · 395

(A)

(B)

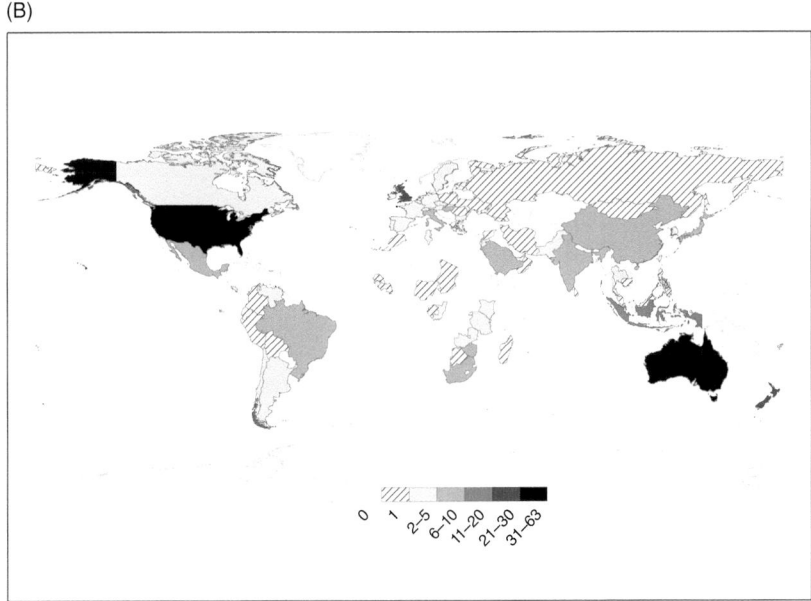

Figure 13.2 Geographic distribution of conservation translocations from the IUCN Conservation Translocation Specialist Group global conservation translocation case studies based on (a) 99 species of plants and (b) 322 species of animals.

13.4.2.2 Future Opportunities

Locke et al. (2019) introduced an implementation concept to deliver global biodiversity conservation and sustainable use. This '3Cs framework' outlines three key conditions:

- C1 – increase conservation efforts to protect and conserve endangered species and ecosystem fragments.
- C2 – establish ecologically representative and well-connected systems of protected areas.
- C3 – retain ecological integrity and global processes such as carbon storage, rainfall generation, and large migrations amongst others.

Conservation translocations have a role to play in supporting such conditions, but more biodiversity surveys and research are needed in countries that may lack necessary funds and expertise. Such foundational work would help to determine where and when conservation translocations could be used in synergy with other strategies such as area-based conservation, invasive non-native species mitigation, and combating of wildlife crime. An important step is to accelerate knowledge exchange to support the building of conservation translocation capacity in countries where such needs and opportunities are outstanding. To this effect the IUCN Conservation Translocation Specialist Group initiated international training programmes in 2016 that have been offered in four countries to date and will increasingly be focused on biodiverse, tropical countries.

13.4.3 Rewilding

Rewilding in its broadest form is about restoring or increasing 'wildness'. That said, exactly why, how, and where this should occur is debated in the context of evolving concepts regarding definitions and purpose.

13.4.3.1 What Aspects are Contentious for Conservation Translocations?

It is clear that rewilding and conservation translocations intersect in a positive way, but the nature of this alignment is difficult to articulate; whereas the four different forms of conservation translocation are clearly defined (IUCN, 2013; Armstrong et al., 2019), rewilding definitions have been changing since the term was introduced thirty years ago. Passive or ecological rewilding need not involve human-mediated translocation; for example, abandoned farms have gradually reverted to more natural conditions over time (Corlett, 2016; Seddon, this volume). However, the majority of rewilding activities have an active component that includes translocations of single or multiple species with the objectives of wider ecosystem benefit – a subset of these are known as 'trophic rewilding' (Guyton et al., 2020; Seddon, this volume). Although some see rewilding as relatively novel, others suggest that it is a mere subset of restoration ecology (Hayward et al., 2019). Proponents emphasise that restoration of ecological processes is fundamentally important to rewilding, and that interventions may have indirect or unexpected results (Perino et al., 2019). While decision-making in the face of uncertainty is central to planning for specific conservation translocation objectives (Ewen et al., this volume), rewilding goes further in embracing and even celebrating the lack of predictable consequences to allow for future evolutionary potential on an ecosystem level (Corlett, 2016; Pettorelli et al., 2019). One fundamental challenge lies in the introduction of species beyond their indigenous range, where the risk of non-native species effects on biodiversity can be managed under single-species assisted colonisation of conservation translocations (Brodie et al., 2021b), but is potentially impossible to control in a multi-species or even novel ecosystem rewilding context. Finally, conservation translocations embrace active management in terms of moving, supporting, and monitoring released individuals to refine subsequent conservation strategies that ultimately improve species viability or ecological function. In contrast, rewilding approaches see human intervention as sometimes necessary but generally undesirable; the end stages of rewilding certainly aim towards systems that are free of human interference (Corlett, 2016; Carver et al., 2021).

13.4.3.2 Future Opportunities

Despite uncertainties or variation in rewilding approaches, the term and its associated desire to restore wildness has captured the public imagination, especially in the context of severely degraded ecosystems. The reintroduction of species that are wide-ranging, are keystones, or yield important ecological functions has underpinned both the origin and the current direction of the rewilding movement (Corlett, 2016; Carver et al., 2021). Conservation translocations are frequently referred to as an inspirational source of 'conservation optimism'; as such they serve as potential sentinels not only for future ecological benefits but also for further capturing public support for responsible rewilding actions. Although reintroductions and reinforcements of many species, including marine plants and algae, are pursued primarily to restore ecological function (Swan et al., 2016), conservation translocations have sometimes been seen as focused on single species, but increased recognition of ecosystem-level contributions with responsible rewilding initiatives could help to garner even greater support and interest in the future (Gaywood and Stanley-Price, this volume). Ecological replacements – the intentional movement and release of species beyond indigenous range to perform specific ecological functions such as those of extinct species – may help to advance rewilding objectives while already falling under established IUCN guidelines (IUCN, 2013; Jones et al., this volume).

13.5 Beyond the Range

Conservation translocations are traditionally known for reintroductions, but challenges and opportunities are changing to the extent that releases within indigenous range are not always preferable or necessarily most feasible.

13.5.1 Extinct in the Wild

'A taxon is "Extinct in the Wild" when it is known only to survive in cultivation, in captivity or as a naturalized population (or populations) well outside the past range'(IUCN Standards and Petitions Committee, 2019). 'Extinct in the Wild' species normally exist only under human care in zoos or botanical gardens, although by the IUCN definition they could also be in nature but well outside of their indigenous range.

13.5.1.1 What Aspects Are Contentious for Conservation Translocations?
Aspects relating to the recovery and classification of 'Extinct in the Wild' species are contentious in several ways. The survival of these animal species depends almost entirely on zoos, but some find the idea of confining animals for ultimate conservation purposes such as reintroductions problematic under the 'compassionate conservation' umbrella (Wallach et al., 2018). Indeed, zoos and botanical gardens represent a last stand for 'Extinct in the Wild' species, and a lack of resources in such organisations could ultimately lead to species extinction (Trask et al., 2020). Moreover, the suitability of 'Extinct in the Wild' species for release may dwindle over time due to factors such as inbreeding depression, evidenced in the sihek (Guam kingfisher *Todiramphus*

cinnamominus; Trask et al., 2021), or rapid loss of behavioural suitability of nearly extinct Vancouver Island marmots *Marmota vancouverensis* (Dixon-McCallum et al., 2021; Ewen et al., this volume). While some species may suddenly become extinct in the wild, such as the recent Loa water frog *Telmatobius dankoi* extirpation due to new dams in Chile, others have been maintained under human care for decades.

Recovery of 'Extinct in the Wild' species should first and foremost seek to re-establish populations within indigenous range if factors leading to their original decline can be addressed, such as in the case of the Przewalski's horse *Equus ferus przewalskii* (Kaczensky et al., 2017).

A fundamental issue why many species remain 'Extinct in the Wild' is that the threats causing their original extirpation are difficult to remove or mitigate, such as in the case of the Guam kingfisher which was driven to near extinction by the invasive brown tree snake *Boiga irregularis*. Assisted colonisation beyond indigenous range may be helpful as it is used increasingly for animals and especially plants (Hällfors & Dalrymple, this volume) to give species a foothold in the wild for eventual return to their indigenous range, or to help establish populations of permanent conservation value. Nevertheless, the trade-off of risks surrounding the potential invasiveness of such release candidates has prompted calls for global policy (Brodie et al., 2021a), which in turn have prompted debate (Brodie et al., 2021b; Ricciardi & Simberloff, 2021). Finally, as per the IUCN Red List definition, populations that are created for conservation purposes outside of indigenous range, such as for the Guam rail *Hypotaenidia owstoni*, currently are not counted in terms of assessing IUCN species status.

13.5.1.2 Future Opportunities

Well-planned conservation translocations rarely have unintended consequences (Novak et al., 2021), and while assisted colonisation has already prevented the extinction of some species, damaging consequences for recipient ecosystems from these activities are extremely rare (Brodie et al., 2021a). Increased opportunities in policy and practice to allow for experimental releases beyond indigenous range could open the door for 'Extinct in the Wild' species that otherwise may be managed to extinction in the institutions where they remain. Such releases would not only need to consider potential impacts on endemic flora and fauna, but would also need to carefully establish trade-offs between the benefits of newly established populations and the risk that donor populations under human care could go extinct (Trask et al., 2020). Moreover, if refinements to the Red List categories, or alignment for considerations under IUCN Green Status assessments (Grace et al., 2021), officially integrated the benefit of conservation-planned extralimital populations, then down-listing to 'Critically Endangered' status would be facilitated. This in turn could help create increasing support and momentum for 'Extinct in the Wild' species in general. Because 'Extinct in the Wild' species are among the most endangered on the planet, showcased progress could help to spur public engagement, confidence, and support that could go well beyond translocations to conservation in general on a global level.

13.5.2 Climate Change Adaptation

Climate change adaptation encompasses management strategies to lessen the impacts of climate change (Owen, 2020). For conservation translocations, the primary questions are when and where translocations should occur in response to current or anticipated climate change.

13.5.2.1 What Aspects Are Contentious for Conservation Translocations?
Moving species in response to climate change is primarily contentious because risks are elevated if species are moved beyond their indigenous range, and these risks must be weighed against the counter-risks of inaction. The Bramble Cay melomys *Melomys rubicola* went extinct due to habitat loss and storm surges attributed to climate change; relocating the species from this small island to suitable mainland habitat could have prevented its extinction (Butt et al., 2020). Casazza et al. (2021) assessed potential range losses due to climate change of 188 endemic vascular plant species in Italy. Under pessimistic scenarios, they determined that up to 95 per cent of taxa would experience range losses of greater than 30 per cent. For species categorised as Vulnerable or Endangered under the IUCN Red List, 32–35 assisted colonisation attempts would be required per species to compensate for such range loss (Casazza et al., 2021).

Assisted colonisation is primarily contentious because of risks that species could become invasive (Ricciardi & Simberloff, 2021). Nevertheless, Butt et al. (2020) asked why assisted colonisation is not conducted more frequently than it has been to address clear climate change threats. They suggest that, aside from concerns about invasiveness, several factors might be at play:

1. Time frame: the effects of climate change may be perceived as being distant in the future, and hence may not necessitate action now.
2. Socio-political aspects: decision makers may be reluctant to spend significant resources on actions that could lead to societal opposition.
3. Uncertainty: practitioners may feel that it is difficult to adapt global climate change models to the local level, and species distribution models may not adequately predict ecological interactions that could affect assessments of future habitat suitability.
4. Fear of failure: attempting novel approaches and failing in the short term might be seen as more risky than being blamed sometime in the future for damages that would arise from inaction in the long term (Butt et al., 2020).

We suggest that another aspect of climate change adaptation could be controversial. While all eyes are on the controversies around moving species beyond indigenous range, it could, and perhaps should, be contentious to reintroduce species into their native range if future climate change scenarios deem conditions there to be unsuitable.

13.5.2.2 Future Opportunities
Assisted colonisation for climate change adaptation has already occurred with tree species in North America over the last two decades. The case of the torreya pine *Torreya taxifolia* has attracted a lot of attention because translocations to suitable sites were conducted by dedicated conservation groups in the alleged absence of government leadership (Brodie et al., 2021a). While scientific debate continues regarding such activities, another knowledge gap is the lack of government policy nationally or globally. Building upon the Council of Europe's Bern Convention recommendation to apply the IUCN conservation translocation guidelines (IUCN, 2013; Trouwborst et al., this volume) for climate-triggered translocations, there are now calls to develop international policy for assisted colonisation under the UN Convention on Biological Diversity (Brodie et al., 2021a). Confidence in this conservation tool should grow as the body of well-conducted research increases. This could be done by concentrating on species that could be translocated at small scales, have little invasion risk, have robust monitoring protocols, and promote

political or public support (Butt et al., 2020). Such trials are already conducted for animals such as the western swamp turtle *Pseudemydura umbrina*, which is otherwise likely to go extinct due to climate change in Australia (Mitchell et al., 2016).

13.5.3 Assisted Colonisation of Humans

Earth could be affected by a cataclysmic event such as an asteroid collision or by other disasters initiated by unsustainable activities such as climate change and a proliferation of diseases, for which the COVID-19 pandemic may have been a prelude. Stephen Hawking suggested that humans should look at colonising other planets, 'We will not establish self-sustaining colonies in space for at least the next hundred years, so we have to be very careful in this period.' This sentiment is echoed by the entrepreneur Elon Musk, who told a press conference in 2013, 'Either we spread Earth to other planets, or we risk going extinct. An extinction event is inevitable and we're increasingly doing ourselves in' (Rincon, 2018).

13.5.3.1 What Aspects Are Contentious for Conservation Translocations?
Humans colonising other planets and moons to decrease our extinction risk would constitute an 'assisted colonisation'. Related actions could result in intentional or accidental releases of other organisms. *Homo sapiens* has been a notorious coloniser, wiping out disadvantaged groups such as Neanderthals. In more recent times colonisers have threatened or eliminated many indigenous societies. Biodiversity has also suffered massive extinctions, especially because of releases of non-native species. The Israeli lunar lander Beresheet crashed on Earth's moon in April 2021. Its payload was composed of artificial tree resin with hair follicles and blood samples, some dehydrated tardigrades, and samples from major holy sites, such as the Bodhi tree *Ficus religiosa* in India. A few thousand extra dehydrated tardigrades were sprinkled onto tape that was attached to the lunar library. Tardigrades can stay viable for long periods in such dehydrated states – as such, unintentional human-facilitated translocation of species beyond Earth has begun. Increasing numbers of unmanned missions to Mars are now on the quest to discover other life forms. It is only a matter of time before humans travel to Mars or other planets while carrying food crops or animals for

eventual subsistence. And it is also only a matter of time until we discover other living beings. In humanity's inherent desire to 'go where no one has gone before', the possibility exists that true to thousands of years of past exploration, initial excitement will overwhelm careful consideration regarding the risks and benefits surrounding 'releases' of our own species onto other planets.

13.5.3.2 Future Opportunities

As missions to other planets gather pace, the question might arise: are we ready – technologically and socially – to live together peacefully on other planets? While Stephen Hawking and Elon Musk inadvertently promote such human assisted colonisation, some argue that humanity is not yet morally and ethically prepared (Billings, 2019). Just as the IUCN Guidelines for Reintroductions and Other Conservation Translocations are applicable to all species, and as taxon-specific guidelines have been created in concert, similar guidelines should be developed for humanity. As in those for other species, thoughtful definition of value-based goals should be considered in light of risks and benefits that can be evaluated and adaptively managed over time. This could help to yield successful human establishment, while reducing potentially negative impacts on life forms that may actually call those planets 'home'.

13.6 Respect, Kindness, and Collaboration to Advance the Future of Conservation Translocations

The world is faced with immense and dynamic challenges for the shared future of nature and humanity. Conservation is among the most transdisciplinary issues of our time, and this yields a diverse richness of perspectives, tools, and resources. Solutions to many challenges are not easy to develop or agree upon, and this complexity in turn creates debate or contention on approaches and outcomes. Controversies may be uncomfortable, but we believe that they are the breeding ground for fruitful and courageous innovation. We believe that such progress must be achieved with respect for differing perspectives, kindness in conduct with one another, and collaboration to have far-reaching impacts on a global scale. We have limitless hope and confidence that such approaches will yield profound benefits that will address the challenges of the present and emerging issues of the future.

13.7 Key Messages

- The use of conservation translocations as a transdisciplinary conservation tool to prevent extinction, recover populations, and restore ecological function is on the rise.
- The growing impact of reintroductions, reinforcements, assisted colonisations, and ecological replacements can be attributed to a number of factors, including an escalation of benefits for species, ecosystems, and human communities driven by bold innovations and courageous ambitions of the global conservation translocation community.
- The inclusion of diverse philosophical perspectives combined with increased need, interest, scope, and policy alignment has driven a broadening of novel approaches, innovations, and tools; however, associated aspects can be contentious.
- To advance the conservation impact of conservation translocations, we have grouped eleven of these contentious issues into three broad categories – genes, species, and ecosystems – and then reframed them as growth opportunities.
- Contentious issues can create conflict, but we suggest that identifying common ground on agreed conservation values, negotiating with respectful kindness, and advancing progress through collaboration will enable powerful advancements for effective conservation translocations in the future.

Acknowledgements

We are extremely grateful to Aisling Rayne and Lacey Hebert for their tremendous help in literature alignment and formatting, and to Typhenn Brichieri-Colombi for assistance in terms of map creation.

References

Allendorf, F. W., Hohenlohe, P. A. & Luikart, G. (2010) Genomics and the future of conservation genetics. *Nature Reviews Genetics*. 11, 697–709.

Aridi, R. (2020) Scientists Cloned an Endangered Wild Horse Using the Decades-Old Frozen Cells of a Stallion. *Smithsonian Magazine* [Online]. Available from: www.smithsonianmag.com/smart-news/save-endangered-wild-horse-species-scientists-cloned-stallion-using-its-decades-old-frozen-cells-180976069/ [Accessed 28 June 2021].

Armstrong, D. P., Seddon, P. J. & Moehrenschlager, A. (2019) Reintroduction. In Fath, B. D. (ed.) *Encyclopedia of Ecology*, 2nd ed, vol. 1. Oxford, Elsevier, pp. 458–466.

Baker, L. (2017) Translocation biology and the clear case for compassionate conservation. *Israel Journal of Ecology & Evolution*. 63, 52–60.

Bekoff, M. (2013) *Ignoring Nature No More: The Case For Compassionate Conservation*. Chicago, IL, University of Chicago Press.

Bennett, J. R., Maloney, R. F., Sleeves, T. E., Brazill-Boast, J., Possingham, H. P. & Seddon, P. J. (2017) Spending limited resources on de-extinction could lead to net biodiversity loss. *Nature Ecology & Evolution*. 1, 4.

Billings, L. (2019) Colonizing other planets is a bad idea. *Futures*. 110, 44–46.

Brodie, J. F., Lieberman, S., Moehrenschlager, A., et al. (2021a) Global policy for assisted colonization of species. *Science*. 372, 456–458.

Brodie J. F., Lieberman, S., Moehrenschlager, A., et al. (2021b) Assisted colonization risk assessment—Response. *Science*. 372, 925–926.

Butt, N., Chauvenet, A. L. M., Adams, V. M., et al. (2020) Importance of species translocations under rapid climate change. *Conservation Biology*. 35, 775–783.

Carver, S., Convery, I., Hawkins, S., et al. (2021) Guiding principles for rewilding. *Conservation Biology*. 35, 1882–1893.

Casazza, G., Abeli, T., Bacchetta, G., et al. (2021) Combining conservation status and species distribution models for planning assisted colonisation under climate change. *Journal of Ecology*. 109, 2284–2295.

Cleves, P. A., Tinoco, A. I., Bradford, J., Perrin, D., Bay, L. K. & Pringle, J. R. (2020) Reduced thermal tolerance in a coral carrying CRISPR-induced mutations in the gene for a heat-shock transcription factor. *Proceedings of the National Academy of Sciences of the United States of America*. 117, 28899–28905.

Connolly, N. (2009) Dutch keep seized orchids. Queen Elizabeth Botanic Park, Grand Cayman Islands [Online]. Available from: www.botanic-park.ky/dutch-keep-seized-orchids/ [Accessed 22 July 2021].

Corlett, R. T. (2016) Restoration, reintroduction, and rewilding in a changing world. *Trends in Ecology & Evolution*. 31, 453–462.

D'Cruze, N. & Macdonald, D. W. (2016) A review of global trends in CITES live wildlife confiscations. *Nature Conservation*. 15, 47–63.

Dessalvi, G., Borgo, E. & Galli, L. (2021) The contribution to wildlife conservation of an Italian Recovery Centre. *Nature Conservation*. 3, 1–20.

De Woody, J. A., Harder, A. M., Mathur, S. & Willoughby, J. R. (2021) The long-standing significance of genetic diversity in conservation. *Molecular Ecology*. 30, 4147–4154.

Diaz, S., Settele, J., Brondizio, E. S., et al. (2019) Pervasive human-driven decline of life on Earth points to the need for transformative change. *Science*. 366, 1327–1333.

Dixon-MacCallum, G. P., Rich, J. L., Lloyd, N., Blumstein, D. T. & Moehrenschlager, A. (2021) Loss of predator discrimination by critically endangered Vancouver Island marmots within five generations of breeding for release. *Frontiers in Conservation Science*. 2. https://doi.org/10.3389/fcosc.2021.718562

Ewen, J. G., Canessa, S., Converse, S. J. & Parker, K. A. (2023) Decision-making in animal conservation translocations: biological considerations and beyond. In Gaywood, M. J., Ewen, J. G., Hollingsworth, P. M. and Moehrenschlager, A. (eds.) *Conservation Translocations*. Cambridge, Cambridge University Press.

Fitzpatrick, S. W., Bradburd, G. S., Kremer, C. T., Salerno, P. E., Angeloni, L. M. & Funk, W. C. (2020) Genomic and fitness consequences of genetic rescue in wild populations. *Current Biology*. 30, 517–526.

Forsdick, N. J., Cubrinovska, I., Massaro, M. & Hale, M. L. (2021) Microsatellite genotyping detects extra-pair paternity in the Chatham Island Black Robin, a highly inbred, socially monogamous passerine. Emu - Austral Ornithology. 1, 143–150.

Frankham, R., Ballou, J. D., Ralls, K., et al. (2017) *Genetic Management of Fragmented Animal and Plant Populations Introduction*, New York, Oxford University Press.

García-Dorado, A. & Caballero, A. (2021) Neutral genetic diversity as a useful tool for conservation biology. *Conservation Genetics*. 22, 541–545.

Gaywood, M. J. & Stanley-Price, M. (2023) Moving species: reintroductions and other conservation translocations. In Gaywood, M. J., Ewen, J. G., Hollingsworth, P. M. and Moehrenschlager, A. (eds.) *Conservation Translocations*. Cambridge, Cambridge University Press.

Germano, J. M., Field, K. J., Griffiths, R. A., et al. (2015) Mitigation-driven translocations: are we moving wildlife in the right direction? *Frontiers in Ecology and the Environment*. 13, 100–105.

Grace, M. K., Akçakaya, H. R., Bennett, E. L., et al. (2021) Testing a global standard for quantifying species recovery and assessing conservation impact. *Conservation Biology*. 35, 1833–1849.

Guyton, J. A., Pansu, J., Hutchinson, M. C., et al. (2020) Trophic rewilding revives biotic resistance to shrub invasion. *Nature Ecology & Evolution*. 4, 712–720.

Hällfors, M. & Dalrymple, S. E. (2023) Assisted colonisation and ecological replacement. In Gaywood, M. J., Ewen, J. G., Hollingsworth, P. M. and Moehrenschlager, A. (eds.) *Conservation Translocations*. Cambridge, Cambridge University Press.

Hamilton, J. A. & Miller, J. M. (2016) Adaptive introgression as a resource for management and genetic conservation in a changing climate. *Conservation Biology*. 30, 33–41.

Hamilton, T. M., Canessa, S., Clark, K., et al. (2020) Applying a values-based decision process to facilitate comanagement of threatened species in Aotearoa New Zealand. *Conservation Biology*. 12, 24–36.

Hanson, M., Hollingshead, N., Schuler, K., Siemer, W. F., Martin, P. & Bunting, E. M. (2019) Species, causes, and outcomes of wildlife rehabilitation in New York State. *bioRxiv*. 8, 184–197.

Hansson, B., Morales, H. E. & Van Oosterhout, C. (2021) Comment on "Individual heterozygosity predicts translocation success in threatened desert tortoises". *Science*. 372, 234–238.

Harrington, L. A., Moehrenschlager, A., Gelling, M., Atkinson, R. P. D., Hughes, J. & Macdonald, D. W. (2013) Conflicting and complementary ethics of animal welfare considerations in reintroductions. *Conservation Biology*. 27, 486–500.

Harrington, L. A., Lloyd, N. & Moehrenschlager, A. (2023) Animal welfare, animal rights, and conservation translocations: moving forward in the face of ethical dilemmas. In Gaywood, M. J., Ewen, J. G., Hollingsworth, P. M. and Moehrenschlager, A. (eds.) *Conservation Translocations*. Cambridge, Cambridge University Press.

Hayek, T., Stanley Price, M. R., Ewen, J. G., Lloyd, N., Saxena, A. & Moehrenschlager, A. (2016) *An Exploration of Conservation Breeding and Translocation Tools to Improve the Conservation Status of Boreal Caribou Populations in Western Canada*. Calgary, Alberta, Canada, Centre for Conservation Research, Calgary Zoological Society.

Hayward, M. W., Callen, A., Allen, B. L., et al. (2019) Deconstructing compassionate conservation. *Conservation Biology*. 33, 760–768.

Herse, M. R., Lyver, P. O., Scott, N., et al. (2020) Engaging Indigenous Peoples and Local Communities in environmental management could alleviate scale mismatches in social–ecological systems. *BioScience*. 70, 699–707.

Herse, M. R., Tylianakis, J. M., Scott, N. J., et al. (2021) Effects of customary egg harvest regimes on hatching success of a culturally important waterfowl species. *People and Nature*. 3, 499–512.

Hill, R., Adem, Ç., Alangui, W. V., et al. (2020) Working with Indigenous, local and scientific knowledge in assessments of nature and nature's linkages with people. *Current Opinion in Environmental Sustainability*. 43, 8–20.

Hoffmann, A. A., Weeks, A. R. & Sgro, C. M. (2021) Opportunities and challenges in assessing climate change vulnerability through genomics. *Cell*. 184, 1420–1425.

Iacona, G., Maloney, R. F., Chades, I., Bennett, J. R., Seddon, P. J. & Possingham, H. P. (2017) Prioritizing revived species: what are the conservation management implications of de-extinction? *Functional Ecology*. 31, 1041–1048.

IUCN (2013) *Guidelines for Reintroductions and Other Conservation Translocations. Version 1.0*. Gland, Switzerland, IUCN Species Survival Commission.

IUCN (2016) *IUCN SSC Guiding Principles on Creating Proxies of Extinct Species for Conservation Benefit*. Gland, Switzerland, IUCN Species Survival Commission.

IUCN Standards and Petitions Committee (2019) *Guidelines for Using the IUCN Red List Categories and Criteria* [Online], Version 14. Available from: www.iucnredlist.org/resources/redlistguidelines [Accessed 21 June 2021].

Jamieson, I. G. & Lacy, R. C. (2012) Managing genetic issues in reintroduction biology. *Reintroduction Biology*. 6, 441–475.

Johnson, J. A., Altwegg, R., Evans, D. M., et al. (2016) Is there a future for genome-editing technologies in conservation? *Animal Conservation*. 19, 97–101.

Johnson, P. J., Adams, V. M., Armstrong, D. P., et al. (2019) Consequences matter: compassion in conservation means caring for individuals, populations and species. *Animals*. 9, 8–15.

Jones, C. J., Tatayah, V., Moorhouse-Gann, R., Griffiths, C., Zuël, N. & Cole, N. (2023) Slow and steady wins the race: using non-native tortoises to rewild islands off Mauritius. In Gaywood, M. J., Ewen, J. G., Hollingsworth, P. M. and Moehrenschlager, A. (eds.) *Conservation Translocations*. Cambridge, Cambridge University Press.

Kaczensky, P., Burnik Šturm, M., Sablin, M. V., et al. (2017) Stable isotopes reveal diet shift from pre-extinction to reintroduced Przewalski's horses. *Scientific Reports*. 7, 5950–5959.

Kardos, M. & Shafer, A. B. A. (2018) The peril of gene-targeted conservation. *Trends in Ecology & Evolution*. 33, 827–839.

Kardos, M., Armstrong, E., Fitzpatrick, S. W., et al. (2021) The crucial role of genome-wide genetic variation in conservation. *Proceedings of the National Academy of Sciences.* 118, e2104642118.

Kosch, T. A., Silva, C. N. S., Brannelly, L. A., et al. (2019) Genetic potential for disease resistance in critically endangered amphibians decimated by chytridiomycosis. *Animal Conservation.* 22, 238–250.

Kyriazis, C. C., Wayne, R. K. & Lohmueller, K. E. (2021) Strongly deleterious mutations are a primary determinant of extinction risk due to inbreeding depression. *Evolution Letters.* 5, 33–47.

Lewis, B., Griffiths, R. A. & Wilkinson, J. W. (2017) Population status of great crested newts (*Triturus cristatus*) at sites subjected to development mitigation. *Herpetological Journal.* 27, 133–142.

Lindsey, P. A., Miller, J. R. B., Petracca, L. S., et al. (2018) More than $1 billion needed annually to secure Africa's protected areas with lions. *Proceedings of the National Academy of Sciences of the United States of America.* 115, E10788–E10796.

Liu, H., Ren, H., Liu, Q., Wen, X. Y., Maunder, M. & Gao, J. Y. (2015) Translocation of threatened plants as a conservation measure in China. *Conservation Biology.* 29, 1537–1551.

Locke, H., Ellis, E. C., Venter, O., et al. (2019) Three global conditions for biodiversity conservation and sustainable use: an implementation framework. *National Science Review.* 6, 1080–1091.

Miller, J. M., Poissant, J., Hogg, J. T. & Coltman, D. W. (2012) Genomic consequences of genetic rescue in an insular population of bighorn sheep (*Ovis canadensis*). *Molecular Ecology.* 21, 1583–1596.

Mitchell, N., Rodriguez, N., Kuchling, G., Arnall, S. & Kearney, M. R. (2016) Reptile embryos and climate change: modelling limits of viability to inform translocation decisions. *Biological Conservation.* 204, 134–147.

Mittermeier, R. A., Turner, W. R., Larsen, F. W., Brooks, T. M. & Gascon, C. (2011) Global biodiversity conservation: the critical role of hotspots. In Zachos, F. E. and Habel, J. C. (eds.) *Biodiversity Hotspots.* Berlin, Springer.

Nason, S. E., Lloyd, N., Kelly, C. D., Brichieri-Colombi, T., Dalrymple, S. E. & Moehrenschlager, A. (2021) Maximizing the effectiveness of qualitative systematic reviews: a case study on terrestrial arthropod conservation translocations. *Biological Conservation.* 254, 8.

Neaves, L. E., Ogden, R. & Hollingsworth, P. M. (2023) Genomics and conservation translocations. In Gaywood, M. J., Ewen, J. G., Hollingsworth, P. M. and Moehrenschlager, A. (eds.) *Conservation Translocations.* Cambridge, Cambridge University Press.

Novak, B. J., Phelan, R. & Weber, M. (2021) US conservation translocations: Over a century of intended consequences. *Conservation Science and Practice.* 3, 19.

Overbeek, A., Galla, S., Brown, L., et al. (2020) Pedigree validation using genetic markers in an intensively-managed taonga species, the critically endangered kakī (*Himantopus novaezelandiae*). *Notornis.* 67, 709–716.

Owen, G. (2020) What makes climate change adaptation effective? A systematic review of the literature. *Global Environmental Change-Human and Policy Dimensions.* 62, 132–145.

Pelai, R., Hagerman, S. M. & Kozak, R. (2021) Whose expertise counts? Assisted migration and the politics of knowledge in British Columbia's public forests. *Land Use Policy*. 103, 287–296.

Perino, A., Pereira, H. M., Navarro, L. M., et al. (2019) Rewilding complex ecosystems. *Science*. 364, 351–563.

Pettorelli, N., Durant, S. M. & Du Toit, J. T. (2019) Rewilding: a captivating, controversial, twenty-first-century concept to address ecological degradation in a changing world. In: Pettorelli, N., Durant, S. M. and Du Toit, J. T. (eds.) *Rewilding*. Cambridge, Cambridge University Press, pp. 1–11.

Phelan, P. & Ribbons, A. (2021) To Restore Biodiversity, Embrace Biotech's 'Intended Consequences'. *Scientific American* [Online]. Available from: www.scientificamerican.com/article/to-restore-biodiversity-embrace-biotechs-lsquo-intended-consequences-rsquo/ [Accessed June 2021].

Phelan, R., Baumgartner, B., Brand, S., et al. (2021a) Intended consequences statement. *Conservation Science and Practice*. 1–3.

Phelan, R., Kareiva, P., Marvier, M., Robbins, P. & Weber, M. (2021b) Why intended consequences? *Conservation Science and Practice*. 3, 1–3.

Piaggio, A. J., Segelbacher, G., Seddon, P. J., et al. (2017) Is it time for synthetic biodiversity conservation? *Trends in Ecology & Evolution*. 32, 97–107.

Powell, W. A., Newhouse, A. E. & Coffey, V. (2019) Developing blight-tolerant American chestnut trees. *Cold Spring Harbor Perspectives in Biology*. 11, 16–24.

Ralls, K., Ballou, J. D., Dudash, M. R., et al. (2018) Call for a paradigm shift in the genetic management of fragmented populations. *Conservation Letters*. 11, 1–6.

Ralls, K., Sunnucks, P., Lacy, R. C. & Frankham, R. (2020) Genetic rescue: a critique of the evidence supports maximizing genetic diversity rather than minimizing the introduction of putatively harmful genetic variation. *Biological Conservation*. 251, 1–8.

Randall, L., Lloyd, N. & Moehrenschlager, A. (2018) *Guidelines for Mitigation Translocations of Amphibians: Applications for Canada's Prairie Provinces*. Calgary, Alberta, Canada, Centre for Conservation Research, Calgary Zoological Society.

Rayne, A., Byrnes, G., Collier-Robinson, L., et al. (2020) Centring Indigenous knowledge systems to re-imagine conservation translocations. *People and Nature*. 2, 512–526.

Reid, A. J., Eckert, L. E., Lane, J. F., et al. (2021) "Two-Eyed Seeing": an Indigenous framework to transform fisheries research and management. *Fish and Fisheries*. 22, 243–261.

Ricciardi, A. & Simberloff, D. (2021) Assisted colonization risk assessment. *Science*. 372, 925.

Rincon, P. (2018) Stephen Hawking's warnings: What he predicted for the future. [Online]. Available from: www.bbc.com/news/science-environment-43408961 [Accessed 28 June 2021].

Robinson, J. A., Brown, C., Kim, B. Y., Lohmueller, K. E. & Wayne, R. K. (2018) Purging of strongly deleterious mutations explains long-term persistence and absence of inbreeding depression in Island Foxes. *Current Biology*. 28, 3487–3499.

Rodriguez, J. P. (2017) The difference conservation can make: integrating knowledge to reduce extinction risk. *Oryx*. 51, 1–2.

Rodríguez, J.P., Sucre, B., Mileham, K., et al. (2022) Addressing the Biodiversity Paradox: mismatch between the co-occurrence of biological diversity and the human, financial and institutional resources to address its decline. *Diversity*. 14, 708.

Samuel, M. D., Liao, W., Atkinson, C. T. & Lapointe, D. A. (2020) Facilitated adaptation for conservation – can gene editing save Hawaii's endangered birds from climate driven avian malaria? *Biological Conservation*. 241, 1–9.

Sandler, R. L., Moses, L. & Wisely, S. M. (2021) An ethical analysis of cloning for genetic rescue: Case study of the black-footed ferret. *Biological Conservation*. 257, 109–118.

Seaborn, T., Andrews, K. R., Applestein, C. V., et al. (2021) Integrating genomics in population models to forecast translocation success. *Restoration Ecology*. 29, 112–124.

Seddon, P. J. (2017) The ecology of de-extinction. *Functional Ecology*. 31, 992–995.

Seddon, P. J. (2023) The role of conservation translocations in rewilding and de-extinction. In Gaywood, M. J., Ewen, J. G., Hollingsworth, P. M. and Moehrenschlager, A. (eds.) *Conservation Translocations*. Cambridge, Cambridge University Press.

Seddon, P. J., Griffiths, C. J., Soorae, P. S. & Armstrong, D. P. (2014a) Reversing defaunation: restoring species in a changing world. *Science*. 345, 406–412.

Seddon, P. J., Moehrenschlager, A. & Ewen, J. (2014b) Reintroducing resurrected species: selecting deextinction candidates. *Trends in Ecology & Evolution*. 29, 140–147.

Shaw, J. (2018) Why is biodiversity so important? [Online]. Available from: www.conservation.org/blog/why-is-biodiversity-important [Accessed May 2021].

Silcock, J. L., Simmons, C. L., Monks, L., et al. (2019) Threatened plant translocation in Australia: a review. *Biological Conservation*. 236, 211–222.

Soorae, P. S. (ed.) (2008) *Global Re-introduction Perspectives: Re-introduction Case-studies from around the Globe*. IUCN/SSC Re-introduction Specialist Group, Abu Dhabi, UAE.

Soorae, P. S. (ed.) (2010) *Global Re-introduction Perspectives: Additional Case-studies from around the Globe*. IUCN/ SSC Re-introduction Specialist Group, Abu Dhabi, UAE.

Soorae, P. S. (ed.) (2011) *Global Re-introduction Perspectives: 2011. More Case Studies from around the Globe*. IUCN/SSC Re-introduction Specialist Group, Gland, Switzerland and Environment Agency, Abu Dhabi, UAE.

Soorae, P. S. (ed.) (2013) *Global Re-introduction Perspectives: 2013. Further Case Studies from around the Globe*. IUCN/SSC Re-introduction Specialist Group, Gland, Switzerland and Environment Agency, Abu Dhabi, UAE.

Soorae, P. S. (ed.) (2016) *Global Re-introduction Perspectives: 2016. Case-studies from around the Globe*. IUCN/SSC Re-introduction Specialist Group, Gland, Switzerland and Environment Agency, Abu Dhabi, UAE.

Soorae, P. S. (ed.) (2018) *Global Reintroduction Perspectives: 2018. Case Studies from around the Globe*. IUCN/SSC Re-introduction Specialist Group, Gland, Switzerland and Environment Agency, Abu Dhabi, UAE.

Steeves, T. E., Johnson, J. A. & Hale, M. L. (2017) Maximising evolutionary potential in functional proxies for extinct species: a conservation genetic perspective on de-extinction. *Functional Ecology.* 31, 1032–1040.

Swan, K. D., McPherson, J. M., Seddon, P. J. & Moehrenschlager, A. (2016) Managing marine biodiversity: the rising diversity and prevalence of marine conservation translocations. *Conservation Letters.* 9, 239–251.

Swan, K. D., Lloyd, N. A. & Moehrenschlager, A. (2018) Projecting further increases in conservation translocations: a Canadian case study. *Biological Conservation.* 228, 175–182.

Taylor, H. R., Dussex, N. & Van Heezik, Y. (2017) De-extinction needs consultation. *Nature Ecology & Evolution.* 1, 111–112.

Teixeira, J. C. & Huber, C. D. (2021a) The inflated significance of neutral genetic diversity in conservation genetics. *Proceedings of the National Academy of Sciences of the United States of America.* 118, 10.

Teixeira, J. C. & Huber, C. D. (2021b) Authors' Reply to Letter to the Editor: Neutral genetic diversity as a useful tool for conservation biology. *Conservation Genetics.* 22, 547–549.

Tengo, M., Hill, R., Malmer, P., et al. (2017) Weaving knowledge systems in IPBES, CBD and beyond-lessons learned for sustainability. *Current Opinion in Environmental Sustainability.* 26, 17–25.

Trask, A., Canessa, S., Moehrenschlager, A., Newland, S., Medina, S. & Ewen, J. (2020) Extinct-in-the-wild species' last stand. *Science.* 369, 516–516.

Trask, A. E., Ferrie, G. M., Wang, J., et al. (2021) Multiple life-stage inbreeding depression impacts demography and extinction risk in an extinct-in-the-wild species. *Scientific Reports.* 11, 234–244.

Trouwborst, A., Blackmore, A., Blyth, S., Fleurke, F., McCormack, P. & Gaywood, M. J. (2023) Conservation translocations and the law. In Gaywood, M. J., Ewen, J. G., Hollingsworth, P. M. and Moehrenschlager, A. (eds.) *Conservation Translocations.* Cambridge, Cambridge University Press.

Tunstall, T., Kock, R., Vahala, J., et al. (2018) Evaluating recovery potential of the northern white rhinoceros from cryopreserved somatic cells. *Genome Research.* 28, 780–788.

Turner, W. R., Brandon, K., Brooks, T. M., Costanza, R., Da Fonseca, G. a. B. & Portela, R. (2007) Global conservation of biodiversity and ecosystem services. *Bioscience.* 57, 868–873.

Undin, M., Lockhart, P. J., Hills, S. F. K. & Castro, I. (2021) Genetic rescue and the plight of Ponui hybrids. *Frontiers in Conservation Science.* 1, 1–13.

UNEP (2006) Summary of the second global biodiversity outlook. *Conference of the Parties to the Convention on Biological Diversity.* Curitiba, Brazil, UNEP.

Verschuuren, B., Mallarach, J-M., Bernbaum, E., et al. (2021) *Cultural and Spiritual Significance of Nature. Guidance for Protected and Conserved Area Governance and Management.* Best Practice Protected Area Guidelines Series No. 32. Gland, Switzerland, IUCN, xvi + 88 pp.

Wallach, A. D., Bekoff, M., Batavia, C., Nelson, M. P. & Ramp, D. (2018) Summoning compassion to address the challenges of conservation. *Conservation Biology.* 32, 1255–1265.

Westbrook, J. W., Zhang, Q., Mandal, M. K., et al. (2020) Optimizing genomic selection for blight resistance in American chestnut backcross populations: a trade-off with American chestnut ancestry implies resistance is polygenic. *Evolutionary Applications*. 13, 31–47.

Whiteley, A. R., Fitzpatrick, S. W., Funk, W. C. & Tallmon, D. A. (2015) Genetic rescue to the rescue. *Trends in Ecology & Evolution*. 30, 42–49.

Wisely, S. M., Ryder, O. A., Santymire, R. M., Engelhardt, J. F. & Novak, B. J. (2015) A road map for 21st century genetic restoration: gene pool enrichment of the black-footed ferret. *Journal of Heredity*. 106, 581–592.

Part IV
Case Studies

14 · *Reintroduction of the Endemic Plant* Manglietiastrum sinicum *(Magnoliaceae) to Yunnan Province, China*

WEIBANG SUN, LEI CAI, AND PETER M. HOLLINGSWORTH

14.1 Background

Huagaimu *Manglietiastrum sinicum* is endemic to south western China where it occurs in subtropical broadleaved evergreen forests and is classed as a plant species with extremely small populations (PSESP) (Law, 1979; Figlar & Nooteboom, 2004; Xia et al., 2008; Sun et al., 2012; Ma et al., 2013; Sun et al., 2019a, 2019b). As an endemic species restricted to the south eastern province of Yunnan, huagaimu has been proposed as a priority species for national protection in China (Anon, 1999), and it has been categorised as Critically Endangered on the *China Species Red List* (Wang & Xie, 2004), the *Red List of Magnoliaceae* (Cicuzza et al., 2007; Rivers et al., 2016), and the *Threatened Species List of China's Higher Plants* (Qin et al., 2017). Huagaimu was listed in 2010 as one of 62 plant species with very small population sizes in Yunnan Province, and as one of the 120 Chinese PSESPs requiring urgent conservation interventions (Sun et al., 2019b; Yang et al., 2020). The latest population survey recorded only 52 individuals in the wild from eight localities (Wang et al., 2016). Land use change to commercial forestry and agriculture, and wider habitat destruction leading to habitat fragmentation, have been invoked in driving the observed reduction in population sizes (Chen et al., 2016). An additional anthropogenic pressure is seed harvesting for propagation and cultivation as the species is valued for ornamental plantings.

Detailed studies on the biology of huagaimu (Chen et al., 2016; Chen, 2017) showed that limitations in pollination and seed dispersal may constrain the viability of existing populations. Huagaimu is self-compatible with hermaphrodite flowers; nocturnal beetles are observed

Figure 14.1 Fruits of huagaimu (photo: Ye Chen). (A black and white version of this figure will appear in some formats. For the colour version, please refer to the plate section.)

to visit flowers and result in successful pollination (Chen et al., 2016). Seed set was significantly increased by hand pollination, suggesting pollinator limitation on seed production (Chen et al., 2016). When seeds are produced, they have a red fleshy aril, consistent with bird dispersal (Figure 14.1), but field observers did not record visiting birds, suggesting that habitat fragmentation may have impacted on disperser abundance (Chen et al., 2016).

To support the conservation of this species, a programme of research, sampling, *ex situ* cultivation, and translocations has been undertaken at the Kunming Botanical Garden (KBG) of the Kunming Institute of Botany (KIB), Chinese Academy of Sciences.

14.2 Methods

Seeds of huagaimu were collected in 2004 and 2005 from four individual trees from the type locality (Fadou village, Xichou County). Seedlings propagated from these seeds were cultivated and maintained in KBG's nursery until they were ready for translocation (Figure 14.2). Using this

Reintroduction of *Manglietiastrum sinicum* · 417

Figure 14.2 Propagation of huagaimu in the nursery of Kunming Botanical Garden, Kunming Institute of Botany (photo: Yuan Zhou). (A black and white version of this figure will appear in some formats. For the colour version, please refer to the plate section.)

source material, conservation translocations were undertaken at four sites in two main areas.

1. In November 2007 and July 2008, KBG and Fauna and Flora International (FFI), in collaboration with the local forestry station, planted 200 two-year-old saplings at each of two reintroduction sites (Shangchang and Xiaoqiaogou) in Fadou Township, Xichou County, in south east Yunnan (400 saplings in total). All samples were planted, mapped, and measured at the same time. Local participants were trained and paid a stipend by the government to carry out the translocations, scientific management, dynamic monitoring, data acquisition, and archiving.
2. In July 2010, the State Forestry Administration of China (SFA), the Yunnan Provincial Forestry Department of China (YPFD), Botanic Gardens Conservation International (BGCI), KBG, and Maguan Forestry Bureau jointly established two translocation sites at Laoqing and Guiniu of Guipishan in Maguan County, Yunnan. At Laoqing,

100 individuals were reintroduced, and at Guipishan, an existing population of three individuals was reinforced with a further 100 individuals. The translocated material consisted of three-year-old saplings.

A total of 74 translocated plants (Shangchang 15, Xiaoqiaogou 19, Laoqing 20, Guiniu 20) were sampled in 2015 for genetic analysis. These samples were screened for genetic variation with 20 highly polymorphic microsatellite markers, and compared with samples from six wild populations (n = 40 total) and two *ex situ* populations (n = 38 total), one at KBG, and one at South China Botanical Garden (SCBG).

14.3 Outcome

Following nearly ten years of monitoring and management by KBG and partners, data acquisition and analyses were carried out on the four sites in May 2018.

1. Individuals reintroduced in 2007 and 2008 to the two sites in Xichou County had survival of 100% in the first year, but plants gradually died off in subsequent years, leading to survival rates of 20% or less (Table 14.1). At Shangchang, the survival rate was 20%, with an average plant height of 4.14 m (representing an average height increase of 3.24 m), and an average base diameter of 4.72 cm (representing an average increase of 3.46 cm). At Xiaoqiaogou, the survival rate was only 2%. Here the average surviving plant height was 2.25 m (Table 14.1).
2. Individuals reintroduced/reinforced in 2010 at sites in Maguan County had a much higher survival rate. At Laoqing, the survival rate was 75%, with the average plant height 3.31 m (Table 14.1). At Guiniu of Guipishan, the survival rate was 74%, and the average plant height was 3.23 m (Table 14.1).

A careful retrospective examination of the reasons for a better outcome for the Maguan translocations compared to the Xichou sites suggested that horticultural practices associated with the outplanting may have been important. The teams involved in the Maguan translocations included professional horticulturists, with greater experience of plant care and cultivation. As an example, when plants that died at the Xichou sites were dug up and examined, many still had the original plastic packaging material around their roots which would have restricted their ability to become established.

Table 14.1. *Survival rate and plant measurements from four translocated populations of huagaimu from Yunnan province in 2018.*

	Xichou – Shangchang	Xichou – Xiaoqiaogou	Maguan – Laoqing	Maguan – Guipishan
Survival rate	20%	2%	75%	74%
Average plant height (increase)	4.14 m (3.24 m)	2.25 m (1.47 m)	3.31 m (1.97 m)	3.23 m (1.68 m)
Average base diameter (increase)	4.72 cm (3.46 cm)	3.78 cm (2.26 cm)	5.50 cm (3.77 cm)	4.49 cm (2.75 cm)

Genetic diversity analysis recovered 111 alleles from 20 microsatellite loci from the 40 sampled individuals from the wild populations, with an average of 5.55 alleles per locus. Samples from the four translocated populations contained 89 alleles, with an average of 4.45 alleles per locus (Chen, 2017). A similar level of diversity was detected (83 alleles total) in the two *ex situ* populations (KBG, SCBG), with an average of 4.15 per locus (Chen, 2017).

Further details of the translocations and their monitoring are reported by Sun et al. (2019b).

14.4 Discussion and Future Recommendations

A key factor influencing the success of the huagaimu translocation was the horticultural practices associated with the initial plantings in Xichou County. There was a major improvement in outcomes in the Maguan translocations, associated with the availability of experienced horticulturists with knowledge of propagation and cultivation of the focal species. Linking horticultural skills to future translocation projects for this species is important, and there is a more general point about enhancing capacity in conservation horticulture to support plant species translocations in China and elsewhere, especially where projects involve volunteers and citizen scientists.

Despite the small number of founder individuals for the translocated populations (four open pollinated mother trees from Xichou County), there was no evidence of a major genetic bottleneck, with the translocated populations having about 80 per cent of the allelic diversity of a sample of 40 of the 52 remaining known wild individuals. Nevertheless, future translocation projects would benefit from sampling from different populations to maximise representativeness of the remaining wild individuals and to maintain as much of the remaining species-wide genetic variation as possible.

The successful recovery of huagaimu will take time. The evidence of pollinator and seed dispersal limitations in natural populations illustrates the challenges to existing populations spreading and colonising new sites. Further translocations will likely be required, along with long-term maintenance and monitoring of sites, and studies of species biology to inform management to enhance population regeneration and expansion. This will require ongoing funding and project support to continue the cooperation with local partners and communities near the translocation sites.

14.5 Summary

- In total, 600 two-year-old and three-year-old saplings of the critically endangered Chinese endemic huagaimu were translocated into the species' natural range.
- There were very low survival rates (2 per cent and 20 per cent respectively) of translocated individuals at two sites in Xichou County, with a much higher survival rate of 74 per cent at two sites in Maguan County.
- The over-riding determinant of translocation success is considered to be the difference in horticultural practices used in the different translocations, with much more favourable outcomes associated with greater horticultural training and expertise.

References

Anon. (1999) List of National Key Protected Wild Plants (First Group). *The Order of National Forestry Bureau and Agriculture Ministry of China.* 4, 2–13.

Chen, Y. (2017) Conservation biology of *Manglietiastrum sinicum* Law (Magnoliaceae), a plant species with extremely small populations. Kunming Institute of Botany, Chinese Academy of Sciences, University of Chinese Academy of Sciences.

Chen, Y., Chen, G., Yang, J., et al. (2016) Reproductive biology of *Magnolia sinica* (Magnoliaecea), a threatened species with extremely small populations in Yunnan, China. *Plant Diversity.* 38, 253–258.

Cicuzza, D., Newton, A. & Oldfield, S. (2007) *The Red List of Magnoliaceae.* Cambridge, Lavenham Press.

Figlar, R. B. & Nooteboom, H. P. (2004) Notes on Magnoliaceae IV. *Blumea - Biodiversity, Evolution and Biogeography of Plants.* 49, 87–100.

Law, Y. W. (1979) A new genus of Magnoliaceae from China. *Acta Phytotaxonomica Sinica*. 17, 72–74.

Ma, Y. P., Chen, G., Grumbine, R. E., et al. (2013) Conserving plant species with extremely small populations (PSESP) in China. *Biodiversity and Conservation*. 22, 803–809.

Qin, H. N., Yang, Y., Dong, S. Y., et al. (2017) Threatened species list of China's higher plants. *Biodiversity Science*. 25, 696–744.

Rivers, M., Beech, E., Murphy, L. & Oldfield, S. (2016) *The Red List of Magnoliaceae Revised and Extended*. Richmond, Botanic Gardens Conservation International.

Sun, W. B., Zhou, Y., Li, X. Y., et al. (2012) Population reinforcing program for *Magnolia sinica*, a critically endangered endemic tree in southeast Yunnan province, China. In Maschinski, J. and Haskins, K. E. (eds.) *Plant Reintroduction in a Changing Climate*. Washington, DC, Island Press, pp. 65–69.

Sun, W. B., Ma, Y. P. & Blackmore, S. (2019a) How a new conservation action concept has accelerated plant conservation in China. *Trends in Plant Science*. 24, 4–6.

Sun, W. B., Yang, J. & Dao, Z. L. (2019b) *Study and Conservation of Plant Species with Extremely Small Populations (PSESP) in Yunnan Province, China*. Beijing, China, Science Press.

Wang, B., Ma, Y. P., Chen, G., et al. (2016) Rescuing *Magnolia sinica* (Magnoliaceae), a critically endangered species endemic to Yunnan, China. *Oryx*. 50, 446–449.

Wang, S. & Xie, Y. (2004) *China Species Red List (Vol 1)*. Beijing, China, Higher Education Press.

Xia, N. H., Liu, Y.H. & Nooteboom, H. P. (2008) Magnoliaceae. In Wu, Z. Y., Raven, P. H. and Hong D. Y. (eds.) *Flora of China*, vol. 7. Beijing, China, Science Press & St. Louis, MO, Missouri Botanical Garden Press, pp. 48–91.

Yang, J., Ca, L., Liu, D. T., et al. (2020) China's conservation program on Plant Species with Extremely Small Populations (PSESP): progress and perspectives. *Biological Conservation*. 244, 108535.

15 · *Applying Adaptive Management to Reintroductions of Pyne's Ground-Plum* Astragalus bibullatus

MATTHEW A. ALBRECHT

15.1 Background

Practitioners face many uncertainties when designing reintroduction programmes. Determining the most suitable recipient sites, appropriate genetic source material, and the best management options can be challenging when the biology of the focal species is poorly understood, as is often the case in reintroductions with rare plants. Adaptive management (AM) can overcome these challenges and help reduce uncertainties in reintroduction programmes through an iterative process of structured decision-making and a continual cycle of learning by doing (McCarthy et al., 2012). Key aspects of AM are defining the problem, developing hypotheses or predictions, implementing management manipulations or experiments, monitoring outcomes, re-evaluating hypotheses, and then incorporating new information into the reintroduction programme.

This case study illustrates how sequential experiments applied in an AM framework improved reintroduction outcomes of the federally endangered and long-lived perennial herb, Pyne's ground-plum *Astragalus bibullatus*. To reduce the extinction risk of Pyne's ground-plum, a programme was initiated in 2000 to reintroduce populations to unoccupied and protected sites within its indigenous range – the Stones River Watershed in Central Tennessee, USA (U.S. Fish and Wildlife Service, 2011). Pyne's ground-plum is a narrow endemic of limestone glades, which are rocky grasslands with thin soils. It is presently known from a few small populations in restricted patches of suitable habitat within the south eastern portion of the watershed. Although most populations are protected, they often occur adjacent to roads or rights-of-way where they are potentially threatened by human activities.

Historical observations over the last century indicate that the species was once more widespread in the watershed, including the northern portion where it is now extirpated. Human population growth and change in land use have resulted in extensive habitat loss and degradation throughout the watershed.

At the outset of the reintroduction programme, we knew little about the biology and optimal habitat requirements of the species. Pyne's ground-plum – a short-statured, rosette-forming perennial – initiates growth and flowering in early spring and sets seed in early summer. However, most naturally occurring populations are small and rarely set seed, but do contain moderate levels of genetic diversity (Baskauf & Snapp, 1998). Populations vary in the amount of woody plant cover and habitat openness, although the response of this species to environmental variation was poorly understood at the beginning of the reintroduction programme. Given the large amount of uncertainty about the biology of Pyne's ground-plum and the conditions required for reintroduced populations to persist, we used an AM approach to test different habitat conditions and management manipulations over time to gain information needed to improve future outcomes.

15.2 Methods

The first reintroduction experiment tested the effects of population genetic source and seedling transplant season in five habitat patches at a protected site. Because the remaining natural populations of Pyne's ground-plum showed genetic admixture and low genetic differentiation (Baskauf & Snapp, 1998), the reintroduction programme opted to mix source material to maximise genetic variation. However, concern over whether some population sources or genotypes might have differential adaptation to novel reintroduction sites and the optimal timing to transplant seedlings created uncertainty in the programme. To reduce uncertainty, the first reintroduction used near-equal representation of maternal lines from three remnant population sources and varied the seasonal (spring and fall) timing of outplanting (Albrecht & McCue, 2010). *Ex situ* propagated seedlings were outplanted into five mesic habitat patches (within 500 m of each other) with the goal of establishing a meta-population. After six years, demographic monitoring revealed that transplant vital rates (i.e. survival, growth, and flowering) varied little between the three population sources. Interactions between population donor source and habitat patch for most vital rates were non-significant, providing no clear evidence of differential source adaptation. Overall, survival and flowering rates were greater for autumn transplants than spring transplants and demographic rates varied among habitat patches, with only a few supporting seedling recruitment. After 10 years, recruitment was not sufficient to replace mortality of the original founders, resulting in the extirpation of all five patches.

Although the first reintroduction attempt in 2001 failed to create persistent populations, the study design provided support for using mixed-population sources and transplanting in autumn. Upon reviewing monitoring outcomes, integrating new knowledge of the study system, and considering demographic and genetic theory, the recovery team developed three non-mutually exclusive hypotheses to explain why the first reintroduction failed. First, initial founder sizes (30–40 plants per patch) were lower than those that typically support persistent plant reintroductions (Albrecht & Maschinski, 2012). We predicted that increasing founding population sizes would improve population viability. Second, shortly after outplanting, some reintroduced plants were browsed by small-vertebrate herbivores and were subsequently caged to prevent further damage; however, the subsequent effect on plant vital rates remained unknown since caging was not experimentally manipulated. We predicted that caging would improve vital rates since a single herbivory event can have long-lasting consequences on perennial herbs. Finally, ongoing studies of natural populations revealed that Pyne's ground-plum requires high light levels for growth and reproduction but can persist for decades in shaded conditions as small, non-reproductive individuals and in a soil seed bank (Albrecht et al., 2016). We hypothesised that the woody encroached and fire-suppressed locations in the first reintroduction reduced population growth.

Applying the AM cycle, Albrecht and Long (2019) tested these three predictions simultaneously with experiments replicated at three locations in 2012, nearly 10 years after the initial reintroduction attempt failed. At the first reintroduction site, we hypothesised that two mesic locations where the species had completed its lifecycle but failed to persist could support a persistent population with a larger initial population size, herbivore exclusion, and mechanical thinning of woody vegetation. The third location was a fire-maintained, open, and xeric site, which was the hypothesised optimal habitat. At all sites, we outplanted and caged all transplants from multiple source populations in autumn using initial founder sizes (168 plants) nearly four times greater than the first reintroduction (Figure 15.1).

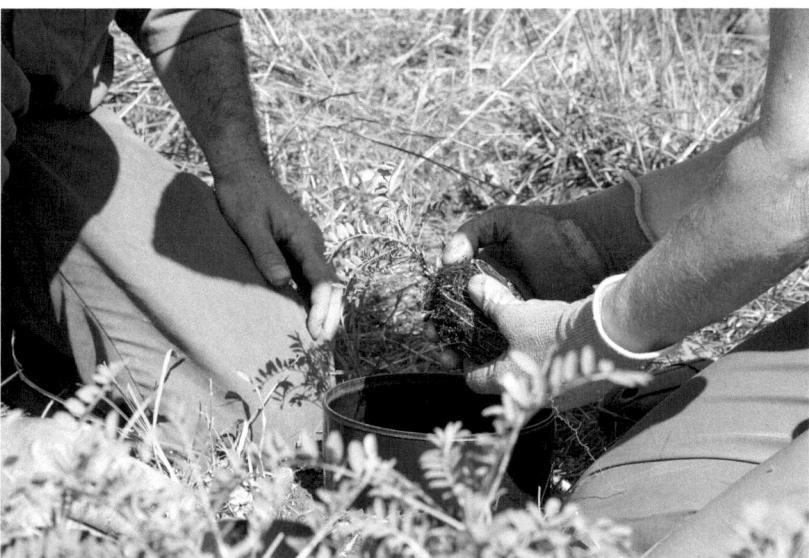

Figure 15.1 Conservation staff and volunteers carefully transplant a Pyne's ground-plum seedling for a reintroduction experiment. Seedlings were propagated from seed that had been stored in the Missouri Botanical Garden's frozen seed bank (photo: M. Albrecht). (A black and white version of this figure will appear in some formats. For the colour version, please refer to the plate section.)

15.3 Outcomes

After four years, vital rates at the two mesic sites were similarly low to the first reintroduction, despite herbivore exclusion, larger founder sizes, and less woody vegetation than the previous attempt. Yet survival, growth rates, and reproduction were much higher at the xeric location (Figure 15.2). At the xeric location, the greatest flowering and seed production occurred in caged transplants, indicating that herbivores can severely limit fitness. Applying this new information to subsequent reintroductions (2016–2017) at three additional xeric grasslands resulted in 90 per cent survival and substantial seed production over a three-year period when plants were caged (Figure 15.3).

15.4 Discussion and Future Recommendations

By applying an AM framework, we continuously gained new information through experimentation, careful monitoring and documentation, and re-evaluating hypotheses. These guided subsequent decision-making in

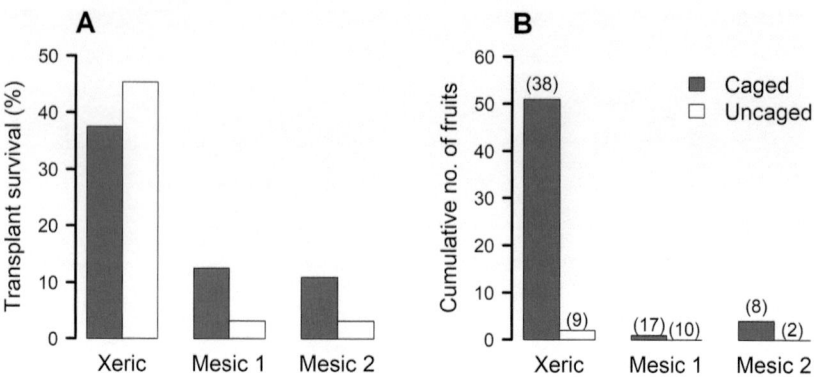

Figure 15.2 (A) Cumulative survival and (B) cumulative number of fruits produced by caged and uncaged Pyne's ground-plum transplants over a four-year period (2013–2017) in xeric and mesic reintroduction sites. Half of the 168 plants initially reintroduced at each site in 2012 were caged. Numbers in parentheses represent the cumulative number of plants that flowered during this period.

Figure 15.3 A reintroduced Pyne's ground-plum plant (four years old) with fruits that had been caged at the beginning of the reintroduction experiment (photo: M. Albrecht). (A black and white version of this figure will appear in some formats. For the colour version, please refer to the plate section.)

the reintroduction programme. Sequential experiments showed that successful reintroduction of Pyne's ground-plum required xeric, open sites and exclusion of vertebrate herbivores. Due to multiple failed attempts after habitat manipulations, reintroductions to mesic sites were discontinued. Upon further review, failed reintroductions occurred at locations with slightly wetter soils compared to natural populations, due to subtle differences in hydrology. The variable cover of woody vegetation in natural populations made it challenging to determine optimal habitat conditions, a common problem with rare plants known from only a few locations. We suspect the high cover of woody vegetation in some natural populations threatens their long-term persistence, and we propose future vegetation thinning and prescribed fire management.

We recommend continued application of AM in the Pyne's ground-plum reintroduction programme. Long-term monitoring will be required to determine whether open, xeric locations can support seedling recruitment and persistence. Future reintroductions should explore the role of soil microbes in enhancing performance, and incorporate breeding system or pollinator manipulations to understand the cause of low seed set in some plants and locations.

15.5 Summary

- Translocations of other rare plants could similarly benefit from applying AM.
- When conducted as experiments and placed in an AM framework, failed reintroductions can provide valuable information that can be applied to subsequent reintroductions.
- Spanning nearly two decades, the Pyne's ground-plum reintroduction programme also highlights the need for long-term institutional commitments, experimentation, and follow-through to meet conservation goals.

References

Albrecht, M. A. & Long, Q. G. (2019) Habitat suitability and herbivores determine reintroduction success of an endangered legume. *Plant Diversity*. 41, 109–117.

Albrecht, M. A. & Maschinski, J. (2012) Influence of founder population size, propagule stages, and life history on the survival of reintroduced plant populations. In Maschinski, J. and Haskins, K. E. (eds.) *Plant Reintroduction in a Changing Climate: Promises and Perils*. Washington, DC, Island Press, pp. 171–188.

Albrecht, M. A. & McCue, K. A. (2010) Changes in demographic processes over long time scales reveal the challenge of restoring an endangered plant. *Restoration Ecology.* 18, 235–243.

Albrecht, M. A., Becknell, R. E. & Long, Q. (2016) Habitat change in insular grasslands: Woody encroachment alters the population dynamics of a rare ecotonal plant. *Biological Conservation.* 196, 93–102.

Baskauf C. J. & Snapp, S. (1998) Population genetics of the cedar-glade endemic *Astragalus bibullatus* (Fabaceae) using isozymes. *Annals of the Missouri Botanical Garden.* 85, 90–96.

McCarthy, M. A., Armstrong, D. P. & Runge, M. C. (2012) Adaptive management of reintroduction. In Ewen, J. G., Armstrong, D. P., Parker, K. A. and Seddon, P. J. (eds.) *Reintroduction Biology: Integrating Science and Management.* Hoboken, NJ, Blackwell Publishing Ltd., pp. 256–289.

U.S. Fish and Wildlife Service (2011) Recovery plan for *Astragalus bibullatus* (Pyne's Ground-plum). Atlanta, GA.

16 · *Five Reasons to Consider Long-Term Monitoring: Case Studies from Bird Reintroductions on Tiritiri Matangi Island*

DOUG P. ARMSTRONG, ELIZABETH H. PARLATO, AND JOHN G. EWEN

16.1 Background

One question that inevitably arises in translocation programmes is how long to continue monitoring. The dynamics of a translocated population can be roughly divided into three stages: establishment, growth, and regulation (Sarrazin, 2007). Short-term monitoring can address questions at the establishment stage – for example, the effect of alternative release strategies on post-release survival (Batson et al., 2015). Consequently, research at the establishment phase has dominated the translocation literature (Taylor et al., 2017). Medium-term monitoring is clearly needed to allow population growth to be estimated (Armstrong et al., 2008; Converse & Armstrong, 2016) and to facilitate adaptive management to improve the growth rate if needed (McCarthy et al., 2012). However, it is less clear whether intensive monitoring should be extended beyond the time frame needed to be confident that the population is growing. Such long-term monitoring diverts resources, so it is important to consider its likely benefits and whether these outweigh the costs (Buxton et al., 2020).

Here we consider the benefits that have arisen from intensive long-term monitoring of two forest bird species reintroduced to Tiritiri Matangi, a 220 ha New Zealand island that is gradually being restored through regeneration of native forest and reintroduction of multiple taxa.

16.2 Methods

Toutouwai (North Island robin, *Petroica longipes*) and hihi (stitchbird, *Notiomystis cincta*) were reintroduced in 1992 and 1995, respectively, and initially subject to two-year monitoring programmes. The hihi population is also subject to continuous management, including supplementary feeding, provision of nest boxes, and control of nest mites in these boxes. The monitoring for both populations included regular re-sighting surveys to estimate survival, intensive monitoring of individual breeding pairs, and colour-banding of chicks in the nest. Extending this monitoring to five years was sufficient to be confident these populations had a high probability of persistence (Armstrong et al., 2002). However, in both cases intensive monitoring was continued for more than 20 years, with breeding pairs or females closely monitored through the breeding season (September–February), most chicks colour-banded, and at least two re-sighting surveys conducted (at start and end of breeding season). So what led to this long-term monitoring and what did it achieve?

16.3 Outcomes

16.3.1 Informing Management Decisions for the Focal Population

The most obvious impetus to continue monitoring is to provide information needed for ongoing management. For example, both populations quickly became source populations for further translocations, with the toutouwai population harvested five times from 1999 to 2016 to establish six further populations and the hihi population harvested ten times from 2005 to 2020 to establish or reinforce seven populations (Figure 16.1). These harvesting decisions were made based on predicted demographic impacts on the source population as well as benefits at the new site, so they required ongoing data collection to do the modelling necessary to make those predictions (e.g. Panfylova et al., 2019). The populations' responses to harvesting then made it possible to validate and improve the models (Figure 16.1).

16.3.2 Predicting the Future Viability of the Focal Population

Ongoing data collection also makes it possible to assess the need for interventions such as genetic rescue (Bell et al., 2019) to avoid future declines due to inbreeding depression. For Tiritiri Matangi toutouwai, long-term data on pedigrees as well as survival and reproduction made it possible to quantify the degree of inbreeding depression and project its likely impact on population growth over the next 150 years, allowing evidence-based assessment of the need for genetic rescue (Armstrong

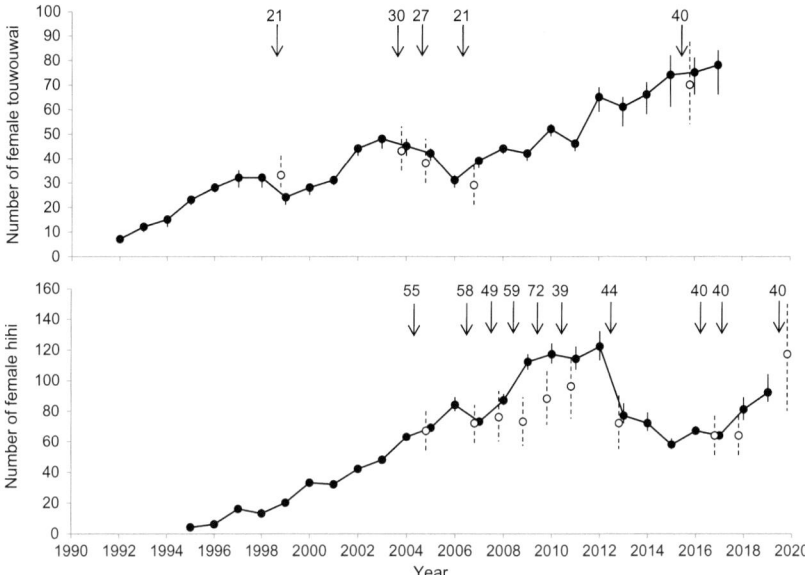

Figure 16.1 Application of population models for guiding harvesting of toutouwai (New Zealand robin) and hihi populations on Tiritiri Matangi for translocation to other sites. Lines and filled circles show the estimated numbers of females at the start of each breeding season, with error bars showing 95% credible intervals. Arrows show numbers harvested, and open symbols show median pre-harvest projections from population models, with error bars showing 95% prediction intervals. Model parameters were updated as data were added, and for toutouwai the form of density dependence in juvenile survival was updated based on responses to harvests. The inaccuracy of pre-harvest projections for hihi in 2009 and 2010 resulted from the population being underestimated at that time due to some females remaining undetected for several years.

et al., 2021a). These projections also required inferences about density-dependent regulation in relation to habitat regeneration that were only possible with long-term data.

16.3.3 Informing Site Selection for Further Reintroductions

Long-term data also make it possible to measure how populations and species respond to changes in environmental conditions over time and space. Chauvenet et al. (2013) analysed survival and reproduction rates of Tiritiri Matangi hihi in relation to temperature and rainfall, and used the results to assess site suitability for the species both now and under future climate change scenarios. Consequently, the Hihi Recovery Group is now considering introducing the species to sites south of its historic range, as these sites are predicted to become suitable for hihi whereas the sites with the largest current populations are predicted to become unsuitable by 2050 (Chauvenet et al., 2013).

16.3.4 Predicting Population Dynamics at Other Sites

For both hihi and toutouwai, the detailed analyses possible with the Tiritiri Matangi datasets have greatly facilitated construction of models for making predictions at new reintroduction sites. These analyses not only inform the appropriate model structure, such as with respect to sex and age classes, but also values for parameters that can only be estimated from long-term datasets. For example, the estimated random annual variation (environmental stochasticity) in hihi survival and reproduction rates at Tiritiri Matangi and other sites has recently been used to account for annual variation when making projections for the recently reintroduced population at Rotokare Scenic Reserve (Parlato et al., 2021).

16.3.5 Improving Our Understanding of the Dynamics of Reintroduced Populations

Long-term data are needed in order to model the full range of factors affecting the dynamics of reintroduced populations and therefore to identify the key factors affecting those dynamics. The long-term data for Tiritiri Matangi toutouwai show that survival and/or reproduction rates have been affected by inbreeding depression and random annual variation, as well as age and sex differences, density dependence, and post-release effects. However, the key factors found necessary to capture the population's dynamics over 26 years (Armstrong et al., 2021b) were identified in the first five years (Armstrong et al., 2002), giving reassurance that useful models can be created from medium-term data.

16.4 Discussion and Recommendations

The five benefits listed reflect our experience with these particular species and do not capture all potential benefits of long-term studies. In particular, we have not attempted to measure ecosystem effects such as those revealed from long-term monitoring of the reintroduction of grey wolves *Canis lupus* to Yellowstone National Park (Ripple & Beschta,

2012). It is therefore useful for reintroduction practitioners to carefully consider the potential benefits of long-term monitoring in terms of their objectives. Although long-term studies have tended to arise opportunistically rather than through careful planning (Miao et al., 2009), reintroduction and other conservation translocation practice could be improved through a more proactive approach.

We therefore recommend that translocation practitioners consider a structured decision-making approach (Gregory et al., 2012) whereby they attempt to predict how well management alternatives will achieve their objectives, including a clear representation of uncertainty in these predictions. Monitoring primarily serves to reduce uncertainty in how well our objectives are being met and is therefore important for refining management. It is therefore good to try to predict explicitly how such learning will improve management when making decisions about what type of monitoring to do and how long to monitor. This can potentially be done through formal value of information analysis (Canessa et al., 2015), a practice very much at the cutting edge of reintroduction biology (see Ewen et al., this volume).

16.5 Summary

Long-term monitoring of reintroduced populations may:

- Inform ongoing management decisions for the focal population, such as supplementary feeding or harvesting.
- Make it possible to predict the future viability of small populations, and therefore the need for genetic management.
- Inform site selection for further reintroductions, for example by showing how survival and reproduction rates vary with climatic conditions.
- Facilitate our ability to ability to predict population dynamics at new reintroduction sites, for example by informing values for parameters that can be estimated only from long-term data.
- Improve our understanding of the dynamics of reintroduced populations, so future monitoring and management can focus on the key factors affecting persistence.

References

Armstrong, D. P. & Seddon, P. J. (2008) Directions in reintroduction biology. *Trends in Ecology and Evolution.* 23, 20–25.

Armstrong, D. P., Davidson, R. S., Dimond, W. J., et al. (2002) Population dynamics of reintroduced forest birds on New Zealand islands. *Journal of Biogeography.* 29, 609–621.

Armstrong, D. P., Parlato, E. H., Egli, B., et al. (2021a) Using long-term data for a reintroduced population to empirically estimate future consequences of inbreeding. *Conservation Biology.* 35, 859–869.

Armstrong, D. P., Parlato, E. H., Egli, B., et al. (2021b). Capturing the dynamics of small populations: a retrospective assessment using long-term data for an island reintroduction. *Journal of Animal Ecology.* 90, 2915–2927.

Batson, W. G., Gordon, I. J., Fletcher, D. B. & Manning, A. D. (2015) Translocation tactics: a framework to support the IUCN Guidelines for wildlife translocations and improve the quality of applied methods. *Journal of Applied Ecology.* 52, 1598–1607.

Bell, D. A., Robinson, Z. L., Funk, W. C., et al. (2019) The exciting potential and remaining uncertainties of genetic rescue. *Trends in Ecology and Evolution.* 34, 1070–1079.

Buxton, R. T., Avery-Gomm, S., Lin, H., Smith, P. A., Cooke, S. J. & Bennett, J. R. (2020) Half of resources in threatened species conservation plans are allocated to research and monitoring. *Nature Communications.* 11, 4668.

Canessa, S., Guillera-Arroita, G., Lahoz-Monfort, J., et al. (2015) When do we need more data? A primer on calculating the value of information for applied ecologists. *Methods in Ecology and Evolution.* 6, 1219–1228.

Chauvenet, A. L. M., Ewen, J. G., Armstrong, D. P. & Pettorelli, N. (2013) Saving the hihi under climate change: a case for assisted colonization. *Journal of Applied Ecology.* 50, 1330–1340.

Converse, S. J. & Armstrong, D. P. (2016) Demographic modeling for reintroduction decision-making. In Jachowski, D., Millspaugh, J., Angermeier, P. and Slotow, R. (eds.) *Reintroduction of Fish and Wildlife Populations.* Oakland, CA, University of California Press, pp. 123–146.

Ewen, J. G., Canessa, S., Converse, S. J. & Parker, K. A. (2023) Decision-making in animal conservation translocations: biological considerations and beyond. In Gaywood, M. J., Ewen, J. G., Hollingsworth, P. M. and Moehrenschlager, A. (eds.) *Conservation Translocations.* Cambridge, Cambridge University Press.

Gregory, R., Failing, L., Harstone, M., Long, G., McDaniels, T. & Ohlson, D. (2012) *Structured Decision Making: A Practical Guide to Environmental Management Choices.* Chichester, Wiley-Blackwell.

McCarthy, M. A., Armstrong, D. P. & Runge, M. C. (2012) Adaptive management of reintroduction. In: Ewen, J. G., Armstrong, D. P., Parker, K. A. and Seddon, P. J. (eds.) *Reintroduction Biology: Integrating Science and Management.* Oxford, Wiley-Blackwell, pp. 257–289.

Miao, S. L., Carstenn, S. & Nungesser, M. (eds.). (2009) *Real World Ecology: Large-Scale and Long-Term Case Studies and Methods.* New York, Springer.

Panfylova, J., Ewen J. G. & Armstrong, D. P. (2019) Making structured decisions for reintroduced populations in the face of uncertainty. *Conservation Science and Practice.* 1, e90.

Parlato, E. H., Ewen, J. G., McCready, M., Gordon, F., Parker, K. A. & Armstrong, D. P. (2021) Incorporating data-based estimates of temporal variation into projections for newly monitored populations. *Animal Conservation*. 24, 1001–1012.

Ripple, W. J. & Beschta, R. L. (2012) Trophic cascades in Yellowstone: the first 15 years after wolf reintroduction. *Biological Conservation*. 145, 205–213.

Sarrazin, F. (2007) Introductory remarks: a demographic frame for reintroductions. *Ecoscience*. 14, iv–v.

Taylor G., Canessa, S., Clarke, R. H., et al. (2017) Is reintroduction biology an effective applied science? *Trends in Ecology and Evolution*. 32, 873–880.

17 · *Multiple Reintroductions to Restore Ecological Interactions in a Defaunated Tropical Forest*

MARCELO LOPES RHEINGANTZ,
ALEXANDRA DOS SANTOS PIRES, AND
FERNANDO A. S. FERNANDEZ

17.1 Background

Worldwide, defaunation leads not only to global and local extinctions but also to the severe loss of ecological interactions (Dirzo et al., 2014).

Reintroductions can be used to restore populations and are also a main tool of the rewilding movement, allowing restoration of ecosystem function (IUCN, 2013). Refaunation (Oliveira-Santos & Fernandez, 2010) and trophic rewilding (Svenning et al., 2016; Seddon, this volume) have been proposed as the cheapest, safest, and most efficient rewilding approaches to restore tropical ecosystems (Galetti et al., 2017).

The Atlantic Forest is a biodiversity hotspot and a priority biome for trophic rewilding (Galetti et al., 2017), prompting the initiation of the REFAUNA programme (Fernandez et al., 2017). Our first target area is Tijuca National Park (TNP), a 3,953 ha reserve within Rio de Janeiro city that suffered extensive deforestation and hunting in the past. In the nineteenth century forest cover was restored but the fauna remained impoverished. In the 1970s there was a multi-species reintroduction effort at TNP (Coimbra-Filho et al., 1973) and at least two species established populations: boa constrictor *Boa constrictor* and channel-billed toucan *Ramphastos vitellinus*. The isolation of TNP means that it cannot be naturally recolonised by most missing species but this makes it a good natural laboratory for studying the benefits and risks of refaunation efforts.

Our goal is to restore TNP as far as possible to pre-colonial conditions and to achieve self-regulating systems by restoring key missing ecological interactions. To accomplish this, we needed to identify the missing species, evaluate the feasibility of reintroductions, translocate individuals, and monitor their fate and the restoration of ecological interactions.

17.2 Methods

In 2008, Ivandy Castro Astor, former TNP director, drew co-author A. S. P.'s attention to the dozens of *Joannesia princeps* seeds (known as 'cutieira' or agouti tree) piled on the ground. They realised that the only species able to disperse them in the area, the red-rumped agouti *Dasyprocta leporina*, was missing. We also noted that other vertebrates included in the TNP management plan were missing – some of which have important ecological roles – such as carnivores, peccaries, reptiles, birds of prey, and primates.

As a response to Ivandy Castro Astor's observations, red-rumped agouti reintroduction started in 2010. Following this, Macedo (2017) plotted mammalian extinctions throughout the Brazilian Atlantic Forest, helping us to understand better which species were missing. Galetti et al. (2017) then described trophic rewilding priorities in Neotropical forests. These ideas were consolidated in a workshop in 2018 to define TNP reintroduction priorities, involving researchers, environmental agencies, park managers, and *ex situ* managers. We used the following criteria: historical record, ecological role, stock availability, available resources at TNP, potential hunting/removal pressure, management conditions, and social support. It was decided to reintroduce generalists of low trophic level first (agoutis, folivore-frugivore primates, tortoises, and others), followed by low-level specialists, and then species of higher trophic levels (native Felidae, Mustelidae, and birds of prey) only when prey abundance could sustain predator populations (Galetti et al., 2017).

17.3 Outcomes

17.3.1 Red-Rumped Agoutis

Red-rumped agoutis (Figure 17.1A) are scatter-hoarding rodents that disperse large seeds at long distances, equivalent to bigger-sized animals. Individuals from a semi-captive population were captured and transported to Rio Zoo facilities. They were examined, quarantined, and tagged and then transported to TNP acclimatisation pens. Thirty-one animals were released between 2010 and 2014, using a delayed release protocol. We monitored individuals by radio tracking (Cid et al., 2014). In 2014, we started to monitor population trends through the use of camera trapping (Kenup et al., 2018).

There was a high mortality rate during quarantine and acclimatisation. However, a population was successfully established in TNP (Kenup

Figure 17.1 Species released at Tijuca by REFAUNA: (A) red-rumped agouti; (B) brown howler; (C) yellow-footed tortoise (photos: Marcelo Rheingantz). (A black and white version of this figure will appear in some formats. For the colour version, please refer to the plate section.)

et al., 2018). The population was estimated to be around 30–40 individuals in 2018 within the surveyed area, and all individuals present since 2016 have been wild-born (Kenup et al., 2018).

Agoutis interacted with at least 23 plant species, hoarding three (*J. princeps*, *Astrocaryum aculeatissimum*, and *Sterculia chicha*). They are the only frugivores able to disperse seeds larger than 30 mm in diameter in TNP, and the seeds of *A. aculeatissimum* and *J. princeps* are only being buried in areas with agoutis (Zucaratto, 2013; Mittelman et al., 2020).

17.3.2 Brown Howlers

Brown howlers *Alouatta guariba* (Figure 17.1B) are folivore-frugivores that have small home ranges for a primate, can disperse seeds, and are charismatic. Six adults, three captive-born and three confiscated from the illegal wildlife trade, were quarantined, health checked, and tagged at Rio de Janeiro Primatology Center (CPRJ) before being sent to the TNP acclimatisation pen. These individuals were released using a delayed release protocol between 2015 and 2017. They have been monitored by radio tracking and active searching.

Two released males were returned to captivity due to their abnormal interaction with visitors. One couple formed in the wild and had three offspring, one each year; they form the single remaining group. A second female was found dead and the other male has been missing since 2017. Transmitters led to serious wounds in the ankles of the males (Genes et al., 2019a). The population remains in the establishment phase and requires additional releases. Unfortunately, this has not been possible due to a regional yellow fever outbreak.

Brown howlers interacted with 60 plant species: for several plants, they became their only disperser in TNP (Genes et al., 2019b). Twenty-one dung-beetle species interacted with brown howler faeces, most of them tunnellers, which can act as second dispersers (Genes et al., 2019b).

17.3.3 Yellow-Footed Tortoises

Yellow-footed tortoises *Chelonoidis denticulatus* (Figure 17.1C) are ecosystem engineers and generalist feeders that can also disperse seeds over long distances. Adults do not have natural predators at TNP and they have a long lifespan. Animals were moved from several captive populations to the Wildlife Screening Center of the Brazilian Institute of the Environment and Renewable Natural Resources in Rio de Janeiro

(CETAS-IBAMA-RJ) in 2019, where they were examined and tagged. Twenty-eight animals were released in 2020 using two different methods, immediate and six-month delayed release, to provide an experimental evaluation of the two methods. Animals were radio tracked once a week.

All released tortoises were still alive three months after release. Some animals dispersed more than 1.3 km from the release area. The monitoring of tortoises was temporarily stopped due to COVID-19. We will return to track them and release more as soon as possible.

Tortoises consumed parts of at least five native plant species, but only the restart of the monitoring will allow us to determine their role in restoring ecological interactions.

17.4 Discussion and Future Recommendations

TNP is a severely defaunated forest; only 11 of the 33 medium and large-sized mammal species that originally occurred there in pre-colonial times were present in 2010 (Macedo, 2017). Many other species are also missing. The REFAUNA programme has achieved promising results, but so far we have released only three species. Other species will be reintroduced, with the aim of restoring complementary ecological interactions. The next candidate species include the blue-and-yellow macaw *Ara ararauna*, jacutinga *Pipile jacutinga*, iguana *Iguana iguana*, red-browed amazon *Amazona rhodocorytha*, and lesser grison *Galictis cuja*. Apart from the grison, all are low-trophic level generalists with non-redundant ecological roles. All of them fit the trophic rewilding criteria (Galetti et al., 2017).

Besides further releases, we will also continue to monitor the species already reintroduced and their interactions. We need to understand how to expand the red-rumped agouti distribution within TNP. In 2018, we started an experiment (delayed versus immediate release, with or without supplementary feeding) in another sector of TNP to define the best reintroduction protocol for the species and to establish a new population. The brown howlers will need population reinforcement, and further releases are planned.

Rewilding programmes should identify key species missing in an area, and their ecological roles, in order to assess which reintroductions can be most effective in restoring function. The refaunation of TNP has been encouraging; we have evidence that the reintroductions carried out are recovering lost ecological interactions, even with a limited number of released individuals. Populations of all species are still well below their respective carrying capacities but hopefully, as they increase and occupy all parts of TNP, they will restore much of the ecosystem functionality in this impoverished Atlantic Forest reserve. The hope is that more conservation translocations will be used as a tool to mitigate the effects of defaunation throughout this threatened biome.

17.5 Summary

- Refaunation can be used to restore empty habitats in the Atlantic Forest.
- TNP has suitable habitat for all ongoing reintroductions.
- At TNP, some plants are only dispersed/consumed by reintroduced species.
- We need to continue monitoring the recovery of ecological interactions and population establishment.

References

Cid, B., Figueira, L., Mello, A., Pires, A. S. & Fernandez, F. A. S. (2014) Short-term success in the reintroduction of the red-humped agouti *Dasyprocta leporina*, an important seed disperser, in a Brazilian Atlantic Forest reserve. *Tropical Conservation Science*. 7, 796–810.

Coimbra-Filho, A. F., Aldrighi, A. D. & Martins, H. F. (1973) Nova contribuição ao restabelecimento da fauna do Parque Nacional da Tijuca. *Brasil Florestal*. 4, 7–25.

Dirzo, R., Young, H. S., Galetti, M., Ceballos, G., Issac, N. J. B. & Collen, B. (2014) Defaunation in the Anthropocene. *Science*. 25, 401–406.

Fernandez, F. A. S., Rheingantz, M. L., Genes, L., et al. (2017) Rewilding the Atlantic Forest: restoring the fauna and ecological interactions of a protected area. *Perspectives in Ecology and Conservation*. 15, 308–314.

Galetti, M., Pires, A. S., Brancalion, P. H. S. & Fernandez, F. A. S. (2017) Reversing defaunation by trophic rewilding in empty forests. *Biotropica*. 49, 5–8.

Genes, L., Cezimbra, T., Moreira, S. B., Pissinatti, A. & Rheingantz, M. L. (2019a) Getting along with radio-telemetry: effects on howler monkeys (*Alouatta guariba clamitans* Atelidae-Primates) welfare and monitoring effectiveness. *Boletim da Sociedade Brasileira de Mastozoologia*. 80, 39–42.

Genes, L., Fernandez, F. A. S., Vaz de Mello, F. Z., da Rosa, P., Fernandez, E. & Pires, A. S. (2019b) Effects of howler monkey reintroduction on ecological interactions and processes. *Conservation Biology*. 33, 88–98.

IUCN (2013) *Guidelines for Reintroductions and Other Conservation Translocations. Version 1.0*. Gland, Switzerland, IUCN Species Survival Commission.

Kenup, C. F., Sepulvida, R., Kreischer, C. & Fernandez, F. A. S. (2018) Walking on their own legs: unassisted population growth of the agouti *Dasyprocta leporina*, reintroduced to restore seed dispersal in an Atlantic Forest reserve. *Oryx*. 52, 571–578.

Macedo, L. (2017) What are we missing? Historical loss of populations and landscape use by mammals in the Brazilian Atlantic Forest. PhD Thesis, Universidade Federal do Rio de Janeiro, Brazil.

Mittelman, P., Kreischer, C., Pires, A. S. & Fernandez, F. A. S. (2020) Agouti reintroduction recovers seed dispersal of a large-seeded tropical tree. *Biotropica*. 52, 766–774.

Oliveira-Santos, L. G. R. & Fernandez, F. A. S. (2010) Pleistocene rewilding, frankenstein ecosystems, and an alternative conservation agenda. *Conservation Biology*. 24, 4–5.

Seddon, P. J. (2023) The role of conservation translocations in rewilding and de-extinction. In Gaywood, M. J., Ewen, J. G., Hollingsworth, P. M. and Moehrenschlager, A. (eds.) *Conservation Translocations*. Cambridge, Cambridge University Press.

Svenning, J. C., Pedersen, P. B. M., Donlan, C. J., et al. (2016) Science for a wilder Anthropocene: synthesis and future directions for trophic rewilding research. *PNAS*. 113, 898–906.

Zucaratto, R. (2013) Os frutos que as cutias comiam: recrutamento da palmeira *Astrocaryum aculeatissimum* na ausência de seu principal dispersor de sementes. MSc dissertation, Universidade Federal Rural do Rio de Janeiro, Seropédica, Brazil.

18 · *Bringing Jaguars and Their Prey Base Back to the Iberá Wetlands, Argentina*

EMILIANO DONADIO, TALÍA ZAMBONI, AND SEBASTIÁN DI MARTINO

18.1 Background

Large carnivores play vital roles in ecosystems. However, they are rapidly declining in distribution and abundance. As they vanish, so do the key ecological processes that depend on them (Estes et al., 2011). Therefore, maintaining functioning large carnivore populations is a major goal of many conservation efforts – to the extent that the reintroduction of these species is gaining acceptance worldwide (Linnell & Jackson, 2019).

Jaguars *Panthera onca* illustrate this pattern. In Argentina, habitat loss and hunting are thought to have reduced populations to fewer than 200 wild jaguars subsisting on less than five per cent of their historical geographic range (Di Bitetti et al., 2016). Indeed, most regions of Argentina were stripped of this apex predator, including the Iberá wetlands, a protected area in Corrientes province, where the last jaguars were recorded in the 1950s. In Iberá, the main prey of jaguars also declined or disappeared, mostly because of commercial hunting and habitat loss to agriculture. Consequently, Iberá posed a daunting challenge: restoring a large carnivore and its prey base while halting current threats in a protected area that lacked strong enforcement.

18.2 Methods

In 1983, the Iberá wetlands were designated a provincial reserve. At that time, the site featured multiple land uses, including cattle ranching, rice farming, and forestry, as well as unauthorised livestock activities and poaching. Its total area was 1.3 million ha, of which 553,000 ha were public land and 750,000 ha private land. In 2009, our organisation, Fundación Rewilding Argentina (FRA), worked with the provincial government to raise the legal status of the wetlands from provincial reserve to that of a provincial park; this change increased the level of protection of the area. Additionally, FRA purchased 159,800 of the 750,000 ha of private land, restored it, and donated it to the Argentine Park Service, who designated these 159,800 ha a national park in 2018. FRA's aims were to establish an area that would support a viable population of jaguars and sufficient wild prey, thereby minimising conflicts with humans.

In 2010, we conducted a habitat suitability analysis for jaguars that evaluated prey and habitat availability and potential threats. The study concluded that Iberá could support up to 90 jaguars with minimal conflict with humans (De Angelo, 2011). Following this, we began to obtain jaguars for reintroduction from two sources: (1) animals that were born in the wild, illegally captured, and held by other parties, but then seized by authorities and donated to FRA; and (2) animals born in our *ex situ* breeding centre. Releases of animals from both sources are underway and involve a similar process. Individuals deemed suitable for release are first held in 1.5 ha pens and fed live prey whilst avoiding contact with humans. After about 18 months, these animals are moved to a 30 ha pen, where they are exposed to most prey and habitats that they will encounter in the park. Animals are finally released once they can kill large prey and show clear elusive behaviours in the presence of humans. We deploy Iridium GPS collars on all individuals for close monitoring.

In parallel, FRA has begun boosting the diversity and abundance of prey for these jaguars. Working in close association with provincial and federal authorities, land protection aimed at the recovery of caimans *Caiman yacare*, capybaras *Hydrochoerus hydrochaeris*, brownish brockets *Mazama gouazoubira*, and marsh deer *Blastocerus dichotomus* has been enhanced. However, other prey species in Iberá were nearly or locally extinct. Therefore, we are carrying out a multi-species conservation translocation programme. Currently, we are reinforcing populations of pampas deer *Ozotoceros bezoarticus* by translocating wild individuals from private lands (Figure 18.1), and reintroducing giant anteaters *Myrmecophaga tridactyla*, mostly by raising and releasing orphan cubs that had been taken illegally from the wild and then seized by authorities. Collared peccaries *Pecari tajacu*, provided by zoos and wildlife shelters, have also been released. All conservation translocations are made following the IUCN guidelines (IUCN, 2013).

Social acceptance is key to the success of reintroductions and other conservation translocations, especially those aiming to restore large

Figure 18.1 Conservation translocations of wild pampas deer from neighbouring private lands, where the deer's habitat is being modified by pine plantations, to restored grasslands inside the protected area. This resulted in two thriving populations of this threatened native species (photo: Rafael Abuín Aido – Fundación Rewilding Argentina). (A black and white version of this figure will appear in some formats. For the colour version, please refer to the plate section.)

carnivores that present a high risk of conflict (Linnell & Jackson, 2019). Therefore, we launched surveys to understand people's perceptions of jaguars and their knowledge of the reintroduction project. We carried out regional (i.e. towns and cities of Corrientes province) and local (i.e. towns near the Iberá wetlands) surveys. Results from these surveys enhanced our communication strategy for the project. We also held meetings with local, provincial, and national authorities. We found that these meetings helped to develop trust and transparency. Authorities and regional leaders were also involved in our communication strategy (see Section 18.3).

18.3 Outcome

Currently, 712,800 ha of the Iberá wetland are protected under a strict conservation designation, providing a large core area where a population of jaguars can thrive. Six jaguars, three males and three females, born in the wild in Brazil and Paraguay that had been captured and subsequently seized by Brazilian authorities were donated to the project. We successfully bred captive jaguars for the first time in 2018. As a result, two female siblings born in the centre were raised without any contact with humans. One of the sisters was released in 2021 and subsequently gave birth. The two females from Brazil gave birth in 2020. These females, each with two four-month-old cubs, and the male also from Brazil were released during 2021. The three Paraguayan jaguars other jaguars are scheduled to be released in 2023.

The enforcement of legal protection within the federal and provincial reserves in Iberá has boosted numbers of at least three important prey species: marsh deer, capybaras, and caimans. Indeed, current marsh deer and capybara densities are higher than those reported for the productive Pantanal, where jaguar populations thrive (Ávila, 2017). Furthermore, translocations of pampas deer were successful; currently, two populations thrive in the park. Likewise, two populations of giant anteaters and two populations of collared peccaries have established. Founding groups of anteaters and peccaries are currently being established in three new locations through additional conservation translocations.

The province-wide survey of stakeholders showed that 95 per cent of the interviewees (n = 433) supported jaguar reintroduction. Interestingly, 65 per cent (n = 15) of interviewed cattle ranchers also supported the project (Caruso and Jiménez Pérez, 2013). The second survey, conducted only in towns near the reintroduction area, showed high levels of support, on condition that animals were kept in semi-captivity (i.e. large fenced corrals; Zamboni, 2015). This strong support was in part triggered by the expectation that jaguars might become an attraction for tourists, thereby boosting the emerging ecotourism industry in the region. Subsequently, we launched an outreach programme emphasising the message that individuals would be released, not held in corrals, and jaguars would be key players in consolidating a new economy based on nature restoration. Finally, as the time to release the jaguars approached, we published a series of short videos in which the governor and provincial representatives of Corrientes, as well as mayors of the main towns around Iberá, ranchers, and cultural and religious provincial leaders, gave explicit support to the programme. Although we did not evaluate quantitatively the impact of these activities, we found no negative responses to the release of the first jaguars. Furthermore, some neighbouring communities that historically relied on livestock production are currently embracing ecotourism as an alternative economic activity.

18.4 Discussion and Future Recommendations

We expect to release a total of eight to 20 adult jaguars (using a ratio of three females to each male) within a period of three years. We will evaluate space use, trophic ecology, and the putative effects of jaguars on ecological communities. We will use camera traps to identify wild-born jaguars and estimate demographic parameters to monitor the trajectory of the population over time. We will monitor population genetics using non-invasive sampling techniques. Demographic and genetic parameters will be used to define the need to reinforce the population with more individuals. Iridium GPS collars will allow us to determine whether to recapture animals when they disperse into risky unprotected areas, or pose an unacceptable risk to livestock, or both. Individuals that conflict repeatedly with humans might be captured and kept at the breeding centre. We will continue to evaluate the attitudes of local people towards jaguars while we provide nearby communities with effective tools to develop an economy based on wildlife observation.

18.5 Summary

- In Argentina, jaguars have been lost from most of their range, but efforts are being made to reintroduce them in the Iberá wetlands, where they went extinct in historical times.
- Before the reintroduction of jaguars, we strengthened the protection status of the reserve, restored the jaguars' prey base, communicated to a broad audience the goals of the project, and are transforming the local economy from extraction based to ecotourism based.
- In 2021, after ten years of preparation, we released three females, two of them with two four-month-old cubs each, and a male.
- Communication campaigns, meetings with all stakeholders, including local communities and decision makers, and a strategy to convert the local economy into ecotourism were key to building strong support for the return of the jaguar.
- Future actions will include releasing eight to 20 individuals within the next three years, monitoring demographic and genetic parameters of the population, maintaining ongoing work with local communities leading to the establishment of a robust ecotourism economy, and close engagement with the public as jaguars disperse into new areas where predation on livestock might occur.

References

Ávila, B. (2017) Evaluación de un método de monitoreo aéreo de fauna mediante fotografía en los Esteros del Iberá (Corrientes, Argentina). Tesis de Maestría, Centro de Zoología Aplicada, Universidad Nacional de Córdoba, Córdoba, Argentina, 68 pp.

Caruso, F. & Jiménez Pérez, I. (2013) Tourism, local pride, and attitudes towards the reintroduction of a large predator, the jaguar *Panthera onca* in Corrientes, Argentina. *Endangered Species Research*. 21, 263–272.

De Angelo, C. (2011) Evaluación de la aptitud del hábitat para la reintroducción del yaguareté en la cuenca del Iberá. Report to The Conservation Land Trust, Corrientes, Argentina, 71 pp.

Di Bitetti, M. S., De Angelo, C., Quiroga V., et al. (2016) Estado de conservación del jaguar en la Argentina. In Medellín, R. A., de la Torre, A., Zarza, H., Chávez, C., et al. (eds.) *El jaguar en el siglo XXI: La perspectiva continental*. Mexico City, Ediciones Científicas Universitarias, Universidad Nacional Autónoma de México, pp. 447–478.

Estes, J. A., Terborgh J., Brashares J. S., et al. (2011) Trophic downgrading of planet Earth. *Science*. 333, 301–307.

IUCN (2013) *Guidelines for Reintroductions and Other Conservation Translocations: Version 1.0*. Gland, Switzerland, IUCN Species Survival Commission.

Linell, J. D. C. & Jackson, C. R. (2019) Bringing back large carnivores to rewild landscapes. In Pettorelli, N., Durant S. M. and du Toit, J. T. (eds.) *Rewilding*. Cambridge, Cambridge University Press, pp. 248–279.

Zamboni, T. (2015) Percepción de actores locales sobre el yaguareté (*Panthera onca*) y su potencial reintroducción en Iberá, Corrientes, Argentina. Tesis de Maestría, Instituto Internacional en Conservación y Manejo de Vida Silvestre, Universidad Nacional de Costa Rica, Heredia, Costa Rica, 86 pp.

19 · *The Return of the Eurasian Beaver to Britain: The Implications of Unplanned Releases and the Human Dimension*

ROISIN CAMPBELL-PALMER, ANDREW BAUER, SIMON JONES, BEN ROSS, AND MARTIN J. GAYWOOD

19.1 Background and Current Status

By the end of the nineteenth century the Eurasian beaver *Castor fiber* was on the verge of extinction (Nolet & Rosell, 1998). Today, through hunting regulations, protective legislation permitting natural expansion, and translocations, there are now thought to be over 1.5 million animals across its natural range (Halley et al., 2020).

19.2 Methods

The Eurasian beaver became extinct on the island of Britain several centuries ago. Assessment of the desirability and feasibility of reintroducing Eurasian beavers began in Scotland in the 1990s. This was prompted by the biodiversity and wider environmental benefits that beavers bring through their ecosystem engineering activities (Brazier et al., 2021; Stringer & Gaywood, 2016). After much debate and changes in political situations, the scientifically monitored, licensed, five-year 'Scottish Beaver Trial' reintroduction started in 2009 in the small, remote catchment of Knapdale on the west coast of Scotland (Jones & Campbell-Palmer, 2014). Around this time reintroduction feasibility studies had also been completed for England and Wales (Gurnell et al., 2008; Jones et al., 2011).

The popular desire to return Eurasian beavers also resulted in the establishment of several private collections in Britain, from some of which animals escaped. Animals may also have been released without authorisation into the surrounding countryside. By the end of the Scottish Beaver Trial in 2014, the beavers resulting from unauthorised releases and escapes in the east of Scotland far outnumbered the official release. After much controversy, debate, and research the Scottish Government announced in 2016 that it was minded to allow both populations to remain, and Eurasian beavers became formally protected in May 2019. This was a historic turning point, the first government-approved reintroduction of a previously extinct mammal in the UK (Gaywood, 2018). In November 2021, the Scottish Government further announced a wider programme of beaver restoration to begin from 2022.

Additional unauthorised escapes have occurred elsewhere in Britain, for example on the River Otter in Devon. This became the first (retrospectively) licensed trial in England from 2015 to 2020 after a public campaign resulted in the backtracking of government body attempts to trap and remove the animals. There is now also a growing drive to establish fenced Eurasian beaver projects to demonstrate beaver benefits, particularly in England. In summer 2022 the UK Government also announced legislation which will establish a legal mechanism to protect beavers and manage their release and control in England. Beavers have, haphazardly, now returned to parts of Britain, some via small-scale official release projects but mainly through either accidental escapes or deliberate unofficial releases. Estimates suggest that around 2,000 individuals could now be present in Britain.

19.3 Outcomes: Population Restoration and the Consequences of Unofficial Releases

Concerns over beaver restoration are varied but generally relate to: socio-economic costs, especially to agriculture; risks to public and animal health; non-native introductions (beaver species and pathogens); diversion of limited conservation resources from existing biodiversity needs; polarisation of views; and limited genetic diversity. The associated issues are complex and varied but we have summarised the key topics here, with a focus on those that apply to unofficial releases:

1. **Sourcing and genetic make-up:** Historically, population 'bottle-necking' has led to a huge reduction of genetic diversity of the species compared to historical levels. The exact composition of the founders of many unauthorised populations remain unknown, but most appear to have originated from Bavaria and Norway, and small numbers from Poland/Lithuania (Campbell-Palmer et al., 2020). The presence of

small, reintroduced populations with restricted genetic flow across the geographically isolated island of Britain means that long-term, planned genetic management may be required. (Ritchie-Parker at al., 2022).

2. **Health status and risk assessment**: Whilst robust health screening was applied for the official Eurasian beaver release process in Scotland (Goodman et al., 2012), this was not the case for any unauthorised releases and escapes. Concerns over lack of traceability and unknown health status for the unofficial populations meant that significant financial resources were needed to run a retrospective trapping and health screening programme (Campbell-Palmer et al., 2021). Screening of wild beavers in Britain has not revealed any significant pathogens to date, although this has been described as a matter of luck rather than the result of any responsible process (Campbell-Palmer et al., 2021). Unauthorised releases have also resulted in beaver persecution and welfare challenges in some places.

3. **Risk of introducing non-native, invasive species:** The North American beaver, *C. canadensis*, is remarkably similar in appearance, behaviour, and ecology: although hybridisation is not an issue, both species are hard to discriminate without more invasive testing. The current spread of *C. canadensis* in parts of Europe, including Finland and Russia, is of great concern, requiring significant conservation resources for eradication (Dewas et al., 2012). Public objections to lethal control can limit its use, even for invasive, non-native species. North American beavers would likely survive well in British landscapes. There is no evidence to date of this species being present in the UK, either through accidental (through lack of species identification expertise) or deliberate introduction (Campbell-Palmer et al., 2020).

4. **Lack of engagement with stakeholders and associated negative impacts:** As significant modifiers of freshwater ecosystems, beaver activities can be challenging in modern, heavily modified landscapes. Tayside beavers appeared on some of the most intensively farmed land for food production in Scotland, without any prior engagement with land managers or other local stakeholders, so conflict was inevitable and damaging. Therefore, the sectors likely to be most affected by beaver presence were given no say in their restoration on Tayside. This resulted in significant tensions and mistrust that had to be addressed in order to move forward. To some, endorsing Eurasian beaver reintroduction was seen as legitimising wildlife crime and potentially encouraging further releases. Similarly, management used to mitigate beaver impacts, particularly by land owners in

low-gradient, prime agricultural areas, has caused controversy with some conservation groups and the wider public; for example, 39 licences were issued in 2021 for lethal control and dam removal on Tayside, with a reported 87 beavers shot and 33 translocated (NatureScot, 2020). Despite the prior development of a management framework including lethal control, these figures increased polarisation between some stakeholder groups and created additional challenges in moving forward strategically. The mistrust resulting from unauthorised and unilateral action can persist for a long time and compromise future conservation work. Although challenging, a compromise agreement was eventually reached between the main parties. The importance of a multi-stakeholder forum to develop management approaches cannot be understated and should be used for other areas and species when appropriate.

19.4 Discussion and Future Acceptance

The return of the Eurasian beaver to Britain has been influenced by many diverse drivers and factors. The question is, if the unofficial releases and escapes of beavers had not taken place in one of the most intensively farmed and productive parts of Scotland, would we by now have had a planned, managed roll-out of formal and less opposed reintroductions in more appropriate catchments? However, that did not happen. Instead the history, and likely future, of Eurasian beaver restoration, is a complex mix of conservation, scientific research, politics, and social attitudes. Whilst an increasingly urbanised human population appears to grow further away from being physically connected to nature, our national love of natural history, fed by the media, has seen very high levels of public support for iconic species conservation projects such as the white-tailed eagle *Haliaeetus albicilla* reintroduction. The majority of the public do not experience direct impacts, which are shouldered by a tiny minority of land managers who may understandably feel ignored or criticised for highlighting the conflicts that beavers can create

The experiences gained through the reintroduction of high-profile species, such as Eurasian beaver, informed the production of the Scottish Code for Conservation Translocations (National Species Reintroduction Forum, 2014) by stakeholders from across the conservation, environmental, and land and water management sectors. The Code is based on IUCN Guidelines and promotes best practice and the consideration of legal, biological, and, importantly, socio-economic factors (Gaywood & Stanley-Price, this volume).

The appearance of beavers outside of official translocations has fast-tracked the retrospective formation of multi-stakeholder forums and the development of appropriate beaver management systems. These have been challenging because they start at points where relationships and trust may have already been damaged. Eurasian beavers have certainly raised a number of land management issues, crossing multiple statutory body remits – including animal health, water management, and wildlife management. 'Retrofitting' management has also proved to be very expensive. Such processes have been complicated by the legal status of the animals, public reaction, and the varying opinions of different land managers. However, the work of such forums is generally constructive. They provide valuable opportunities for a range of interest groups and statutory bodies to work together using a holistic approach to agree shared goals and understand and address wider issues highlighted by beaver presence, such as flood alleviation, managing run-off from intensive farming along riparian habitats, and wetland conservation. The urgent need to develop a more strategic, national approach to Eurasian beaver restoration, with the involvement of wide stakeholder representation, resulted in the production of Scotland's Beaver Strategy (Copsey, 2022). This includes a vision statement for 2045, and planned action from 2022 to 2032.

Ultimately, the power and authority to sanction species reintroductions rest with the UK's devolved political administrations. Parliaments, when setting policy and legislation, must balance scientific evidence with varied public and stakeholder opinions and issues (the use of structured decision-making tools for conservation translocations is described by Ewen et al., this volume). One person's trail-blazing conservation project is another's foolhardy rewilding venture that can damage property and livelihoods, and trust between the land management and wider conservation communities, thereby damaging prospects of future species reintroductions and other ecosystem restoration activities. The added complexity is that the impact of Eurasian beavers' engineering abilities means that active management, including lethal control, maybe needed to enable coexistence with humans in the modern world. Therefore, if we wish to live with Eurasian beavers again, we may need to accept that at times we will also need to actively manage them. This is not an easy position for a politician to sell or communicate to many members of the voting public. Politicians must measure the environmental, social, economic, *and political* costs and benefits of supporting or opposing beavers.

19.5 Summary

- The Eurasian beaver has returned to Britain, presenting fundamental challenges and opportunities for all involved.
- The species will inevitably expand throughout British freshwater systems and provide significant benefits.
- Unofficial releases have presented challenges in terms of sourcing and genetics, health status and disease risks, the risk of introducing the non-native North American beaver species, and the lack of engagement with communities and resulting conflict.
- Agreed approaches require development using multi-stakeholder forums to recognise and promote benefits whilst sensitively managing the impacts of beavers' activities on people's livelihoods.

References

Brazier, R. E., Puttock, A., Graham, H. A., Auster, R. E., Davies, K. H. and Brown, C. M. (2021) Beaver: Nature's ecosystem engineers. *Wiley Interdisciplinary Reviews: Water.* 8, e1494.

Campbell-Palmer, R., Senn, H., Girling, S., et al. (2020) Beaver genetic surveillance in Britain. *Global Ecology and Conservation.* 24, E01275.

Campbell-Palmer, R., Rosell, F., Naylor, A., et al. (2021) Eurasian beaver (*Castor fiber*) health surveillance in Britain: assessing a disjunctive reintroduced population. *Veterinary Record.* 188, e84.

Copsey, J. A. (ed.) (2022) Scotland's Beaver Strategy 2022-2045. IUCN SSC Conservation Planning Specialist Group.

Dewas, M., Herr, J., Schley, L., et al. (2012) Recovery and status of native and introduced beavers *Castor fiber* and *Castor canadensis* in France and neighbouring countries. *Mammal Review.* 42, 144–165.

Ewen, J. G., Canessa, S., Converse, S. J. & Parker, K. A. (2023) Decision-making in animal conservation translocations: biological considerations and beyond. In Gaywood, M. J., Ewen, J. G., Hollingsworth, P. M. and Moehrenschlager, A. (eds.) *Conservation Translocations.* Cambridge, Cambridge University Press.

Gaywood, M. (2018) Reintroducing the Eurasian beaver *Castor fiber* to Scotland. *Mammal Review.* 48, 48–61.

Gaywood, M. J. & Stanley-Price, M. (2023) Moving species: reintroductions and other conservation translocations. In Gaywood, M. J., Ewen, J. G., Hollingsworth, P. M. and Moehrenschlager, A. (eds.) *Conservation Translocations.* Cambridge, Cambridge University Press.

Goodman, G., Girling, S., Pizzi, R., et al. (2012) Establishment of a health surveillance program for the reintroduction of the Eurasian beaver (*Castor fiber*) into Scotland. *Journal of Wildlife Disease.* 48, 971–978.

Gurnell, J., Gurnell, A. M., Demeritt, D., et al. (2008) The feasibility and acceptability of reintroducing the European beaver to England. Sheffield, Natural England/People's Trust for Endangered Species.

Halley, D., Savelijev, A. P. & Rosell, F. (2020) Population and distribution of beavers *Castor fiber* and *Castor canadensis* in Eurasia. *Mammal Review*. 51, 1–21.

Jones, A., Halley, D., Gow, D., Branscombe, J. & Aykroyd, T. (2011) Welsh Beaver Assessment Initiative Report: an investigation into the feasibility of reintroducing European beaver (*Castor fiber*). Wildlife Trusts Wales.

Jones, S. & Campbell-Palmer, R. (2014) The Scottish Beaver Trial: The story of Britain's first licensed release into the wild. Edinburgh, Scottish Wildlife Trust and Royal Zoological Society of Scotland.

National Species Reintroduction Forum (2014) The Scottish Code for Conservation Translocations. Inverness, Scottish Natural Heritage.

NatureScot (2022) *Summary of Beaver Populations and Licence Returns Covering the Period 1st January to 31st December 2021*. Available from: www.nature.scot/doc/summary-beaver-populations-and-licence-returns-covering-period-1st-january-31st-december-2021 [Accessed 28 August 2022].

Nolet, B. A. & Rosell, F. (1998) Comeback of the beaver *Castor fiber*: an overview of old and new conservation problems. *Biological Conservation*. 83, 165–173.

Ritchie-Parker, H., Ball, A., Campbell-Palmer, R., Taylor, H. & Senn, H. (2022) Genetic diversity analysis of beaver (*Castor fiber*) in England. Natural England Research Report NECR433, Peterborough.

Stringer, A. P. & Gaywood, M. J. (2016) The impacts of beavers *Castor* spp. on biodiversity and the ecological basis for their reintroduction to Scotland, UK. *Mammal Review*. 46, 270–283.

20 · *The Role of Community Engagement in Conservation Translocations: The South of Scotland Golden Eagle Project (SSGEP)*

CATHERINE BARLOW

20.1 Background

The South of Scotland Golden Eagle Project (SSGEP) is an exciting, ambitious, and collaborative project between land managers and conservationists working to increase the breeding population and range of golden eagles *Aquila chrysaetos* in southern Scotland. Identified by the Scottish Government as a priority for its biodiversity conservation programme (Scottish Government, 2020), SSGEP offers a ground-breaking opportunity to galvanise community interest and support for Scotland's iconic bird.

Once widespread, the population of golden eagles in the south of Scotland is now tiny and fragmented. The SSGEP's legacy will be a healthier and viable population of golden eagles in the south of Scotland, enjoyed and supported by local communities and land managers, and forming the basis for greatly enhanced ecotourism opportunities and wider economic development. SSGEP launched in December 2017 with the first three eagles released in summer 2018. The six-year delivery phase is planned for completion in 2024.

20.2 Partnership Approach

The partnership involves Scottish Land & Estates representing landowning interests, the Royal Society for the Protection of Birds (RSPB Scotland) representing conservation interests, the statutory bodies Scottish Forestry (SF) and NatureScot, and the Southern Uplands

Partnership (SUP). SUP, based near Selkirk in the Scottish Borders, is host partner of SSGEP and key to promoting the project as integral to the communities and people of southern Scotland.

The project follows the established translocation protocols set out in the Scottish Code for Conservation Translocations (National Species Reintroduction Forum, 2014; Gaywood & Stanley-Price, this volume). This has involved locating nests containing twin golden eagles in the Scottish Highlands and Islands, translocating one chick from each pair to pens in the south of Scotland, and then releasing the young birds after a few weeks once they are ready (SSGEP, 2021). These releases therefore represent a reinforcement by SSGEP of the small and isolated population of golden eagles in the south of Scotland (three pairs in 2021). The vital part of the SSGEP will be to build collaboration between land owners, their employees, and local conservationists who need to work together to deliver its objectives. In previous projects of this type, landowning and land management interests have not been centrally involved in their design, management, and implementation, although they have been heavily involved in some aspects of project delivery. Through this novel, integrated approach we shall build a firm, loyal buy-in from landowning and land managing interests and create a platform for future collaborative working to improve the long-term prospects of golden eagles in the south of Scotland. SSGEP will seek to break down barriers of suspicion and mistrust that might presently exist. We aim to involve both local raptor experts and local landowners in the monitoring and wider work of the project. Volunteers and interested members of the public will also be asked to assist with various aspects of the project delivery.

20.3 Community Engagement

Education and community outreach work is vital to the lasting success of this project and is delivered by two full-time 'Community Outreach Officers' who work with schools, local communities, tourism bodies, and others, all with an interest in the success of the SSGEP. Land manager events will be targeted to promote beneficial land and forestry management operations for golden eagles.

Through the first three years of the 'Eagle Schools' programme we have reached over 3,000 primary aged children from 11 schools across the south of Scotland. The programme consists of 10 engaging, eagle-themed lessons covering everything from eagle ecology, habits, and current challenges to the bird's historical significance to the people of

Scotland. The sessions have been enthusiastically received by students and staff alike, and many schools are planning to run 'Eagle Schools' annually as part of the curriculum. Schools in the south have also been 'twinned' with schools in Sutherland and the Highlands from where the birds have been sourced. Students exchange letters and stories about 'their' eagles as well as other wildlife. Eagle School culminates in the participating schools opening their doors to the local community with the students presenting their learning to the people of their town or village. These school 'Eagle Days' have been vital for reaching these individuals within the communities who might not have otherwise engaged with the project.

Another successful element to our outreach work has been a partnership with Scouting Scotland that resulted in the creation of an occasional 'Eagle Champions' badge. Through a series of tasks, challenges, and educational sessions (including an exciting tree climb to achieve an eagle's eye view of the Southern Uplands), around 100 young scouts across the south are working towards their Eagle Champions badge so far (Figure 20.1).

As well as working to engage and inspire young people, we also reach out to community groups across the south. Talks, presentations, and raptor identification courses have all been well received, giving the team the opportunity to address personally any concerns or difficult questions that may arise. Overall, the general feeling garnered from these presentations has been one of excitement and a desire to have an increased chance of sighting a magnificent iconic raptor over their local landscape. In all these different ways, during the first two years of SSGEP we have engaged with over 11,000 members of communities across the south of Scotland and beyond.

The COVID-19 pandemic was a significant challenge for the project; the 2020 translocation season was cancelled and all face-to-face public engagement by the SSGEP team was halted. The team switched to increasing opportunities for online engagement through the project website (www.goldeneaglessouthofscotland.co.uk) while planning ahead for the anticipated lifting of restrictions. A series of project blogs (www.goldeneaglessouthofscotland.co.uk/blog) involving a wide variety of stakeholders, experts, and volunteers and online resources (www.goldeneaglessouthofscotland.co.uk/outreach) encouraged engagement despite COVID-19 restrictions.

In partnership with a local estate we have created an 'Eagle information point' called 'The Eyrie' at Philiphaugh Waterwheel Café at the

The South of Scotland Golden Eagle Project · 459

(a)

(b)

Figure 20.1 Young scouts (a) meeting a golden eagle and (b) climbing trees to get an eagle's eye view of the landscape, to earn their 'Eagle Champions' badges (photo: Phil Wilkinson/SSGEP). (A black and white version of this figure will appear in some formats. For the colour version, please refer to the plate section.)

gateway to the Yarrow Valley. Informative media and video are included in a visitor's experience; these also create opportunities for trained volunteers to become involved with the project. It is hoped this could be the first of many partnerships with local communities and landowners to encourage visitor interest in golden eagles across the south of Scotland.

Moffat also hosted an Eagle Festival celebrating golden eagles and encouraging nature based tourism in the area in September 2021. A week-long programme of events including an environmental street fair, informative talks every evening from a wide range of countryside stakeholders, a family fun day, a hill run through Eagle Country, live music, eagle-themed menus at local restaurants, and more allowed us to not only engage more effectively with a wider range of the community but also benefit the town of Moffat economically. By working in partnership with Moffat and, in the future, other towns across the south to develop their role as ecotourism destinations, we hope to create a lasting legacy of enthusiasm, excitement, and welcome for the golden eagles in the south of Scotland. Our intention is to evaluate the socio-economic impacts of these activities and the wider engagement towards the end of the project

20.4 Building Trust

The relationship between the game industry and conservation has a long and often difficult history – not constrained to eagles but birds of prey in general. At the project's conception, the intention was to develop an equal partnership of land management sectors and conservationists that would foster a positive and productive relationship in the future. We hope that by building trust between conservationists and land management practitioners we will reduce or prevent conflict and promote a more positive relationship with golden eagles. It was agreed to employ a dedicated member of staff to liaise directly with landowners, gamekeepers, and shooting interests – to inform, reassure, and advise them on how best to accommodate golden eagles. We feel that this 'relationship' is vital to the future of golden eagles in the south of Scotland. At the end of SSGEP we also hope to have a network of landowners, their employees, and interested members of the public who have a shared passion for 'their' local golden eagles and a wish to see a thriving population of these birds in the south of Scotland. Stakeholders in the project area will be communicating between themselves through sharing information and monitoring the outcome for future breeding pairs of golden eagles.

20.5 Looking Ahead

The area where we are focusing our project is often perceived as a 'poor relation' when it comes to high-profile conservation projects in Scotland, which are more often directed to areas perceived to have higher biodiversity value, for example Scotland's National Parks or the Highlands and Islands. Public attitudes towards birds of prey and their conservation needs in the south of Scotland have changed. A number of other bird of prey populations, including those of the buzzard *Buteo buteo* and goshawk *Accipiter gentilis*, have recovered in this part of Scotland. However, this will not happen for golden eagles unless they are given a helping hand. The first birds released as part of SSGEP are thriving and interacting with naturally fledged young in the south of Scotland. We hope to add to this population in coming years.

20.6 Summary

- We are using communication and engagement with stakeholders to attempt to foster a positive view of eagles.
- We are exploring the opportunities for nature based tourism to highlight the benefits of eagles.
- We are taking a partnership approach with equal input from land management sectors and conservationists.

References

Gaywood, M. J. & Stanley-Price, M. (2023) Moving species: reintroductions and other conservation translocations. In Gaywood, M. J., Ewen, J. G., Hollingsworth, P. M. and Moehrenschlager, A. (eds.) *Conservation Translocations*. Cambridge, Cambridge University Press.

National Species Reintroduction Forum (2014) *The Scottish Code for Conservation Translocations*. Inverness, Scottish Natural Heritage.

Scottish Government (2020) *2020 Challenge for Scotland's Biodiversity: A strategy for the conservation and enhancement of biodiversity in Scotland*. Edinburgh, Scottish Government.

SSGEP (2021) *All about the project: an introduction to the project aims and objectives*. Available from: www.goldeneaglessouthofscotland.co.uk/about/the-project [Accessed 25 March 2021].

21 · *The European Native Oyster and the Challenges for Conservation Translocations: The Scottish Experience*

CASS BROMLEY AND DAVID W. DONNAN

21.1 Background

Ostrea edulis, the European native or flat oyster (hereafter 'native oyster', Figure 21.1) has a natural range in the north east Atlantic from Norway to the north coast of Africa and into the Mediterranean and Black Sea. Evidence from shell middens indicates that harvesting of the species has taken place since the Neolithic, cultivation since around 95 BCE, and commercial fishing since the Middle Ages.

The Industrial Revolution brought demand for cheap food, and the native oyster became popular with the growing urban populations. It is believed that oyster beds throughout European waters occupied substantially larger areas than is the case today. A nineteenth century map shows extensive offshore banks in the North Sea (Olsen, 1883). This illustrates general locations, whereas the actual footprint of the beds was probably far more complex. In the Firth of Forth, Scotland, oyster grounds occupied an area larger than the adjacent city of Edinburgh, and the 'Oyster Ground' in the North Sea is estimated to have covered 20,000–25,000 km^2.

In the nineteenth century, there was a tendency to view fish and shellfish stocks as 'inexhaustible'. Not only were the oysters fished for human consumption, they were also used for the replenishment of depleted oyster grounds elsewhere. This activity, together with the effects of pollution, disease, and other factors, resulted in the collapse of populations within a relatively short space of time, particularly in the deeper water beds and, for example, the Firth of Forth.

The decline of Scotland's native oysters is similar to what has happened to the species throughout European waters. Native oysters

(a)

(b)

Figure 21.1 Native oysters in (a) the Firth of Clyde and (b) the Loch Sween Marine Protected Area (photos: (A) NatureScot; (B) David Donnan/NatureScot). (A black and white version of this figure will appear in some formats. For the colour version, please refer to the plate section.)

are classed as a Threatened and/or Declining Species and Habitat by European legislation (the OSPAR Convention) (OSPAR, 2009), and this treatment as a conservation priority is reflected in their status as a Priority Marine Feature in Scottish waters. In terms of restoring the species, Scottish waters divide roughly into two situations. On the east coast, where native oysters have been extirpated, the situation is relatively straightforward. Projects such as the Dornoch Environmental Enhancement Project (DEEP) have selected sites with evidence of previous populations where the environment remains capable of supporting reintroduced oysters and are addressing biosecurity issues for translocating stock. On the west coast and the islands, the picture becomes more complex. Here, there are fragmented populations, mainly in the many sea lochs (University Marine Biological Station Millport, 2007). Rather than full restoration, any intervention is more likely to involve reinforcement/enhancement to increase extant populations.

Although restoration and enhancement are at an early stage, experience to date has confirmed that biosecurity is the overarching consideration and is especially important where there are already native oysters present. Movements of shellfish carry inherent risks of transporting disease-causing organisms, predators, and pests, as well as a variety of non-native species (see Carlton & Ruiz, 2005). As native oyster populations collapsed, other oyster species were imported as alternatives. Commercial translocations from overseas also brought non-native species that are harmful to both the native oyster beds and wider marine communities, for example the American oyster drill *Urosalpinx cinerea* and slipper limpet *Crepidula fornicata* (Hancock, 1969). Since its introduction to European waters in 1979, the parasite *Bonamia ostreae* has caused mortalities of more than 90 per cent in some populations of the native oyster (Culloty & Mulcahy, 2007).

Early restoration efforts involving translocations were largely commercially driven, to regenerate declining fisheries. In contrast, contemporary native oyster restoration is driven mainly by biodiversity and ecosystem goals, including: recognising the importance of complex habitat for biodiversity; restoring the native oyster's range; enhancing resilience to climate change; and providing ecosystem services such as water quality enhancement and carbon sequestration. The growing interest in restoration has resulted in new networks being established to facilitate knowledge transfer and develop best practice (e.g. the Native Oyster Restoration Alliance (NORA) and the UK & Ireland Native Oyster Network).

Successful native oyster restoration faces a number of uncertainties and challenges, the most important of which are the availability of biosecure stock in sufficient quantity, and mitigating disease and non-native species risks during translocation. Other issues include developing restoration projects in the context of the increasingly complex mix of users, activities, and stakeholders of coastal waters. The oyster restoration networks therefore have a crucial role to play in developing generic guidance and standards to meet these challenges (see, for example, Preston et al., 2020; zu Ermgassen et al., 2020).

21.2 Methods

Two main methods are currently employed for native oyster conservation translocation. First, there is the translocation of oysters to a site where they are then kept in containment, for example in bespoke structures or in baskets, and sometimes utilising existing infrastructure such as marina pontoons. The oysters in the containers may be juveniles being grown prior to seabed deployment, or mature individuals forming a broodstock for the purpose of providing larvae to enhance local recruitment. Second, oysters may be translocated directly onto the seabed, often in conjunction with the deposit of shell material (cultch) as a substrate for the settlement of future larvae.

A key component of these translocations is a workable biosecurity protocol addressing both diseases of oyster and non-native species (Preston et al., 2020). This protocol must be capable of operating feasibly at a scale relevant to the numbers of oysters being translocated. The biosecurity considerations also extend to the deposit of cultch, usually marine shell obtained as a bivalve fishery by-product. This shell material has the potential to contain passenger organisms and therefore must be treated or left to weather until it becomes biologically inert.

The availability of biosecure stock and cultch for restoration projects remains a significant bottleneck. The natural populations of native oysters are therefore of great importance as a potential source of broodstock. However, native oyster populations have a high conservation status (OSPAR, 2009; Donnan et al., 2016) and it is critical that restoration activity (including collection for broodstock) does not present a risk to their conservation. Other challenges include preserving the genetic integrity of populations and their local adaptations. Reliance on hatchery-produced stock using small numbers of parents can create genetic bottlenecks and introduce genetic drift or homogenise populations.

21.3 Outcome

The desired outcomes for native oyster translocation in Scotland are restoring the species to its former range on the east coast (i.e. the re-establishment of viable, self-sustaining populations where they have been extirpated) and enhancing or strengthening extant populations of native oysters on the west coast in areas where they still exist. With current projects at an early stage, it is not possible to assess success at this point. However, experience so far points to the value of adequately incorporating the following elements that have consequences for planning, funding, and time scales:

- Carrying out adequate reviews and surveys to understand the receiving environment, particularly where extant populations of oysters are present, is critical for setting meaningful objectives for the project (including, for example, informing an objective decision on whether to proceed in any given area).
- Establishing an adequate supply of biosecure oysters with the necessary genetic provenance is a major priority to support conservation translocations for this species.
- Monitoring to assess success and biosecurity is fundamental but may not be given adequate resources and long-term commitment if funding is only confirmed for short time scales.

Consequently, the combination of oyster biology and practical considerations means that projects need to factor in realistic time scales of at least 20–25 years from inception to gaining a reliable indication that the restored population has become self-sustaining.

21.4 Discussion and Future Recommendations

Restoration involving conservation translocation is not the only possible action that can be taken to conserve native oysters. Where there is already a population present, it may be more appropriate to focus on protection from pressures rather than undertake restoration with the risk of an adverse outcome through the introduction of disease or non-native species.

Where it is deemed appropriate to carry out conservation translocations, there is no 'one-size-fits-all' approach; each proposal needs to be assessed on its own merits. Requirements will vary depending on location including, for example, existing activities, the extent of suitable habitat, and relevance to protected areas.

It is important for restoration practitioners to be aware of the risks and to ensure that rigorous biosecurity protocols are built into project planning at every stage of the process. There is a need to improve knowledge of extant native oyster populations to determine where enhancement may be best employed. Although native oysters are a natural component of north east Atlantic ecosystems, in many locations they have been absent for over 100 years. Projects need to be mindful of this and also of setting realistic time scales for work.

21.4.1 Recommendations

- Marine management frameworks vary between jurisdictions. Therefore, practitioners should seek advice from the relevant competent authorities and nature agencies as early as possible in the design phase of projects.
- Projects need to set clear objectives and realistic, long-term time scales from the outset.
- Objectives should be informed by good baseline data for the receiving site and a programme of monitoring. Both are essential for adaptive management, particularly when a project has connectivity to a protected area.
- Project plans should be built around conservation translocation best practice, for example the IUCN (2013) guidelines, and national codes (e.g. National Species Reintroduction Forum, 2014), and implement workable biosecurity protocols.
- Large-scale conservation translocations should not be carried out until issues surrounding biosecure, reliable sources of oysters have been addressed.

21.5 Summary

- Restoration of the European native oyster is underway across Europe, and Scotland provides a good example of the current experience and lessons to be learnt for future restoration work.
- Translocations of shellfish carry inherent risks that must be taken into account when considering whether this is the most appropriate conservation action to take.
- Marine restoration projects require strict biosecurity protocols to address disease and non-native species risks relevant to both species' translocation and associated operations.

- Native oyster restoration is a long-term commitment: projects need to organise appropriate surveys and consultation periods before starting work, together with long-term monitoring after the initial deployment phase.

References

Carlton, J. T. & Ruiz, G. M. (2005) The magnitude and consequences of bioinvasions in marine ecosystems: implications for conservation biology. In Norse, E. A. and Crowder, L. B. (eds.) *Marine Conservation Biology: The Science of Maintaining the Sea's Biodiversity*. pp. 123–148.

Culloty, S. C. & Mulcahy, M. (2007) *Bonamia ostreae* in the native oyster *Ostrea edulis*: a review. Marine Environment and Health Series, No. 29. Marine Institute.

Donnan, D. W., Manson, F. J. & Macdonald, I. (2016) Native oyster. In Gaywood, M. J., Boon, P. J., Thompson, D. B. A. and Strachan, I. M. (eds.) *The Species Action Framework Handbook*. Battleby, Perth, Scottish Natural Heritage. Available from: www.nature.scot/species-action-framework-handbook [Accessed 15 June 2021].

Hancock, D. A. (1969) Oyster pests and their control. Ministry of Agriculture, Fisheries and Food Laboratory Leaflet (New Series) No. 19, Lowestoft.

IUCN (2013) *Guidelines for Reintroductions and Other Conservation Translocations. Version 1.0*. Gland, Switzerland, IUCN Species Survival Commission.

National Species Reintroduction Forum (2014) *The Scottish Code for Conservation Translocations*. Inverness, Scottish Natural Heritage.

Olsen, O. T. (1883) *The Piscatorial Atlas of the North Sea, English and St. George's Channels, Illustrating the Fishing Ports, Boats, Gear, Species of Fish (How, Where, and When Caught), and Other Information Concerning Fish and Fisheries*. London, Taylor and Francis.

OSPAR (2009) *Background Document on* Ostrea edulis *and* Ostrea edulis *Beds*. OSPAR Commission Biodiversity Series. Available from: www.ospar.org/documents?v=7183 [Accessed 8 June 2022].

Preston, J., Gamble, C., Debney, A., Helmer, L., Hancock, B. & zu Ermgassen, P. S. E. (eds.) (2020) *European Native Oyster Habitat Restoration Handbook*. London, The Zoological Society of London.

University Marine Biological Station Millport (2007) *Conservation of the native oyster* Ostrea edulis *in Scotland*. Battleby, Perth, Scottish Natural Heritage Commissioned Report, no. 251. Available from: www.nature.scot/naturescot-commissioned-report-251-conservation-native-oyster-ostrea-edulis-scotland [Accessed 7 July 2021].

zu Ermgassen, P., Bonačić. K., Boudry, P., et al. (2020) Forty questions of importance to the policy and practice of native oyster reef restoration in Europe. *Aquatic Conservation: Marine and Freshwater Ecosystems*. 30, 2038–2049.

22 · *Slow and Steady Wins the Race: Using Non-native Tortoises to Rewild Islands off Mauritius*

CARL G. JONES, VIKASH TATAYAH, ROSEMARY MOORHOUSE-GANN, CHRISTINE GRIFFITHS, NICOLAS ZUËL, AND NIK COLE

22.1 Introduction

In this study, we use an extant species to fulfil the ecological roles of an extinct one in order to restore species interactions and natural processes, and to nurture a more resilient ecosystem. A replacement species must be ecologically similar and, if possible, should be taxonomically closely related. Here we consider the use of non-native tortoises to fulfil the ecological function of the extinct *Cylindraspis* tortoises from Mauritius and to help restore plant communities. These studies were conducted on Ile aux Aigrettes, a 26 ha coralline islet about 850 m off the coast of Mauritius, and on Round Island, a rugged 219 ha island 22.4 km north of Mauritius (Tatayah et al., 2018).

The endemic tortoises in the genus *Cylindraspis* were herbivores and keystone species in the ecology of the Mascarene Islands (Mauritius, Réunion, and Rodrigues). They would have been important grazers, browsers, tramplers, seed dispersers, and nutrient cyclers (Griffiths, 2014). There was a radiation of five species across the three Mascarene Islands, with two species in Mauritius, the domed *C. inepta* and a saddle-backed *C. triserrata*. Tortoises were found in the highest densities in the more open forest around the coast (Cheke & Hume, 2008).

The tortoises available as ecological replacements, the radiated tortoise *Astrochelys radiata* (Madagascar) and Aldabra giant tortoise *Aldabrachelys gigantea* (Seychelles), were believed to be the most appropriate based on genetic and morphological traits (Griffiths et al., 2012; Griffiths, 2014).

22.2 Endemic Tortoise and Plant Co-evolution

There was assumed co-evolution between tortoises and the native vegetation (Eskidsen et al., 2004; Griffiths et al., 2010; Griffiths, 2014). The tortoises, as well as the endemic, extinct, Mauritius sheldgoose *Alopochen mauritiana* (Hume, 2017), probably shaped vegetation communities by grazing in open, flat, lowland areas. This plant community was characterised by a mosaic of palms and screw-pines with some hardwoods, and open areas that were likely composed of a community of grasses, sedges, and herbaceous plants. These latter areas would probably have been closely cropped, with patches of tussock grasses such as the Mascarene endemic *Chrysopogon argutus* forming a palm savannah (Vaughan & Wiehe, 1937). This tussock grass is heliophilous, has adaptations to avoid being grazed, and provides important habitat for invertebrates, reptiles (including juvenile tortoises), and nesting seabirds. By the twentieth century, the only remnant of this community was on Round Island (Vaughan & Wiehe, 1937).

In 1986, introduced rabbits *Oryctolagus cuniculus* were eradicated from Round Island. In the absence of their grazing pressure, non-native grazing-intolerant and invasive grasses (e.g. *Cenchrus echinatus* and *Chloris barbata*) flourished and the native tussock-forming *C. argutus* declined (North et al., 1994; Bullock et al., 2002). It was suggested that without grazers to restore the grazing climax plant community, several plant species could become extinct (Jones, 2002; Griffiths et al., 2010).

22.3 Tortoise Studies on Ile Aux Aigrettes

In the 1990s, the Mauritian Wildlife Foundation established Aldabra giant tortoises on Ile aux Aigrettes to study how they interacted with the vegetation (Tatayah et al., 2018). Feeding observations and DNA meta-barcoding found that the tortoises fed on native and non-native plants (Moorhouse-Gann, 2017). The tortoises consumed the seeds of large-seeded natives such as *Eugenia lucida* and the ebony tree *Diospyros egrettarum*, and dispersed whole seeds in their faeces. The ebony seeds from tortoise faeces show enhanced germination rate (29.0% versus 1.8%) and time (52.3 versus 82.4 days) compared to the seeds in whole fruit (Griffiths et al., 2011; Griffiths, 2014). Before the introduction of tortoises, recruitment was limited as few seeds were dispersed beyond the shadow of adult trees.

22.4 Tortoise Studies on Round Island

Tortoises were introduced to Round Island in 2007 to see if they would restore the grazing climax vegetation community and be effective seed dispersers for the native, endemic screw-pines, palms, and hardwood trees. Twelve sub-adult Aldabra giant tortoises (about eight years old) and 12 adult radiated tortoises were kept in enclosures for 11 months to investigate how they interacted with the vegetation. Following release, they grazed on most non-native plants and significantly reduced vegetation cover, height, and seed production, reflecting what was seen in the enclosure study (Griffiths et al., 2013). They generally avoided feeding on native leaves, although the Aldabra giant tortoises readily consumed the fallen fruits of the screw-pine *Pandanus vandermeeschii* and the palm *Latania loddigesii*. Improved germination and seedling establishment of dispersed seeds was observed (Griffiths, 2014). The tortoises created grazed areas, like the tortoise lawns on Aldabra (Merton et al., 1976), although many species hypothesised as components of a Mauritian grazing climax vegetation community remain absent or rare (Moorhouse-Gann et al., 2021). The native tussock grass *C. argutus* that was not grazed, increased in abundance having benefited from reduced competition from the fast-growing non-native grasses (Griffiths et al., 2013).

Both tortoise species tended to keep to the flatter areas of the island, on the summit and on some of the gentler slopes. Consequently much of the island was not used. During 2016–2018, when the population was more than 650 mostly juvenile and sub-adult individuals, they occupied about 20 per cent of the island. The larger tortoises moved more during the dry season, whereas smaller individuals moved more during the wet season and travelled more over varied terrain and on the bare slopes. The tortoises were unable to traverse rocky steps more than 15–20 cm, and it was predicted that 90% of the tortoises could reach 60% of the island, but only 70% could reach a minimum of 85% of the island (Stephani, 2017).

Aldabra giant tortoises are better adapted to living on Round Island than the radiated tortoise, which has a high-domed carapace and high centre of gravity and is not adapted for moving over rough ground and steep slopes. We have focused upon the Aldabra giant tortoise for our studies because it is ecologically more generalised and more effective at spreading the seeds of palms and screw-pines and controlling non-native vegetation (Gerlach, 2014; Griffiths, 2014). Since December 2010, we have steadily increased the number of Aldabra giant tortoises: in

Figure 22.1 A two-year-old tortoise, hatched on Round Island, being assessed for microchip insertion and individual recognition (photo: Nik Cole). (A black and white version of this figure will appear in some formats. For the colour version, please refer to the plate section.)

2020 there were 683 (95% confidence interval: 677–689) of the 691 individuals released since 2008. A further 31 tortoises, known to have hatched on the island since 2016, have survived to greater than two years of age (Figure 22.1), with many smaller unidentifiable hatchlings being found.

22.5 Discussion

These studies are in their early stages and it will be decades before we see the full impact of the tortoises upon the plant and animal communities, with the likely emergence of yet-unknown interactions, as well as differences across spatial and temporal scales. We demonstrate the potency of using ecological replacements to reactivate ecological processes and functions, such as dispersing the seeds of threatened endemic trees and palms, and the control of some invasive plants (Figure 22.2). The recovery of a grazing climax community and the restoration of tussock grasses are being witnessed, with plans to reintroduce into the grazed areas herbaceous species that were once probably part of this community.

Using Non-native Tortoises to Rewild Islands · 473

Figure 22.2 Surveying the impact of tortoise herbivory in exclosure (pictured) and control plots on Round Island (photo: Nik Cole). (A black and white version of this figure will appear in some formats. For the colour version, please refer to the plate section.)

The Mascarene Islands have been radically altered by extinctions and there are opportunities to develop a community of proxy species. The tortoises we have used are primarily grazers: a browsing saddle-backed tortoise that would likely be a suitable replacement for *C. triserrata* is the Española tortoise *Chelonoidis hoodensis* from the Galápagos Islands.

There are other species that may be used to replace extinct taxa that performed important ecological functions. Mauritius had flightless rails that would have been scavengers and predators of invertebrates and small reptiles, including an undescribed species which was a derivative of the white-throated rail *Dryolimnas cuvieri* from Madagascar. This was similar to the still extant flightless Aldabra rail *D. (cuvieri) aldabranus* (Hume, 2017), which would be an appropriate ecological replacement to trial (Jones, 2008).

Seabirds bring nutrients from the marine environment to the terrestrial, and replacing lost seabirds is a priority. The Mascarene Islands supported a population of the Mascarene booby *Papasula* sp., last recorded in 1832 (Hume, 2017). These were closely allied to the Abbott's booby *Papasula abbotti*, which would be a suitable replacement.

The reinstatement of missing ecosystem interactions by the tortoises illustrates that highly degraded ecosystems may be rebuilt using ecological replacements. Further research is essential to understand the role of other potential ecological replacements in shaping the Mauritian ecosystems.

22.6 Summary

- The radiated tortoise and Aldabra giant tortoise were used as ecological replacements for the extinct Mauritian tortoise *Cylindraspis inepta*. Aldabra giant tortoises were better adapted as seed dispersers and grazers, and have become the species of choice. It is intended to remove the radiated tortoises from Round Island.
- It is suggested there was an open grazing climax plant community maintained by tortoises with heliophilous native plants that have adaptations to avoid being grazed or browsed and respond to grazing with a prostrate growth form. The Aldabra giant tortoises fed on fallen fruits and spread the seeds of hardwood trees, screw-pines, and palms in their droppings that subsequently demonstrated enhanced germination and growth rates.
- Preliminary work was done on Ile aux Aigrettes before the release on Round Island.
- The tortoises have established grazed areas colonised by the native tussock grass *Chrysopogon argutus* that the tortoises do not graze; *C. argutus* is therefore benefiting from reduced competition from the fast-growing non-native grasses.

References

Bullock, D. J., North, S. G., Dulloo, M. E. & Thorsen, M. (2002) The impact of rabbit and goat eradication on the ecology of Round Island, Mauritius. In Veitch, C. R. and Clout, M. N. (eds.) *Turning the Tide: The Eradication of Invasive Species*. Gland, Switzerland, ISSG, SSC, World Conservation Union, pp. 53–63.

Cheke, A. & Hume, J. (2008) *Lost Land of the Dodo*. London, T & A. D. Poyser.

Eskildsen, L. I., Olesen, J. M. & Jones, C. G. (2004) Feeding response of the Aldabra giant tortoise (*Geochelone gigantea*) to island plants showing heterophylly. *Journal of Biogeography*. 31, 1785–1790.

Gerlach, J. (ed.) (2014) *Western Indian Ocean Tortoises. Ecology, Diversity, Evolution, Conservation, Palaeontology*. Manchester, SIRI Scientific Press.

Griffiths, C. J. (2014) Rewilding in the Western Indian Ocean. In Gerlach, J. (ed.) *Western Indian Ocean Tortoises. Ecology, Diversity, Evolution, Conservation, Palaeontology*. Manchester, SIRI Scientific Press, pp. 325–349.

Griffiths, C. J., Jones, C. G., Hansen, D. M., et al. (2010) The use of extant non-indigenous tortoises as a restoration tool to replace extinct ecosystem engineers. *Restoration Ecology*. 18, 1–7.

Griffiths, C. J., Hansen, D. M., Jones, C. G., Zuël, N. & Harris, S. (2011) Resurrecting extinct interactions with extant substitutes. *Current Biology*. 21, 762–765.

Griffiths, C. J., Zuël, N., Tatayah, V., Jones, C. G., Griffiths, O. & Harris, S. (2012) The welfare implications of using non-native tortoises as ecological replacements. *PLoS ONE*. 7, e39395.

Griffiths, C. J., Zuël, N., Jones, C. G., Ahamud, Z. & Harris, S. (2013) Assessing the potential to restore historic grazing ecosystems with tortoise ecological replacements. *Conservation Biology*. 27, 690–700.

Hume, J. P. (2017) *Extinct Birds*, 2nd ed. London, Christopher Helm.

Jones, C. G. (2002) Reptiles and amphibians. In Perrow, M. R. and Davy, A. J. (eds.) *Handbook of Ecological Restoration. Volume 1. Principles of Restoration*. Cambridge, Cambridge University Press, pp. 355–375.

Jones, C. G. (2008) *Practical conservation on Mauritius and Rodrigues, steps towards the restoration of devastated ecosystems*. In Cheke, A. and Hume, J. (eds.) *Lost Land of the Dodo*. London, T & A. D. Poyser, pp. 226–259.

Merton, L. F. H., Bourn, D. M. & Hnatiuk, R. J. (1976) Giant tortoise and vegetation interactions on Aldabra atoll – part 1: inland. *Biological Conservation*. 9, 293–304.

Moorhouse-Gann, R. J. (2017) Ecological replacement as a restoration tool: disentangling the impacts and interactions of Aldabra giant tortoises (*Aldabrachelys gigantea*) using DNA metabarcoding. PhD thesis, Cardiff University.

Moorhouse-Gann, R. J., Vaughan, I. P., Cole, N. C., et al. (2021). Impacts of herbivory by ecological replacements on an island ecosystem. *Journal of Applied Ecology*, 00, 1–17.

North, S. G., Bullock, D. J. & Dulloo, M. E. (1994) Changes in the vegetation and reptile populations on Round Island, Mauritius, following eradication of rabbits. *Biological Conservation*. 67, 21–28.

Stephani, A. (2017) The dispersal potential of Giant Aldabra Tortoises, Aldabrachelysgigantea. MSc thesis, University of Zurich, Switzerland.

Tatayah, V., Zuël, N., Cole, N. C., Griffiths, C. & Jones, C. G. (2018) Introduction to Ile aux Aigrettes, Mauritius of the Aldabra giant tortoise as an ecological replacement for the extinct Mauritius tortoise. In Soorae, P. S. (ed.) *Global Reintroduction Perspectives: 2018. Case Studies from around the Globe*. Gland, Switzerland, IUCN/SSC Reintroduction Specialist Group, pp. 87–91.

Vaughan, R. E. & Wiehe, P. O. (1937) Studies on the vegetation of Mauritius: 1. A preliminary survey of the plant communities. *Journal of Ecology*. 25, 289–343.

23 · Assisted Colonisation as a Conservation Tool: Tasmanian Devils and Maria Island

CAROLYN HOGG AND PHIL WISE

23.1 Background

The Tasmanian devil *Sarcophilus harrisii* is the world's largest marsupial carnivore, found only on the island state of Tasmania, Australia (Figure 23.1). Devils are listed as endangered due to significant population declines as a result of an infectious cancer, devil facial tumour disease (DFTD; Figure 23.2).

In disease-free areas (north west and south west Tasmania) devils live for five to six years (Guiler, 1978) and breed between two and five years of age (Guiler, 1970). Older individuals are more susceptible to DFTD and so, in diseased areas, lifespan is reduced to between three to four years of age with females commencing their first breeding season at age one (Jones et al., 2008). Although devils produce multiple offspring at one time, each female has only four teats in her pouch meaning she can produce up to four joeys per year (Guiler, 1970).

The Save the Tasmanian Devil Program (STDP, established in 2003) is the Australian and Tasmanian government response to DFTD. A zoo-based insurance population commenced in 2006 to preserve the species (Jones et al., 2007). By 2012, the population was comprised of devils housed in intensive zoo-based facilities (1–2 animals per enclosure) or group housing (4–15 animals per enclosure), and on an island (Hogg et al., 2017). As part of this meta-population, devils were translocated to Maria Island National Park in a series of assisted colonisation events in 2012 and 2013, with further reinforcements in 2017 and 2019 (Table 23.1).

The purpose of Maria Island was to establish a disease-free devil population living under wild conditions to complement the intensive management of the zoo-based insurance population. Prior to the translocation to Maria Island, captive devils had not been released except for a small number of hand-raised orphans (Sinn et al., 2014). In 2015, the

Figure 23.1 A disease-free Tasmanian devil (photo: Carolyn Hogg). (A black and white version of this figure will appear in some formats. For the colour version, please refer to the plate section.)

Maria Island population was repurposed to provide devils to genetically augment existing wild diseased populations under the trial Wild Devil Recovery Project (Fox & Seddon, 2019).

Here we present a short summary and associated references for further information on the Maria Island devil project.

23.2 Methods

23.2.1 Study Site

Maria Island is a 115 km^2 national park located off the east coast of Tasmania. During initial assessments, the devil carrying capacity on the island was estimated to be 100–120 individuals.

23.2.2 Release and Harvest Events

Since 2012, there have been a number of different release and harvest events to and from Maria Island (Table 23.1). Each release is based on monitoring data from the previous releases, adapting releases to learnings from previous ones. The same can be said for harvest events, which account for a range of factors from demographic stability to maintaining

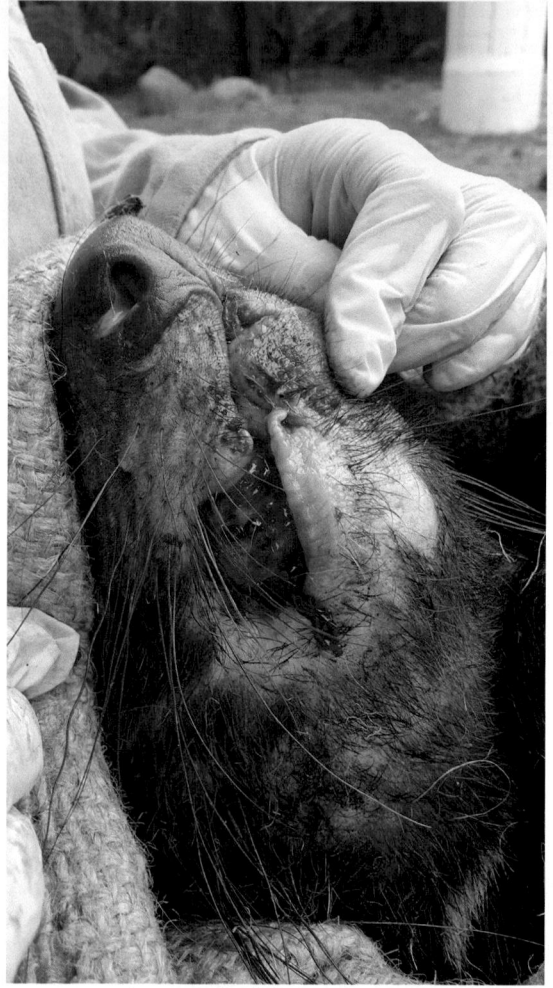

Figure 23.2 Tasmanian devil with devil facial tumour disease (photo: Carolyn Hogg). (A black and white version of this figure will appear in some formats. For the colour version, please refer to the plate section.)

genetic diversity of the population as well as operational constraints (Hogg et al., 2020).

23.2.3 Monitoring

Monitoring has changed over time. Before the first release, island-wide monitoring of other species was undertaken using camera traps located

Table 23.1. *Tasmanian devil release to and harvest events from Maria Island. Note, some females are removed during the breeding season with their pouch young (PY). M, males; F, females.*

Year	Release to Maria	Harvest from Maria
2012	7M; 8F	
2013	8M; 5F	
2014	0	0
2015	0	0
2016	0	14M; 2F
2017	8F	19M; 14F (38PY)
2018	0	16M; 13F (22PY)
2019	2M; 6F (12PY)	4M; 6F (11PY)
2020	0	4M; 6F (10PY)
Total	17M; 27F (12PY)	57M; 41F (81PY)

in a randomised grid across the island, in addition to other survey techniques. These cameras showed occupancy of all species on the island. Remote microchip readers were installed to monitor devils and replenished with fresh bait/fish oil. Initially these were used occasionally, however in 2018 a regular replenishment regime of every three months commenced.

Monitoring trips using PVC pipe traps occurred monthly for three months after the first two releases (2012, 2013). After this, monitoring trips occurred quarterly in 2014 and 2015 and were then reduced to six-monthly in July 2015 and continue on this schedule. All released devils were microchipped and had ear biopsies collected. Any new devils trapped on the island are microchipped, biopsied, and have their overall body condition checked. DNA is extracted from the ear biopsies and analysed using reduced representation sequencing (see Hogg et al., 2020, McLennan et al., 2020).

Each January an annual survey is undertaken to determine population size. This is based on a trap-line running north to south consisting of 40–60 traps over six trapping nights. Since 2016, harvest and/or release events have occurred in June each year.

23.3 Outcome

After the initial assisted colonisation events, the population grew rapidly and reached over 100 individuals by mid-2014. Due to the low population densities early on, the first cohort of island-born females showed some precocial breeding (age 1), which contributed to the rapid population growth. Genetic data have been used to reconstruct the island-born pedigree since the project outset. This significant investment has paid dividends for the management of the island. Initial analysis showed that males from the 2013 release cohort barely contributed to any of the offspring born on the island compared to the 2012 males (McLennan et al., 2018), leading to a skew in genetic representation on the island.

Due to the significant population growth in the early years, the island reached carrying capacity in 2014–2015. Contraception was trialled as a method for reducing over-represented bloodlines. The trial was successful in females, with contraception being effective for one to two breeding seasons depending on when it was used (Cope, 2018).

Since 2016, Maria Island has also been used as a source population for the trial Wild Devil Recovery Project (see Hogg et al., 2020 for details). It should be noted that for devils, Maria Island is the only source population for mainland Tasmania translocations at this time. This means that intensive genetic monitoring is essential to ensure the maintenance of the island devil population's genetic integrity.

The impacts of devils on the island ecosystem are varied, with minimal impact to some mammal species and possible benefits for other species such as long-nosed potoroos *Potorous tridactylus* and eastern pygmy possum *Cercartetus nanus*; others have significantly decreased in numbers, such as short-tailed shearwaters *Ardenna tenuirostris* and little penguins *Eudyptula minor*. Many of these changes in species' numbers and distribution were anticipated during the initial assessment of the island ecosystem prior to the devils' release. Changes in the ecosystem were deemed to be an acceptable risk by the STDP and the Tasmanian Parks & Wildlife department (Wise et al., 2019).

23.4 Future Recommendations

Maria Island was initially established to house disease-free devils as part of the insurance population. Due to a large number of unknowns regarding disease transmission (other than DFTD), the 2012 release cohort was only from Tasmanian-based facilities. Unbeknownst at the time, a significant number of first release cohort individuals were related due to their original founding grandparents being related (Hogg et al., 2018). This was further exacerbated by housing their parents in group housing where genetic assignment was difficult due to a limited microsatellite set (Farquharson et al., 2019). Pedigree reconstruction showed that the 2012 cohort contributed significantly to the island population, and rectifying these over-represented bloodlines has been an ongoing and extensive task. *Our advice to others using conservation translocations (e.g. assisted colonisation, reintroductions) is ensure that your founding population is as unrelated as possible.*

Undertaking an assisted colonisation to ensure that there was a disease-free population for a species facing an infectious disease was no small task. The Maria Island project benefited greatly from communication between all stakeholders and investment from all concerned, including the Tasmanian government agencies (STDP, Parks & Wildlife Service), the zoo industry, and academia. *The project is the success it is today due to the stakeholder network that was established from the outset, where each party brings their strengths and skill sets* (see Hogg et al., 2019 for details regarding governance and stakeholder engagement).

As with any translocation, post-release monitoring was essential. For this situation, ongoing genetic monitoring has been particularly critical in understanding the reproductive variance on the island. Without this monitoring a species suffering from inherently low genetic diversity would have lost even more diversity and become a poor source population. Through the selective contraception trial and annual harvesting (since 2016) we have removed over-represented individuals from the population while releasing other genetically complementary individuals. As a result, the Maria Island population now represents some of the most genetically diverse, disease-free devils in the world (McLennan et al., 2020). This has made it a preferred source population for reinforcing wild devil populations on mainland Tasmania (Hogg et al., 2020). *We recommend that monitoring programmes for island conservation projects budget for ongoing genetic analysis, particularly if the island is to be used as a source population.*

23.5 Summary

- Tasmanian devils are endangered due to an infectious clonal cancer that has reduced populations by up to 80 per cent since it first arose in 1996.
- As part of a management strategy for the species, an island population was established through an assisted colonisation event on Maria Island National Park. The original purpose of the Maria Island population was to establish and maintain a disease-free population of devils. The

island is now used as a source site for trial releases of devils to mainland Tasmania populations.
- The 2012 release cohort to the island had a high degree of relatedness. However, through dedicated management strategies, including contraception and selective harvesting, this situation has been rectified and the Maria Island population now represents a genetically diverse group.
- Monitoring, using traditional methods of trapping and camera traps, in addition to genetic monitoring, has been essential to the establishment and maintenance of the Maria Island population.

References

Cope, H. R. (2018) Contraception as a wildlife conservation tool. PhD thesis, The University of Sydney, Australia.

Farquharson, K. A., Hogg, C. J. & Grueber, C. E. (2019) A case for genetic parentage assignment in captive group housing. *Conservation Genetics*. 20, 1187–1193.

Fox, S. & Seddon, P. J. (2019) Wild devil recovery: managing devils in the presence of disease. In Hogg, C. J., Fox, S., Pemberton, D. and Belov, K. (eds.) *Saving the Tasmanian Devil: Recovery through Science-Based Management*. Melbourne, Australia, CSIRO Publishing.

Guiler, E. (1970) Observations on the Tasmanian Devil, *Sarcophilus harrisii* (Marsupialia: Dasyuridae) II. Reproduction, breeding and growth of pouch young. *Australian Journal of Zoology*. 18, 63–70.

Guiler, E. R. (1978) Observations on the Tasmanian devil, *Sarcophilus harrisii* (Dasyuridae: Marsupialia) at Granville Harbour, 1966-75. *Papers and Proceedings of the Royal Society of Tasmania*. 112, 161–188.

Hogg, C. J., Lee, A. V., Srb, C. & Hibbard, C. (2017) Metapopulation management of an Endangered species with limited genetic diversity in the presence of disease: the Tasmanian devil *Sarcophilus harrisii*. *International Zoo Yearbook*. 51, 137–153.

Hogg, C. J., Wright, B., Morris, K. M., et al. (2018) Founder relationships and conservation management: empirical kinships reveal the effect on breeding programs when founders are assumed to be unrelated. *Animal Conservation*. 22, 348–361.

Hogg, C., Fox, S., Pemberton, D. & Belov, K. (2019) *Saving the Tasmanian Devil: Recovery through Science-Based Management*. Melbourne, Australia, CSIRO Publishing.

Hogg, C., McLennan, E., Wise, P., et al. (2020) Preserving the demographic and genetic integrity of a single source population during multiple translocations. *Biological Conservation*. 241, 108318.

Jones, M. E., Jarman, P. J., Lees, C. M., et al. (2007) Conservation management of tasmanian devils in the context of an emerging, extinction-threatening disease: devil facial tumor disease. *Ecohealth*. 4, 326–337.

Jones, M. E., Cockburn, A., Hamede, R., et al. (2008) Life-history change in disease-ravaged Tasmanian devil populations. *Proceedings of the National Academy of Sciences of the United States of America*. 105, 10023–10027.

McLennan, E. A., Gooley, R. M., Wise, P., Belov, K., Hogg, C. J. & Grueber, C. E. (2018) Pedigree reconstruction using molecular data reveals an early warning sign of gene diversity loss in an island population of Tasmanian devils (*Sarcophilus harrisii*). *Conservation Genetics*. 19, 439–450.

McLennan, E. A., Grueber, C. E., Wise, P., Belov, K. & Hogg, C. J. (2020) Mixing genetically differentiated populations successfully boosts diversity of an endangered carnivore. *Animal Conservation*. 23, 700–712.

Sinn, D. L., Cawthen, L., Jones, S. M., Pukk, C. E. & Jones, M. E. (2014) Boldness towards novelty and translocation success in captive-raised, orphaned Tasmanian devils. *Zoo Biology*. 33, 36–48.

Wise, P., Peck, S., Clarke, J. & Hogg, C. J. (2019) Conservation introduction of Tasmanian devils to Maria Island: a managed response to DFTD. In Hogg, C. J., Fox, S., Pemberton, D. and Belov, K. (eds.) *Saving the Tasmanian Devil: Recovery through Science-Based Management*. Melbourne, Australia, CSIRO Publishing.

Index

Abies fraseri, 261
Accipiter gentilis, 461
acclimatisation, 136, 282, 289, 437–439
Acer campestre, 247
Acer lobelii, 247
Acinonyx jubatus jubatus, 79, 101
acoustic anchoring, 118
Acrocephalus familiaris kingi, 337
adaptation
   local, 214–215, 222, 276, 284–287, 292, 340
adaptive
   management, 24, 62, 130, 199, 422–427, 429
   potential, 225–226, 271, 273, 276, 281, 284, 287–295, 383, 397
   variation, 273, 292
adder
   common European, 158, 160, 162
adenovirus, 162
Aedes aegypti, 164
agouti
   red-rumped, 436–441
agouti tree, 437
agriculture, 21, 28, 92, 244, 264, 359, 415, 443, 450–451, 453
Agrilus planipennis, 243, 259, 263
albatross, 87
   black-footed, 336
   Laysan, 336
Aldabrachelys gigantea, 20, 337, 469–474
Allee effect, 115, 220
Alopochen mauritiana, 470
Alouatta guariba, 439
Amazona rhodocorytha, 440
Amorpha herbacea var. *crenulata*, 218–231
animal rights, 7, 180–205, 310, 386–388
Anisogramma virgultorum, 254
Anoplophora glabripennis, 243

ant
   red barbed, 158
anteater
   giant, 21, 444
Anthochaera phrygia, 158
anthropause, xxiii, 6
Anthropocene, 3
Aotearoa. *See* New Zealand
apex consumers, 11, 223–225, 335, 357, 371, 443
Apicystis bombi, 170
Apteryx mantelli, 16
aquatic ecosystems, 32, 164, 223, 261, 381
   freshwater, 14, 32, 53, 55, 289, 449–454
   marine, 13, 28, 45, 92, 97, 306, 462–468, 473
Aquila chrysaetos, 31, 307, 456–461
Aquila heliacea adalberti, 19
Ara ararauna, 440
Ara macao, 303
Ardenna tenuirostris, 480
Argentina, 21, 335, 443–447
Arvicola amphibius, 159
ash
   common, 20, 243, 246, 258–259, 261
ash dieback fungal disease, 20, 243, 246, 258, 261
aspen
   European, 286
Aspergillus spp., 162
assisted colonisation, 9, 15, 17–19, 34, 40, 46–67, 83–86, 99, 244–245, 331–347, 375, 397, 399–403, 432, 476–482
assisted migration. *See* assisted colonisation
Astacus spp., 17
Astragalus bibullatus, 220, 422–427
Astrochelys radiata, 469–474
Athene cunicularia, 199
aurochs, 20–22, 364

Australia, 10, 16, 21, 158, 212, 224, 258, 261, 282, 289–290, 307, 389, 394, 476–482
  legislation, 85–88, 98
Austria, 165, 286
avian
  malaria, 16, 155–156, 292
  poxvirus, 155–156

back-breeding, 364
bacteria, 151, 162, 164, 168, 244, 247
bandicoot
  golden, 16
banksia
  feather leaf, 224
Banksia brownii, 224
Bateson's Cube, 202
Batrachochytrium dendrobatidis, 152, 172, 338, 385
Baylisascaris procyonis, 153
beak and feather disease virus, 122
bear
  grizzly, 307
beaver, 31
  Eurasian, 8, 10–11, 28, 56, 92, 154, 159, 278, 305, 449–454
  North American, 278, 451
beech
  American, 260
beech bark disease, 260
beetle, 415
  Asian longhorned, 243
  bark, 262
  dung, 439
  heather, 248
  mountain pine, 263
  Rocky Mountain pine, 262
behavioural conditioning, 112–114, 136–138, 182, 185–188, 190, 201–202, 387, 444
Belgium, 158, 165
benign introduction. *See* assisted colonisation
Bern Convention, 79, 81–93, 96, 101–102, 104, 401
bettong
  brush-tailed, 21
  burrowing, 282
  eastern, 10
Bettongia gaimardi, 10
Bettongia lesueur, 282
Bettongia penicillata, 21

Betula pendula, 254
Betula pubescens, 254
biodiversity
  crisis, xxiii–xxiv, 6, 13, 34, 204, 388
  hotspot, 436
  loss, 347, 387
  maintenance, 332
  paradox, 393–396
  restoration, 12, 20, 27, 80–81, 392
  threats to, 13–17, 30–31, 89, 93, 103, 397
  value, 461
biological control, 7, 55, 334
biosecurity, 33, 85, 159, 168–174, 244–264, 462–468
biosphere, 52
biotechnology, 22, 292–294, 364–366, 384–386
birch
  downy, 254
  silver, 254
bison
  American, 31, 308
  European, 21, 335
  steppe, 21, 359
Bison bison, 31, 308
Bison bonasus, 21, 335
Bison priscus, 21
Blastocerus dichotomus, 444
Boa constrictor, 436
Boiga irregularis, 399
Bombina bombina, 286
Bombina variegata, 130
Bombus subterraneus, 158, 170–174
Bonamia ostreae, 464
Bos primigenius, 20–22, 364
Brachylophus vitiensis, 281
Bradybatus kellneri, 246
Brazil, 182, 394, 436–441, 445–446
Brazilian Atlantic Forest, 436–441
brocket
  brownish, 444
browsing, 424, 469
bryophyte, 212, 214
buffalo
  African, 91
Bufo periglenes, 338
bulgy-eye disease, 248
bumblebee
  short-haired, 158, 170–174
bunting
  cirl, 153, 158

bunyavirus, turtle, 162
bushbuck
  Cape, 89
Buteo buteo, 461
butterfly
  de Prunner's ringlet, 19
  Esper's marbled white, 277
  marbled white, 336
  Provence chalkhill blue, 19
  small skipper, 337
buzzard, 461

caiman, 444
Caiman yacare, 444
Cairngorms, 14, 18
Caladenia hastata, 223
Callaeas wilsoni, 337
Calluna vulgaris, 248
Cambodia, 308
camel
  dromedary, 16
Camelus dromedarius, 16
Canada, 16, 136–138, 308, 389
Candida spp., 162
canine
  distemper virus, 153
  hepatitis virus, 153
  herpesvirus, 153
  parvovirus, 153
Canis lupus, 10, 83, 153, 304–308, 310, 335, 370, 432
capacity building, 25–27, 396, 419
capercaillie, 7
Capra ibex, 281, 290
captive animals
  adaptation to captivity, 113, 129, 282–283
  disease risk, 16–17, 55, 150, 161, 188
  release of, 24, 112–114, 118, 136–138, 181, 188, 192, 194, 390–391, 440
captive breeding, 85, 111–112, 159, 182, 193, 283, 290, 387, 398–400, 444
capybara, 444
carnivores, 56, 173, 304–309, 354, 387, 443–447, 476–482
Carpathian Biodiversity Protocol, 81
carrying capacity, 51, 309, 441, 480
Carterocephalus palaemon, 158
Castanea dentata, 22, 97, 242, 385
Castor canadensis, 278, 451
Castor fiber, 8, 10–11, 28, 56, 92, 154, 159, 278, 305, 449–454

cat
  domestic, 16, 278
Centauria corymbosa, 226
Central America, 80, 242
Ceratotherium simum cottoni, 386
Cercartetus nanus, 480
Cervus nippon taiouanus, 308
Chad, 282
cheetah
  Asiatic, 79
  Southern African, 79, 101
Chelonoidis abingdonii, 19, 338
Chelonoidis denticulatus, 439–440
Chelonoidis hoodensis, 20, 473
Chelonoidis triserrata, 473
chestnut
  American, 22, 97, 242, 385
chestnut blight, 242, 385
Chile, 399
China, 222, 230, 389, 415–420
  legislation, 98
Chlamydia spp, 162
chocolate tree, 214
Christmas Island, 280
Chrysopogon argutus, 470
chytrid fungus, 152, 172, 338, 385
Cicerbita alpina, 13
Ciconia ciconia, 198
Circus cyaneus, 158, 167–169
CITES, 99, 103, 390
citizen science, 62, 231, 419
climate change, 9, 14–15, 44, 60, 85–86, 212, 221, 331, 333, 339–341, 393, 464
  adaptation to, 15, 83, 215, 220, 285–286, 400–402
  mitigation, 165, 255–264, 385, 393
  targets, 32
climate matching, 285
cloning, 22, 293–294, 364–365, 384–386
clover
  Florida prairie, 217
coastal dunes, 215
coccidia, 169
coccobacillus, 164–170
coffeeberry, 246
cognitive bias, 109–110
collaboration, 33, 52, 202, 306, 314, 389–390, 403, 456
compassionate conservation, 181–183, 200, 204, 386–388, 398

competition, 21, 55, 112, 168, 190, 221, 224, 228, 345, 471
condor
 California, 8, 156
conflict resolution, 317–324
Connochaetes spp., 91, 153
conservation introduction, 5–6, 15, 60, 83–84, 88, 95, 101–102, 331–347
conservation optimism, xxiii, 6, 397
conservation programme
 appraisal process, 25, 52
 cost-benefit analysis, 205, 294–295
 failure, 4, 8, 27, 34, 52, 55, 62, 66, 182, 258, 401
 feasibility, 28, 51–52, 109, 280, 307, 449
 motivations, 7, 32–33, 43–67
 objectives, 5, 120–122, 387, 467
 personnel, 30, 68, 231
 planning, 4–6, 30, 109–110
 prioritisation, 12, 15
 risk assessment, 92, 103, 150, 152, 156–172, 200, 244–248, 343–344, 451, 464
 strategies, 122
 success, 69
 threat assessment, 46, 60
 trade-offs, 399
contraception, 479–481
Convention on the Conservation of Migratory Species of Wild Animals (CMS), 83–84, 87, 93, 97
copepod, 287
coral reefs, 10, 33, 385
Coregonus albula, 95
Coronella austriaca, 158, 173
Costa Rica, 303, 338
crane
 Eurasian, 158
 whooping, 129
crayfish
 European species, 17
 Murray, 307
 North American signal, 17
Crepidula fornicata, 464
cricket
 British field, 158
 wart-biter, 158
CRISPR, 22, 273, 292–293, 384
Crithidia bombi, 170
crocodile
 Orinoco, 310

Crocodylus intermedius, 310
Crocuta crocuta, 153
Cryphonectria parasitica, 242
Cryptoblepharus egeriae, 280
cryptogam, 15, 18
Cryptosporidium baileyi, 248
Cuora trifasciata, 158, 162
Cyanoramphus novaezelandiae, 114
Cyathostoma spp., 167–168
Cylindraspis inepta, 469–474
Cylindraspis triserrata, 469–474
Cynomys ludovicianus, 10, 197
Cynomys spp., 371

Dalea carthagenensis var. *floridana*, 217
Dasyprocta leporina, 436–441
decision tree, 130, 226
decision-making
 evidence-based, 24
 structured, 16, 19, 96, 110, 120–126, 202–204, 392, 453
 support tools, 9, 25–27, 30
 trade-offs, 123, 202
 transparency, 30
 uncertainty, 24, 34, 112, 120, 123–130, 197, 397
Decticus verrucivorus, 158
deer
 marsh, 444
 Pampas, 21, 444
 sika, 308
de-extinction, 22, 293–294, 358–359, 364–366, 370, 384–386
deforestation, 436
Delissea waianaeensis, 220
Dendroctonus ponderosae, 262–263
Dendrophylax lindenii, 391
desman
 Pyrenean, 19
destination site
 selection, 46, 116–118, 214, 254, 424
diet analysis, 275, 279–280, 289
Diospyros egrettarum, 20, 470
Diplacus aurantiacus, 246
disease,
 epidemic, 153–164, 168, 173, 242
 in plants, 242–244
 infectious, 149–156
 management, 122, 250
 mitigation, 255, 393
 non-infectious, 156, 161, 163

disease (cont.)
  resistance, 215, 274, 280–281, 283, 292
  risk, 55, 152, 161, 165, 215, 248
  risk analysis, 149–150, 249–255
  risk assessment, 172, 334, 344
  risk management, 150, 165–174
  screening, 161, 169–171, 224
  surveillance, 150, 165, 170–174
  susceptibility, 119, 248, 250, 258, 282, 476
  transmission, 16–17, 33, 44, 55, 149–156, 480
  zoonotic, 16, 56, 163–164
dispersal
  ability, 15, 51, 54, 117–118, 339, 440
  benefits of, 26
  risks of, 26, 60, 108, 115–118, 188, 447
DNA barcoding. See meta-barcoding
dormouse
  hazel, 159
Drepanis coccinea, 292
Dutch elm disease, 242

eagle
  golden, 31, 307, 456–461
  Spanish imperial, 19
  white-tailed, 28–29, 31, 96, 159, 307–308, 310, 452
Earth BioGenome Project, 291
ebony tree, 20, 470
Echinococcus multilocularis, 56
ecological
  barrier, 152–164, 248, 254, 367
  integrity, 354, 396
  niche, 53, 64, 160, 228, 338–339
  proxy. See ecological replacement
  replacement, 19–21, 40, 46–67, 83–84, 131, 331–347, 355, 358, 368, 398, 469–474
  substitute. See ecological replacement
  surrogate. See ecological replacement
ecosystem
  dynamics, 275, 345–346
  engineer, 10–11, 21–22, 31, 358, 371, 439, 449
  function, 10–13, 19–21, 45, 131, 245, 331, 338–339, 341–347, 355, 436–441, 469–474
  modification, 12, 56, 451
  processes, 6, 243, 259–260, 322, 342, 345–346, 397
  resilience, 31, 35, 345, 365

  restoration, 6, 10–13, 31, 35, 80–81, 131, 245, 371
  services, 27, 31, 56, 243, 259, 356, 393, 464
Ectopistes migratorius, 22, 366
eDNA analysis, 273, 289, 295
egret
  snowy, 8
Egretta thula, 8
elephant, 89, 359
  African, 91
  Sri Lankan, 308
Elephas maximus maximus, 308
Emberiza cirlus, 153, 158
emerald ash borer, 243, 259, 263
Emydidae, 163
Emys orbicularis, 129, 279
Endangered Species Act (ESA), 231
endemism
  disease, 164
  species, 86, 132, 136, 218, 281, 399, 415, 469
England, 7, 19, 158–159, 162, 164–174, 307
Entamoeba invadens, 162
environmental
  impact assessment, 93
  impact mitigation, 21, 23
  stochasticity, 226, 432
Epidalea calamita, 158
Equus ferus przewalskii, 22, 281, 385, 399
Equus hemionus, 289
Erannis defoliaria, 260
Erebia triaria, 19
Estonia, 196, 261
ethical dilemmas, 183, 200–202
ethics
  animal, 181, 184, 200–202, 387
  conservation, 182
  environmental, 181, 387
  of assisted colonisation, 346
  of cloning, 293–294
  of gene editing, 292–293
Euastacus armatus, 307
Eudyptula minor, 480
Eugenia lucida, 470
European Union, 78, 81, 97
  legislation, 99–105
evergreen forest, 262, 415
evolutionary
  potential. See adaptive potential
  rescue, 273, 283–285, 333

Index · 489

ex situ conservation, 46, 51, 58, 83, 158–159, 200, 213, 282–283, 416
exit strategies, 46–67, 86, 122, 344
expert
   elicitation, 127, 129
   judgement, 18, 30
   knowledge, 52, 122, 132, 200
Extinct in the Wild, 111, 201, 212, 398–400
extinction, 3, 20, 26–27, 51, 212, 342, 365, 437

Fagus grandifolia, 260
Falco peregrinus, 15, 369
falcon
   peregrine, 15, 369
Felis catus, 16, 278
Felis silvestris, 278
ferret
   black-footed, 385
financial costs of translocations, 12, 24, 26, 109, 231, 307, 389, 394
fir
   Fraser, 261
fish, 32, 53, 55, 96–97, 283
fisheries, 92, 464
   illegal, 14
   management, 32, 51, 53, 273–275, 294
fitness, 64, 226, 258, 273, 276, 280, 284, 287, 383–384
Flavocetraria nivalis, 18, 337
folivory, 439
forestry, 32–33, 92, 245, 264, 415, 443, 457
Formica rufibarbis, 158
fox
   red, 16
France, 158, 167, 244
Francisella tularensis, 164–170
Frangula californica, 246
Fraxinus excelsior, 20, 243, 246, 258–259, 261
French Polynesia, 158
frog
   green and golden bell, 389
   Loa water, 399
   mountain chicken, 158
   northern leopard, 195
   pool, 158, 161, 163, 172
   southern corroboree, 385
frugivory, 439
Fundación Rewilding Argentina, 21, 443
fungi, 64–65, 94–95, 151, 214, 221–223
Fusarium spp., 162

Galápagos Islands, 19–20, 323, 338–339
Galemys pyrenaicus, 19
Galictis cuja, 440
gecko
   giant wall, 280
gene
   drive, 292–293
   editing, 22, 292–294, 384–386
   flow, 26, 214, 226, 273, 283, 290, 333, 451
genetic
   augmentation, 17, 477
   bottleneck, 227, 276, 419, 450, 465–466
   differentiation, 226, 281, 423
   diversity, 26, 51, 65, 217, 254, 275–276, 282, 284–285, 290, 383, 450, 480
   drift, 273, 282, 465
   engineering, 22, 364–366
   management, 225, 283, 285, 291, 451
   mixing, 287–295
   monitoring, 27, 46–67, 130, 199–200, 229–231, 257, 276, 283, 289–291, 345, 423, 429–433, 447, 481
   rescue, 65, 273, 276, 283–285, 384, 430
   restoration, 284–285
   screening, 275, 277, 418
   swamping, 274, 284, 286
genetically modified organisms (GMOs), 97, 293
genomics, 277–291, 383–384
Geoemydidae, 163
Germany, 158, 260, 278, 286
giant tortoise, 19–20, 337, 368, 371
   Aldabra, 20, 337, 469–474
   domed, 469–474
   Española, 20, 473
   Pinta Island, 19, 338
   saddle-backed, 469–474
Gopherus agassizii, 23
Gopherus polyphemus, 389
goshawk, 461
grassland, 21, 32, 222, 230, 245, 422, 445
grazing, 13–14, 27, 252, 359, 367–368, 469–474
Great Plains, 31
grison
   lesser, 440
ground-plum
   Pyne's, 422–427
Grus americana, 129
Grus grus, 158

Gryllus campestris, 158
guppy
  Trinidadian, 284–285
Gymnogyps californianus, 8, 156

habitat
  assessment, 188, 228
  complexity, 21, 464
  concept, 116
  connectivity, 116–118
  degradation, 88, 221, 366, 393, 423
  destruction, 415
  enhancement, 188
  fragmentation, 15, 88, 415
  loss, 9, 13, 15, 393, 423, 443
  management, 6, 32, 213
  quality, 115–116, 228
  restoration, 27, 32, 51, 245, 308, 431
  suitability, 401, 431
  suitability analysis, 444
  suitability modelling, 18, 53, 228
Haliaeetus albicilla, 28–29, 31, 96, 307–308, 310, 452
Haliaeetus leucocephalus, 159
harrier
  hen, 158, 167–169
Hawaii, 212, 228, 281, 292, 336
heather
  common, 248
Hemileuca spp., 277
hemlock
  eastern, 260–263
herbivores, 20, 65, 220–221, 223–225, 307, 357, 365, 424–427
hihi, 127–128, 132–136, 429–433
hippopotamus, 91
Hippopotamus amphibius, 91
honeycreeper, 292
honeyeater
  Regent, 158
Hong Kong, 158, 163
horse
  Przewalski's, 22, 281, 385, 399
horticulture, 217, 231, 245, 415–420
huagaimu, 415–420
human dimensions, 12, 27–34, 91, 303–322
  culture, 28, 31, 65, 313, 387, 392
  health, 26
  livelihoods, 11, 26, 56, 65, 303
  relationship with nature, xxiii, 27, 30, 346, 356

religion, 7, 28, 31, 86
rights, 52
well-being, 26, 31, 65, 231, 303
human–wildlife
  coexistence, 181, 305–306
  conflict, 30, 89–94, 304, 308, 390
Hungary, 165, 277
hunting, 54, 88, 436, 443, 449
  illegal, 310
  trophy, 103, 388
hyaena
  spotted, 153
hybridisation, 19, 26, 55, 88, 274, 278, 287–295, 383
Hydrochoerus hydrochaeris, 444
Hymenoscyphus fraxineus, 20, 243, 246
Hypotaenidia owstoni, 399

Iberá wetlands, 21, 443–447
ibex
  Alpine, 281, 290
iguana
  Fijian crested, 281
  green, 440
Iguana iguana, 440
illegal wildlife trade, 390, 439, 444
in situ management. *See* post-release support
inbreeding depression, 58, 225–226, 271, 274, 276, 281–282, 398, 430
India, 79, 101
Indigenous Peoples, xxiii, xxv, 12, 30, 86–87, 309, 346, 356, 392–393
individual
  identification, 188, 194, 229, 430
  variation, 115, 197
  well-being, 181, 386
influence diagram, 123–126
injury, 156, 185, 387, 390
insurance population, 85, 476–477
introduced species, 15, 33, 450
  risk assessment, 19, 55
introgression, 274, 277–278, 286, 383, 385
invasive species, 16, 33, 55, 60, 80, 89, 221, 422, 451
  control, 6, 51, 293, 334, 451
  disease risk, 17, 26
  legislation, 93–97
  management, 7
  risk assessment, 93, 344–345
Ireland, 308, 310
iridovirus, 162

islands, 16, 117, 212, 371
Isoodon auratus, 16
Isospora normanlevinei, 153
Italy, 129, 165, 244, 390, 400
IUCN, 4, 7, 108, 177
   Conservation Translocation Specialist Group, 9, 25, 66, 120, 184, 363, 396
   Green Status, 400
   Red List, 66, 390, 400
   Species Survival Commission, 9
   translocation guidelines, xxiv, 9, 25, 31, 43, 67, 79, 84, 109, 120, 249, 271, 307, 390
Ixodae spp., 164

Jacquemontia reclinata, 215
jacutinga, 440
jaguar, 21, 335, 443–447
Joannesia princeps, 437
juniper, 249, 252–255
Juniperus communis, 249, 252–255

kakapo, 7, 333
kākāriki, 114
keystone species, 10, 31, 45, 335, 341–342, 358, 361, 368–371, 397, 469
kingfisher
   Guam, 111, 159–160, 398
kite
   red, 156, 172
kiwi
   North Island brown, 16
koala, 280
kokako, 337

Lacerta agilis agilis, 158
Lagomorpha, 164
Lagorchestes hirsutus, 16
land
   development, 22–24
   management, 10, 12
   ownership, 11, 33, 68, 92, 309, 318, 451, 460
   stewardship, 12
   use, 13–14, 22, 28, 33, 309, 415, 423, 443
Lasiorhinus latifrons, 21
Latania loddigesii, 471
leadplant
   crenulate, 218–231
legal considerations, 33–34, 44, 57, 251
   barriers, 33, 57, 84–101, 346, 391
   context, 78–80
   obligations, 80–83, 182
   permissions, 22, 28, 57, 59, 68, 84, 88–105, 307
   support, 33, 80–84
Leontopithecus rosalia, 182
Leptodactylus fallax, 158
lichen, 53, 214, 261
   crinkled snow, 18
limpet
   slipper, 464
lion
   African, 91, 153
Lithobates pipiens, 195
lizard
   sand, 158
Lochmaea suturalis, 248
Loxodonta africana africana, 91
Lutra lutra, 82
Lycaon pictus, 91
Lymantria dispar subsp. *dispar*, 243
lynx
   Canada, 182
   Eurasian, 82, 307
   Iberian, 19, 82, 309
Lynx canadensis, 182
Lynx lynx, 82, 307
Lynx pardinus, 19, 82, 309

macaw
   blue-and-yellow, 440
   scarlet, 303
Macronesia, 280
mainland island, 16, 133
mala, 16
maladaptation, 225, 283
mammoth
   woolly, 359
Mammuthus primigenius, 359
managed relocation. *See* assisted colonisation
Manglietiastrum sinicum, 415–420
maple
   field, 247
   Lobel's, 247
Margaritifera margaritifera, 14
Maria Island, 17, 290, 337, 476–482
marmot
   Vancouver Island, 136–138, 198, 399
Marmota vancouverensis, 136–138, 198, 399
Marssonina betulae, 254
marten

marten (cont.)
  pine, 159
Martes martes, 159
Mascarene Islands, 20, 469–474
Mauritius, 20, 122, 469–474
Mazama gouazoubira, 444
media, 28, 31, 231, 310, 334
megafauna, 307, 359–360
Melanargia galathea, 336
Melanargia russiae, 277
Meleagris gallopavo silvestris, 163
melomys
  Bramble Cay, 400
Melomys rubicola, 400
meta-barcoding, 273–274, 277–281, 289–291, 475
meta-population, 24, 111, 230, 423, 482
Mexico, 8
microbiome, 280–281, 289, 365
microsite, 214
microsporidian, 170
migratory species, 83–84, 87–91, 98
millerbird, 337
Milvus milvus, 96, 156, 172
mink
  European, 196
miromiro
  North Island, 108
mitigation translocation, 22–24, 45–67, 388–390
Mohoua albicilla, 115
monitoring, 27, 50, 66, 130, 231, 257, 276, 342, 345–346
monkey
  brown howler, 439
monkey flower
  orange sticky, 246
Montserrat, 158
moorland, 245
mosquito, 164
moth
  Barberry carpet, 158
  bog buck, 277
  gypsy, 243
  mottled umber, 260
  winter, 260
multi-species translocation, 21–22, 368, 436, 443–447
Muscardinus avellanarius, 159
mussel
  freshwater pearl, 14
Mustela lutreola, 196
Mustela nigripes, 385

mutualism, 21, 223, 275, 280, 308, 436–441
Mycobacteria, 162
mycorrhizae, 221–223, 246, 280
Myrmecophaga tridactyla, 21, 444

Namibia, 79
nature-based solutions, 33
nematode, 170, 243
Neotoma floridana, 153
Nepal, 159
New Zealand, 7, 16, 31, 108, 111, 114–115, 132–136, 155, 424–427
newt
  great crested, 23, 389
Norway, 8, 159, 164, 304, 450
Nosema bombi, 170
Notiomystis cincta, 127–128, 132–136, 429–433
nyala, 89

oak
  sessile, 251, 260
  valley, 285
Oman, 8, 369
Oncorhynchus gorbuscha, 287
Oncorhynchus tshawytscha, 283
Operophtera brumata, 260
Ophiostoma novo-ulmi, 242
orchid, 216, 222–223
  ghost, 391
Oryctolagus cuniculus, 95, 470
oryx
  Arabian, 8, 369
  scimitar-horned, 282
Oryx dammah, 282
Oryx leucoryx, 8, 369
Ostrea edulis, 462–468
Otospermophilus beecheyi, 164
otter
  Eurasian, 82
outbreeding depression, 58, 225, 271, 274, 276, 286–295, 384
Ovis canadensis, 282, 290
owl
  burrowing, 199
oyster, 32
  European native, 462–468
oyster drill
  American, 464
Ozotoceros bezoarticus, 21, 444

Pacifastacus leniusculus, 17

Index · 493

Paecilomyces spp., 162
palm, 471
Palmyra Atoll, 159
Pandanus vandermeeschii, 471
panther
  Florida, 283
Panthera leo, 91, 153
Panthera onca, 21, 335, 443–447
Panthera tigris, 308
parakeet
  Mauritius 'echo', 122
parasite, 55, 119, 149, 160
Pareulype berberata, 158
parrot
  red-browed amazon, 440
Partulidae, 158
pathogen. *See* disease
peatland, 241, 245, 247
Pecari tajacu, 21, 444
peccary
  collared, 21, 444
Pelophylax lessonae, 158, 161, 163, 172
penguin
  little, 480
Penicillium spp., 162
persecution, 54, 62, 156, 193, 304, 310, 451
personality, animal. *See* individual variation
pesticide, 16, 103, 223
petrel, 91
Petrogale lateralis pearsonii, 21
Petroica longipes, 117, 429–433
Petroica macrocephala toitoi, 108
Petroica traversi, 115
Phascolarctos cinereus, 280
Philesturnus carunculatus carunculatus, 155
Philesturnus rufusater, 111
Phoebastria immutabilis, 336
Phoebastria nigripes, 336
Phyllostegia kaalaensis, 281
physiology, 118–120, 200
Phytophthora austrocedri, 249, 251–252
Phytophthora cactorum, 246
Phytophthora cinnamomi, 242, 261
Phytophthora infestans, 242
Phytophthora quercina, 253
Phytophthora ramorum, 242, 246
Phytophthora spp., 244–245, 251–255, 258
Phytophthora tentaculata, 246
pigeon
  passenger, 22, 366
pine
  torreya, 401
  Wollemi, 254
Pipile jacutinga, 440
plague, crayfish, 17
plant louse, 224
Plant Species with Extremely Small Populations (PSESP), 415–420
Plasmodium elongatum, 155
Platysternon megacephalum, 158
poaching, 51, 66, 304, 310, 443
Podocnemis expansa, 310
Poecilia reticulata, 284–285
poisoning, 16, 156, 172, 201
Poland, 159, 165, 450
politics, 28, 31, 452
pollination, 22, 27, 45, 56, 65, 223, 275, 280, 415
pollution, 14–16, 60, 156, 263, 462
Polyommatus hispanus, 19
popokatea, 115
population
  augmentation. *See* reinforcement
  connectivity, 26
  control, 94
  decline, 31, 45–46, 59, 172, 342, 462
  demographics, 51, 64
  dynamics, 432
  enhancement, 464. *See* assisted colonisation
  establishment, 24, 55
  growth, 26, 63, 69, 116, 228, 282, 430, 480
  management, 430
  persistence, 24, 69, 119, 383, 430
  resilience, 26, 58
  re-stocking. *See* reinforcement
  restoration, 5, 32
  self-sustainability, 69
  structure, 26, 277–279
  supplementation. *See* reinforcement
  viability, 5, 30, 46, 65, 281, 397, 415
  viability analysis, 51–52, 61
Populus tremula, 286
Porphyrio hochstetteri, 16
Portugal, 82, 309
possum
  brush-tailed, 16, 289
  eastern pygmy, 480
post-release support, 69, 114, 116, 120, 194, 228
  disease management, 122, 165, 430
  feeding. *See* supplementary feeding
  nest box provisioning, 430
  shelter, 119, 197
  threat management, 220

494 · Index

post-release support (cont.)
  trade-offs, 198–199
potato blight, 242
potoroo
  long-nosed, 480
Potorous tridactylus, 480
prairie dog, 371
  black-tailed, 10, 197
precautionary principle, 102–104
predation, 45, 55, 65, 117, 119, 201, 223–225, 280, 357, 365
predator control, 124–125, 185, 188
problem animals, 7, 67, 89–94, 308
Procyon lotor, 153
propagation, plant, 99, 251–254
protected area, 6, 57, 82, 88–89, 94, 244, 308, 354, 360, 394
protozoan, 151, 170
Pseudemydura umbrina, 85, 337, 402
Pseudophryne corroboree, 385
Psittacula eques, 122
public
  awareness, 3, 56
  engagement, 6, 30, 56, 306–317, 457–460
  health, 56, 450
  opposition, 6, 11, 26, 56, 304, 401
  outreach, 31, 56, 307, 457–460
  support, 231, 308, 397, 437, 444
publication bias, 25, 27
Puma concolor couguar, 283

quarantine, 112, 159–160, 170–174, 187, 193, 254, 256, 437
Quercus lobata, 285
Quercus petraea, 251, 260

rabbit
  European, 95, 470
rabies, 56
raccoon
  common, 153
radio tracking, 137, 200, 437, 440
rail
  Guam, 399
Ramphastos vitellinus, 436
ranavirus, 172
Ranoidea aurea, 389
rat
  black, 16
  central rock, 16

Rattus rattus, 16
reciprocal translocations, 65
refaunation, 436–441
reforestation, 32
reinforcement, 7, 40, 46–67, 81–83, 132–136, 276, 283–285
reintroduction, 4, 7–9, 24, 31, 40, 46–67, 81–83, 87, 418, 422, 429–433, 449
release cohort
  age, 116, 129–130, 137–138, 188, 195, 227
  cohort size, 61, 114–115, 118, 226, 424
  genetic diversity, 58, 271, 420
  propagule size (plants), 227
  selection, 215, 280, 282
  sex ratio, 115–116
  source, 112–114, 190, 282
release site. *See* destination site
release strategy, 188, 194
  delayed, 113, 118, 188, 195, 198, 281, 437, 440, 444
  hard release, 61
  immediate, 113, 440
  soft release, 61
  stepping-stone, 137–138
  timing, 194–195, 423
reovirus, 162
reproductive isolation, 340
reptilian paramyxovirus (PMV), 162
restoration ecology, 10, 213, 334–335, 366–367, 397
rewilding, 11–13, 20, 31–32, 131, 154, 308, 335, 354–372, 396–398, 436
  ecological, 360–361, 370, 396
  passive, 359–360, 370, 396
  Pleistocene, 359
  trophic, 357–359, 371, 397, 436–441
rhinoceros
  greater one horned, 159
  northern white, 385
Rhinoceros unicornis, 159
rinderpest virus, 153, 173
riparian habitat, 453
robin
  Chatham Island black, 115
  North Island, 117
Rodentia, 164
Round Island, 20, 469–474

saddleback
  North Island, 111
  South Island, 155

sage, 246
Salix lapponum, 13
salmon
　Chinook, 283
　pink, 287
Salvelinus fontinalis, 289
Salvia spp., 246
Sarcophilus harrisii, 17, 85, 283, 337, 476–482
Sarcoptes scabiei, 153
Scandinavia, 7, 304
Sciurus carolinensis, 154
Sciurus vulgaris, 7, 153–164
Scotland, 7, 28–29, 31, 56, 159, 278, 305, 307, 449–454, 456–468
　legislation, 57, 94–99
Scottish Code for Conservation Translocations, 25, 29, 52, 67, 249, 452, 457
screw-pine, 471
seagrass beds, 32, 371
seed bank, 215–216, 425
seed dispersal, 19, 308, 338, 345, 365, 415, 436–441, 469–474
shearwater
　short-tailed, 480
sheep
　bighorn, 282, 290
sheldgoose
　Mauritius, 470
sihek, 111, 159–160, 398
skink
　blue-tailed, 280
skipper
　chequered, 158
snail
　Partula, 158
snake
　boa constrictor, 436
　brown tree, 399
　smooth, 158, 173
social science, 30, 305, 310–313
sociality, 60, 115–124
socio-economic considerations, 26–33
　benefits, 27, 29–30
　compensation for impacts, 28
　costs, 304
　monitoring, 65–66
　risks, 4, 17
source population
　genetics, 285, 290

selection, 46, 52–59, 188, 225–226, 275–276, 281–283, 292, 424
South Africa, 242
　legislation, 88–94
Spain, 82–83, 167, 244
species
　analogue. *See* ecological replacement
　decline, 45–46, 59, 172, 342
　distribution, 26, 54, 102, 289
　distribution modelling, 53, 339, 401
　extirpation, 44–45, 116, 342
　management, 10, 51
　range, 46, 66, 340
　recovery, 5–6, 9, 26, 44, 80–81
　recovery plans, 85–86, 114, 212
　viability, 46, 391
Sphaerularia bombi, 170
Sphagnum spp., 247–248
Sphenodon punctatus, 337
squirrel
　Californian ground, 164
　grey, 154
　red, 7, 153–164
squirrelpox virus, 152–164, 173
Sri Lanka, 308
stakeholder
　conflicts, 304, 317–324, 451
　engagement, 28–31, 56, 61–62, 67, 86–87, 171, 205, 305–306, 308, 312, 343, 451, 481
starvation, 119, 201
steppe, 359
stitchbird. *See* hihi
stork
　white, 198
stress, 118–120, 152, 155, 168–169, 185, 187–188, 194–197, 200, 387
　in plants, 215–217
Strigops habroptilus, 7, 333
subspecific substitution. *See* ecological replacement
supplementary feeding, 114, 118–119, 188, 197–199, 430, 440
Sweden, 8, 158, 165, 170, 304, 310
SWOT analysis, 54–55
symbiosis, 280–281
Syncerus caffer, 91
synthetic biology, 384–386

Taiwan, 244, 308
takahe, 16

tamarin
  golden lion, 182
tamarisk, 57
Tamarix spp., 57
Tarentola gigas, 280
Tasmania, 17, 85–86, 290, 476–482
Tasmanian devil, 17, 85, 283, 290, 337, 476–482
Tasmanian devil facial tumour disease (DFTD), 17, 283, 476
taxon substitution. *See* ecological replacement
taxonomic uncertainty, 101–102, 275, 278
Telmatobius dankoi, 399
Tephrosia angustissima var. *corallicola*, 217
territoriality, 60, 282
Tetrao urogallus, 7
Theobroma cacao, 214
thistle
  alpine blue sow-, 13
Thymelicus sylvestris, 337
tick, 164
tīeke
  North Island, 111
  South Island, 155
tiger, 308
Tigriopus californicus, 287
Tijuca National Park, 436–441
Tiritiri Matangi Island, 108, 134, 424–427
toad
  European fire-bellied, 286
  Monte Verde golden, 337
  natterjack, 158
  yellow-bellied, 130
Todiramphus cinnamominus, 111, 159–160, 399
tomtit, 108
Torreya taxifolia, 334, 336, 401
tortoise. *See also* giant tortoise
  gopher, 389
  Mojave desert, 23
  radiated, 469–474
  western swamp, 85, 337
  yellow-footed, 439–440
toucan
  channel-billed, 436
tourism, 27, 31, 308, 446, 456, 460
toutouwai, 117, 429–433
Tragelaphus angasii, 89
Tragelaphus sylvaticus, 89
transportation, 151, 159, 188, 251, 257, 387
Trichosurus vulpecula, 16, 289
Trioza barrettae, 224

Triturus cristatus, 23, 389
trophic
  cascading, 11, 223, 357
  downgrading, 11
  interactions, 275, 279–280, 355–358, 371
trout
  brook, 289
trypanosome, 170
Tsuga canadensis, 260–263
tuatara, 337
tularaemia, 164–170
Tunisia, 282
turkey
  Eastern wild, 163
turtle
  big-headed, 158
  European pond, 129, 279
  golden coin, 158, 162
  Orinoco, 310
  western swamp, 402

Ukraine, 277
United Arab Emirates, 283
United Kingdom, 15, 19–20, 23, 153–164, 249, 253–255, 389
United Nations
  Convention on Biological Diversity, 80, 93–96, 102–104, 401
  Decade on Ecosystem Restoration, 35, 355, 381
  Environment Programme (UNEP), 98
United States, 8, 10, 16, 97, 218, 261, 278, 282, 304, 394, 422
Urosalpinx cinerea, 464
Ursus arctos horribilis, 307

vaccination, 153
value of information analysis, 129–130
Vancouver Island, 399
vendace, 95
Venezuela, 310
Vietnam, 158
Vipera berus, 158, 160, 162
vole
  European water, 159
Vulpes vulpes, 16

wallaby
  black-footed, 21
weather, 63, 187, 217, 431
weevil, 246

welfare, animal, 6, 26, 33, 61, 84, 100, 112–114, 118–120, 183–205, 386–388, 451
wetland, 10, 21, 85, 443–447, 453
wild ass, 289
wild dog, 91
wildcat, 301
wildebeest, 91
 western white-bearded, 153
wildlife rehabilitation, 7, 158–159, 390–391
wildlife trade, illegal, 51, 66
willow
 woolly, 13
wolf
 grey, 10, 82, 153, 304–308, 310, 335, 370, 432
Wollemia nobilis, 254

wombat
 southern hairy-nosed, 21
woodland, 20, 32, 245, 252, 254, 290
woodrat
 eastern, 153
World Organization for Animal Health, 156
World Wide Fund for Nature (WWF), 79

Xylella fastidiosa, 244

Yellowstone National Park, 10, 153, 304, 335, 370, 432

Zoological Society of London (ZSL), 156–158
Zyzomys pedunculatus, 16